# Using the Agricultural, Environmental, and Food Literature

# BOOKS IN LIBRARY AND INFORMATION SCIENCE

## A Series of Monographs and Textbooks

FOUNDING EDITOR

**Allen Kent**

*School of Library and Information Science*
*University of Pittsburgh*
*Pittsburgh, Pennsylvania*

1. Classified Library of Congress Subject Headings: Volume 1, Classified List, *edited by James G. Williams, Martha L. Manheimer, and Jay E. Daily*
2. Classified Library of Congress Subject Headings: Volume 2, Alphabetic List, *edited by James G. Williams, Martha L. Manheimer, and Jay E. Daily*
3. Organizing Nonprint Materials, *Jay E. Daily*
4. Computer-Based Chemical Information, *edited by Edward McC. Arnett and Allen Kent*
5. Style Manual: A Guide for the Preparation of Reports and Dissertations, *Martha L. Manheimer*
6. The Anatomy of Censorship, *Jay E. Daily*
7. Information Science: Search for Identity, *edited by Anthony Debons*
8. Resource Sharing in Libraries: Why · How · When · Next Action Steps, *edited by Allen Kent*
9. Reading the Russian Language: A Guide for Librarians and Other Professionals, *Rosalind Kent*
10. Statewide Computing Systems: Coordinating Academic Computer Planning, *edited by Charles Mosmann*
11. Using the Chemical Literature: A Practical Guide, *Henry M. Woodburn*
12. Cataloging and Classification: A Workbook, *Martha L. Manheimer*
13. Multi-media Indexes, Lists, and Review Sources: A Bibliographic Guide, *Thomas L. Hart, Mary Alice Hunt, and Blanche Woolls*
14. Document Retrieval Systems: Factors Affecting Search Time, *K. Leon Montgomery*
15. Library Automation Systems, *Stephen R. Salmon*
16. Black Literature Resources: Analysis and Organization, *Doris H. Clack*
17. Copyright–Information Technology–Public Policy: Part I–Copyright–Public Policies; Part II–Public Policies–Information Technology, *Nicholas Henry*
18. Crisis in Copyright, *William Z. Nasri*
19. Mental Health Information Systems: Design and Implementation, *David J. Kupfer, Michael S. Levine, and John A. Nelson*

*ADDITIONAL VOLUMES IN PREPARATION*

# Using the Agricultural, Environmental, and Food Literature

edited by

**Barbara S. Hutchinson**
*University of Arizona*
*Tucson, Arizona, U.S.A.*

**Antoinette Paris Greider**
*University of Kentucky*
*Lexington, Kentucky, U.S.A.*

MARCEL DEKKER, INC.          NEW YORK · BASEL

ISBN: 0-8247-0800-8

This book is printed on acid-free paper.

**Headquarters**
Marcel Dekker, Inc.
270 Madison Avenue, New York, NY 10016
tel: 212-696-9000; fax: 212-685-4540

**Eastern Hemisphere Distribution**
Marcel Dekker AG
Hutgasse 4, Postfach 812, CH-4001 Basel, Switzerland
tel: 41-61-260-6300; fax: 41-61-260-6333

**World Wide Web**
http://www.dekker.com

The publisher offers discounts on this book when ordered in bulk quantities. For more information, write to Special Sales/Professional Marketing at the headquarters address above.

Current printing (last digit):
10 9 8 7 6 5 4 3 2 1

**PRINTED IN THE UNITED STATES OF AMERICA**

# Preface

Since the publication of G. P. Lilley's reference work *Information Sources in Agriculture and Food Science* in 1981, we have been witness to incredible changes both in the area of agricultural research and in our ability to disseminate the results of that research. Advances have been made in the areas of plant and animal genetics, in animal and human nutrition and health, and in our understanding of the effects of human and climatic actions on the land and its natural resources. So, too, have we become more efficient and productive in bringing safe and healthy food to people around the world. These developments have been made possible through the efforts of a strong and growing worldwide network of agricultural researchers and organizations, and by significant and often dramatic improvements in technology. Ever more powerful computers, continued refinements in miniaturization, the development of the Internet, and expanded communications capabilities have created a foundation from which we are now able to link people, information, data, and decision-making tools on a 24/7 basis, from nearly any location, and as conveniently as in the palm of our hands.

Although strongly related to many other disciplinary areas, the applied nature of agriculture ensures its close ties to both human systems and technological developments. If the results of research are to be put to practical use for the benefit of all, a symbiotic relationship must exist among the scientific, educational, and information technology communities. In fact, another striking characteristic of the past decade has been an increased collaboration among scientists, teachers, extension personnel, technicians, and information specialists to make key information easily and widely accessible. Also involved in this mix have been the policymakers who fund the programs for bringing all these resources to bear on the challenging human and environmental issues of the day. By working together, people from these diverse backgrounds have made great strides in developing distributed and broad-based information systems such as the Agriculture Network Information Center (AgNIC) and the National Biological Information Infrastructure (NBII) through which significant content and learning tools have been brought to the public for the first time. Digital library projects and similar state-of-the-art initiatives also have provided testbeds for accessing specialized resources such as geospatial information, images, video, and other nontextual formats. This combination of trends has changed the face of the information industry and the nature of scientific communication and, as such, has influenced the content of this work.

The book's 16 chapters reflect not only the extent of the agricultural discipline but the variety of resources available for the study of these wide-ranging subjects. Although generally following a common structure, authors were given the latitude to accommodate the particular vagaries of their area of interest. Thus, certain chapters highlight resources and aspects that are unique to a topic. This can be seen most notably in Chapters 7, 14, 15, and 16, which include a variety of resources covering diverse subtopics. In addition, the inclusion of World Wide Web sites have been handled slightly differently. Most of the authors include a separate list for significant Web sites, while others have incorporated URLs for selected sites within the categories covered. URLs are also included for electronic journals if they are available free of charge. URLs for publishers are provided in appropriate sections.

As with any work of this sort, there are many people who contributed time and expertise to ensure its successful completion. First, of course, are the chapter authors. We thank them all for their patience and perseverance, and for their willingness to spend long hours on compiling and editing their contributions. In addition, we express thanks to the many outside reviewers who provided comments and suggestions for improvements to the various chapters. Some of these people are noted at the end of the chapters, but we also want to mention a few others here. In particular, we thank Tom Greider, Susan Moody, Don Post, and Anne Hedrich for their timely help in reviewing manuscripts. We especially thank Heather Severson, Carla Casler, and Susan Moody for their unwavering assistance with many of the more tedious last-minute tasks.

It is our sincere hope that *Using the Agricultural, Environmental, and Food Literature* will provide the basis for further advancements in agriculture and its related fields, and a deeper understanding of their essential role in determining quality of life for people throughout the world. The book is intended to serve both as an entry point into the discipline for those new to the field and as a guide for those wishing to expand their knowledge beyond their own areas of expertise. We firmly believe that universal access to accurate and current information is crucial for achieving a more equitable and stable world for ourselves and for the generations to come.

May we all eat well and tread lightly.

*Barbara S. Hutchinson*

# Contents

# Contributors

**Robert S. Allen**   Agricultural, Consumer and Environmental Sciences Library, University of Illinois at Urbana-Champaign, Urbana, Illinois, USA

**Francine Bernard***   Library Services, Food Research and Development Centre, Agriculture and Agri-Food Canada, Quebec, Canada

**Carla Long Casler**   Arid Lands Information Center, Office of Arid Lands Studies, The University of Arizona, Tucson, Arizona, USA

**Kathleen Ann Clark†**   Life Sciences Library, Purdue University, West Lafayette, Indiana, USA

**Antoinette Paris Greider**   Agricultural Information Center, University of Kentucky, Lexington, Kentucky, USA

**Anita L. Hayden**   Arid Lands Information Center, Office of Arid Lands Studies, University of Arizona, Tucson, Arizona, USA

**Barbara S. Hutchinson**   Arid Lands Information Center, Office of Arid Lands Studies, University of Arizona, Tucson, Arizona, USA

---

\* *Current affiliation:* Soils and Crops Research and Development Centre, Agriculture and Agri-Food Canada, Quebec, Canada

† *Current affiliation:* Biotechnology Librarian, University Library, University of Illinois at Urbana-Champaign, Urbana, Illinois, USA

**Jodee L. Kawasaki**   Reference Team, Montana State University, Bozeman, Montana, USA

**Louise Letnes**   Waite Library, Department of Applied Economics, University of Minnesota, St. Paul, Minnesota, USA

**Tim McKimmie**   University Library, New Mexico State University, Las Cruces, New Mexico, USA

**Sheila D. Merrigan**   Cooperative Extension, University of Arizona, Tucson, Arizona, USA

**Heather K. Moberly**   Veterinary Medicine Library, Oklahoma State University, Stillwater, Oklahoma, USA

**Elaine A. Nowick**   Branch Services, University of Nebraska, Lincoln, Nebraska, USA

**Mary Anderson Ochs**   Albert R. Mann Library, Cornell University, Ithaca, New York, USA

**Amy L. Paster**   Life Science Library, The Pennsylvania State University, University Park, Pennsylvania, USA

**Mary E. Patterson**   Engineering and Computer Science Library, Cornell University, Ithaca, New York, USA

**M. Louise Reynnells**   Rural Information Center, National Agricultural Library, ARS-USDA, Beltsville, Maryland, USA

**Patricia J. M. Rodkewich**   Magrath Library, University of Minnesota, St. Paul, Minnesota, USA

**Karl R. Schneider**   National Agricultural Library, ARS-USDA, Beltsville, Maryland, USA

**Gretchen Stephens**   Veterinary Medical Library, Purdue University, West Lafayette, Indiana, USA

**Irwin Weintraub**   Brooklyn College, Brooklyn, New York, USA

**Sue Wilkinson**   Information Service, Food Marketing Institute, Washington, D.C., USA

# Using the Agricultural, Environmental, and Food Literature

# 1

## An Introduction to the Literature and General Sources

**Antoinette Paris Greider**
University of Kentucky, Lexington, Kentucky, USA

### DEFINING AGRICULTURE

When we study agriculture, just what is it that we are studying? To look for a definition, one need go only as far as the agencies that serve the discipline. The Web site of the Food and Agriculture Organization (FAO) of the United Nations lists programs such as production agriculture, economics, fisheries, and nutrition. Along with these more traditional programmatic areas are newer areas of interest dealing with topics such as biotechnology and ethics in food and agriculture; gender and food security; rural youth; organic agriculture; and biological diversity. The United States Department of Agriculture (USDA) Web site lists program areas for food safety; natural resources and environment; rural development; food, nutrition and consumer services as prominently as farm and foreign agricultural services; and marketing and regulatory programs.

The diverse fields of study in agriculture are also illustrated through the curriculums of institutions of higher education. Wageningen University is part of Wageningen University and Research Centre (Wageningen UR), which is a strategic alliance between Wageningen University and the Dutch Agricultural Research Institute, DLO. The graduate programs offering Ph.Ds at that institution include Experimental Plant Sciences; Production Ecology and Resource Conservation; Food Technology, Agrobiotechnology, Nutrition and Health Sciences; Animal Sciences; Environment and Climate Research; and Social Sciences.

This shows that agriculture encompasses much more than the basic production of food. Graduates from agricultural programs find employment in industry, government, and academia, as well as self-employment. The various areas of modern-day agriculture are an integral part of modern-day society.

## COMPLEXITY OF THE LITERATURE

The complexity of the discipline of agriculture is reflected in its literature. Agriculture has its roots in the basic sciences and must be used in conjunction with the basic science literature. The student in entomology, for example, must also be well versed in zoology and the student in agricultural economics must also know basic economics to be able to carry out their research. In a sense, agriculture encompasses all of the literature in the pure and applied sciences as well as the social sciences.

Agriculture as a science in the United States dates back to the late nineteenth century with the creation of the agricultural experiment stations and the formation of the land-grant colleges. The formal separation between agriculture and the ''pure'' sciences remains but the research boundaries have blurred. So too have the literature boundaries. It is no longer possible to scan the major agricultural journals and have all of the information you need. With areas such as biotechnology and natural resource management, agriculture crosses over into every other discipline and it is necessary to depend on secondary sources to provide a snapshot of the literature. The successful information seeker uses all information tools at his/her disposal including the traditional literature, World Wide Web, conferences, e-mail, and peer interaction to name a few. There are literally millions of sources available that a searcher must wade through to teach, conduct research, or help the general farming population. The amount of information can be overwhelming and daunting to search.

## THE COMPLEXITY OF SEARCHING THE LITERATURE

Knowing what to ask for and how to ask for it is the key to successful searching. Precision searching means knowing the vocabulary of the subject being searched. The ease of searching among the various agricultural disciplines varies with the discipline. Searching in natural resource management is more difficult than searching in plant sciences because of the interdisciplinary nature of the vocabulary. Searching for a species of plant is not difficult because all you have to put in is the plant's common and scientific name. Searching for an ecological topic is much more difficult because of the nondescriptive and broad vocabulary used in the discipline. The plant is a tangible entity while ecology is a concept. A great deal of thought and some research of the vocabulary are required to execute a successful electronic search in most of the subjects. The intellectual effort is at the keyboard as the machine will only give the searcher what they ask for. Being able to articulate the search problem in the appropriate vocabulary is the most difficult part of the literature searching process.

Once the topic is defined it is necessary to choose the correct source to search for information. Determining the subject area that has the appropriate source to search is the most difficult part of searching for information. It is particularly difficult in agriculture because of the scope of the subject. The appropriate source could be something that is not even traditional agriculture. For example, when searching a fluid mechanics question in agricultural engineering, the appropriate sources could be those that cover agricultural engineering, such as CAB Abstracts followed by the engineering sources. To compound the problem, many of the sources are available in multiple formats and structures. For example, the AGRICOLA Database from 1970 to 2000 is available in print as the *Bibliography of Agriculture* and in multiple electronic formats depending on the electronic service available. It is available through Dialog and FirstSearch on a pay as you go searching plan and on various other electronic services and products that can be purchased on subscription. It is available through SilverPlatter, Ovid, directly from the National Agricultural Library (NAL), and now through ProQuest. All have the same data but are searched differently. Knowing what is available and how to search a particular system for a given subject area is another key to successful information gathering.

The World Wide Web further complicates the information gathering process. The Web is a vehicle providing access to secondary information sources such as bibliographic databases such as AGRICOLA as well as to provide primary information. Searching the Web is not an easy task and it is impossible to search the entire Web. At best, any given search engine covers 40% of the Web and many of the various search engines cover different topical areas. A Web search can yield both valuable information along with a great deal of information not of interest. Precision searching on the Web is more difficult but there are techniques to improve search results. Learning how to use one search engine well is an excellent investment of time. Most of the major search engines have help screens and spending 30 minutes to learn those features can save hours of searching time.

The Web provides access to millions of Web sites and a major challenge is to determine the validity of the content on each site. If the Web site creator is unknown, then more searching is necessary to determine its credibility. One method is to look at the extension in the address. The extension will tell you the originating country of the Web site or what type of agency has provided the information. For example, .gov means the Web site was done in an agency of the U.S. government while .nl means that the Web site is from a source in the

Netherlands. Home pages from unknown sources should not be used without validating the information.

Not all information on the Web is free. Many Web-based services charge a subscription fee and limit users to one institution. Government-produced secondary sources such as AGRICOLA and Medline are available through commercial vendors for a fee and free through their administering government agencies. There are also many electronic journals available through the Web but the majority of them do not have free access to their content. Most scholarly journals require a subscription, as they do in print, but many provide free access to the abstracts of the articles contained in each issue. Current awareness services are available through many publishers. These services allow the user to set up an information profile that is stored centrally on the publisher's computer. The profile is run at a specified time (depending on the publication schedule of the information source) and articles of interest are automatically sent to the user's e-mail address. Information on these services can be found at the various publishers' Web sites. Access to the full articles generally is available for a fee.

The advent of electronic journals has provided a number of enhanced services through the databases. Current awareness services are available through many of the major database producers. For example, SilverPlatter has a feature in its software that allows users to save a valid search on the SilverPlatter server. The search is automatically run each time the specified database is updated and the results are sent to the subscribers' e-mail boxes. This is most useful in agriculture because it allows the searcher to go across traditional disciplines in choosing databases. If the library subscribes to all of the necessary databases then the searcher can set up a subject or journal title search that covers subjects in the pure, applied, and social sciences along with the humanities if necessary.

Many of the electronic services now offer "hooks to holdings" and article delivery. "Hooks to holdings" provide a link to the local library catalog for sources the institution owns. By clicking a tag that indicates to check the catalog, a window opens and the location of the source indicated comes up. If the library subscribes to the article linking service, then any e-journals that the library subscribes to are linked to the database and only a keystroke away from the information seeker. As with many of the databases, the enhancements are available for a fee so availability varies from institution to institution.

In this book we are doing more than providing access to sources in agriculture but in essence we are redefining the agricultural literature. The subject breakdown reflects the broad scope of agriculture in today's society. The scope of coverage was outlined by the editors but each chapter author was given only general guidelines on how to organize their chapter and given the freedom to alter that organization if they felt it necessary. Authors were asked to discuss how to begin searching for their topics in the literature and to select the most prominent resources available to the searcher as of the year 2000. They were asked to include only those sources that would be the most useful to a broad audience in their subject area. For this reason, the chapters that follow are intended to be a selective resource and not an exhaustive treatment of the various subject areas. It is intended to provide a starting point for searching and not a step-by-step guide. Each discipline within agriculture has many subdisciplines that could not be explored here. It is recommended that a librarian who is a subject specialist be consulted to provide assistance in constructing an in-depth search on a given topic.

The Web resources described here are accompanied by an URL. URLs are not static and do change. The date after the URL (in parentheses) for Web sites listed is the last date that the Web site was checked and the URL was valid at that time. You can do some problem solving with URLs by determining what might have changed. For instance you are looking for the *Journal of Food Science* and it has the following URL: http://www.acs.jfsc/contents/html. If you cannot find the Web page you are looking for, begin dropping off elements on the right end of the URL. By putting in http://www.acs.jfsc you may find the journal. If that does not work, go to the main Web page (http://www.acs) to see if URL has been changed for the journal. In some instances you may have to do a Web search for the publisher to see if their Web site still exists or if they have revamped their entire home page.

The book is not intended to be an exhaustive work on all sources in the agricultural literature. Due to page limitations and the time involved, choices were made on what resources to include. However, this work does provide some pointers on how to deal with the traditional literature. The focus is not only on the sources but on how to search for additional sources. As such major secondary sources are described so as to give the user the tools needed for locating a wealth of other pertinent sources. To quote Dr. Samuel Johnson, ''Knowledge is of two kinds. We know a subject ourselves, or we know where we can find information upon it.'' It is hoped this will provide users with the guidance they need to locate the information that is essential for building and strengthening the world's agricultural sector and for ensuring the sector's viability for future generations.

## GENERAL SOURCES

There are some resources so broad in their coverage of agricultural topics that they would have needed to be described in each chapter. To conserve space they are included here. Further discussion of using these information tools is available in the various subject chapters with an emphasis on searching the topic being covered in the chapter. The descriptions have been contributed by various authors whose names appear at the end of the abstract.

AgNIC (http://www.agnic.org) (cited August 15, 2001). Beltsville, MD: U.S. National Agricultural Library. The Agriculture Network Information Center (AgNIC) is a voluntary alliance of the NAL, land-grant universities, and other agricultural organizations, in cooperation with citizen groups and government agencies. AgNIC focuses on providing agricultural information in electronic format over the World Wide Web using the Internet. One of the objectives of AgNIC is that member participants take responsibility for small vertical segments of agricultural information (including basic, applied, and developmental research; extension; and teaching activities in the food, agricultural, renewable natural resources, forestry, and physical and social sciences) and develop Web sites and reference services in specific subject areas (Introduction, AgNIC Alliance Governance Document). Subject pages include traditional agricultural subjects such as plant sciences and extension as well as pages on consumer and family studies; earth and environmental sciences; and people and organizations. The site also includes a calendar of agriculturally related conferences, meetings, and seminars; has a search feature with a thesaurus; and allows the user to ask a question. (A.P. Greider)

AGRICOLA. 1970– . Beltsville, MD: U.S. National Agricultural Library. AGRICOLA (AGRIcultural OnLine Access), titled CAIN (Cataloging and Indexing) during the years 1970–75, covers all fields of the broadly defined term *agriculture*. It includes bibliographic records for all materials added to the collection of the NAL, plus materials submitted by cooperating institutions. AGRICOLA is the best resource for locating publications of the USDA and the U.S. land-grant institutions. Ninety percent of the items in AGRICOLA are journal articles or book chapters. The remaining 10% include monographs, microforms, audiovisuals, and so forth. Since 1985 a controlled vocabulary has been used to assign descriptors to most records in AGRICOLA. The vocabulary comes from either (or both) the *Library of Congress Subjects Headings* (Library of Congress 2000) or the *CAB Thesaurus* (Wightman 1999). In addition to the subject headings used, subject category codes are added for each record. AGRICOLA is available to the public through the World Wide Web at http://www.nal.usda.gov/ag98. It is also available online and on CD-ROM through several vendors. Older issues (pre 1970) are available in paper format under the title *Bibliography of Agriculture* (Pat Rodekwich and Louise Letnes).

## Related Sources to AGRICOLA

List of Journals AGRICOLA is Indexed in (http://www.nalusda.gov/indexing/ljiarch.htm) (cited November 20, 2000). 1995– . Beltsville, MD: U.S. National Agricultural Library. Provides two lists of the journals indexed in a specific year: one list is arranged alphabetically by the full title of the journal, the other list is arranged alphabetically by the abbreviated name of the journal. Lists of the journals that have been added or subtracted from the List of Journals Indexed in in a particular year are provided as well. Not all journals indexed in AGRICOLA are abstracted, so there is a separate list of journals that are abstracted (Kathleen Clark).

Subject Category Codes and Scope Notes. (http://www.agnic.org/cc/hv.html) (cited August 15, 2001). Beltsville, MD: U.S. National Agricultural Library. Subject Category Codes (SSC) and Scope Notes are alphanumeric designations for the broad subject areas in agriculture. The codes are assigned by the indexers to records to aid in subject searching (A.P. Greider).

*Agricultural Information Centers: A World Directory.* 2000. J.S. Johnson, R.C. Fisher, and C.B. Robertson, eds. Twin Falls, ID: IAALD. The directory is arranged alphabetically by country and then by city and parent institution and provides descriptions of approximately 3,900 agricultural information resource centers in 180 countries. All entries include scope and language of library collection, databases maintained, search services provided, name and telephone number of the director, and e-mail addresses and Web sites where available. The one-volume work includes institution, city, and subject indexes (Irwin Weintraub).

AGRIS. 1975– . Rome: AGRIS Coordinating Centre. AGRIS is an electronic database that covers worldwide literature dealing with all aspects of agriculture. The print version of AGRIS is *Agrindex* (1975– ). AGRIS is updated monthly. The records for AGRIS are input by 161 national and 31 international/intergovernmental centers, which submit about 14,000 items per month. Thus, AGRIS provides good coverage of the major journal literature as well as of regional publications not elsewhere indexed—the so-called "grey literature." All AGRIS records are assigned subject headings from *AGROVOC*, a multilingual agricultural thesaurus. AGRIS records are also assigned AGRIS categories. The AGRIS centers collect bibliographic references (to date, about 3 million) to either conventional (such as journal articles, books) or nonconventional materials (such as theses, reports) not available through normal commercial channels. AGRIS encourages the exchange of information among developing countries, whose literature is often not covered by other international systems. AGRIS, *AGROVOC* (in print), and the *AGRIS Subject Categories* (not in print) are freely available on the Internet from the FAO at http://www.fao.org/agris/. Additionally, several vendors offer subscriptions to AGRIS, such as SilverPlatter and DIALOG (Kathleen Clark with added information by Sheila Merrigan and Tim McKimmie).

## Related Sources to AGRIS

AGROVOC (http://www.fao.org/agrovoc/) (cited August 21, 2001). Rome: FAO. Searchable thesaurus for the controlled terms used in the AGRIS database. Allows the searcher to look up the term in English, French, Spanish, or Portuguese and provides the term in all languages on each record.

AGRIS/Caris Subject Category Codes (http://www.fao.org/scripts/agris/c-categ.htm) (cited August 21, 2001). Rome: FAO.    Similar to the Subject Category Codes used in AGRICOLA. Allows the searcher to search broad subject categories such as plant sciences with a subject code.

BIOSIS Previews (online). 1969– . Philadelphia, PA: BIOSIS. Available on CD-ROM and through various vendors on the World Wide Web.    BIOSIS Previews covers both the biological and medical literature worldwide and provides them in a structured database. Approximately 550,000 items are added annually to this database, which contained 13 million citations at the end of 1999. The database provides enhanced access through the use of codes to pull out broad subject areas. The print counterparts to BIOSIS Previews includes *Biological Abstracts* that provides access to the journal literature in the life sciences and *Biological Abstracts/RRM* (reports, reviews, meetings) that covers meeting papers, symposia, conference proceedings, workshops, books, literature reviews, reports, patents, and CD-ROM and other software. The database contains over 2 million records and is updated monthly. A complete description of BIOSIS products and services can be found at www.biosis.org.

## Print Counterparts

*Biological Abstracts*, v. 1– 1926– . Philadelphia, PA: BIOSIS.

*Biological Abstracts/RRM*, v. 1– 1967– . Philadelphia, PA:BIOSIS. (Sheila Merrigan, Tim McKimmie, and A.P. Greider).

CAB Abstracts. 1973– Wallingford, UK: CABI.    Provides comprehensive, worldwide coverage of all aspects of agricultural, forestry, veterinary medicine, and biological information. Over 14,000 serials and journals in over 50 languages are scanned, as well as books, reports, and other publications. It is updated monthly and available online from 1973 to the present. CAB Abstracts is the aggregate, electronic version of nearly 50 printed indexes including primary and secondary review journals, some of which go back as far as 1913. This database cumulates 45 individual abstract journals covering most branches of agriculture. CAB uses a controlled vocabulary that is outlined in the *CAB Thesaurus* to index each record. In addition, each record is assigned a CABICODE that allows limiting keyword searches to records in a specific subject area. Combining these CABICODEs with descriptors allows great precision in searching. CAB Abstracts is available online through various vendors and is also available on CD-ROM. Access to CAB Abstracts is available electronically from a variety of vendors including Dialog, Ovid, SilverPlatter, and CAB.

## Print Counterparts

*Animal Breeding Abstracts*, 1933–
*Dairy Science Abstracts*, 1939–
*Forest Products Abstracts*, 1978–
*Forestry Abstracts*, 1939/40–
*Helminthological Abstracts*, 1932–
*Horticultural Abstracts*, 1931–
*Index Veterinarius*, 1933–
*Nutrition Abstracts and Reviews*, 1931–
*Plant Breeding Abstracts*, 1930–
*Review of Applied Entomology*, 1913–
*Soils and Fertilizers*, 1938–
*Veterinary Bulletin*, 1931–
*Weed Abstracts*, 1952–
*World Agricultural Economics and Rural Sociology Abstracts*, 1959–

(Kathleen Clark with added information by Pat Rodkewich, Louise Letnes, and A.P. Greider).

## Related Sources to CAB

*CAB Thesaurus.* 1999. Wallingford, UK; New York: CAB International. In an alphabetic, hierarchical structure, provides the subject headings that are applied to records in CAB Abstracts as well as AGRICOLA. Beyond its utility as a guide to the CAB Abstracts and subject headings, this book is a useful tool for locating alternative keywords when searching for agricultural literature in any database (Kathleen Clark).

*CABI* Codes (diskette). 1994. Wallingford, UK: CAB International. ISBN: 851989659. CABICODES are derived from the classification schemes used in the AGRICOLA and AGRIS databases with changes made to allow for different subject code emphasis. These codes enable searchers to locate general subjects by using a code consisting of two identical alphabetic characters and three digits. There are approximately 250 codes and a listing of codes with no scope notes can be found at http://www.cabi.org/publishing/products/dbmanual/cabicode/List.asp.

CSA (online). Bethesda, MD: CSA. Formerly called Cambridge Scientific Abstracts (CSA), this data file is an electronic collection of over 50 databases in the life sciences. CSA provides one search interface to search all data files. The data files are organized by broad subject and can be selected as a group or searched individually. Content coverage includes life sciences; environmental sciences and pollution; aquatic sciences and fisheries; biotechnology; engineering; computer sciences; materials science; sociology; and linguistics. The search interface provides both easy and advanced searching. Printed counterparts are available for many of the databases. Coverage dates vary for each database (A.P. Greider).

CARIS (http://www4.fao.org/caris/). Rome: FAO. Current Agricultural Research Information System (CARIS) was created by the FAO in 1975 to identify and facilitate the exchange of information about current agricultural research projects being carried out in or on behalf of developing countries. CARIS identifies projects dealing with all aspects of agriculture: plant and animal production and protection, post harvest processing of primary agricultural products, forestry, fisheries, agricultural engineering, natural resources and the environment as related to agriculture, food and human nutrition, agricultural economics, rural development, agricultural administration, legislation, education, and extension. Some 137 national and 19 international and intergovernmental centers participate in CARIS (Sheila Merrigan and Tim McKimmie).

CRIS (http://cristel.nal.usda.gov:8080/). 1975– . Beltsville, MD: USDA Cooperative State Research, Education, and Extension Service, Science and Education Resources Development. The Current Research Information System (CRIS), established by the USDA, serves as the USDA documentation and reporting system for publicly supported agricultural, food and nutrition, and forestry research in the United States that is ongoing or recently completed. Projects are conducted or sponsored by USDA research agencies, state agricultural experiment stations, the state land-grant university system, other cooperating state institutions, and participants in a number of USDA research-grant programs. Similar to the FAO's CARIS database, CRIS is of interest to field agronomists because it may point to research in progress but not yet published. To maximize search efficacy, use the Manual of Classification of Agricultural and Forestry Research, which is a type of subject categorization used for CRIS (Kathleen Clark).

## Related Sources to CRIS

Manual of Classification of Agricultural and Forestry Research Classifications used in CRIS (http://cristel.nal.usda.gov:8080/star/manual.html). 1993. Washington, D.C.: USDA Current Research Information System, Cooperative State Research Service. Provides the taxonomy used in CRIS to categorize research projects. The manual is available for download from the Web site (Kathleen Clark).

Current Contents (online). Philadelphia, PA: Institute for Scientific Information. Current Contents is a current awareness service from Institute for Scientific Information that provides access to complete bibliographic information for over 8,000 journals and 2,000 books. Print and diskette copies of Current Contents cover the following areas: agriculture, biology, and environmental sciences; arts and humanities; clinical medicine; engineering; computing and technology; life sciences; physical, chemical, and earth sciences; business collection; and electronics and telecommunications collections. The online file merges all topics into one database. Electronic files for Current Contents allow the user to set up auto-alert profiles that provide users with e-mail lists of the new citations that have been added to the database each week. Current Contents is available through the Institute for Scientific Information and several other vendors and coverage varies with the service (Kathleen Clark and A.P. Greider).

Dissertation Abstracts (http://wwwlib.umi.com/dissertations/gateway). Ann Arbor, MI: Proquest Information and Learning. Dissertation Abstracts contains more than 1.6 million entries and is the single, authoritative source for information about doctoral dissertations and master's theses. The database represents the work of authors from over 1,000 graduate schools and universities. Forty-seven thousand new dissertations and 12,000 new theses are added to the database each year. Citations for dissertations published from 1980 to the present and for master's theses from 1988 to the present include abstracts. The most current two years of citations and abstracts in the Dissertation Abstracts database are available online free of charge. Titles published since 1997 are available in PDF digital format and have previews that are 24 pages long. Institutions with a subscription to Dissertation Abstracts have access to dissertations back to 1861. Full dissertations can be ordered for a fee. Dissertation Abstracts is also available in print with the same coverage (Sheila Merrigan and Tim McKimmie).

FirstGov (http://www.first.gov.gov) (cited August 23, 2001). 2000– . Washington, DC: U.S. Office of First Gov, Office of Government Policy, General Services Administration. Official U.S. government portal to 30 million pages of government information, services, and online transactions. FirstGov features everything including an index to various topics, forms to complete some of your government transactions online, and links to state and local governments. The Web site has a basic search engine that allows searching the federal government, a particular state government, all states in one search, or a combination of both federal and state governments. The search engine provides an advanced feature for more precise searching. It is the aim of the Web site to give the user secure and seamless access to government publications and services.

*Guide to Sources for Agricultural and Biological Research.* 1981. J.R. Blanchard and L. Farrell, eds. Berkeley, CA: University of California Press. Describes and evaluates the important sources of information for the fields of agriculture and biology with major emphasis on agriculture and related subjects. The references are arranged by broad subject and then by format within the subject and works published prior to 1958 are generally not included unless important for retrospective searching. A brief narrative precedes each subject section of short, descriptive, sequentially numbered annotations. There is an author index, title index, and subject index at the end of the volume. Subject categories include agriculture and biology in general, plant sciences, crop protection, animal sciences, physical sciences, food sciences and nutrition, environmental sciences, social sciences, and computerized databases for bibliographic searching. Though somewhat dated, this work and its previous edition, *The Literature of Agricultural Research* (1958) provide good information on many of the standard sources in the field (A.P. Greider).

*Information Sources in Agriculture and Food Science*. 1981. G.P. Lilly. London, ed.: Butterworths. Part of the series *Butterworth's Guides to Information Sources*, this work concentrates on information sources in agriculture only. The volume is organized around format and provides a series of bibliographic essays written by different information specialists from the United Kingdom. There is a combined title and subject index at the end of the work. While not as useful as the Blanchard and Farrell guide as a quick reference tool, this book does provide a thoughtful discussion on the types of sources used by agricultural scientists (A.P. Greider).

*International Union List of Agricultural Serials*. 1990. Wallingford, Oxon, UK; New York: CAB International. Lists the serials indexed in AGRICOLA, AGRIS, and CAB Abstracts. Although out of date, the list is still useful for deciphering abbreviated journal titles. In addition to the title(s) of the journals, it provides the name and location of the publisher, beginning volumes and dates, earlier or later titles, language of text, ISSN, frequency of publication, and which of the three databases index the journal (Kathleen Clark).

*Journal Citation Reports*. 1973– . Philadelphia, PA: Institute for Scientific Information. Also available online. A bibliometric analysis of journals in the Institute of Scientific Information database. The journal titles are taken from Science Citation and Social Sciences Citation data files. The report does not follow a standard format throughout, but basically includes what journals have been cited, how often, and how frequently. It also includes both the citing and cited journal. The 1988 volume has two physical volumes. The first contains cited journals, where they have been cited, and how many times. The second volume has citing journals and includes various rankings such as impact factor and immediacy index. The product is also available electronically for a fee (A.P. Greider).

Web of Science. 1945– . Philadelphia, PA: Institute for Scientific Information. A relatively new product, Web of Science provides a user-friendly online search interface to the three ISI citation indexes, the Science Citation Index, the Social Sciences Citation Index, and the Arts & Humanities Citation Index. It provides the most current index to over 8,000 journals of the world's leading scholarly research journals in the sciences, social sciences, and arts and humanities, as it is updated weekly. About 1,000 of the journals covered are in agriculture, biology, or the environmental sciences. So, while the agricultural coverage in Web of Science is not as complete as in AGRICOLA or CAB Abstracts, certainly the most important agricultural journals are well represented. In addition to the capability to search for topics by keywords, Web of Science has a unique "cited reference" feature for finding articles that have cited a given article. This mimics the natural way scientists seek additional material on a topic, that is, by following the cited or citing literature. Abstracts are provided for most of the recent articles. Although Web of Science is primarily a database of journal citations, "cited materials" may be books, dissertations, government reports, in-press articles, or any other material. As there are many errors in the ways literature is cited, this database may be used to track down the "correct" citation for errant cites. Although the journals indexed in Science Citation Index are leading scholarly and technical journals, the citation part of the index includes all of the references in each article. These references include a great deal of gray literature (Kathleen Clark with additional information by Amy Paster and Heather Moberly).

WorldCat (Available through FirstSearch on the World Wide Web). Dublin, OH: OCLC Inc.   This database contains over 41 million records of all materials cataloged by OCLC member libraries. The database includes all formats of information and has entries dating back as early as 1,000 BC. Access to records in the database is broad. In addition to the standard author, title, and keyword, users can also search by series, ISBN and ISSN numbers, and many other elements. This product is available at most institutions through FirstSearch (A.P. Greider).

# 2

# Agricultural Economics

**Louise Letnes and Patricia J. M. Rodkewich**
University of Minnesota, St. Paul, Minnesota, USA

"Economics is a study of how people, individually or in groups, allocate scarce resources among competing wants to maximize satisfaction over time for each and for the group."—Sjo, 1976

Economics is a social science. All the social sciences focus on the study of human behavior, either individually or in groups. Economics focuses on the human behaviors involved with how people make their living, how they produce goods and services, how they exchange these goods and services, and how they acquire wealth. The discipline of economics has developed a body of principles and theories that clarify complex relationships and provide methods that can be applied to solve specific problems. Economists who apply these tested economic principles and theories to solve real-world problems in a specific economic sector are called applied economists. Applied economists who work with agriculture and related agricultural economic problems are agricultural economists.

In the early history of the United States about 90% of the population were farmers and agriculture was the most important sector in the economy. The focus of agriculture was almost completely on the family farm and the crops and livestock it produced and used. Each farmer was an independent operator and for the most part produced enough to feed, clothe, and shelter his own family with little left over to trade and, actually, with little need to trade. The Industrial Revolution changed everything. In a long and slow transformation workers left farms and moved to live and work in towns where factories were being built. These factories began to produce goods that farmers wanted and needed to buy and in turn the workers needed the food the farmers produced. Farm production increased and farmers were able to raise more food than they needed for their own use and now they had a market for their extra crops. This new system required a "middle man" operation because every farmer could not contact every consumer. This "middle man" would buy from farmers and sell to consumers or, often, add value to the product and then sell to consumers.

All of these changes demanded a new name, *agribusiness*, and a new definition. *Agribusiness* is:

> ". . . the sum total of all operations involved in the manufacture and distribution of farm supplies, production operations of the farm, and the storage, processing and distribution of farm commodities and items made from them." (Davis, 1957, 3)

The agricultural sector had transformed into the agribusiness sector. It was still the job of the agricultural economist to apply the principles and theories of economics but the scope of possible research had become much broader. Some examples of research topics examined by agricultural economists are conservation costs and returns, water needs and costs, weather and its effects, production, transportation, government policy, laws and regulations, biotechnology results and affects, new crops, agricultural trade, rural growth, rural businesses, farm inputs, markets, futures, labor, migrant workers, land costs, fuel costs, taxes, inheritance problems, consumer preferences, and marketing studies. Agricultural economists often serve governmental bodies seeking to make policy decisions by completing economic studies that predict outcomes of various actions based on the economic models used in the study. Governments rely on the research of agricultural economists to assist them in formulating food and agriculture policy decisions for their countries.

The research interests of the agricultural economists have continuously expanded. Environmental economics and natural resources economics, which might be viewed as separate areas of research, actually overlap in many ways with agricultural economics and are rapidly becoming part of the core research interests of agricultural economists. International development, which usually takes, at least in its early stages, the form of agricultural and rural development, is also a natural expansion of research for agricultural economists.

## THE LITERATURE OF AGRICULTURAL ECONOMICS

Every body of literature has its traditions, its own way of doing things. The scholarly works of one academic field are presented many times in a different way than are the scholarly works of another field. Agricultural economists are quite traditional in their literature, publishing articles in scholarly journals and scholarly books, except for the practice of sending out working papers—works in progress circulated for review and comment before publication.

The traditional arbitrator of scholarly published journal literature has long been and continues to be the professional academic societies. The societies have developed a rigorous review and editing system that is followed before a paper is accepted and published. Respected scholars who are members of the societies volunteer to review papers as part of their contribution to their organizations. This insures that the standards of the society are upheld.

Book literature quality rests on the reputation of the author and also the publisher. Book literature does not undergo the same peer review system that journal articles do, but instead relies on a very structured editing system by a highly qualified editor. There is a review system, but it takes place after the book is published rather than before. Many of the professional journals publish reviews of new books.

Agricultural economists share the tradition of working papers with many of the social scientists. The working paper is a form of an informal peer review process. Authors use the suggestions and comments returned from their peers to strengthen their papers before they are submitted to a professional journal. Most, if not all, university economics departments (theoretical and applied) sponsor working-paper series. The agricultural economics working papers have been an elusive literature until recently when they began to be collected, indexed, and available full text on the Internet.

A large part of an agricultural economist's work has been to synthesize scholarly information so that others can apply the information. To do so, the agricultural economists follow the traditions of agricultural scientists and use state agricultural extension and experiment station publications to bring research results to the persons needing it the most, the rural community. Many also publish in the trade literature read by farmers and others working in agribusiness. The U.S. Department of Agriculture (USDA) has published much research in agricultural economics and they also have provided many statistical resources for the economist's use.

It is necessary to remember that the scope of agricultural economics is broad and the search for its literature is just as broad. An agribusiness study might be indexed in the business indexes rather than the agricultural indexes. Environmental or conservation studies with an economic emphasis might appear in the biology literature rather than the agricultural. The key to locating all interesting research is to think and search broadly.

Those interested in the history of this discipline might wish to consult one or both of two works by Henry C. Taylor. In the first work, Taylor traces and describes the beginnings of a new discipline named agricultural economics (Taylor, 1905). The second work, published much later, describes the achievements in agricultural economics from 1840–1932 (Taylor, 1952).

## ABSTRACTS AND INDEXES

The field of agricultural economics is served by four major indexes: AGRICOLA, CAB Abstracts, AGRIS, and EconLit. The interdisciplinary nature of research sometimes, or even often, requires that other subject-specific indexes beyond the agriculture or economics indexes normally searched by agricultural economists be used if the researcher wishes to be comprehensive.

AGRICOLA, CAB Abstracts, and AGRIS are agriculture databases that include agricultural economics literature. The three databases overlap somewhat in coverage. One study (Thomas, 1990) found that of 11,619 titles indexed by at least one of the three indexes, only 450 were in all three indexes. Another brief study (Nixon, 1995) directly investigated the overlap in agricultural economics literature. It found little or no overlap between AgECONCD (a subset of CAB Abstracts) and AGRICOLA. An earlier study (Farget, 1984) compared 15 European and North American databases, including AGRICOLA, AGRIS, and CAB Abstracts by searching 9 socioeconomic agricultural topics. CAB Abstracts fared the best both in number and quality of references retrieved.

AGRICOLA covers all fields of agriculture, including agricultural economics. AGRICOLA is the best resource for locating publications of the USDA and the U.S. land-grant institutions. In the early 1970s the American Agricultural Economics Association and the Economic Research Service (ERS) of the USDA jointly created the American Agricultural Economics Documentation Center (AAEDC). The goal of the AAEDC was to increase the indexing of agricultural economics literature in AGRICOLA. Records with abstracts were added for all agricultural economics journal articles as well as many materials authored by agricultural economists in land-grant institutions and government agencies in the United States and Canada. The project ended in 1985 with the U.S. National Agricultural Library (NAL) taking over responsibility for adding these materials to AGRICOLA. The inclusion of abstracts on the records was halted at that time.

Since 1985 a controlled vocabulary has been used to assign descriptors to most records in AGRICOLA. The vocabulary comes from either (or both) the *Library of Congress Subjects Headings* (2000) or the *CAB Thesaurus* (1999). Checking these vocabulary lists is important. A search for *Agricultural Policy* as a subject term in AGRICOLA turns up no citations. A check of the *CAB*

*Thesaurus* shows that the used subject heading is *Agriculture and State*, a phrase that will lead to many hits.

Searching keywords combined with a subject category code will direct the search to appropriate citations. For example, enter the category code E110 and *policy* and *Canada* as keywords. The category code E110 (Land Development, Land Reform, and Utilization) limits the search to that category, which means the search for *policy* and *Canada* are returned if they fit into that category. Subject category codes used for agricultural economics include:

A500: Agricultural Research and Methodology
C210: U.S. Extension Services
D500: Laws and Regulations
E100: Agricultural Economics (General)
E110: Land Development, Land Reform, and Utilization (Macroeconomics)
E130: Agricultural Production (Macroeconomics)
E200: Farm Organization and Management (Microeconomics)
E300: International Agricultural Development Aid Programs
E310: U.S. Food and Nutrition Programs
E400: Cooperatives
E500: Rural Sociology
E550: Rural Development
E560: Rural Community Public Services
E700: Distribution and Marketing of Agricultural Products
E710: Grading, Standards, Labeling
E720: Consumer Economics
P200: Water Resources and Management
X100: Mathematics and Statistics
X700: Economics and Management
X800: Social Sciences, Humanities, and Education

A complete list of these codes is found on this Web site: http://www.agnic.org/cc. AGRIS indexes international information on all aspects of agriculture. The AGRIS category codes for agricultural economics include:

E10: Agricultural Economics and Policies
E11: Land Economics and Policies
E12: Labour and Employment
E13: Investment, Finance, and Credit
E14: Development Economics and Policies
E16: Production Economics
E20: Organization, Administration/Management of Agricultural Enterprises or Farms

E21: Agro-Industry
E40: Cooperatives
E50: Rural Sociology
E51: Rural Population
E70: Trade, Marketing, and Distribution
E71: International Trade
E72: Domestic Trade
E73: Consumer Economics
E90: Agrarian Structure

The complete list of subject categories for AGRIS is found on the AGRIS Web page in the Documentation Tools section: http://www.fao.org/agris.

CAB Abstracts indexes agricultural literature, including agricultural economics and rural development. CAB uses a controlled vocabulary that is outlined in the *CAB Thesaurus* (Wightman, 1999) to index each record. It is essential to use the *CAB Thesaurus* when doing a comprehensive search in CAB Abstracts. The following example from the thesaurus shows the term, *Agricultural Insurance*, and the additional terms that broaden or narrow, or lead to a related term connected to the topic. BT (broader term) expands the search, NT (narrower term) tightens the search, and RT (related term) leads to other terms:

Agricultural Insurance
BT1  insurance
NT1  animal insurance
NT1  crop insurance
NT1  hail insurance
NT1  livestock insurance
RT    agricultural disasters
RT    cooperative insurance
RT    fire insurance
RT    social insurance

In addition, each record is assigned a CABICODE that allows limiting keyword searches to records in a specific subject area. Combining these CABICODEs with descriptors allows great precision in searching. CABICODES for agricultural economics include:

EE100: Economics (General)
EE110: Agricultural Economics
EE120: Policy and Planning (General)
EE130: Supply, Demand, and Prices
EE140: Input Supply Industries
EE145: Farm Input Utilization
EE150: Environmental Economics

EE160: Land Use and Valuation
EE165: Agricultural Structure and Tenure System
EE170: Water Resources, Irrigation, and Drainage Economics
EE200: Farming Systems and Management
EE300: Cooperatives
EE350: Industry and Enterprises
EE450: Development Aid, Agencies, and Projects
EE500: Food Policy, Food Security, and Food Aid
EE520: Food Industry
EE600: International Trade
EE700: Distribution and Marketing of Products
EE720: Consumer Economics
EE730: Transport
EE800: Investment, Finance, and Credit
EE900: Labour and Employment
EE950: Income and Poverty

The complete list of CABICODEs is found online at http://www.cabi.org/WHATSNEW/Content/Cabicodes.htm.

The agricultural economics section of CAB Abstracts is available separately on a CD-ROM titled AgECONCD covering the years 1973 to the present. The print publication of the agricultural economics section of CAB Abstracts is titled *World Agricultural Economics and Rural Sociology Abstracts* (1958– ).

EconLit. 1969– . Philadelphia, PA: American Economic Association. EconLit (Economic Literature Index) is the major index for economic literature. Agricultural economists use EconLit to investigate economic theory and its applications, which they then apply to the specific agricultural economics problem they are researching. More than 400 economics journals, plus collective volumes, books, dissertations, and some working papers are indexed and abstracted. These citations are also included in the print quarterly journal, *Journal of Economic Literature*. Included in EconLit is the Cambridge University Press' *Abstracts of Working Papers in Economics* (*AWPE*).

EconLit uses the *Journal of Economic Literature* classification system to provide subject category codes for its citations. A complete list is found on the Web at http://www.econlit.org/econlit/elhomsub.html. Using these codes in conjunction with relevant keywords is the most productive use of the database. While all sections of EconLit are of potential interest to agricultural economists, the Q1 section is specifically dedicated to the discipline:

Q1: Agriculture
Q10: General
Q11: Aggregate Supply and Demand Analysis; Prices

    Q12: Micro Analysis of Farm Firms, Farm Households, and Farm Input
       Markets
    Q13: Agricultural Markets and Marketing; Cooperatives; Agribusiness
    Q14: Agricultural Finance
    Q15: Land Ownership and Tenure; Land Reform; Land Use; Irrigation
    Q16: Research and Development; Agricultural Technology; Agricultural
       Extension Services
    Q17: Agriculture in International Trade
    Q18: Agricultural Policy, Food Policy
    Q19: Other

EconLit is available in print as the *Index of Economic Articles in Journal and Collective Volumes*, online as the Economic Literature Index, and on CD-ROM.

## Other Indexes

Although not as central as the previous indexes, the following indexes are also valuable for research in agricultural economics.

ABI Inform. 1971– . Ann Arbor, MI: Bell and Howell.    ABI Inform is a business and management database, abstracting more than 1,000 international serials and providing online full-text articles from more than 600 of its serials. ABI Inform indexes journals, newspapers, and magazines. ABI Inform uses a controlled vocabulary and category codes. Included in the vocabulary terms are *agribusiness*, *agricultural banking*, *agricultural commodities*, *agricultural co-operatives*, *agricultural economics*, *agricultural lending*, *agricultural policy*, *agricultural subsidies*, *farm loans*, *farm price supports*, and many more. The category code for agriculture is 8,400 (Agriculture Industry). ABI Inform is available online and on CD-ROM.

AgEcon Search: Research in Agricultural and Applied Economics (http://agecon. lib.umn.edu). 2000. St. Paul, MN: University of Minnesota, University of Minnesota Libraries.    This Web-based service provides indexing to over 3,000 full-text reports in agricultural economics. The reports include working-paper series from many United States and some international departments of agricultural economics at major universities. Also included in the service are conference papers—mainly from the annual meetings of the American Agricultural Economics Association and the Western Agricultural Economics Association. Other organizations such as the International Agricultural Trade Research Consortium have included their papers on AgEcon Search. Users may search the entire database by keyword or author, or may limit their search to a particular institution. Abstracts are provided for most entries. AgEcon Search was developed at the University of Minnesota with the support of the American Agricultural Economics Association, the Farm Foundation, and the USDA's Economic Research Service.

*F&S Index: Europe.* 1978– ; *F&S Index: International.* 1967– ; *F&S Index: United States.* 1960– . Foster City, CA: Gale Group. The *F&S Indexes* contain citations to international literature on industries and companies from business and trade magazines and newspapers as well as government and international agency reports. The indexes are organized by Standard Industrial Classification (SIC) codes. Agricultural trade publications are among those indexed. This is a good database for information on agribusiness. The *F&S Indexes* are available in print, online, and on CD-ROM.

IDEAS (Internet Documents in Economics Access Service) (http://ideas.uqam. ca/). Montreal, Quebec, Canada: University of Quebec. IDEAS is an Internet service that indexes and provides access to economics working papers. IDEAS uses the data from the RePEc (http://repec.org) database that includes over 130 archives of working papers from organizations such as the National Bureau of Economic Research and the U.S. Federal Reserve Banks. Other collections of economics working papers are included such as EconWPA (Economics Working Paper Archive) (http://econwpa.wustl.edu) and WoPEc: Electronic Papers in Economics (http://netec.mcc.ac.uk/WoPEc.html). The database contains information on nearly 70,000 working papers with about one-third of them linked to the full text of the papers. Some of the papers have *Journal of Economic Literature* codes (http://www.econlit.org/econlit/elhomsub.html) and can be browsed by these codes or searched by keyword. The *Journal of Economic Literature* code for Agricultural and Natural Resources Economics (Q) lists many papers of interest to agricultural economists, but a search of the database by keyword will retrieve many more items.

Lexis-Nexis Academic Universe. Bethesda, MD: Congressional Information Service. The Congressional Information Service established a Universe Library in 1998 consisting of Statistical Universe, Congressional Universe, History Universe, State University, and Lexis-Nexis Academic Universe. The Lexis-Nexis Academic Universe provides news, business, legal, and reference information to university libraries and individuals. Lexis-Nexis is a powerful option for searching for business (including agribusiness) information. Lexis-Nexis indexes a wide range of business periodicals, trade literature, newspapers, and government publications, many of them with links to the full text of the article. Lexis-Nexis Academic Universe is available online and through the Internet by contract.

PAIS International. 1972– . New York, NY: Public Affairs Information Service, Inc. PAIS International indexes and abstracts a wide variety of U.S. and international material in the areas of public affairs, public policy, and general social sciences. Agriculture, environment, international relations, business, and economics are just a few of the subjects covered that make this a rich resource for agricultural economists. Another important aspect of this database is the many formats of material included—journal articles, books, government documents, congressional reports, reports of public and private agencies, conference proceedings, and so forth. Materials are indexed with a controlled vocabulary developed by PAIS. PAIS International is available in print (*PAIS Bulletin*, *PAIS Foreign Language Index*, and *PAIS International in Print*), as well as online and in CD-ROM.

Social Science Research Network (http://www.ssrn.com).2000. Social Science Electronic Publishing, Inc.   Social Science Research Network (SSRN) is a commercial Internet service that disseminates social science research abstracts of working papers and journal articles. Some of the abstracts have links to an electronic library of full-text papers. A subset of SSRN is the Economics Research Network that includes abstracting journals in several subdisciplines of economics, including agricultural and natural resource economics, as well as international trade and development economics. There is a fee for subscribing to the abstracting journals, but access to the database is free.

Social Sciences Citation Index; Science Citation Index. 1973– . Philadelphia, PA: Institute for Scientific Information.   Social Sciences Citation Index is a large multidisciplinary database that indexes those journals it has determined to be core journals in each social sciences field. Social Sciences Citation Index covers most of the list of core journals listed in the journals section of this chapter, but makes no pretense of providing the coverage of economics offered by indexes such as EconLit. A weakness in the searching of Social Sciences Citation Index is that there is no thesaurus of terms to be used as a guide for searching, no category codes to use, and no abstracts to search. Its strengths lie in the fact that you are doing a search across all social science fields and this allows the search to stray outside the strictly economics field into others that might produce unexpected results. One study (Ekwuzel and Saffran, 1985) found that Social Sciences Citation Index covers less than half of EconLit's journals. It would not be a substitute for EconLit, but instead, a good complement.

Social Science Citation Index has a companion index for the sciences titled Science Citation Index. It indexes the core journals in science and technology fields. Because it is very probable that articles on the economics of ''anything'' can be published in the ''anything'' journal as well as the economic journal, it is useful to do the same search in the Science Citation Index as well as the Social Sciences Citation Index. That search will pick up the economics of crop production found in Crop Science or the economics of hog lots in Animal Science.

The unique aspect of the citation indexes is that they not only index the new articles in the core journals, they also index the citations at the end of each of those articles. This is useful in two ways. First, it leads to the possibility of other references that might be valuable, and second, a search can be done to see who else has cited an article. If the researcher knows of a particularly useful article, say, the classic work in a subject, the database can be searched to find out who else has cited that classic work. The indexes are available in paper, online (Web of Science), and on CD-ROM.

## BIBLIOGRAPHIES

Bibliographies, selected and annotated by scholars in the field, are good places to begin a search for the scholarly literature of a subject. Often they can be located by searching library catalogs using subject-appropriate keywords connected to an additional keyword, *bibliography*. A search of the indexes, again using subject-appropriate keywords with the additional term, *review*, often retrieves special articles called review articles. Review articles are very useful because the writer, who is an expert in the field, produces a history of the subject to date, and identifies the researchers and the articles that have brought it to this point. Attached to review articles are extensive bibliographies of all the important works to date. Several books that review agricultural economics literature are mentioned in the core literature section of this chapter.

Widespread electronic access to databases and library catalogs has made it possible for almost anyone to quickly create their own extensive bibliography tailor-made to their own needs by searching appropriate indexes and library catalogs. These bibliographies would be in the category of ''quick and dirty'' lists of possibly useful materials, not selective lists. The NAL has published a series of AGRICOLA searches titled *Quick Bibliography Series*. These searches can be easily updated as the search strategy is included in the publication. Many of these bibliographies are focused on topics related to the economics of agriculture.

Dissertations are also an excellent source of bibliographies in specialized subject areas. A major review of the literature of the subject of the thesis is a requirement for acceptance of the thesis. Information on locating dissertations is provided later in this chapter.

Agricultural economics department publication lists are important resources useful for keeping up with new publications by research faculty. The AGECONdotCOM Library contains a list of Departments of Agricultural Economics (http://www.aeco.ttu.edu/Links/acadepts.htm). Most departments link to a list of their department and faculty publications from their home pages. A good example is the Giannini Foundation of Agricultural Economics list of agricultural economics publications (Dote, 2000) from the University of California.

## BIOGRAPHIES

Searches of various databases and online library catalogs can be done to retrieve biographies of agricultural economists. CAB Abstracts, AGRICOLA, and Library

of Congress Subject Headings all use the term *Agricultural Economists* as
a subject heading. Pairing this term with *Biographies* in CAB Abstracts and
AGRICOLA, or *Biography* in online library catalogs will retrieve information
on individual agricultural economists. In addition, *The New Palgrave: A Dictionary of Economics* (Eatwell, 1987) contains about 700 biographies of economists.

*Agricultural and Veterinary Sciences International Who's Who*. 1994. 5th edition.
New York, NY: Stockton Press, 1,157 p.    This directory contains an alphabetical
listing of notable persons involved with agriculture. Each listing includes a brief
biographical sketch with an address. At the end of the volume, names are listed
alphabetically by country and subdivided by subject category. One of the subject
categories under each country is *agricultural economics.*

*American Men and Women of Science: A Biographical Directory of Today's
Leaders in Physical, Biological and Related Sciences*. 1998. 20th ed. New Providence, NJ: R.R. Bowker, 8 vols., 8,500 p.    This directory includes listings with
brief biographical information. One of the categories of interest is agricultural
economics. This directory does not cumulate from edition to edition, so it is
necessary to check older editions for biographies of economists who are deceased.

*Who's Who in Economics*. Blaug, M., ed. 1999. 3rd ed. Northampton, MA: Edward Elgar, 1,235 p.    This alphabetical list of economists who are most often
cited in the journal literature has a principal field of interest at the back of the
book, which includes agricultural economists and natural resources economists.
A brief biography of each economist, including career history, professional affil-
iations and awards, brief statements of principal contributions, and bibliographies
is available. The dictionary covers 1700–1996 and is indexed by place of birth
and residence.

## CORE LITERATURE

There have been numerous studies done through the years detailing the history
of scholarly work in the discipline of agricultural economics. By reviewing the
publication output of researchers, it is possible in retrospect to identify changes
in research direction and concentration that have resulted in new approaches to
solving the economic problems of agriculture. These historical reviews have produced lists of the most significant scholarly work published during the nearly
100 years of the discipline's existence. Two of the studies (Martin, 1977–92)
and (Olsen, 1991), are interesting because, although they are not years apart in
publication time, they are years apart in methodology. The Martin volumes are
extensive studies that rely on a scholar's approach to the field and personal knowledge of all work that has gone on before to identify the significant contributions
to the literature. The Olsen book relies on a system of citation analysis to deter-

mine the most important publications based on the number of times they are cited.

Another method of looking at important literature in a field is to examine materials awarded prizes by professional associations. For example, the American Agricultural Economics Association provides an annual award for outstanding journal articles, published research reports, theses, and publications of enduring quality. The awards are listed each year in the final issue of the *American Journal of Agricultural Economics*.

## Monographs

Gardner, B.L. and R.C. Rausser, eds. 2001. *Handbook of Agricultural Economics*. 2 vols. Amsterdam, The Netherlands: Elsevier Science.   This work contains a set of commissioned articles that survey the field of agricultural economics. The publication is in two volumes, 1A: Agricultural Production, and 1B: Marketing, Distribution and Consumers.

Martin, L.R., ed. 1977–92. *A Survey of Agricultural Economics Literature*. 4 vols. Minneapolis, MN: University of Minnesota Press.   This work, commissioned by the American Agricultural Economics Association, is a comprehensive survey of agricultural economics literature. The volumes cover the literature from the 1940s to the 1970s and review monographs, journal articles, and document literature. The literature was selected and the chapters written by outstanding scholars. Volume 1 covers the traditional fields of agricultural economics; volume 2 covers quantitative methods; volume 3 covers welfare, rural development, and natural resources; and volume 4 covers agriculture in economic development.

Olsen, W.C. 1991. *Agricultural Economics and Rural Sociology: The Contemporary Core Literature*. Ithaca, NY: Cornell University Press, 346 p.   Olsen's book is the most comprehensive guide to core books and monographs for agricultural economics. Through an involved citation analysis, Olsen compiled the top-ranked monographs for developed countries and then did the same for developing countries. Also listed in Olsen's book is a compilation of 11 classic monographs determined by a vote of the Fellows of the American Agricultural Economics Association in 1987 (p. 334).

Sondag, P., G. Dote, and L. Letnes. 1991. *Core Monographs in Agricultural Economics: Survey Report*. Michigan State University, Department of Agricultural Economics, Staff Paper 91–49. East Lansing, MI: 16 p.   This article reports the results of a 1990 survey of agricultural economists to identify core monographs in the field of agricultural economics. Agricultural economics faculty were asked to list titles that were ''classic, core, basic, must-have'' titles for agricultural economics. The list in the report consists of 35 titles that were cited more than once by survey respondents. A full list of cited titles (160) is also included.

## Core Journal Articles

Fox, K.A. and D.G. Johnson. 1969. *Readings in the Economics of Agriculture.* The Series of Republished Articles on Economics, v. 13. Homewood, IL: Richard D. Irwin, 517 p. This volume is one of a series sponsored by the American Economics Association to republish and provide an overview of major studies published in journals or conference/symposium volumes from 1945–66. This volume surveys the literature of agricultural economics.

Peters, G.H., ed. 1995. *Agricultural Economics.* International Library of Critical Writings in Economics. Brookfield, VT: Edward Elgar, 633 p. This volume is a collection of important journal articles in the field of agricultural economics. Most of the articles are from the 1980s and 1990s, but some were published in the 1960s and 1970s. The articles are divided into sections: Dominant Issues in Agricultural Economics, Agricultural Policy Analysis, Agricultural Policy Studies, and Issues Relating to Developing Countries.

## Keeping Up with New Publications

Keeping up with new monographs that originate from so many sources is difficult. New books are regularly reviewed in some journals (often long after they are published) and some have a list of newly received (waiting to be reviewed) books. Some journals contain publishers' advertisements for new books. A discussion and list of the core journals of agricultural economics is found later in this chapter. Journals that feature book reviews and/or lists of new books are identified in this list. See the article by Clark and Mai (2000) for more agricultural economics journals that include reviews.

## DICTIONARIES AND ENCYCLOPEDIAS

Dowdy, G.T. and L.W. Garnett. 1966. *Dictionary of Agricultural Economics and Related Terms.* Tuskegee Institute, AL: Tuskegee Institute, 90 p. While somewhat dated, this short volume does a good job of defining words and/or phrases frequently found in agricultural economics publications.

Eatwell, J., M. Milgate, and P. Newman. 1987. *The New Palgrave: A Dictionary of Economics.* 4 vols. New York, NY: Stockton Press. More like an encyclopedia than a dictionary, the nearly 2,000 articles included are detailed, and written by subject experts. Each article includes an extensive bibliography and refers to related articles within the publication. The article on agricultural economics covers the history of the discipline and describes the development of many subfields such as marketing and supply/demand analysis. Other areas covered are agricultural growth and population change, agriculture and economic development, agricultural supply, free trade and protection, and futures markets. Biographies of many economists are included in the dictionary.

Food and Agriculture Organization of the United Nations. 1992. *Glossary of Terms for Agricultural Insurance and Rural Finance*. FAO Agricultural Services Bulletin 100. Rome, Italy: FAO, 155 p. This work is targeted at researchers, teachers, and practitioners. Two main fields are covered—rural finance and agricultural insurance. The purpose of the publication is to ease dialog and comparisons between countries using these services.

Fuell, L.D., D.C. Miller, and M. Chesley. 1988. *Dictionary of International Agricultural Trade*. Agriculture Handbook (USDA) no. 411. Washington, DC: U.S. Foreign Agricultural Service, 96 p. Su DOCS no.: A 1.76 no. 411. This publication is useful for those studying international agricultural trade. Included are definitions of terms related to commodities, policies, programs, transportation, and storage. Appendices cover conversion factors for various commodities.

Heil, S. and T.W. Peck, eds. 1998. *Encyclopedia of American Industries*. 2nd ed. 2 vols. Detroit, MI: Gale Research. These volumes first provide industry snapshots, followed by information on organization and structure; background and development; current conditions; industry leaders; information sources; workforce descriptions; overviews of the relationship of the U.S. industry to the rest of the world; and information about research and technology in the world. The publication covers food industries such as meat-packing plants, fluid milk, grain-milling products, sugar beets, wheat, rice, beef cattle feedlots, hogs, farm-management services, and general farms. The volumes include many graphs and charts.

Hinkelman, E.G. 1999. *Dictionary of International Trade: Handbook of the Global Trade Community: Includes 12 Key Appendices*. 3rd ed. Novato, CA: World Trade Press, 412 p. This dictionary is more than an ordinary dictionary as it includes maps, currencies, weights and measures, a list of Web resources, and a list of reference resources on international trade.

Lipton, K.L. 1991. *Agriculture, Trade, and the GATT: A Glossary of Terms*. Agriculture Information Bulletin (USDA) no. 625. Washington, DC: USDA. Economic Research Service, 58 p. Su DOCS no.: A 1.75 no. 625. Terms associated with the General Agreement on Tariffs and Trade (GATT), plus other terms related to agriculture programs, food assistance, trade and development, and conservation policies are defined.

Lipton, K.L. 1995. *Dictionary of Agriculture: From Abaca to Zoonosis*. Boulder, CO: Lynne Rienner Publishers, 345 p. The emphasis of this dictionary is agricultural policy and economics. Also included is a listing of major U.S. agricultural and trade legislation, provisions of North American Free Trade Agreement (NAFTA) affecting agriculture, and a guide to weights and measures.

Lipton, K.L. and S.L. Pollack. 1989. *A Glossary of Food and Agricultural Policy Terms, 1989.* Agriculture Information Bulletin no. 573. Washington DC: USDA, Economic Research Service, 46 p. Su DOCS no.: A 1.75 no. 573. "This glossary is designed to serve as a practical guide to the many terms associated with food and agricultural policies and programs. Embodied in a myriad of complex programs, these policies cover agricultural commodities, international trade and development, domestic and international food assistance, and conservation. The glossary includes program descriptions, terms used in implementing the programs, and the Federal agencies involved" (p. i).

Magill, F.N. 1991. *Survey of Social Science*: *Economics Series*. 5 vols. Pasadena, CA: Salem Press. Magill's work includes sections on agricultural economics, international trade, and other topics of interest to agricultural economics. Each signed article is followed by a bibliography of suggested readings. A subject index provides the reader with cross-references to related sections of interest.

Schapsmeier, E.L. and F.H. Schapsmeier. 1976. *Encyclopedia of American Agricultural History*. Westport, CT: Greenwood Press, 467 p. This encyclopedia includes definitions of agricultural programs, legislation, and organizations. The items are arranged alphabetically with special topic indexes at the end of the volume bringing terms and phrases on particular issues together.

Womach, J. and C. Canada. 2000. *Agriculture*: *A Glossary of Terms, Programs, Laws and Websites*. Huntington, NY: Nova Science Publishers, 247 p. Nearly 2,000 terms and phrases related to agriculture are defined in this work. The emphasis of the glossary is on agricultural policy. Also included are acronyms, agencies, programs, laws, and Web sites connected to agriculture. An earlier version of this work (1997) is available as a Congressional Research Report 97-905 ENR on the Internet at http://www.cnie.org/nle/AgGlossary/AgGlossary.htm.

World Bank. 1994. *Annotated Glossary of Terms Used in the Economic Analysis of Agricultural Projects*; *Taken from J. Price Gittinger, Economic Analysis of Agricultural Products*. Washington, DC: World Bank, 134 p. This glossary contains the definitions of words frequently used in agricultural project analysis. The terms are defined in Spanish and in English.

## DIRECTORIES

Some might question the value of directories at a time when so many companies and organizations have Internet sites that probably give more up-to-date information than a directory would. These are valid comments and if someone wanted the address of a particular company and the name of its CEO, the first place to look probably would be the Internet. But if the question to be answered is one asking for a list of all the cooperatives based in Minnesota, a directory would be the place to go. The value of the old-fashioned paper directory is that it brings together similar organizations or companies then divides them by state or region, by product, or by size, etc.

American Agricultural Economics Association, 1987– . *Directory and Handbook.* Ames, IA: The Association. Published annually. This alphabetical list of agricultural economists includes their names, contact information, and specialty listing. Also included in the directory is a list of agricultural economics departments in the United States and Canada.

American Economic Association, 1999. *1997 Survey of Members Including Classification Listings.* In *American Economic Review* 87(6): 31–674. The American Economics Association occasionally publishes a membership survey as a special edition of its journal, the *American Economic Review.* In this directory, members are listed alphabetically with a brief biography. This is followed by a classification of members by fields of specialization and academic affiliation. Many agricultural economists are included in this directory. This directory is also available on the Web at http://www.eco.utexas.edu/AEA.

American Society of Farm Managers and Rural Appraisers. 1929– . *Membership Directory.* Denver, CO: The Society. Published annually. "The mission of the American Society of Farm Managers and Rural Appraisers is to represent professionals in financial analysis, valuation, and management of agricultural and rural resources" (1997 annual). This directory contains a list of their membership.

Baumann, S.C., compiler. 1995. *The Economics Institute Guide to Graduate Study in Economics and Agricultural Economics in the United States of America and Canada.* 9ᵗʰ ed. Boulder, CO: The Economics Institute, 372 p. This comprehensive directory provides a list of 170 graduate training programs in economics and agricultural economics at universities in the United States and Canada. Information is included on size, location, strengths, and degree requirements. In addition, each entry contains a list of faculty members with information on their fields of specialization and professional experience.

International Association of Agricultural Economists. 1999. *IAAE Membership Directory 1996–1998.* Oak Brook, IL: The Association, 67 p. The Association updates this directory every three years, providing contact information for each member by country.

USDA, Cooperative State Research, Education, and Extension Service (http://www.pwd.reeusda.gov). *Cooperative State Research, Education, and Extension Service Online Directory of Professional Workers in Agriculture.* This Web publication replaces the former *Directory of Professional Workers in State Agricultural Experiment Stations and Other Cooperating State Institutions*, published as *Agriculture Handbook no. 305*. The last print publication was issued in 1994. This database lists all of the agricultural economists (plus workers from other disciplines) employed at U.S. state agricultural experiment stations and the USDA.

USDA, Economic Research Service (http://www.ers.usda.gov/AboutERS/ specialists). 2000. About ERS: Subject Specialists. The ERS has created an Internet directory to provide a list of ERS specialists. The listing includes e-mail addresses and phone numbers of these specialists. The list is arranged by subject matter.

USDA, Foreign Agricultural Service. 1998. *Information Sources on International Agricultural Trade*. Washington, DC: The Service, 84 p. Su DOCS no.: A 67.2: In 3/998. This directory provides a listing of countries, commodities, and other subject terms, with experts listed for each. The experts are from inside and outside of the Foreign Agricultural Service (FAS). The directory is also available on the FAS home page at http://www.fas.usda.gov/scriptsw/fassubj/fassubj-frm.asp.

USDA, Rural Business-Cooperative Service (http://www.rurdev.usda.gov/rbs/ pub/sr22.pdf). 1968– . Directory of Farmer Cooperatives. Washington, DC: The Service. Published annually. Farmer-owned cooperatives in the United States are listed in this directory. The listings are by state and include the type of cooperative as well as contact information.

## GOVERNMENT DOCUMENTS

Government publications are important resources for agricultural economists. Economists use statistical data to find, test, and verify their conclusions and many of the statistics are gathered by government agencies on the national, state, or local level. Intergovernmental agencies are also collectors, publishers, and distributors of a good deal of the statistical data available. Two major agricultural indexes, AGRICOLA and AGRIS, include government publications appropriate for agricultural economics research. Many countries and intergovernmental organizations are now using the Internet to distribute their publications and statistical data. The Internet guides listed in the Web section of this chapter will assist in locating these resources. There are also many government documents listed in the statistics section of this chapter. A new comprehensive guide to U.S. documents that traces series through years of publication has recently been published (Batten 2000).

## GUIDES TO THE LITERATURE

Literature guides are particularly useful for those who have not had much experience using the literature of a particular discipline. These guides are an organized reminder that indexes, handbooks, or other types of works exist that will help in a search for information. Because any guide to the literature of an academic discipline is out of date before it is published, it is necessary to check for additional recently published works in order to stay up to date.

Ben-Zion, B. 1994. *Economic Information: Where to Find It.* Santa Rosa, CA: BZ Publications, 200 p.    This volume is an annotated economics literature guide that includes the major reference tools for economics, a list of major economics journals, and a lengthy list of statistical sources. A section for agricultural economics statistical sources is included as well as a section on statistical methods and interpretation.

Blanchard, J.R. and L. Farrell, eds. 1981. *Guide to Sources for Agricultural and Biological Research.* Berkeley, CA: University of California Press, 735 p.    This is a thorough and comprehensive guide to the literature of agriculture. It has strong social sciences sections including agricultural components of economics, biography, history, geography, legislation, development, education, land reform, and rural sociology. Each section is arranged by type of material and includes annotated citations. The Blanchard volume continues to be useful as the basic guide to resources prior to 1981.

Herron, N.L., ed. 1996. 2$^{nd}$ ed. *The Social Sciences: A Cross-Disciplinary Guide to Selected Sources.* Englewood, CO: Libraries Unlimited, 323 p.    This social science literature guide starts with a general introduction to the literature of the social sciences and provides annotations for general social science reference tools. Chapters on subdisciplines of the social sciences follow. The economics and business chapters will be especially useful for economists. Neither chapter directly addresses agricultural economics or agribusiness, but the work provides lengthy annotations of more general resources.

Li, T. 2000. *Social Science Reference Sources: A Practical Guide.* 3$^{rd}$ ed. Westport, CT: Greenwood Press, 495 p.    Li's work is the most current listing of social science reference sources presently available. It gives a general overview of reference sources, including Internet search engines and Web sites. The second half of the guide is devoted to subdisciplines of the social sciences, including sections for business and economics. These sections list materials by type of publication and contain informative annotations that both describe and compare the resources. The economics resources will be useful for agricultural economists.

Littleton, I.T. 1969. *The Literature of Agricultural Economics: Its Bibliographic Organization and Use.* North Carolina Agricultural Experiment Station. Technical Bulletin no. 191. Raleigh, NC: The Station, 65 p.    This work is dated, but provides a historical perspective of the field. The report is an outcome of thesis research done on indexing and abstracting of U.S. agricultural economics literature.

Olsen, W.C. 1991. *Agricultural Economics and Rural Sociology: The Contemporary Core Literature*. Ithaca, NY: Cornell University Press, 346 p.   This is a comprehensive monograph on the literature of agricultural economics. In this book Bernard Stanton traces the development of agricultural economics in the United States and Margot Bellamy looks at CAB International's services for agricultural economists. Olsen authored the chapter on the characteristics of agricultural economics literature. The bulk of the book consists of Olsen's core lists of monographs, journals, and serials in agricultural economics identified by using citation analysis. Also included is a list of grey literature series in agricultural economics. The final chapter of the book updates the social sciences section of the *Guide to Sources for Agricultural and Biological Research* (Blanchard, 1981). The Olsen book is must reading for anyone involved in collection management of agricultural economics literature.

Schreiber, M.N. 1997. *International Trade Sources: A Research Guide*. Research and Information Guides in Business, Industry, and Economic Institutes 12. New York, NY: Garland Publishing, 327 p. Also issued as *Garland Reference Library of Social Science 1068*.   Schreiber's guide to international trade sources will prove useful to agricultural economists specializing in agricultural trade research. Schreiber provides annotated listings of many resources along with indexes to provide access by author, title, and subject.

## HANDBOOKS AND GUIDES

Handbooks are valuable reference tools. Usually the information in a handbook is written for a particular audience and, very often, the contents provide a wide range of information of interest to that audience. Some examples of the kind of information that might be included are society information, directory of members, statistics, standards, and suppliers. This list of handbooks and guides is short partly because of decisions made to place some works that might be called handbooks in other sections of this chapter, many in the statistical section.

Hoag, D.L. 1999. *Agricultural Crisis in America: A Reference Handbook*. Santa Barbara, CA: ABC-CLIO, 270 p.   ''Seven agricultural crises are examined in this book: (1) Farm and Ranch Survivability, (2) Modernization, (3) Feeding a Growing World, (4) Safe Food and Drinking Water, (5) Stewardship and the Environment, (6) Urbanization and Land Use, and (7) Country and Urban Conflicts.'' After the crises are described in depth, a chronology of events that led to the crises are provided, as well as biographies of persons related to the crises. Also included in the handbook are lists of organizations, resources for further reading, and a glossary of terms.

Rose, F.S., ed. 1999. *Commodity Trading Manual.* Chicago, IL: Chicago Board of Trade, 410 p. This guide gives an introduction and overview of the futures industry. A description of the futures market for each traded commodity is included.

Ulrich, H., ed. 1985. *The Practical Grain Encyclopedia.* Chicago, IL: Commodity Center Corporation, 169 p. This encyclopedia contains a glossary of market terminology, maps of major growing areas, crop schedules, farmer marketing data, a history of USDA programs, and historical charts of prices and inter-commodity spreads.

## JOURNALS

Scholarly research is usually advanced in small incremental steps by individual researchers. New insights and advances in knowledge made by researchers are generally reported to the entire scholarly community in the form of journal articles. The journals that publish these articles serve two purposes—first, reading the journals keeps the researcher informed about advances made by others, and second, the journals offer a place for the researcher's own work to be viewed by others.

Several studies have been undertaken in recent years to attempt to define a core collection of agricultural economics journals. Olsen's study of the core literature of agricultural economics (Olsen, 1991) found that a large number of journals/serials (5,000) are scanned for the agricultural economics sections of CAB Abstracts. Olsen did a citation analysis of these journals to ascertain which titles were the primary/core journals for agricultural economics. This comprehensive list is published in his book on agricultural economics literature (Olsen, 1991, 78). A survey of the journal holdings of agricultural economics department libraries and reading rooms was done to determine the most frequently held agricultural economics journals (Dote, 1996). This survey resulted in a list of 243 journal titles arranged by frequency of ownership. Another citation survey of a predetermined set of 19 agricultural economics journals was done to develop a ranking of those 19 journals (Perry, 1999). In addition to the 19 ranked agricultural economics journals, Perry prepared another list of frequently cited journals not specifically devoted to agricultural economics, but cited by agricultural economics articles. Another rankings study (Barrett, 2000) developed a ranking of journals in 16 subdisciplines of applied economics, including agricultural economics. Another study (Broder, 1984) conducted a survey of agricultural economists to establish a ranking of journals of agricultural economics.

The following list contains the titles of journals specifically devoted to the literature of agricultural economics. Journals from other areas of economics are included if they appeared on more than one of the core agricultural economics lists in the previously mentioned studies. Each journal is marked to designate if it has book reviews (R) or new book lists (L) in each issue.

*Agribusiness* (R)
*Agricultural Economics* (R)
*Agricultural Economics Research* (ceased publication)
*Agricultural Finance Review* (R)
*Agricultural and Resource Economics Review* (formerly *Northeastern Journal of Agricultural and Resource Economics*)
*American Economic Review*
*American Journal of Agricultural Economics* (R)
*Australian Journal of Agricultural and Resource Economics* (formerly *Australian Journal of Agricultural Economics*) (R)
*Canadian Journal of Agricultural Economics* (R)
*Econometrica*
*Economic Development and Cultural Change* (R)
*Economic Journal* (R) (L)
*European Review of Agricultural Economics* (R)
*Food Policy* (R)
*Food Research Institute Studies* (ceased publication)
*Indian Journal of Agricultural Economics* (R) (L)
*International Economic Review* (L)
*Journal of Agribusiness*
*Journal of Agricultural and Applied Economics* (formerly *Southern Journal of Agricultural Economics* (R)
*Journal of Agricultural Economics* (R) (L)
*Journal of Agricultural and Resource Economics* (formerly *Western Journal of Agricultural Economics*)
*Journal of the American Society of Farm Managers and Rural Appraisers*
*Journal of the American Statistical Association* (R)
*Journal of Cooperatives* (formerly *Journal of Agricultural Cooperation*) (R)
*Journal of Econometrics*
*Journal of Economic Literature* (R) (L)
*Journal of Environmental Economics and Management*
*Journal of Futures Markets*
*Journal of International Food and Agribusiness Marketing*
*Journal of Political Economy* (R)
*Land Economics* (R) (L)
*Quarterly Journal of Economics*
*Review of Agricultural Economics* (formerly *North Central Journal of Agricultural Economics*)

*Review of Economics and Statistics*
*Review of Economic Studies*
*Review of Marketing and Agricultural Economics* (merged with *Australian Journal of Agricultural Economics* and renamed *Australian Journal of Agricultural and Resource Economics*)

## PROCEEDINGS

Most professional associations sponsor conferences for their members. The conferences are designed to exchange information and ideas and the papers presented are often published—sometimes in a volume of the society's journal, sometimes in a separate proceedings volume. Some associations post their papers on the Internet. The following list of agricultural economics and related societies gives the Web address of the society/association. Announcement of conferences and publications relating to them are often found at these Web sites.

Agricultural Economics Association of South Africa
http://www.aeasa.org.za
Agricultural Economics Society
http://www.aes.ac.uk
American Agricultural Economics Association
http://www.aaea.org
American Society of Farm Managers and Rural Appraisers
http://www.asfmra.org
Association of Environmental and Resource Economists
http://www.aere.org
Australian Agricultural and Resource Economics Society
http://www.general.uwa.edu.au/u/aares
Canadian Agricultural Economics Society
http://www.caes-scae.org/caes-e.htm
European Association of Agricultural Economists
http://www.lei.dlo.nl/EAAE
Food Distribution Research Society
http://fdrs.ag.utk.edu
International Agricultural Trade Research Consortium
http://iatrcweb.org
International Association of Agricultural Economists
http://www.iaae-agecon.org
International Food and Agribusiness Management Association
http://www.ifama.org
Korean Agricultural Economics Association
http://www.uriel.net/~agrecon

Malaysian Agricultural Economics Association
   http://www.econ.upm.edu.my/~peta
National Agri-Marketing Association
   http://www.nama.org
Northeastern Agricultural and Resource Economics Association
   http://www.narea.org
Russian Independent Agricultural Economics Association
   http://www.user.cityline.ru/~vmlca/naekor/engl.htm
Southern Agricultural Economics Association
   http://www.ag.auburn.edu/saea
Western Agricultural Economics Association
   http://dare.agsci.colostate.edu/thilmany/waea.htm

## STATISTICS

The importance of statistics to all economists cannot be overstated. Most of the research done by economists is quantified by use of statistics. Agricultural economists concentrate on providing solutions to economic problems connected to food and agriculture so they are particularly interested in the statistics related to those areas. Included in this section are the indexes used for finding agricultural statistics, followed by sections on general agricultural statistics, USDA statistics, and international statistics. Because many handbooks contain statistics, they are included in these sections.

### Statistical Indexes

*American Statistical Index.* 1973– . Bethesda, MD: Congressional Information Service. *Index to International Statistics.* 1983– . Bethesda, MD: Congressional Information Service. *Statistical Reference Index.* 1980– . Bethesda, MD: Congressional Information Service.   These three statistical indexes cover three categories of statistical data—U.S. government statistics; international intergovernmental organizations statistics; and state (United States) and private organization statistics. The indexes identify, catalog, announce, describe, and index thousands of statistical publications and, in addition, provide a microfiche copy of most of these works. Each publication is described in detail and indexed by subject, name, and category. All three are available in print, online, and on CD-ROM. The three separate indexes have been merged into a single database and this database is available as part of the Web-based Statistical Universe. Many of the publications are available full text through this database.

### Statistical Handbooks, Guides, and Data Sources

AAEA Task Force on Commodity Costs and Returns. 1998. *Commodity Costs and Returns Estimation Handbook.* Ames, IA: American Agricultural Economics

Association, 530 p. ''This handbook's purpose is to gather in one place information on estimating costs of and returns to agricultural enterprises'' (p. xi).
The monograph covers topics such as government programs participation, inputs,
machinery and building costs, land, labor, farm overhead, international comparisons, and data sources. The report also includes a glossary. The report is available
on the Web at http://waterhome.brc.tamus.edu/care/Aaea.

*Agricultural Finance Databook.* 1978– . Washington, DC: Board of Governors
of the Federal Reserve System, Division of Research and Statistics. Published
quarterly.   This quarterly databook compiles agricultural finance statistics that
are gathered by the Board of Governors of the Federal Reserve System. The data
includes information on farm loans, agricultural banks, and trends in farm real-
estate values.

BRIDGE Information Systems. Bridge Commodity Research Bureau. 1997– .
*The CRB Commodity Yearbook.* New York, NY: John Wiley and Sons. Published
annually.   This annual guide to commodity exchanges, formerly published by
Knight Ridder Financial Commodity Research Bureau, was purchased in 1997
by Bridge Information Systems. The volume is arranged alphabetically by commodity. An overview of the U.S. and world situations and a list of futures markets
that trade the particular commodity begin each section. The introduction is followed by statistics on production, stocks, supply and utilization, prices, volume,
and futures trading.

Buckley, J., consulting editor. 1996. *Guide to World Commodity Markets*: *Physical, Futures and Options Trading.* 7th ed. London, England: Kogan Page,
619 p.   After an introduction to commodity marketing, this guide describes each
individual commodity market, along with statistics for many countries on production, exports, and imports. Each section has graphs and charts to illustrate the
trends in that particular market. The final sections of the guide consist of directories of commodity markets and commodity market trading members. The guide
includes a glossary of terms.

*Dun & Bradstreet/Gale Group Industry Handbook*: *Construction and Agriculture.*1999– . Detroit, MI: Gale Group. Published annually.   This handbook gathers together data and information from federal and private sources. The agricultural section of this edition begins with an article on the changing structure of
agriculture, an overview of the agricultural situation, and profiles of leading companies in the agribusiness field. The next section includes agricultural industry
statistics arranged by SIC code. Also included are financial norms and ratios for
agricultural commodities, directories, and company rankings.

Friedman, C., ed. 1991. *Commodity Prices*: *A Source Book and Index Providing
References to Wholesale, Retail, and Other Quotations for More Than 10,000
Agricultural, Commercial, Industrial, and Consumer Products.* 2nd ed. Detroit,

MI: Gale Research, 630 p. Finding commodity prices can be a time-consuming process. This guide indexes over 200 trade publications, newsletters, and government documents to assist the user in locating various price series (spot, cash, retail, and futures). The indexing is done alphabetically by commodity name for over 10,000 commodities including agricultural commodities.

Garkey, J. and W.S. Chern. 1986. *Handbook of Agricultural Statistical Data.* American Agricultural Economics Association, Economic Statistics Committee, 139 p. This report was prepared to pull together into one publication, information on the many U.S. agricultural data series from various governmental agencies. The entries are arranged by agency and contain detailed descriptions of the methodology, variables, coverage, time period, and publications related to the data. Many of the data series described in this publication have ceased or merged with other series in the 14 years since its publication, but this remains a valuable source for understanding the time series data used frequently by agricultural economists.

Le Vallee, J. 1999. *Market Information Sources Available Through the Internet*: *Daily to Yearly Market and Outlook Reports*, *Prices*, *Commodities and Quotes.* Michigan State University, Department of Economics and Department of Agricultural Economics, MSU International Development Working Paper 64. East Lansing, MI: 30 p. "This guide may be of use to agricultural economists, market information analysts, librarians, and others who are interested in market information. Although the Web site concentrates partly on grains, many of the same references to Web sites in this virtual library refer to many other types of commodities . . ." (p. 3). The guide is in print and also on the Web at http://www.aec.msu.edu/agecon/fs2/market/contents.htm.

*North American Industrial Classification System.* 1997. Washington, DC: Executive Office of the President, Office of Management and Budget, 1247 p. *Standard Industrial Classification Manual.* 1987. Washington, DC: Executive Office of the President, Office of Management and Budget, 705 p. The North American Industrial Classification System (NAICS) set of codes replaced the SIC codes in 1997. These codes are assigned to classify all federal statistical data by type of activity in order to make data compatible. The NAICS appendices provide a chart that assists in comparing the SIC codes to the NAICS codes. The agriculture codes begin with 11. The NAICS codes are available on the Web at http://www.census.gov/naics.html. The SIC codes are available at http://www.osha.gov/oshstats/sicser.html.

*RMA Annual Statement Studies.* 1923– . Philadelphia, PA: RMA. Published annually. This report gives financial ratios for manufacturing, wholesale, retail, construction, and consumer finance establishments. Agriculture is one of the industries covered, with breakdowns within agriculture by SIC code (i.e., wheat, corn, cotton, beef cattle feedlots, and dairy farms.)

*Soya and Oilseed Bluebook.* 1947– . Bar Harbor, ME: Soyatech. Published annually. The *Soya and Oilseed Bluebook* gathers together statistics from various sources for the soy industry (including canola, corn, cottonseed, palm, soybean, and sunflowers). Statistics on production, trade, processing, and prices are provided for the United States and for the world. There is also a directory of companies, a glossary, and standards.

*Statistical Abstract of the United States.* 1894– . Washington, DC: U.S. Department of Commerce. Published annually. Su DOCS no.: C 3.134. This annual compilation of statistics is a good place to begin a search for U.S. data. This publication contains statistics on all aspects of the U.S. economy, including agriculture. Data are included from both government and nongovernment organizations, with some international data. All tables are well documented and can serve as guides to more statistics on a particular subject. The appendices include a guide to sources of statistics, and an explanation of statistical methodology and reliability. The *Statistical Abstract* is available in print, on CD-ROM, and on the Internet at http://www.census.gov/prod/www/statistical-abstract-us.html. The *Statistical Abstract* is supplemented by an earlier U.S. Census Bureau publication that contains data back to 1610 titled *Historical Abstracts of the United States, Colonial Times to 1970.* (U.S. Bureau of the Census. 1975. Bicentennial Edition. 2 vols. Washington, DC: Government Printing Office. Su DOCS no.: C3.134/2: H62.) *Historical Abstracts* is also available on the Web for the years 1790–1960 at http://fisher.lib.virginia.edu/census/.

U.S. Dept. of Agriculture (USDA) (http://www.usda.gov). The Department of Agriculture is the U.S. government agency charged with keeping and distributing agricultural production, supply, consumption, facilities, costs, and returns statistics. The four agencies within the USDA responsible for collecting the raw data and publishing/distributing the results are the Agriculture Marketing Service (AMS), the Economic Research Service (ERS), the Foreign Agricultural Service (FAS), and the National Agricultural Statistics Service (NASS).

The following are statistical publications from the USDA and its four agencies.

*Agriculture Fact Book.* 1962– . Washington, DC: USDA. Published annually. The *Agriculture Fact Book* is a quick guide to the statistics and programs of the USDA. The guide is written with a lay audience in mind, and is full of facts and figures that can readily be used in presentations. The *Fact Book* contains sections on consumption; structure of agriculture; farm sector economics such as income and balance sheet; rural demographics; and information on the various agencies within the USDA. The *Fact Book* is available in print, and in PDF format on the Web at http://www.usda.gov/news/pubs.

*Agricultural Statistics.* 1936– . Washington, DC: USDA. Published annually. Su DOCS no.: A 1.47.   This annual publication is the place to begin a search for U.S. agricultural statistical data. Included are data on production, prices, trade, farm income, consumption, supplies, and costs and returns. Most of the tables are based on data collected by the USDA. *Agricultural Statistics* is available in print and in PDF format on the Web at http://www.usda.gov/nass/pubs/ agstats.htm. The earliest volume available on the Internet is 1994. The statistical data in each edition generally goes back 10 years.

*Statistical Bulletin.* 1923– . Washington, D.C.: USDA.   The U.S. Department of Agriculture publishes a series of *Statistical Bulletins* that gather data on selected agricultural economics subjects. As can be seen by the sampling of the following titles, most include time series that might be useful in a research study. Other titles in this statistical series can be found in major library catalogs or in the *Catalog of U.S. Government Publications* published by the U.S. Government Printing Office and found online at http://www.access.gpo.gov/su_docs/ locators/cgp/index.html. A few titles in this series include *A Historical Look at Farm Income*; *Weather in U.S. Agriculture, Monthly Temperature and Precipitation by State and Farm Production Region, 1950–90*; *Food Spending in American Households, 1980–92*; *European Agricultural Statistics*; *Foreign Ownership of U.S. Agricultural Land Through December 31, 1997*; and *Agricultural Land Values, Final Estimates 1994–98.*

USDA Agriculture Marketing Service (http://www.ams.usda.gov).   The AMS of the USDA publishes marketing statistics and news covering a wide variety of commodities. The reports include information on prices, volume, quality, condition, and other market data on farm products in specific markets and marketing areas. Some of the commodities covered are cotton, dairy, fruits and vegetables, grains, livestock, poultry, and tobacco. The reports are available in print and at the AMS Web site.

USDA Economic Research Service (http://www.ers.usda.gov).   The ERS provides economic analysis on issues related to agriculture, food, the environment, and rural development to improve public and private decision making. Regular reports on the situation and outlook for various commodities are issued by the ERS. Besides a monthly *Agricultural Outlook* report, the ERS provides reports on agricultural income and finance, aquaculture, cotton and wool, feed, fruit and tree nuts, international agriculture and trade, livestock, dairy, poultry, oil crops, rice, sugar and sweeteners, tobacco, vegetables and specialties, and wheat. Most current ERS publications are available in print and also electronically on the Internet at http://www.ers.usda.gov/Publications. Free e-mail subscriptions are available for many ERS serials through the USDA Web site at Cornell University: http://usda.mannlib.cornell.edu.

USDA Economics and Statistics System (http://usda.mannlib.cornell.edu). 2000. Ithaca, NY: Cornell University, Albert R. Mann Library. A cooperative agreement between the Albert R. Mann Library at Cornell University and three USDA economic agencies (ERS, NASS, and World Agricultural Outlook Board) resulted in an archival Internet site that contains reports and data sets from the USDA. The USDA provides the data while the Mann Library takes responsibility for organizing and archiving the data. Mann Library also provides reference assistance in use of the data through the AgNIC-affiliated Web sites (Lawrence, 1996).

*Agricultural Resources and Environmental Indicators, 1996–97.* 1997. Agriculture Handbook (USDA) no. 712. Washington, DC: U.S. Economic Research Service, 347 p. Su DOCS no.: A 1.76 no. 712. Data and information on the relationships between farming practices, conservation, and the environment in the United States are reported in this publication. The data included in the report cover land, water, production inputs, production management, technology, and conservation/environmental programs. This report is available in print and on the Web at http://www.ers.usda.gov/briefing/arei.

Daberkow, S.G. and L.A. Whitener. 1986. *Agricultural Labor Data Sources: An Update.* Agricultural Handbook (USDA) no. 658. Washington, DC: U.S. Economic Research Service, 25 p. Su DOCS no.: A 1.76 no. 658. "This report identifies, describes, and compares various data sources used to analyze the different components of the agricultural work force, and alerts the reader to differences in data sources that complicate comparisons" (p. iii).

*Farm Business Economics Report, 1996.* 1999. Washington, DC: USDA Economic Research Service, 235 p. This report contains national and state financial summaries and cost of production data. The data is updated on the ERS home page at http://www.ers.usda.gov/briefing/fbe.

*Foreign Agricultural Trade of the United States, FATUS, Calendar Year 1998 Supplement.* 1999. Washington, DC: USDA Economic Research Service, 447 p. Formerly a bimonthly publication with annual summaries, *FATUS* is now published annually with access to current data through the Web. The Web site contains 11 years of export and import information for 211 commodity groups and over 250 countries and regions. Data is from the Census Bureau of the U.S. Department of Commerce. The Web site for *FATUS* is http://www.ers.usda.gov/db/fatus/.

*Products and Services from USDA's Economic Research Service, Annual Issue.* 1999– . Washington, DC: USDA Economic Research Service, Published annually. This catalog consists of a subject listing of the statistical publications of the USDA's Economic Research Service (ERS). Most of these resources are available full-text on the Web and in print format. The catalog can also be found at this Web address: http://www.ers.usda.gov/publications.

PS&D View. 2000. Washington, DC: USDA Economic Research Service. This package includes data on production, supply, and distribution variables for many agricultural commodities, along with software that can be used to graph and display the statistics. Data is included for over 190 countries and regions. It is updated monthly and can be purchased from NTIS or downloaded from this Web site: http://usda.mannlib.cornell.edu/data-sets/international/93002.

Putnam, J.J. and J.E. Allshouse. 1999. *Food Consumption, Prices and Expenditures, 1970–97.* Statistical Bulletin (USDA) no. 965. Washington, DC: U.S. Economic Research Service, 155 p. Su DOCS no.: A 1.34 no. 965. Published annually, this report provides U.S. data by commodity on consumption, prices, and expenditures for food. The report is also on the Web at http://www.ers.usda.gov/epubs/pdf/sb965.

Reinsel, E., compiler. 1993. *Agricultural and Rural Economics and Social Indicators.* Agriculture Information Bulletin (USDA) no. 667. Washington, DC: U.S. Economic Research Service, 61 p. Su DOCS no.: A 1.75 no. 667. "This guide provides descriptions and illustrative samples of available statistics from the agricultural and rural economic and social indicators published by the Economic Research Service. . . . Users will learn what the statistics measure, when and where the information is published, whether the data are available in electronic form, and whom to contact for more information" (p. i).

*Weights, Measures, and Conversion Factors for Agricultural Commodities and Their Products.* 1992. Agriculture Handbook (USDA) no. 697. Washington, DC: U.S. Economic Research Service, 71 p. Su DOCS no.: A 1.76 no. 697. This handbook includes conversion factors as well as weights and measures used for agricultural commodities and their products.

Publications from the USDA Foreign Agricultural Service (http://www.fas.usda.gov). The FAS has primary responsibility for the USDA's overseas programs—market development; international trade agreements and negotiations; and the collection of statistics and market information. It also administers USDA's export credit guarantee and food aid programs and helps increase income and food availability in developing nations by mobilizing expertise for agriculturally led economic growth.

The FAS prepares a number of statistical reports. The reports, issued several times per year, include titles on cotton, dairy, livestock, poultry, grain, oilseeds, sugar, tobacco, tropical products (coffee, cocoa, spices, essential oils), and horticulture. These reports are available in print and on the FAS Web page. Also available on the FAS Web page are country reports prepared by FAS attachés (http://www.fas.usda.gov/scriptsw/attacherep/default.asp). These reports cover nearly 170 countries and provide up-to-date statistics and commentary on the agricultural situation in these particular countries. These reports may be searched by country, topic, and date.

Dolinsky, D. 1994. *Desk Reference Guide to U.S. Agricultural Trade.* Agriculture Handbook (USDA) no. 683. Washington, DC: U.S. Foreign Agricultural Service, 64 p. Su DOCS no.: A 1.76 no. 683. This guide gives an overview of the various aspects of U.S. agricultural trade. It is divided into four parts: agricultural exports, agricultural imports, fish and forest products, and a statistical appendix.

Publications from the USDA National Agricultural Statistics Service (http://www.usda.gov/nass). Each year NASS conducts hundreds of surveys and prepares reports covering every aspect of U.S. agriculture such as production and supplies of food and fiber; prices paid and received by farmers; and farm labor and wages. In addition, NASS's 45 State Statistical Offices (http://www.usda.gov/nass/sso-rpts.htm) publish data about many of the same topics for local audiences. NASS reports are available in print by subscription with the National Technical Information Service (NTIS), but are also available on the Internet through the NASS home page. A free e-mail subscription service is offered for the reports through the USDA Web site at Cornell University (http://usda.mannlib.cornell.edu). The NASS home page offers historical data, state data, graphics, links to the publication *Agricultural Statistics*, as well as links to the *Census of Agriculture* and many statistical bulletins. Historical data is found at http://www.usda.gov/nass/pubs/histdata.htm#sb.

*Census of Agriculture.* 1840– . Washington, DC: USDA National Agricultural Statistics Service. The *Census of Agriculture* is the most comprehensive, detailed source of data about the agricultural situation in the United States. The *Census of Agriculture* includes national-, state-, and county-level data. From 1840 through 1920 the census was taken by the U.S. Bureau of the Census every 10 years. It was changed to every five years between the years 1925 to 1974. Beginning with 1982, the census has been taken in years ending in two and seven. In 1997 responsibility for the *Census of Agriculture* was shifted to the NASS.

The *Census of Agriculture* includes data on topics such as land use, land ownership, irrigation, acres planted and quantity harvested, livestock and poultry production, value of products sold, set-aside acres, government payments, hired farm workers, operator characteristics, production expenses, use of fertilizer and chemicals, machinery and equipment owned, market value of land and buildings, and farm income. Each census is accompanied by a *Guide to the Census of Agriculture and Related Statistics* that discusses the scope of the *Census of Agriculture* and the publications generated from it. Each *Census of Agriculture* offers some special studies. The 1997 *Census of Agriculture* has special volumes on rankings, data by zip code, an agricultural atlas, aquaculture, horticulture, and irrigation. The *Census of Agriculture* is available in print and on CD-ROM. The

1992/1997 volumes are on the Web at http://www.nass.usda.gov/census/census97.

*Guide to Products and Services*. 1999– . Washington, DC: USDA National Agricultural Statistics Service. Published annually.  This report is published annually and lists NASS statistical publications. It gives a description of each data series, the release dates of each issue and ordering information. The entries are arranged by subject categories, with a detailed commodity index to facilitate location of series on individual commodities. This report is also available in PDF format on the Web at http://www.usda.gov/nass/pubs/catalog.htm.

*Major Statistical Series of the U.S. Dept. of Agriculture*. 1990. Agriculture Handbook (USDA) no. 671. Washington, DC: U.S. National Agricultural Statistics Service, 12 vols. Su DOCS no.: A 1.76 no. 671, v. 1–12.  The goal of this handbook series is to help the users of NASS materials understand the concepts and data used in preparing NASS statistical series. This publication is useful in tracking time-series data as they change series titles and data definitions.

### Statistics (International)

The easiest international statistics to find are those gathered and published by international governmental organizations such as the Food and Agriculture Organization of the United Nations (FAO). Most countries do collect some agricultural statistics and publish them in their official documents, and lately have begun to put some of them on the Internet. The Web site Statistical Data Locators (http://www.ntu.edu.sg/library/stat/statdata.htm) maintained at Nanyang Technological University Library in Singapore, has links to data by country, as well as links to data by subject. Another site located at Pennsylvania State University (http://www.libraries.psu.edu/crsweb/docs/statnats.htm) links to the national statistical office in many countries—a good place to begin a search for data from that country. Resources for Economists on the Internet (http://rfe.wustl.edu/sc.html) and WebEc: WWW Resources in Economics (http://netec.wustl.edu/WebEc/WebEc.html) each have links to sources of national and international data. An article by Geraldine Foudy (Foudy, 2000) provides an overview of international statistics on the Web.

AMAD: Agricultural Market Access Database (http://www.amad.org). 2000.  "AMAD results from a co-operative effort by Agriculture and AgriFood Canada, EU Commission—Agriculture Directorate-General, Food and Agriculture Organization of the United Nations, Organization for Economic Cooperation and Development, The World Bank, United Nations Conference on Trade and Development, United States Department of Agriculture—Economic Research Service" (AMAD home page). The database includes a broad set of information on tariffs, imports, world reference prices, and exchange rates.

*Economic Accounts for Agriculture.* 1999. Paris, France: Organization for Economic Co-operation and Development. Farm accounts for structure, levels and trends in Organization for Economic Co-operation and Development (OECD) countries are included in this publication. The 1999 edition contains data for 1985–98. The data is available in print and on diskette.

*FAO Production Yearbook.* 1958– . Rome, Italy: Food and Agriculture Organization of the United Nations. Published annually. Time-series data for many countries contained in this yearbook include land use, irrigation, agricultural population, indices of agricultural production, data for agricultural production by crop, and livestock production. This data is also available in long-time series on the CD-ROM from FAO, titled FAOSTAT and on the Internet at http://apps.fao.org.

*FAO Trade Yearbook.* 1958– . Rome, Italy: Food and Agriculture Organization of the United Nations. Published annually. FAO's annual trade publication contains time-series trade data by commodity and by country. Exports and imports in means of agricultural production (tractors, fertilizers, pesticides) and value of agricultural trade by country are included. The CD-ROM published by FAO titled FAOSTAT also contains this data in long-time series. FAOSTAT is also available on the Web at http://apps.fao.org.

*International Trade Statistics Yearbook.* 1950– . New York, NY: United Nations Department of Economic and Social Affairs, Statistics Division. Published annually. This annual compilation of trade statistics lists by country the value, volume, and price of commodities, including agricultural commodities.

Kurian, G.T. 1997. *Global Data Locator.* Lanham, MD: Bernan Press, 375 p. Kurian describes the most important general global statistical sourcebooks and includes copies of their tables of contents. He then does the same for other topical statistical sourcebooks, including publications related to agricultural economics.

OECD Agricultural Commodities Outlook Database 1999–2004. 1999. Paris, France: Organization for Economic Co-operation and Development. The database contains statistics and projections for production, consumption, trade, stocks, and prices of agricultural commodities for OECD countries plus other selected countries from 1970 to the present. The data is available on CD-ROM and on diskette.

*Oil World Annual.* 1987– . Hamburg, Germany: ISTA Mielke. Published annually. This annual publication includes detailed statistics and forecasts for 45 commodities and over 120 countries. It covers oilseeds, oils and fats, and oilmeals in terms of supply and demand, prices, production, and price forecasts.

Pardey, P.G. and J. Roseboom. 1989. ISNAR Agricultural Research Indicator Series: A Global Database on National Agricultural Research Systems. New

York, NY: Cambridge University Press, 547 p. "The ISNAR (International Service for National Agricultural Research) Agricultural Research Indicator Series is a fully sourced and extensively documented set of research personnel and expenditure indicators for national agricultural research systems (NARS) in 154 developing and developed countries for the 27 years 1960 through 1986 . . . the series represents a major effort to consolidate and completely restructure previously available data compilations . . ."(p. 3).

*Producer Subsidy Equivalents and Consumer Subsidy Equivalents Database.* 1997. Paris, France: Organization for Economic Co-operation and Development. Statistics for 1973–97 in this data set provide a framework for quantifying agricultural output.

*The State of Food and Agriculture.* 1951– . Rome, Italy: Food and Agriculture Organization of the United Nations. Published annually. *The State of Food and Agriculture* is an annual review that covers recent developments in the agricultural situation worldwide. Each edition begins with a general overview that is followed by review of the agricultural situation in individual countries and regions. A diskette titled Time Series for SOFA has been included with the past several years of the publication. The diskette contains time-series data and graphing software for about 150 countries on population, agricultural production, food supply, agricultural trade, farm inputs, and land use. This publication is also available on the Web at http://www.fao.org/docrep/x4400e/x4400e00.htm.

*Statistics on Prices Paid by Farmers for Means of Production.* 1989– . Rome, Italy: Food and Agriculture Organization of the United Nations, Statistics Division. *Statistics on Prices Received by Farmers.* 1982– . Rome, Italy: Food and Agriculture Organization of the United Nations, Statistics Division. The 1997 (3rd) edition of the prices paid report includes data from 126 countries for 10 years on prices paid by farmers for materials used in agricultural production (such as seeds and pesticides), factor services (such as rental and interest payments), and investment goods (such as machinery and draught animals). The 1996 (6th edition) of the prices received publication includes 10 years of data for many countries.

World Development Indicators. 1997– . Washington, DC: World Bank. Published annually. This annual World Bank publication is available in print and on CD-ROM. The print version contains data for the current year with some comparison data. The CD-ROM contains long time-series for the various indicators. Data is included on land use, agricultural inputs, agricultural output and productivity, economy, states and markets, trade, and aid.

*World Development Report.* 1978– . Washington, DC: World Bank. Published annually. Each issue of this annual report from the World Bank begins with an examination of some aspect of the world economic development situation. An appendix listing selected world-development indicators follows. This appendix is also issued separately in an enlarged, more comprehensive format as *World Development Indicators*. The *World Development Report* is available in print and on the Web at http://www.worldbank.org/wdr/2000/fullreport.html.

*World Grain Statistics*. 1991– . London, England: International Grains Council. Published annually.    This annual report includes data on production, trade, consumption, and prices for all major grains and grain-producing countries.

## THESES

Exhaustive literature reviews included in each thesis make a search of Dissertation Abstracts for relevant titles particularly useful to researchers. In addition to Dissertation Abstracts, dissertations can be found in several bibliographic indexes and journals. AGRICOLA, CAB Abstracts, and EconLit index dissertations that fit into their subject areas. To retrieve a list of dissertations on a specific subject, combine appropriate subject keywords or subject category codes with the keywords *thesis* or *dissertation*. Several journals provide annual lists of new dissertation titles. These three journals publish a new dissertations list each year: *American Journal of Agricultural Economics* (United States and Canadian), *Indian Journal of Agricultural Economics* (India), and *Journal of Agricultural Economics* (United Kingdom). The *Journal of Economic Literature* includes dissertations from the United States and Canada listed by subject, with a section for agricultural economics.

## TRADE LITERATURE

The end result of the scholarly research process is to transfer the knowledge gained from the research to the people who will use it in a real-life situation. Trade journals such as *Choices*, a publication of the American Agricultural Economics Association, are written in clear, easily understood language and circulated to agribusinesses, farmers, bankers, politicians, and others interested in the economics of agriculture. Agricultural extension publications also fill the gap between the scholarly publication and the practical application with clearly written documents. Some of the business indexes such as *F&S Index*, ABI Inform, and the main U.S. agriculture index, AGRICOLA, are good resources for this type of literature.

## WORLD WIDE WEB SITES

The Internet may be wonderful and may open new worlds to all, but it isn't easy, it isn't dependable, and it bears watching because it isn't always accurate. URLs seem to change on a whim. Nevertheless, it is here to stay and every day it will play a more important part in everyone's search for information. It is recommended that you bookmark some of the following sites specializing in seeking out the best Internet economic sites and make use of their work.

AgEcon Search AgNIC: Agricultural and Applied Economics Resources on the Internet (http://agecon.lib.umn.edu/AgNIC). 2000. St. Paul, MN: University of Minnesota, University Libraries. This section of the U.S. NAL's AgNIC project (http://www.agnic.org) includes links to reference resources, statistical data and resources, full-text resources, and subject division resources for agricultural economics.

Beddow, J. and T. Debass (http://www.agecon.com). 2000. AGECONdotCOM Virtual Library; The World-Wide Web Virtual Library: Agricultural Economics. This Web directory provides a guide to materials available on the Internet for agricultural economists. The library has sections for academic departments, mailing lists, journals and research, markets, policy, trade, associations, software, extension, and data. This is a great place for agricultural economists to begin their search for relevant materials on the Web.

Canada, C., coordinator (http://cnie.org/NLE/CRSreports/Agriculture/ag-80. cfm). 2000. Agriculture: A List of Websites. Washington, DC: Congressional Research Service. This guide is a sampling of the resources available for agriculture on the Web. The guide is arranged by subject, using headings such as colleges and universities, commodity markets and prices, cooperatives, credit/finance, food assistance, farm inputs, food processing/marketing, international aid and development, international organizations, and international trade.

Frank Beurskens Consulting (http://www.agribiz.com). 1999. AgriBiz. E-Markets, Inc. ''AgriBiz organizes information and resources available on the Internet, specific to the global agricultural trading community. The site is organized into three main sections: Articles and Trade News, Markets and Analysis, and Search and Research.''

Goffe, B., ed. (http://rfe.wustl.edu). 2000. Resources for Economists on the Internet. American Economic Association. This comprehensive guide to the Internet for economics resources is an invaluable resource for academic and practicing economists. Included in the guide are sections for data; links to economists and economics departments; jobs and grants; conferences; associations; news media; other Internet guides; scholarly communication (includes bibliographical databases, working papers, journal indices, online journals, academic publishers, newsletters, and style guides); software; and teaching resources. The user is able to navigate the guide by browsing or by searching.

Links for Agricultural Economists (http://www.aaea.org/info/resources). 2000. Ames, IA: American Agricultural Economics Association. The American Agricultural Economics Association's resource page includes links to academic departments, associations, journals, conferences, and more.

Saarinen, L., maintainer (http://netec.mcc.ac.uk/WebEc). 2000. WebEc: World Wide Resources in Economics. Helsinki, Finland: University of Helsinki, Faculty of Social Sciences. WebEc is a directory of Web resources for academic economists. It lists resources by category, including categories such as International Economics, Economics Data, and Agriculture and Natural Resources. Under the Agriculture category are included links to agricultural economics collections, institutions, publications, data, and communication resources.

## AUTHOR'S NOTE

Selecting the most appropriate reference resources for inclusion in this chapter has been a very difficult task. Our focus was agricultural economics, but we also included some basic resources from other fields of economics. Because of limited space we were able to use only about half of the publications we originally selected. If you are aware of an excellent resource and you do not find it listed here, it is very likely that we had to make a hard choice and one of your favorites lost out. To some extent this is a good thing. This chapter is a selective guide to resources that were available at one point in time and it must always be supplemented and updated by the user. Very often updating it will be a matter of finding the latest edition of the one listed in the chapter. Check the Web resources, the research library catalogs, and the indexes listed to find other online, updated, or new resources.

## ACKNOWLEDGMENTS

The authors wish to thank the members of Agricultural Economics Reference Organization (AERO) for their interest in this work and their input. We are also grateful to Dr. Vernon R. Eidman, Head of the Department of Applied Economics at the University of Minnesota, for his comments and suggestions.

## REFERENCES

AGROVOC: Multilingual Agricultural Thesaurus. 4th ed. Rome, Italy: Food and Agriculture Organization of the United Nations. 1999.

Barrett CB, Olia A, Von Bailey D. Subdiscipline-Specific Journal Rankings: Whither Applied Economics. Applied Economics 32:239–252, 2000.

Batten, Donna, ed. Guide to U.S. Government Publications. Detroit, MI: Gale Group. 2000.

Blanchard, Richard J, Farrell L. eds. Guide to Sources for Agricultural and Biological Research. Berkeley, CA: University of California Press. 1981.

Broder JM, Ziemer RF. Assessment of Journals Used by Agricultural Economists at Land-Grant Universities. Southern Journal of Agricultural Economics 16(1):167–172, 1984.

Clark KA, and Mai B. Locating Book Reviews in Agriculture and the Life Sciences. Science and Technology Libraries 18(4):3–27, 2000.

Davis JH, Goldberg RA. Concept of Agribusiness. Boston, MA: Harvard Business School, Division of Research. 1957.

Dote G, comp. Economic Research of Interest to Agriculture 1997–1999. Berkeley, CA: University of California, Division of Agriculture and Natural Resources, Agricultural Experiment Station. 2000.

Dote G, Letnes L. Scholarly Journals in Agricultural Economics Libraries/Reference Rooms: A Survey. Journal of Agricultural and Food Information 3(3):47–63, 1996.

Eatwell J, Milgate M, Newman P. The New Palgrave: A Dictionary of Economics. 4 vols. New York, NY: Stockton Press. 1987.

Ekwurzel D, Saffran B. Online Information Retrieval for Economists: the Economic Literature Index. Journal of Economic Literature 23(4):1728–1763, 1985.

Farget MA. A Performance Test on Various Bibliographic Databases Relevant to Agricultural Economics. Economie Rurale 160:37–39, 1984.

Foudy G. Wide Wide World of Statistics: International Statistics on the Internet. EContent, June 2000. http://www.ecmag.net/EC2000/foudy6.html. 2000.

Lawrence GW. U.S. Agricultural Statistics on the Internet: Extending the Reach of the Depository Library. Journal of Government Information 23(4):443–452, 1996.

Library of Congress, Cataloging Policy and Support Office. 23rd ed. 5 vols. Library of Congress Subject Headings. Washington, DC: Library of Congress, Cataloging Distribution Service. 2000.

Martin LR, ed. A Survey of Agricultural Economics Literature. 4 vols. Minneapolis, MN: University of Minnesota Press. 1977–1992.

Nixon JM. Review of CAB International's CD ROM Database AgECONCD. Journal of Agricultural & Food Information 3(1):93–97, 1995.

Olsen WC. Agricultural Economics and Rural Sociology: The Contemporary Core Literature. Ithaca, NY: Cornell University Press. 1991.

Perry GM. Using Citations to Evaluate the Quality of Agricultural Economics Journals. Review of Agricultural Economics 19(1):166–176, 1999.

Sjo J. Economics for Agriculturalists: A Beginning Text in Agricultural Economics. Columbus OH: Grid. 1976.

Taylor HC. An Introduction to the Study of Agricultural Economics. New York: Macmillan. 1905.

Taylor HC, Taylor AD. The Story of Agricultural Economics in the U.S., 1840–1932: Men, Service, Ideas. Ames, IA: Iowa State College Press. 1952.

Thomas SE. Bibliographic Control and Agriculture. Library Trends 38:542–561, 1990.

Wightman P, ed. *CAB Thesaurus.* 2 vols. Wallingford, UK: CABI Publishing. 1999.

World Agricultural Economics and Rural Sociology Abstracts. Wallingford, UK: CABI Publishing, 1958.

# 3

# Agricultural and Biosystems Engineering

**Mary Anderson Ochs and Mary E. Patterson**
Cornell University, Ithaca, New York, USA

The discipline of agricultural engineering evolved out of the union of agronomy and existing engineering fields, including mechanical engineering and civil engineering. In 1907, the American Society of Agricultural Engineers, (ASAE) was founded, marking the formal identification of the discipline in the United States, but the roots of agricultural engineering go back much further. Perhaps one might even consider that the first agricultural engineers were the early designers of irrigation systems thousands of years ago (Morgan, 1985, 3–5).

The discipline came into its own, however, in the late nineteenth and early twentieth centuries, when researchers and inventors in Europe and in the United States focused on the mechanization of agriculture, designing equipment that increased productivity on the farm and reduced the backbreaking labor of food production. Introduction of the steam engine and later the gasoline or diesel engine had a dramatic impact on this mechanization process, as did the development of the technology to connect implements directly to the tractor (Morgan, 1985, 11). Traditional areas of research included power and machinery, but also soil and water, structures and environment, and electric power and processing (Hall, 1992, p. vii). Rural electrification provided power for improved dairy operations, crop drying, and pumping (Morgan, 1985, p. 13).

In the post–World War II era, many agricultural engineers focused on developing machinery for harvesting specific crops. Grape pickers, cabbage harvesters, and other equipment were developed and put into use for crop production, harvesting, and storage. Others in the field developed improved facilities and equipment for livestock production. For example, agricultural engineers created improved milking equipment for dairy farms and better air-handling systems for large-scale poultry operations. Agricultural engineers also worked on research to improve irrigation technology.

Another important research area during this era was the application of improved instrumentation to agricultural equipment. Developments in instrumentation, such as magnetic resonance imaging (MRI) technology, durable microprocessors, and global positioning systems have been applied to research in agricultural engineering, resulting in the development of technologies, such as precision agriculture and improved food quality control.

Agricultural safety has traditionally been included within the discipline of agricultural engineering. As farming became more mechanized, it also became more hazardous. Agricultural engineers designed equipment features, standards, and procedures that helped insure the safety of farm workers.

The modern era has seen a shift toward biological systems engineering, first at a macro-biological level and more recently at the molecular level. ASAE is now known as the Society for Engineering in Agricultural, Food, and Biological Systems, reflecting this transformation. At the macro-level, agricultural and biosystems engineers have worked on producing ethanol from corn plants and developing better methods for handling animal waste. At the molecular level, bioengineers are joining forces with biologists, geneticists, and nanotechnologists to work in research areas, such as tissue engineering and development of biosensors.

In the developing countries, research and development is still occurring in some of the earlier research areas, as technological transitions continue to occur. Researchers from around the world are interested in sustainable development, making sure that appropriate technology is being used in ways that benefit current generations without sacrificing the resources available to future generations.

Related to agricultural and biosystems engineering are three other engineering areas: food engineering, forest engineering, and aquacultural engineering. Although these areas will not be incorporated in any great depth in this chapter, it is important to note that they are closely related to the more traditional areas of agricultural engineering.

The field of agricultural and biosystems engineering is evolving rapidly with advances in technology. It began as an interdisciplinary field and has continued to draw from the best of many fields to improve agricultural production and facilitate the protection of the environment. This chapter will outline the critical information resources for agricultural or biosystems engineering. It will not attempt to be comprehensive, but instead provide a selected list of resources essential for the upper-level undergraduate or beginning graduate student in agricultural or biosystems engineering.

## THE LITERATURE OF AGRICULTURAL AND BIOSYSTEMS ENGINEERING

In many ways, the literature of agricultural and biosystems engineering is typical of the literature of the other engineering disciplines. It is focused on the journal literature, with some of the most important journals emanating from several key societies. Handbooks, patents, and standards also play an important role. In the United States, the ASAE plays a critical role in publishing many of these resources for the discipline. Similar organizations exist in other nations, and international organizations, such as the Commission Internationale du Genie Rural (CIGR), also play an important part.

Many fields today are cross-disciplinary, but agricultural and biosystems engineering has been so from the very beginning. The literature cuts across agricultural, biological, and engineering resources, requiring the researcher to be familiar with a variety of resources in both engineering and the life sciences. *The Literature of Agricultural Engineering*, edited by Carl W. Hall and Wallace C. Olsen (1992), provides a detailed and comprehensive overview of the contemporary and historical literature of agricultural engineering, as does Bryan Morgan's *Keyguide to Information Sources in Agricultural Engineering* (1985).

## SEARCHING THE LIBRARY CATALOG

Just as the literature of agricultural and biosystems engineering is varied and interdisciplinary, searching the library catalog for information in this field must be approached with multiple strategies. Some of the most useful Library of Congress subject headings related to agricultural and biosystems engineering are listed in the following text:

> Agricultural Biotechnology
> Agricultural Chemicals
>   Environmental Aspects
> Agricultural Engineering
> Agricultural Implements
> Agricultural Machinery
> Agricultural Mechanics
> Agricultural Physics
> Agricultural Pollution

Agricultural Wastes
  Environmental Aspects
Animal Waste
Bioengineering
Dairy Engineering
Drainage
Electricity in Agriculture
Electronics in Agriculture
Electronics in Biology
Farm Buildings
Farm Engines
Farm Equipment
Farm Mechanization
Farm Tractors
Geographic Information Systems
Irrigation
Precision Farming

Most library catalogs now provide for keyword searching, which allows the combining of terms to pull in cross-disciplinary topics. Using the keyword searching capability to combine the term *engineering* with a variety of other keywords should prove successful. Identifying the correct Library of Congress subject heading can often focus directly on a specific topic. Catalog users should note that Library of Congress subject headings are often slow to respond to new areas of research. Thus keyword searching should be used in searching for "state-of-the-art" topics.

The Library of Congress classification system places agricultural engineering material in several areas within the S and T classification schedules. Listed are some important areas:

| | |
|---|---|
| S671-760.5 | Farm machinery and farm engineering |
| S770-S790.3 | Agricultural structures: Farm buildings |
| SH1-691 | Aquaculture |
| TA164 | Bioengineering |
| TA170-171 | Environmental Engineering |
| TC801-978 | Irrigation engineering: Reclamation of wasteland: Drainage |
| TD920-934 | Rural and farm sanitary engineering |
| TH7005-7699 | Heating and ventilation: Air conditioning |
| TJ1480-1496 | Agricultural machinery: Farm machinery |
| TP248.13-248.65 | Biotechnology |

The diversity of the classification of materials in this field is yet another indicator of the interdisciplinary nature of agricultural and biosystems engineering.

## ABSTRACTS AND INDEXES

Indexes by their nature provide structure and organization for the literature of a subject, and this structure in turn offers the beginning scholar an organized pathway to discover what research has been published in journals and presented at conferences. These same indexes provide the established researcher with a current overview of the research direction of others in similar subject specialties.

The area of biosystems and agricultural engineering is covered in the general agricultural databases such as AGRIS, AGRICOLA, BIOSIS, and CAB Abstracts, and in the major engineering databases as listed in the following text. There are also a number of tools specific to agricultural engineering.

*Agricultural Engineering Abstracts.* 1976– . Wallingford, UK: CAB International. ''A bi-monthly print journal, which keeps readers informed about significant research developments in agricultural engineering and instrumentation.'' The index contains approximately 3,500 abstracts per year derived from the CAB Abstracts database. An Internet version of *Agricultural Engineering Abstracts* is also available. Abstracts are broken out into subcategories of agricultural engineering, including precision agriculture, models, instruments; mechanical power; land improvement; pre- and post-harvest technology; protected cultivation; crop harvesting and threshing; crop handling and transport; livestock buildings and equipment; and aquaculture.

*Biological and Agricultural Index.* Previous title was *The Agricultural Index.* (1916–1964). New York: H. W. Wilson Company. *Biological and Agricultural Index* is perhaps most useful as a tool for undergraduates or for locating citations in the major journals in biology and agriculture. Covers approximately 335 core journal titles, across the fields of biology and agriculture. This index is also available as a database beginning with 1983.

BioSciences Information Service of Biological Abstracts. 1969– . BIOSIS Previews. Philadelphia, PA: BioSciences Information Service of Biological Abstracts. With much of agricultural and biosystems engineering moving toward understanding the interworkings of biological systems, the BIOSIS database increases in its importance for the discipline.

Computerized Agricultural Engineering Index. (http://www.engr.ucdavis.edu/%7Ebae/Outreach/aeindex.html) (cited February 2, 2002). William J. Chancellor, Compiler. A computerized index of technical articles in agricultural engineering periodicals and papers—along with searching software—is available without charge from W.J. Chancellor (wjchancellor@ucdavis.edu). Periodicals include *Agricultural Engineering* (since 1950), *Transactions of the ASAE* (since 1958), *Applied Engineering in Agriculture* (since 1985), *Canadian Agricultural Engineering* (since 1963), *Journal of Agricultural Engineering Research* (since

1956), *Agricultural Mechanization in Asia, Africa and Latin America* (since 1971), *Grundlagen der Landtechnik/Landtechnik* (since 1965), *Transactions of the Chinese Society of Agricultural Machinery* (since 1984), *Journal of the Japanese Society for Agricultural Machinery* (since 1985), *International Agricultural Engineering Journal* (since 1992), *the Journal of Agricultural Machinery* (since 1992), *Misr Agricultural Engineering Journal* (since 1995), *Memoria del Congreso Nacional, Asociacion Mexicana de Ingenieria Agricola* (since 1991). Also included are 3,700 selected (mostly unpublished) ASAE papers (since 1980) and all the papers on the CD-ROMs from the 1997, 1998, and 1999 ASAE summer meetings. Entries stopped with January 2000. An expanded Agricultural Engineering database is also available for searching online using a TELNET session to sweetpea.engr.ucdavis.edu (169.237.204.225). When the prompt "Username" appears, type **SEARCH** (enter). No password is required and there is no charge.

*Engineering Index*. 1884/91– . New York: Engineering Index, Inc. This index covers the literature of various specialties in engineering. The arrangement is by subject only from 1884–1906; then by a classed subject listing, 1907–1918; then by an alphabetical subject index in 1919. Beginning in 1928, an author index is included. In the early volumes, subjects such as tractor engines; tractors—farm; farming, electric power—farms; irrigation—soils; and education—agricultural engineering are included.

Ei Compendex Web. Hoboken, NJ: Engineering Information, Inc. This comprehensive engineering database is the electronic version of *Engineering Index*. The electronic coverage begins in 1970. The database adds about 500,000 records yearly from all areas of engineering. Approximately 2,600 journals, conferences, and technical reports are reviewed; conference literature comprises 20% of the database; and English-language publications represent 90% of coverage. Searching can be done by subject term, author, author affiliation, serial title, and keywords. Ei subject terms such as *agricultural engineering* are subdivided by place, state, country, or continent, or by subjects such as computer applications; *agricultural machinery* has subheadings such as economics or performance; *agricultural products* has subheadings such as fermentation or quality control; *agricultural wastes* has subheadings such as disposal or health hazards; and *agriculture* is subdivided by place or subjects such as energy conservation.

Society of Automotive Engineers (http://www.sae.org/) (cited February 2, 2002). 2001. SAE publications plus standards. Warrendale, PA: Society of Automotive Engineers. Covers SAE technical papers published since 1906. Subjects covered include agricultural machinery, alternate fuels, diesel engines, farm machinery, pumps, refrigeration equipment, tractor design, tractor maintenance, tractor operation, and tractor stability.

## BIBLIOGRAPHIES AND GUIDES TO THE LITERATURE

While no comprehensive guide to the agricultural engineering literature has been published recently, the following sources offer an excellent historical perspective on the literature of the field. Many of the sources described in these volumes are still relevant or have been updated in print or electronic formats.

Blanchard, J., and L. Farrell. 1981. *Guide to Sources for Agricultural and Biological Research*. Berkeley, CA: University of California Press. While somewhat dated, this source provides an extensive listing of standard reference sources in agricultural engineering; irrigation, drainage and land reclamation; water resources; and agricultural chemistry.

Cloud, G.S. 1985. *Selective Guide to Literature on Agricultural Engineering*. College Station, TX: American Society for Engineering Education, Engineering Libraries Division. A list of reference sources for graduate students, undergraduate students, and faculty doing research in agricultural engineering.

Hall, C.W. 1976. *Bibliography of Agricultural Engineering Books*. St. Joseph, MI: American Society of Agricultural Engineers. A comprehensive listing of books in agricultural engineering. Includes English-language materials, as well as books in French, German, Russian, and Spanish. Listed by author with specialized subject index included.

Hall, C.W. 1976. *Bibliography of Bibliographies of Agricultural Engineering and Related Subjects*. St. Joseph, MI: American Society of Agricultural Engineers. A supplement to Hall's *Bibliography of Agricultural Engineering Books*. Hall attempts to list all bibliographies included in indexes, abstracts, and journal references as of December 1975.

Hall, C.W., and W.C. Olsen, eds. 1992. *The Literature of Agricultural Engineering*. Ithaca, NY: Cornell University Press. Hall and Olsen provide a detailed survey of the field of agricultural engineering and the literature of the discipline. They used citation analysis and ranking by scholars to establish the core or most important works in the subject area. Subject specialists have contributed chapters on related disciplines, such as aquaculture, food engineering, and forest engineering.

Morgan, B. 1985. *Keyguide to Information Sources in Agricultural Engineering*. London and New York: Mansell. A comprehensive guide to the literature of agricultural engineering written by a librarian with extensive experience in a library devoted to the field. Morgan's historical overview and discussion of the literature provide excellent background for both librarians and new agricultural engineers, even though the book is over 15 years old.

*Encyclopedia of Agricultural and Food Engineering.* New York: Marcel Dekker, in press.

Farrall, A.W. 1979. *Dictionary of Agricultural and Food Engineering.* Danville, IL: Interstate. Provides short definitions of both common and technical terms related to agricultural and food engineering.

Flickinger, M.C., and S.W. Drew. 1999. *Encyclopedia of Bioprocess Technology: Fermentation, Biocatalysis and Bioseparation.* 5 vols. New York: Wiley. One in a series of Wiley encyclopedias in biotechnology aimed at explaining biological processes related to biotechnology. A wide range of topics, from anaerobes to wastewater treatment, that are relevant to the agricultural or biosystems engineer.

*GTZ Bildworterbuch: System Steinmetz.* 1987. Weikersheim, Germany: Margraf. Contains 6,000 terms in English, Chinese, French, Spanish, and German. Designed to aid in the dialog for bringing agricultural technical innovation to the developing countries.

Montgomery, J.H. 1996. *Groundwater Chemicals: Desk Reference.* 2nd ed. Boca Raton FL: CRC/Lewis Publishers. A handbook outlining properties of hazardous chemicals that may be found as groundwater contaminants.

Parker, S.P., and R.A. Corbitt, eds. 1993. *McGraw-Hill Encyclopedia of Environmental Science and Engineering.* 3rd ed. New York: McGraw-Hill. Provides excellent brief overviews of topics related to agricultural and biosystems engineering, such as groundwater hydrology, river engineering, and saline water reclamation. This volume is a subset of the *McGraw-Hill Encyclopedia of Science and Technology,* which has been updated since 1993.

Spier, R.E. 2000. *Encyclopedia of Cell Technology.* 2 vols. New York: Wiley. Another in the Wiley series of biotechnology encyclopedias. This encyclopedia provides a compilation of information, ideas, procedures and guidelines for animal and plant cell technologies, as well as the transfer of ideas between the two. Topics include bioreactors, measurement of cell viability, and online analysis of animal cell culture, to name just a few.

Stickney, R.R. 2000. *Encyclopedia of Aquaculture.* New York: Wiley. Designed for a wide audience working in areas related to farming of aquatic plants or animals for direct or indirect human consumption. This encyclopedia should be useful to the aquaculture professional, as well as to those wanting to learn more about aquaculture.

Tosheva, T., M. Djarova, and B. Deliiska. 2000. *Elsevier's Dictionary of Agriculture, in English, German, French, Russian, and Latin.* Amsterdam; New York: Elsevier. This polyglot dictionary covers 9,389 terms in English, German, French, Russian, and Latin.

Van der Leeden, F., F.L. Troise, and D.K. Todd. 1990. *The Water Encyclopedia.* 2ⁿᵈ ed. Chelsea, MI: Lewis Publishers. Covers climates, hydrology, surface and ground water, water use, water management, water resources agencies, and legislation. Contains over 600 tables of water-related data.

## HANDBOOKS AND MANUALS

Handbooks for the engineer represent an invaluable reference source, consolidating facts, formulas, and fundamentals for specific topics. Again because of the interdisciplinary nature of agricultural engineering, handbooks from across the field of engineering provide useful information to the agricultural or biosystems engineer. Listed are key reference handbooks, many of which are produced by engineering societies.

American Society of Heating Refrigerating and Air-Conditioning Engineers. 1997. *ASHRAE Handbook*: *Fundamentals*. Atlanta, GA: American Society of Heating Refrigerating and Air-Conditioning Engineers. Published in two editions: inch-pound (I-P) units of measurement and International System of Units (SI). Covers basic principles and includes data for HVAC technology. Summary of updates and changes included in preface. Updated in a four-year cycle; information at http://www.ashrae.org (cited February 3, 2002). Chapters include heat transfer, mass transfer, sound and vibration, environmental control for animals and plants, physiological factors in drying and storing farm crops, air contaminants, odors, refrigerants, thermal and moisture control, ventilation, energy estimation, and duct design. Also used as a textbook.

American Society of Heating Refrigerating and Air-Conditioning Engineers. 1998. *ASHRAE Handbook*: *Refrigeration*. Atlanta, GA: American Society of Heating Refrigerating and Air-Conditioning Engineers. Published in two editions: inch-pound (I-P) units of measurement and International System of Units (SI). Updated in a four-year cycle; information at http://www.ashrae.org (cited February 3, 2002). Includes systems; new refrigerants; cold storage facilities; retail equipment; thermal properties; and freezing of food: meat, fish, eggs, fruit; distribution of refrigerated food; and biomedical applications of cryogenic refrigeration.

American Society of Heating Refrigerating and Air-Conditioning Engineers. *ASHRAE Handbook*: *HVAC Applications*. 1999. Atlanta, GA: American Society of Heating Refrigerating and Air-Conditioning Engineers. Focus is on applications and equipment. Published in two editions: inch-pound (I-P) units of measurement and International System of Units (SI). Updated in a four-year cycle; information at http://www.ashrae.org (cited February 3, 2002). Includes transport refrigeration, food display and equipment, refrigerated food technology and processing, ventilation and distribution, hot water and steam equipment, and compressors, pumps, and motors.

American Society of Heating Refrigerating and Air-Conditioning Engineers. 2000. *ASHRAE Handbook*: *HVAC Systems and Equipment*. Atlanta, GA: American Society of Heating Refrigerating and Air-Conditioning Engineers. Published in two editions: inch-pound (I-P) units of measurement and International System of Units (SI). Updated in a four-year cycle; information at http://www.ashrae.org (cited February 3, 2002). Also available in CD format. Includes system selection and analysis, air distribution, heat pumps and heat recovery, air-handling, industrial drying systems, water systems design, motor controls, and heat exchangers.

Atkinson, B., and F. Mavituna. 1991. *Biochemical Engineering and Biotechnology Handbook*. 2nd ed. New York: Stockton Press. Intended to assist the development of biotechnological processes used to produce commodity chemicals from vegetable substances. Chapters cover properties of microorganisms, process biotechnology, thermodynamic aspects of metabolism, microbial activity, product formation, plant cell culture, animal cell culture, enzymes, industrial microbial processes, and principles of costing.

Bolz, R.E., G.L. Tuve, and Chemical Rubber Company. 1973. *CRC Handbook of Tables for Applied Engineering Science*. Cleveland, OH: CRC Press. A desktop source of numerical data for steam, refrigerants, metals, diffusion constants, permeability, conversion factors, formulas, tables, and properties of common solid materials, such as thermal conductivity of wood. Includes a bibliography of data sources.

Bronzino, J.D., ed. 1995. *The Biomedical Engineering Handbook*. Boca Raton, FL: CRC Press. Chapters include physiologic systems, bioelectric phenomena, biomechanics, biomaterials, biomedical sensors, biomedical signal analysis, imaging, medical instruments and devices, biotechnology, tissue engineering, prostheses, rehabilitation engineering, human performance engineering, physiologic modeling, artificial intelligence, regulations, and organizations. Each chapter includes a bibliography and further readings.

CIGR, the International Commission of Agricultural Engineering, ed. 1999. *CIGR Handbook of Agricultural Engineering*. 5 vols. St. Joseph, MI: American Society of Agricultural Engineers. This five-volume set is designed to cover the major fields within agricultural engineering. Volumes include (1) Land and Water Engineering, (2) Animal Production and Aquacultural Engineering, (3) Energy and Biomass Engineering, (4) Plant Production Engineering, and (5) Agro Processing Engineering. Useful to agricultural engineers and students in developed and developing countries.

Dorf, R.C. 1996. *The Engineering Handbook*. Boca Raton, FL; Piscataway, NJ: CRC Press; IEEE Press. A ready reference for the practicing engineer. Provides underlying theories and concepts with appropriate applications. Each chapter defines terms and includes references, and the further information section includes relevant conferences and journals. Topics include kinematics and mechanisms, structures, fluid mechanics, thermodynamics and heat transfer, water resources, circuits, remote sensing, control systems, food and agricultural engineering, safety, and materials engineering.

Karassik, I.J. et al. 1976. *Pump Handbook*. New York: McGraw-Hill. Theory and performance of centrifugal, displacement, and jet pumps—their controls and valves, maintenance, and test procedures.

Kreith, F., ed. 1998. *The CRC Handbook of Mechanical Engineering*. Boca Raton, FL: CRC Press. Ready reference for the practicing engineering covering the traditional areas of thermodynamics, mechanics, heat and mass transfer, as well as modern manufacturing, robotics, computer and environmental engineering, project management, patent law, and bioengineering.

*Lange's Handbook of Chemistry*. 1999. J.A. Dean, ed. 15th ed. New York: McGraw-Hill. Essential handbook covering mathematics, conversion tables, chemistry, thermodynamics, and physical properties.

*Machinery's Handbook*. 2000. E. Oberg et al., eds. 26th edition. New York: Industrial Press. Contains formulas and tables for sections on mathematics, mechanics, properties of materials, dimensioning, tooling, machining, fasteners, threads, gears, machine elements, and measuring units. Tables for screw threads, nuts and bolts, and washers.

Maidment, D.R. 1993. *Handbook of Hydrology*. New York: McGraw-Hill. Chapters are written by experts in hydrology. The chapters include precipitation, evaporation, infiltration, groundwater flow, runoff, streamflow, and transport of substances dissolved or suspended in flowing water. Main sections are the hydrologic cycle, transport, statistics, and technology (which includes computer programs, data collection, hydrologic design, and groundwater pollution control).

*Marks' Standard Handbook for Mechanical Engineers*. 1996. 10th edition. New York: McGraw-Hill. Essential desk reference for engineers. Chapters include mathematical tables; mechanics of solids and fluids; strength of materials; materials of engineering (wood, cement, water, lubrication); power generation (steam: boilers, engines, turbines; gas turbines; hydraulic turbines); building construction; fans, pumps, and compressors; environmental control; and refrigeration. Chapters are divided into subchapters with an overview of topic, references, and many detailed tables, graphs, illustrations, and charts.

Skalak, R., and S. Chien. 1987. *Handbook of Bioengineering*. New York: McGraw-Hill. Articles on mechanics of soft tissue, properties of bone, skin mechanics, thermal properties of skin, rheology of blood flow, artificial kidney, and biomechanics of the spine. Contains extensive references.

Smith, E.H., ed. *Mechanical Engineer's Reference Book*. 1994. 12th edition. Warrendale, PA: Society of Automotive Engineers. Designed to supply a practicing engineer with the basic principles and has additional references. Chapters cover mechanical-engineering principles; electrical and electronics; microprocessors; computers; computer-integrated engineering systems; design standards; materials properties; mechanics of solids; tribology; power units and transmission; fuels; alternative energy; nuclear, offshore, and plant engineering; manufacturing methods; engineering mathematics; health and safety; units symbols; and constants.

Zwillinger, D. 1996. *CRC Standard Mathematical Tables and Formulae*. Boca Raton, FL: CRC Press. A handbook of basic mathematical reference materials required for the disciplines of mathematics, engineering, and physical sciences.

## MONOGRAPHS AND TEXTBOOKS

In *Keyguide to Information Sources in Agricultural Engineering* (1985, 42), Morgan discusses some of the problems in publishing in the field of agricultural engineering. He cites problems with the size of the market for materials and the varying levels of readers interested in texts, ranging from the farmer to graduate students and faculty. Creating a general textbook that can appeal to a large enough market, yet cover the material in any depth, was difficult then and continues to be problematic. Some texts in agricultural engineering and related areas have been updated, but for many texts, the most recent edition is over five years old.

In the United States, the ASAE has attempted to address this problem by creating a series of monographs and textbooks. Other organizations, such as CIGR and Food and Agricultural Organization of the United Nations (FAO), also produce publications in agricultural engineering. Commercial scientific publishers, such as Delmar Thomson and Interstate, continue to produce monographs and textbooks for agricultural engineering, where the market warrants. Searching techniques such as those outlined in the section on using the Library of Congress subject headings will uncover materials in library catalogs. *Book in Print*, available both in print and as an online database from R. R. Bowker, also provides subject access to materials available for purchase from the publishers.

## JOURNALS

The journal literature is the most important format for scholarly communication for the discipline of agricultural and biosystems engineering. Listed in the following text are key journals in agricultural and biosystems engineering. They have been selected by consulting the Institute for Scientific Information's *Journal Citation Reports*, the core list of serials listed in Hall and Olsen's *The Literature of Agricultural Engineering*, and Bryan Morgan's *Keyguide to Information Sources in Agricultural Engineering*. The list includes journals covering traditional areas of agricultural engineering research, as well as the new research areas in biosystems engineering and biotechnology. This list focuses on English-language titles.

Many general science and agriculture journals, such as *The Journal of Agricultural Science*, and specialized journals in related areas, such as *Journal of Dairy Science*, include research articles in agricultural and biosystems engineering. Those titles have not been included in this list. Also not listed are trade journals that provide an important source of information for the agricultural and biosystems engineer. Trade publications may be published by commercial publishers or trade organizations serving the related industries.

Most of the journals listed are available from the publisher in electronic form.

*Acta Biotechnologica*. 1981– . Berlin: Akademie-Verlag.

*Agricultural Mechanization in Asia, Africa and Latin America*: AMA. 1981– . v. 12– . Tokyo: Farm Machinery Industrial Research Corp. Continues *Agricultural Mechanization in Southeast Asia*.

*Agricultural Systems*. 1976– . Barking, UK: Elsevier Applied Science.

*Agricultural Water Management*. 1976– . Amsterdam: Elsevier.

*Agriculture, Ecosystems & Environment*. 1983– . Amsterdam: Elsevier.

American Society for Microbiology. *Applied and Environmental Microbiology*. 1976– . v. 31– . Washington, DC: American Society for Microbiology.

*Applied Engineering in Agriculture*. 1985– . St. Joseph, MI: American Society of Agricultural Engineers.

*Applied Microbiology and Biotechnology*. 1984– . v. 19– . Berlin: Springer International. Continues *European Journal of Applied Microbiology and Biotechnology*.

*Appropriate Technology*. 1974– . London: Intermediate Technology Publications Ltd.

*Aquacultural Engineering*. 1982– . London: Applied Science.

*Biochemical Engineering Journal*. 1998–. Amsterdam; New York: Elsevier.

*Biodegradation*. 1990– . Dordrecht, The Netherlands; Boston: Kluwer.

*Biomass & Bioenergy*. 1991– . Oxford; New York: Pergamon.

*Bioprocess Engineering*. 1986– . Berlin; New York: Springer.

*Bioresource Technology*. 1991– . Barking, UK: Elsevier Science.

largest voluntary standards development organizations in the world. ASTM develops standard test methods, specifications, practices, guides, classifications, and terminology in 130 areas covering subjects such as metals, paints, plastics, textiles, petroleum, construction, energy, the environment, consumer products, medical services and devices, computerized systems, electronics, and many others. More than 10,000 ASTM standards are published each year in the 73 volumes of the *Annual Book of ASTM Standards* (ASTM home page). ASTM sells copies of its standards on its Web page, and often research libraries will subscribe to the *Annual Book of ASTM Standards*.

European Patents (http://ep.espacenet.com/) (cited February 3, 2002). Munich, Germany: European Patent Office. A source for European patents, both pending and granted, and Japanese patents. The designations are European patent (EP), European Community (EC), world patent (WO). Searching can be done by patent publication number, date, applicant, inventor, assignee, title words, and classification. It is possible to download the patent in English one page at a time, or the entire patent in the original language.

USPTO Web Patent Databases (http://www.uspto.gov/patft/) (cited February 3, 2002). Washington, DC: United States Patent and Trademark Office. Freely accessible full-text database of all U.S. patents including design and plant patents. It is possible to search for patents using a variety of fields such as abstract, assignee, date, inventor, patent number, and class/subclass number, but prior to 1976 searching is possible by patent number or current U.S. classification only. Full text and drawings are available and can be printed. Help screens and software information are available on the Web page.

## SEARCH ENGINES AND IMPORTANT WEB SITES

While most scholarly publications related to agricultural and biosystems engineering are not freely available on the Web, sites similar to the ones listed in the following text provide other types of information. Most of the important societies in the field host Web sites. The Web site for the ASAE is listed as an example. All other related society Web sites are listed under the Societies section. Most departments of agricultural and biosystems engineering in the land-grant institutions host Web sites. The Web site for Cornell University's Department of Biological and Environmental Engineering is listed as an example.

American Society of Agricultural Engineers (also The Society for Engineering in Agricultural, Food and Biological Systems) (http://www.asae.org) (cited February 3, 2002). This organization provides the backbone for the literature of agricultural and biosystems engineering in the United States. Their Web site provides a wealth of information about the literature and the society.

Core Historical Literature of Agriculture (http://chla.mannlib.cornell.edu) (cited February 3, 2002). This site provides searchable full text and scanned page images of historical texts in agriculture. It includes such classics in agricultural engineering as, *Technology on the Farm* published in 1940 by the USDA and *Rudimentary Treatise on Agricultural Engineering* by G.H. Andrews, published in 1852.

Cornell University. Department of Biological and Environmental Engineering (http://aben.cals.cornell.edu) (cited February 3, 2002). The Web site for Cornell's Department of Biological and Environmental Engineering. Provides links to program information for undergraduate and graduate students, as well as research in progress in the department. Most agricultural and biosystems engineering departments in universities have similar Web sites.

Edinburgh Engineering Virtual Library (http://eevl.ac.uk) (cited February 3, 2002). The Edinburgh Engineering Virtual Library (EEVL) is probably the largest free specialized guide to Internet resources in Engineering (Powell, 2001). Although it has no predefined section for agricultural engineering, the search engine allows for searching using keywords, such as *agricultural* or *biosystems*.

Google (http://www.google.com) (cited February 3, 2002). Perhaps the most popular all-purpose Web search engine at present, Google offers powerful searching capabilities of the Web with relevance ranking that appears to work quite well. Try their "I'm feeling lucky" search.

Web-agri (http://www.webagri.com) (cited February 3, 2002). According to the site name page, Web-agri is "the agricultural search engine." Especially useful for locating Web sites for agricultural and biosystems engineering departments in universities.

## SOCIETIES

Societies serve an important role in scholarly communication for the discipline of agricultural and biosystems engineering. Each of these societies produces publications and holds annual and specialty conferences that provide a forum for the exchange of information relating to ongoing research. With the advent of the Internet, most societies have a presence on the Web, and URLs for society Web sites have been included. These Web sites often provide a wealth of information about the society and provide access to their publications, either directly on the Web site or available for purchase. In addition, these Web sites can provide calendar listings of upcoming conferences, job openings, and information for high-school students interested in finding out more about careers in this field.

American Society for Agricultural Engineering (http://www.asae.org) (cited February 3, 2002). The new name for this society is the Society for Engineering in Agriculture, Food and Biological Systems. The history of the society is described at length in Robert E. Stewart's *Seven Decades that Changed America: A History of the American Society of Agricultural Engineers, 1907–1977* (St. Joseph, MI: American Society of Agricultural Engineers, 1979). ASAE is a professional and technical organization dedicated to the advancement of engineering applicable to agricultural, food, and biological systems. Founded in 1907 and headquartered in St Joseph, Michigan, ASAE comprises 9,000 members representing more than 90 countries.

American Society of Civil Engineers (http://www.asce.org/) (cited February 3, 2002). "Founded in 1852, the American Society of Civil Engineers (ASCE) represents more than 123,000 members of the civil engineering profession worldwide, and is America's oldest national engineering society. ASCE's vision is to position engineers as global leaders building a better quality of life. The Society is preparing to celebrate its 150th anniversary in 2002" (home page).

American Society for Engineering Education (http://www.asee.org/) (cited February 3, 2002). According to the ASEE home page, the society is a non-profit, member association, founded in 1893, dedicated to promoting and improving engineering and technology education. Includes a division of agricultural and biological engineering.

American Society of Heating, Refrigerating and Air-Conditioning Engineers (http://www.ashrae.org) (cited February 3, 2002). "ASHRAE, the American Society of Heating, Refrigerating and Air-Conditioning Engineers is an international organization of 50,000 persons with chapters throughout the world. The Society is organized for the sole purpose of advancing the arts and sciences of heating, ventilation, air conditioning and refrigeration for the public's benefit through research, standards writing, continuing education and publications" (ASHRAE Web page).

American Society of Mechanical Engineers International (http://www.asme.org/) (cited February 3, 2002). "Founded in 1880 as the American Society of Mechanical Engineers, today ASME International is a nonprofit educational and technical organization serving a worldwide membership of 125,000. Its mission is to promote and enhance the technical competency and professional well-being of members, and through quality programs and activities in mechanical engineering, better enable its practitioners to contribute to the well-being of humankind" (ASME Web page).

Canadian Society for Engineering in Agricultural, Food and Biological Systems (http://www.csae-scgr.ca/) (cited February 3, 2002). Canada's professional organization for agricultural engineers.

Commission Internationale du Genie Rural (http://www.ucd.ie/cigr/) (cited February 3, 2002). "The International Commission of Agricultural Engineering

(CIGR) was set up by a Constituent Assembly on the occasion of the first International Congress of Agricultural Engineering, held in Liege, Belgium in 1930. It is an international, non-governmental, non-profit organization regrouping, as a networking system, regional and national Societies of Agricultural Engineering as well as private and public companies and individuals all over the world. The main aims of CIGR are to: stimulate the development of science and technology in the field of Agricultural Engineering; encourage education, training, and mobility of young professionals; encourage interregional mobility; facilitate the exchange of research results and technology; represent the profession at a worldwide level; work towards the establishment of new associations, both at national and regional levels, and to the strengthening of existing ones; and perform any other activity that will help to develop Agricultural Engineering and Allied Sciences'' (CIGR Web page).

Institute for Biological Engineering (http://www.ibeweb.org/) (cited February 3, 2002). ''The Institute of Biological Engineering was established to encourage inquiry and interest in biological engineering in the broadest and most liberal manner, and to promote the professional development of its members. Toward that end, the Institute provides a multi-disciplinary forum for exchange of ideas, knowledge and information that spans the many fields where biology and engineering converge'' (IBE Web page).

Institution of Agricultural Engineers (UK) (http://www.iagre.org/) (cited February 3, 2002). ''Located in the UK, the Institution of Agricultural Engineers is the professional body for engineers, scientists, technologists and managers in agricultural and allied industries, including forestry, food processing, and agrochemicals'' (Institution Web page).

Instrumentation, Systems, and Automation Society (http://www.isa.org/) (cited February 3, 2002). ''ISA—The Instrumentation, Systems, and Automation Society is a 39,000 member international, non-profit, technical organization. The Society fosters advancements in the theory, design, manufacture, and use of sensors, instruments, computers, and systems for measurement and control in a wide variety of applications'' (ISA Web page).

National Institute for Farm Safety (http://www.ag.ohio-state.edu/~agsafety/NIFS/nifs.htm) (cited February 3, 2002). ''NIFS is the leading organization dedicated to reducing accidents in agriculture. NIFS is a non-profit, voluntary organization consisting of safety and health professionals, and interested organizations. NIFS was formed in 1962 to provide a structure for the professional development of agricultural safety and health professionals. Its focus is to reduce the agricultural death, injury, and illness rate'' (NIFS Web page).

Society of Automotive Engineers (http://www.sae.org/) (cited February 3, 2002). ''The Society of Automotive Engineers is the professional organization for engineers working with cars, aircraft, trucks, off-highway equipment, engines, materials, manufacturing, and fuels. Their membership numbers nearly 80,000 engineers, business executives, educators, and students from more than 97 countries'' (SAE Web page).

Society for Engineering in Agriculture, Food and Biological Systems (new name for ASAE).   For additional listings for agricultural engineering societies worldwide, consult the directories section of this chapter.

## CONCLUSION

The discipline of agricultural and biosystems engineering has historically been a diverse field, and it continues to expand dramatically as new technologies extend the capabilities of researchers. Problems of the past are being solved in new ways. Agricultural and biosystems engineers develop strategies and processes for applying new technologies that allow for sustainable development. Scott (1992) notes that an ASAE Task Force stated that agricultural and biosystems engineers are the key technologists to bring timely science and engineering to productive and ecologically sustainable agriculture and related biological systems. In order to accomplish this, agricultural and biosystems engineers need access to the key literature of the discipline.

## ACKNOWLEDGMENTS

Many thanks to Dr. Gerald Rehkugler, Dr. Michael Timmons, Dr. Norman R. Scott, and Dr. Eric Hallman, who provided an immense amount of help in constructing this chapter.

## REFERENCES

Hall W, Olsen WC. *The Literature of Agricultural Engineering*. Ithaca, NY: Cornell University Press. 1992.

Morgan B. *Keyguide to Information Sources in Agricultural Engineering*. London: Mansell. 1985.

Powell JH. Virtual Engineering Libraries. Science and Technology Libraries. pp. 105–128, 2001.

Scott NR. Engineering for the world's agricultural, food, and environmental needs for the next century. Agricultural Engineering and Rural Development. Proceedings of the International Conference on Agricultural Engineering. (92-ICAE). October 12–14, International Academic Publishers. Beijing, China. v. 1, 1992. pp 1–7.

# 4

---

# Animal Health and Veterinary Sciences

**Gretchen Stephens**
Purdue University, West Lafayette, Indiana, USA

## RESEARCH IN ANIMAL HEALTH AND VETERINARY SCIENCE

''Animal health remains the primary raison d'être for veterinary medicine.'' (Loew, 1996)

Veterinary medicine began with prehistoric people who lived with animals. The earliest known practitioners of veterinary medicine were Sumerian priest-healers. In Greece and Rome, veterinarians concentrated on equine care for military reasons and secondarily on farm animal care. The *Hippiatrika*, a Byzantine text on equine medicine compiled from Greek and Latin authors, was a basis for scientific study in veterinary medicine during the Middle Ages. By the mid-eighteenth century, intensified animal husbandry combined with concern for the many plagues of the time to lead to the establishment of veterinary schools in Lyons (1762) and Alfort (1765). By 1825, there were 30 veterinary schools in 12 European countries.

In North America, the Boston Veterinary Institute (1854) and a veterinary school in Guelph (1862) were the first to produce graduates. State-established veterinary schools came into being in the United States after passage of the Morrill Land Grant Act in 1862, beginning with the veterinary program at Iowa State

University (1879). Organized veterinary medicine began with the establishment of the U.S. Veterinary Medical Association [now the American Veterinary Medical Association (AVMA)] in 1863.

In 1884, the U.S. Congress established the Bureau of Animal Industry within the USDA to fight outbreaks of bovine pleuropneumonia and other epidemic diseases. The Bureau achieved initial success over pleuropneumonia eight years later. Women entered the profession as early as 1889, with the first female practitioner graduating in 1910. As transportation became mechanized, veterinary emphasis on horses declined. Veterinary services for companion animals and veterinarians' roles in public health issues, such as rabies control and meat/milk safety, grew. The development of pharmaceuticals and biologics for farm livestock became profitable, and their availability transformed veterinary practice efficacy.

During the second half of the twentieth century, major funding opportunities for health-related research and postgraduate training allowed some veterinary schools to expand their research and advanced training. Specialization within the veterinary profession appeared after World War II primarily in North America, and by 1993 had grown to 27 specialty boards, colleges, and subspecialties recognized by the AVMA (Dunlop and Williams, 1996).

Today, veterinary medicine is concerned with the scientific study of animal disease prevention and treatment of diverse species. Although there is still concern with the whole animal, funded research frequently focuses on the molecular level. In addition, with human health linked to animal health and production, particularly in developing countries, veterinary medicine continues to contribute to the maintenance and promotion of public health. In 1999, a Food and Agriculture Organization/Office International des Épizooties/World Health Organization (FAO/OIE/WHO) Study Group on Future Trends in Veterinary Public Health recommended that veterinary public health be defined as "the contribution to the complete physical, mental, and social well-being of humans through an understanding and application of veterinary medical science." With major zoonotic diseases emerging, reemerging or endemic, affecting millions and preventing efficient production of needed proteins (such as animals), veterinary medicine continues to be an important partner in improving overall socioeconomic development (*Weekly Epidemiological Record*, 1999).

For a more extensive review of the history of veterinary medicine consult Dunlop and Williams' *Veterinary Medicine: An Illustrated History* (1996), Karasszon's *A Concise History of Veterinary Medicine* (1988), or one of the other histories noted in this chapter.

The literature of animal health and veterinary science (e.g., veterinary medicine) draws from the literatures of agriculture, medicine, and biology. Over the past 20 years, the following trends in veterinary literature have been observed:

- Growth in the clinical veterinary medicine literature, both books and journals.
- Increase in literature on exotic and wildlife veterinary medicine, as well as the human/animal bond.
- Specialty print indexes being created, recombined, and then replaced by electronic bibliographic databases.
- Selected veterinary texts frequently being published in both print and electronic formats (usually CDs) with a few titles being published in CD format only.
- Specialized resources, such as bibliographies and directories, from government or professional specialty groups moving from print to electronic environments.
- Veterinary research journals by major publishers becoming available in electronic full-text formats.
- Continued importance of the journal literature, print or electronic, in transmitting primary research to the profession.
- Growth in Internet communities and member-only Web sites (e.g., NOAH, VIN) to support practitioners and researchers.
- Compact handbooks and texts frequently used by clinicians potentially moving from print to personal digital assistants (PDAs).

Comments on specific aspects of the veterinary literature appear at the beginning of each of the following sections. Government publications are included in appropriate sections of this chapter but are not highlighted in a separate section. International Standard Serials Numbers (ISSNs) and International Standard Book Numbers (ISBNs) are provided whenever possible. Throughout the chapter, references to Web sites/pages on the World Wide Web include the URL and the date of access in parentheses.

## ABSTRACTS, INDEXES, AND DATABASES

Focusing on the premier quartet of veterinary abstracts, indexes, and databases (CAB Abstracts, Index Veterinarius, Veterinary Bulletin, and www.animalscience.com), this section includes current subject-specific abstract journals and databases as well as retrospective English-language veterinary indexes. Foreign-language veterinary abstract journals and indexes are not included.

In addition, major indexing and abstracting services for both the biosciences literature and the agricultural literature provide selected coverage of the animal health and veterinary science literature as reflected in the Joint Collection Development Policy Statement of the National Library of Medicine, the National Agricultural Library and the Library of Congress on Veterinary Science and Related Subjects (March 6, 1996) that can be seen at the following Web sites: http://www.nlm.nih.gov/pubs/cd_vet_sci.html (cited June 27, 2001); http://www.nalusda.gov/acq/cdvetsci.htm (cited June 27, 2001); and http://lcweb.loc.gov/acq/devpol/vet.html (cited June 27, 2001). The veterinary coverage of such services is briefly discussed in the following text with additional detail available elsewhere in this book.

## Current

CAB Abstracts. 1973– . Wallingford, UK: CAB International. An online bibliographic database compiled from the world's scientific and technical literature, CAB Abstracts includes the most comprehensive access to the veterinary medical literature worldwide. Although full coverage of veterinary journals is a primary focus of this database, conference proceedings and book chapters are indexed if available. Although accessible through many vendors, the database now is available directly to the animal health and veterinary science community as a component of various CABI-created Internet communities such as www. animalscience.com. To optimize searching of this extensive, structured database, consult the *CAB Thesaurus* for relevant term(s) for the topic, i.e., *veterinary medicine*, as well as additional terms that broaden (BT), narrow (NT), or direct one to a related term (RT) applicable to the topic.

    VETERINARY MEDICINE
    BT veterinary science
    RT animal pathology
    RT pharmacology
    RT pharmacy
    RT surgical operations
    RT veterinarians
    RT veterinary education
    RT veterinary entomology
    RT veterinary equipment
    RT veterinary helminthology
    RT veterinary mycology
    RT veterinary parasitology
    RT veterinary practice
    RT veterinary schools
    RT veterinary services

    In addition, each CAB Abstracts record is assigned a CABICODE that assists in limiting keyword searches to records in a specific subject area. Because using CABICODEs in combination with terms from the *CAB Thesaurus* will lead to greater searching precision, CAB revised its CABICODEs relating to animal health and veterinary science as follows:

(Those in italics are new as of March 2000. Those with lines through them were discontinued as of March 2000.)

~~CC720 Veterinary Profession~~

*EE117 Veterinary Economics*

LL060 Draught Animals
LL070 Pets and Companion Animals
*LL075 Sport Animals*
LL080 Zoo Animals
~~LL100 Animal Husbandry (General)~~
LL110 Dairy Animals
LL120 Meat Producing Animals
LL130 Egg Producing Animals
~~LL140 Animal Husbandry (Wool and Other Fibres)~~
*LL145 Wool Producing Animals*
*LL148 Fur-bearing Animals*
~~LL150 Animal Husbandry (Other Products)~~
*LL180 Animal Husbandry and Production*
LL190 Animal Slaughter
~~LL200 Animal Breeding and Genetics~~
~~LL210 Animal Reproduction and Development~~
~~LL220 Animal Genetics~~
*LL240 Animal Genetics and Breeding*
*LL250 Animal Reproduction and Embryology*
LL300 Animal Behaviour
*LL400 Animal Anatomy and Morphology*
LL500 Animal Nutrition (General)
LL510 Animal Nutrition (Physiology)
LL520 Animal Nutrition (Production Responses)
LL600 Animal Physiology and Biochemistry (Excluding Nutrition)
*LL650 Animal Immunology*
LL700 Animal Tissue and Cell Culture
LL800 Animal Health and Hygiene (General)
LL810 Animal Welfare
~~LL820 Parasites, Vectors, Pathogens and Biogenetic Diseases of Animals~~
*LL821 Prion, Viral, Bacterial and Fungal Pathogens of Animals*
*LL822 Protozoan, Helminth, Mollusc and Arthropod Parasites of Animals*
*LL823 Veterinary Pests, Vectors and Intermediate Hosts*
LL860 Non-communicable Diseases and Injuries of Animals
~~LL870 Animal Injuries~~
~~LL880 Animal Treatment and Diagnosis (Non-Drug)~~
*LL882 Veterinary Pharmacology and Anaesthesiology*

*Veterinary Drug File*. 1968– . London: Derwent/Thomson Scientific. Updated 10 times a year.    Formerly known as *VETDOC (1968–1992)*, this bibliographic database covers the world's scientific and veterinary literature on all developments and applications of veterinary drugs, vaccines, and biologicals. It indexes over 1,200 scientific and veterinary journals plus conference proceedings and meeting reports. This resource is available online through DataStar and STN, and in print.

www.animalscience.com (cited June 27, 2001). 2000– . Wallingford, UK: CAB International.    An Internet community covering both animal science and animal health, this site features a comprehensive abstract database with a 25-year archive (CAB Abstracts), linkage to full-text primary journals in ingenta (www. ingenta.com) news, spotlight articles, reviews, conference proceedings, reference material, listservs, alerting services, calendars, bookshops, useful links, and more.

## Retrospective

*Accumulative Veterinary Index*. v. 1–4, 1960/63–1975 with five annual supplements through 1980. Arvada, CO: Index Inc. ISSN: 0567-7033. Annual with five-year accumulations.    Indexes ten North American veterinary journals commonly received by veterinary practitioners.

*Humans & Other Species*: *Information Resources on Their Relationship*. 1997–99. Penryn, CA: Rockydell Resources. ISSN: 1093-8915. Quarterly. Previous titles: *InterActions of Man & Animals* (1990–91); *The Interactions Bibliography* (1992–96).    Covers animal-assisted activities and therapy; attitudes toward animals; the role of pets; impact of human-animal bond on human health; pet loss and counseling; animals in the arts, entertainment, and literature; and the ethics of our relationships to animals.

*Index-Catalogue of Medical and Veterinary Zoology*. v. 1–18, 1932–52. Washington, DC: U.S. Government Printing Office. Supplements (v. 1–24; 1953–82).    Index to the world's literature on parasites and parasitisms of man, of domestic animals, and of wild animals whose parasites may be transmitted to man and domestic animals.

*Quarterly Index*. v. 1–8, 1983–90. Oakdale, CA: Veterinary Interface. ISSN: 0740-2430. Quarterly with annual cumulation.    Includes 19 veterinary journals, focusing on cats, dogs, and birds, as well as domestic and exotic pets.

*Small Animal Abstracts*. v. 1–15, 1975–89. Farnham Royal, UK: Commonwealth Agricultural Bureaux. ISSN: 0306-7580. Quarterly.    Contains 1,200 selected abstracts per year from the CAB Abstracts database covering the literature on cats, dogs, and other pets (excluding laboratory animals).

*Tropical Veterinary Bulletin.* v. 1–18, 1912–30. London: Bureau of Hygiene and Tropical Diseases. ISSN: 0372-2635. Quarterly. Superseded by *Veterinary Bulletin.* Focuses on contagious and transmissible diseases caused by bacterial, filterable viruses, metazoan and protozoan parasites, and mycotic diseases.

*Veterinary Update: Clinical Abstract Service.* 1986–94. Goleta, CA: American Veterinary Publications. ISSN: 1059-8456. Bimonthly or monthly. First published as a monthly loose-leaf abstract service under the title (and editions) *Modern Veterinary Practice Reference and Data Service, Large Animal and Small Animal Editions* (1960–74) and then as *Veterinary Reference Service Update, Large Animal and Pet Practice Editions* (1975–85). *Veterinary Update: Small Animal* briefly continued as *Small Animal Medicine Digest* (1995–96).

## Selective

AGRICOLA (http://www.nal.usda.gov/ag98/) (cited June 28, 2001). 1970– . Beltsville, MD: National Agricultural Library. *Bibliography of Agriculture.* 1942– . Beltsville, MD: U.S. National Agricultural Library. ISSN: 0006-1530. In animal health and veterinary science areas, this bibliographic database and its print equivalent, *Bibliography of Agriculture*, reflect the U.S. National Agricultural Library's (NAL's) commitment to collect comprehensively in all areas dealing with the treatment and health maintenance of animals, as well as their diseases, anatomy, and physiology. In addition, NAL's collecting interests are influenced by the USDA legislative mandate under the 1989 Animal Welfare Act and the establishment of the Animal Welfare Information Center (AWIC) leading to its collecting in all aspects of laboratory animal science. NAL indexes some 40 veterinary journals. This database, which is available free of charge, is described in greater detail elsewhere in this book.

AGRIS (http://www.fao.org/agris/) (cited June 28, 2001). 1975– . Rome: AGRIS Coordinating Centre and Food and Agriculture Organization of the United Nations. *Agrindex.* 1975. Rome, Italy: AGRIS Coordinating Centre and Food and Agriculture Organization of the United Nations. Monthly. Created and maintained by the FAO, this bibliographic database indexes agricultural sciences and technology journals including a number of veterinary-related titles. It is available free of charge and is searchable in English, French, and Spanish. This database is described in greater detail elsewhere in this book.

MEDLINE (http://www.ncbi.nlm.nih.gov/entrez/query.fcgi) (cited June 28, 2001). 1960– . Bethesda, MD: National Library of Medicine. *Index Medicus.* 1960– . Bethesda, MD: National Library of Medicine. ISSN: 0019-3879. Monthly with annual cumulations. In animal health and veterinary science areas, this bibliographic database and its print equivalent, *Index Medicus*, reflect the National Library of Medicine's (NLM's) commitment to collect comprehensively in veterinary science as it relates to human health, biomedical research, and advances in biomedicine. It covers comparative medicine and comparative pathology, zoonoses, veterinary pharmacology, veterinary clinical sciences, primatology, and all aspects of laboratory animal science and welfare. Because of NLM's interest in research at the cellular, molecular, and biochemical level and its application to humans and human health, they also collect animal research literature in such fields as immunology, microbiology, parasitology, and toxicology. NLM does not collect in animal breeding or husbandry unless related to human health or laboratory animal science. The journals indexed include some 70 veterinary titles. MEDLINE is available through various commercial vendors, as well as free of charge through NLM's own PubMed database.

## BIBLIOGRAPHIES

Since the advent of user-accessible bibliographic databases, such as CAB Abstracts or AGRICOLA, the number of animal health and veterinary science bibliographies published has decreased to those focused on subjects that cross multiple disciplines, such as human-animal bond, or those covering an extended time period that frequently predates the databases. Most bibliographic series, such as *Bibliographies* from the U.S. Animal and Plant Health Inspection Services, Veterinary Services, Emergency Programs; *Quick Bibliography Series* by the U.S. NAL; or *Annotated Bibliographies* from the Commonwealth Bureau of Animal Health (UK), have ceased publication except for the work of the Animal Welfare Information Center described in the following text. As print bibliographies have languished, specialized bibliographic databases are developing and subject-specific bibliographies are appearing in Web formats.

If you are seeking information from years before those covered by bibliographic databases, bibliographies remain an excellent resource if one is available on the subject desired. Try local or national online catalogs, WorldCat or RLIN, by searching for the ''(topic) AND bibliographies.'' Given the frequent use of a cataloging note, ''Includes bibliography,'' the results of such a search may or may not be useful. Anyone seeking a bibliography on a given subject also should consult the references found in relevant review articles or dissertations.

Although institutes or schools, especially those in Europe, frequently issue bibliographies of faculty publications, those titles are not included here.

Allen, K.M. 1985. *The Human-Animal Bond: An Annotated Bibliography*. Metuchen, NJ: Scarecrow Press. 246 p. ISBN: 0810817926.   As an overview of the significant literature on the pet as an integral part of society, this bibliography includes materials on animals in their rehabilitative roles with psychiatric patients and other hospitalized and institutionalized individuals. It covers publications about pets as companions to the healthy, pet owners' responsibilities and attitudes, public health and pets, animal control, humane education, and the veterinarian's role in counseling bereaved pet owners. This work is arranged by broad topics with author and title indexes.

Bibliography for Transmissible Spongiform Encephalopathies (http://www. aphis.usda.gov/vs/ceah/cei/bsbibps.pdf) (cited June 14, 2001). 1996. Fort Collins, CO: Centers for Epidemiology and Animal Health. 116 p.   This electronic bibliography covers prion diseases, especially transmissible and bovine spongiform encephalopathies is available. Access requires a Web browser and PDF file reader. Citations are grouped by broad topic subdivided by year (oldest citation is 1979). There is an index to first authors.

Davies, A.S., compiler. 1990. *A Bibliography of Sheep and Goat Anatomy*. Palmerston North, NZ: Veterinary Continuing Education, Massey University. 167 p. no. 128. ISSN: 0112–9643.

Dey, B.P. 1986. *Mycobacterioses in Swine and Their Significance to Public Health*. Beltsville, MD: U.S. Department of Agriculture. 92 p. no. 49. ISSN: 0163–0873.   This monographic review covers the relationship between human mycobacterial infection and the disease in swine caused by certain organisms of the MAIS complex. It includes a 28-page bibliography prepared by Jesse Ostroff of the U.S. NAL.

Dingley, P. 1992. *Historic Books on Veterinary Science and Animal Husbandry*: *the Comben Collection in the Science Museum Library*. London: HMSO. 183 p. ISBN: 012905110.   This is a bibliography of the Norman Comben Collection, a collection of 704 printed books and pamphlets, published between 1514 and the mid-nineteenth century, on farriery, animal husbandry, and veterinary science. Purchased by the Science Museum Library (Great Britain) in 1987, this collection, primarily of English-language works published in Britain, is fully described including date of publication, pagination, and notes.

Dubey, J.P. and A. Towle. 1986. *Toxoplasmosis in Sheep*: *A Review and Annotated Bibliography*. St. Albans, UK: Commonwealth Institute of Parasitology. Miscellaneous Publ., no. 10. 152 p. ISBN: 851985629.   The first section of this publication consists of a critical review of current research on toxoplasmosis in sheep and an annotated bibliography of the relevant literature published between 1950 and 1985. Bibliographic entries are classified under eight subject headings and then listed by year of publication; then within each year alphabetically by author. Annotations were taken from abstracts found in *Protozoological Abstracts* (1976–85) and in *Veterinary Bulletin* (1950–75). Author and subject indexes complete this work.

Friedman, R. 1987. *Animal Experimentation and Animal Rights*. Phoenix, AZ: Oryx Press. Oryx Science Bibliographies, v. 9. 75 p. ISBN: 0897743776.   This annotated bibliography of some 245 recent publications covers the "pro-rights" and the "pro-experimentation" viewpoints of using animals in research.

Hails, M.R. 1983. *Plant Poisoning in Animals: A Bibliography from the World Literature, 1960–1979.* Farnham Royal, UK: Commonwealth Agricultural Bureau. 158 p. ISBN: 0859185254. Hails, M.R. 1986. *Plant Poisoning in Animals: A Bibliography from the World Literature, 1980–1982.* Farnham Royal, UK: CAB International. 6,923 p. ISBN: 0851985785. Hails, M.R. 1994. *Plant Poisoning in Animals: A Bibliography from the World Literature, 1983–1992.* Farnham Royal, UK: CAB International 282 p. ISBN: 0851989160. The first volume incorporates material from T.D. Crane's "Plant Poisoning in Animals: A Bibliography" (*Veterinary Bulletin* 43: 165–77; 231–49, 1973) and covers 3,200 references to poisonous plants that have an adverse effect on the health of animals; those containing factors that limit or inhabit optimum animal performance; and those containing antifertility substances. Drawn from *Index Veterinarius* and other relevant abstract journals of CAB International, entries are arranged taxonomically into plant families with author, geographical, and two subject (plant species and animals poisoned) indexes. The second volume is a supplement to the 1983 edition, including some 875 annotated citations for the period 1980–82. Volume three supplements both of the earlier editions with an additional 875 annotated citations from CAB Abstracts.

International Greyhound Research Database (http://nga.jc.net/agc/database/index.html) (cited June 14, 2001). 2001. 6th ed. Corvallis, OR: Oregon State University College of Veterinary Medicine. ISSN: 1057–5464. Begun in 1990, this research bibliography on greyhounds focuses on clinical veterinary research concerning the breed, institutional research germane to the racing greyhound industry, and medical research where greyhounds are used as a model for human disease. Citations to journal articles, books, or proceedings are divided into fourteen categories: muscle, orthopedics, physiology, drugs, nutrition, parasites, reproduction, injuries/repair, training/management, genetics, behavior, performance, and veterinarians' duties. Citations marked "Confidential" are available only to racing commissioners and racing chemists in the Association of Official Racing Chemists. This title is also available on CD-ROM or through the Internet.

Jorgensen, R.J. and C.P. Ogbourne. 1985. *Bovine Dictyocauliasis: A Review and Annotated Bibliography.* Farnham Royal, UK: Commonwealth Agricultural Bureaux. Miscellaneous Publ., no. 8. 104 p. ISBN: 0851985521. This work is a well-illustrated, critical review of current knowledge on a ubiquitous and at times endemic parasite of cattle. It includes an annotated bibliography of some 383 relevant references, covering 1968 to 1984, compiled from *Helminthological Abstracts*. Bibliographic entries are arranged by broad subjects followed by author and subject indexes.

Kellert, S.R. and J.K. Berry. 1985. *Bibliography of Human/Animal Relations*. Lanham, MD: University Press of America. 200 p. ISBN: 0819149586. Addressing the social, economic, political, cultural, and psychological aspects of man's interdependence with the nonhuman world, this bibliography cites scientific or academic materials on the interaction of humans and animals, rather than the biological or ecological characteristics of animal species. The 3,861 citations listed are arranged alphabetically by author's name with appropriate subject labels. Years covered appear to be primarily the 1960s to early 1980s yet references from as early as 1897 are found. Included is a list of journals and major databases searched in preparing the publication, and a broad keyword index based on the subject labels previously mentioned.

Macdonald, A.A. and N. Carlton, compilers. 2000. *A Bibliography of References to Husbandry and Veterinary Guidelines for Animals in Zoological Collections*. London: Federation of Zoological Gardens of Great Britain and Ireland. 611 p. Drawing from published works primarily from 1980 through 2000, this bibliography focuses on husbandry and management of captive (wild) animals. It is arranged alphabetically by author and subdivided into categories: general topics, invertebrates, amphibia, reptiles, birds, fish, and mammals. Seen as a "work in process," the authors intend to update it regularly. It lacks an index to authors, species, and subjects.

Mathias-Mundy, E. and C.M. McCorkle. 1989. *Ethnoveterinary Medicine: An Annotated Bibliography*. Ames, IA: Technology and Social Change Program, Iowa State University. 199 p. ISBN: 0945271166. Published as number six in the series, *Bibliographies in Technology and Social Change*, this title contains abstracts from around the world giving indigenous peoples' animal-care treatments, otherwise known as ethnoveterinary medicine. A new edition of this title is expected in 2001.

Rookmaaker, L.C. 1983. *Bibliography of the Rhinoceros: An Analysis of the Literature on the Recent Rhinoceroses in Culture, History, and Biology*. Rotterdam: A.A. Balkema; Salem, NH: MBS. 292 p. ISBN: 906191261X. This analysis of the literature on the rhinoceros from 1500 through the first half of 1982 covers all aspects, whether cultural, ethnological, historical, art-related, parasitological, or biological. Of special interest is the section of biological literature on the family Rhinocerotidae and the five living species, which includes references to the morphology and anatomy of each species, as well as food, reproduction, parasites, and veterinary studies.

Saint-Martin, G. 1990. *Bibliographie sur le Dromadaire et le Chameau*. 2nd ed. Maisons-Alfort: Institut d'Elevage et de Médecine Vétérinaire des Pays Tropicaux. 2 vols. 824 p. ISBN: 2859851674 (set). Covering the dromedary and the camel and their diseases, this bibliography includes citations in English, French, German, Italian, and Russian.

Thompson, R.C.A., and C.E. Allsopp. 1988. *Hydatidosis: Veterinary Perspectives and Bibliography*. Wallingford: CAB International. 246 p. ISBN: 0851986102. Beginning with a critical review of veterinary research on hydatidosis, or echinococcosis, a parasitic disease of major health and economic impact on livestock, this bibliography includes 591 citations, from the years 1976 through 1986, drawn primarily from *Helminthological Abstracts, Series A*.

U.S. Animal Welfare Information Center, National Agricultural Library (http://www.nal.usda.gov/awic/pubs/awicdocs.htm) (cited June 15, 2001). Quick Bibliography Series (ISSN: 1052-5378), Special Reference Briefs (ISSN: 1052-536X), *AWIC* Resource Series (ISSN: 1082-9644), and AWIC Series (ISSN: 1049-202X). Beltsville, MD: National Agricultural Library. A selected number of the bibliographies in these series, usually compiled using the AGRICOLA database, are relevant to veterinary medicine. Selected titles in Spanish as well. A list of titles is available on the AWIC Web site.

Van Der Westhuizen, E. 1993. *Ostrich Bibliography*. Onderstepoort, South Africa: University of Pretoria, Academic Information Service, Veterinary Science Library. 316 p. ISBN: 0869799258. A growing interest in ostriches worldwide led to the compilation of this bibliography. Drawing from CAB Abstracts (1972–93), BIOSIS (1969–93), AGRICOLA (1970–93), and other relevant bibliographies, regional resources, and the reference collections of ostrich experts, references are arranged alphabetically by author under main subject headings such as behavior, feedlots, and parasites. A descriptor list and an author list facilitate finding references with multiple authors or descriptors. A 1994 update (71 p.) to this bibliography was published in June of 1995 and includes earlier references omitted from the 1993 edition.

Van Der Westhuizen, E. 1994. *African Rhinoceros Bibliography*. Pretoria: University of Pretoria, Academic Information Service, Veterinary Science Library. 245 p. ISBN: 0869799746. Compiled for the Symposium on Rhinos as Game Ranch Animals, the Wildlife Group of the South African Veterinary Association, and the Wildlife Research Programme of the Faculty of Veterinary Science, University of Pretoria, this bibliography is arranged alphabetically by author under main subject headings such as immunology, physiology, or poaching. Drawing references from various sources including CAB Abstracts (1972–94), BIOSIS (1969–94), Zoological Record (1978–94), and the *Index to South African Periodicals*, each reference is assigned a record number that allows for cross-indexing within author list and descriptor list appendices.

## BIOGRAPHICAL DIRECTORIES AND BIOGRAPHIES

This section describes biographical directories that provide individual profiles of veterinarians. Titles covered range from the latest *Who's Who in Veterinary Science and Medicine* to a biographical-bibliographical dictionary of veterinarians written by Georg W. Schrader in 1863. The author found Robert B. Slocum's *Biographical Dictionaries and Related Works* (2ⁿᵈ ed., 1986) to be particularly useful in initially identifying a number of unique veterinary titles not widely held in this country, but those titles have not been included here.

Individual biographies, autobiographies, correspondence, reminiscences, assorted tales, and other life experiences written by or about individual veterinarians have appeared over the years, but the popularity of James Herriot's books in recent years has greatly encouraged the publication of similar works by other veterinarians. Titles can be identified through the print or online catalogs of the U.S. Library of Congress, the U.S. NLM, the U.S. NAL, the British Museum, and the U.K. Royal College of Veterinary Surgeons, or through local library catalogs by doing a subject-heading search for "Veterinarians—biographies" or a keyword search for "veterinarians and biographies." Note that in countries or locations where veterinarians are ethically bound to refrain from advertising, a veterinarian's real name may be disguised. For example, James Herriot's real name was James Alfred Wight.

*Agricultural and Veterinary Sciences International Who's Who*. 1994. 5ᵗʰ ed. Harlow, Essex: Longman. 1,157 p. ISBN: 0582256666; 1561591211. Containing biographical profiles for over 10,800 senior agricultural and veterinary scientists in some 132 countries worldwide, this resource is arranged alphabetically by surname. The subject index lists scientists first by country, then by broad subject categories such as veterinary medicine. Initially published as *Who's Who in World Agriculture: A Biographical Guide in the Agricultural and Veterinary Sciences* (1ˢᵗ–2ⁿᵈ eds., 1979–85).

*American Men and Women of Science: Agricultural, Animal and Veterinary Sciences*. 1974. Edited by the Jaques Cattell Press. New York: R. R. Bowker. 832 p. ISSN: 0094-5110, ISBN: 0835207153. Selected from the *American Men and Women of Science*, 12ᵗʰ ed. (8 vols.), this biographical directory includes some 14,200 men and women for the U.S. and Canada. For more current biographical profiles, see *American Men and Women of Science* (20ᵗʰ ed., 1998 or more recent edition).

Drum, S. and H.E. Whiteley, 1991. *Women in Veterinary Medicine: Profiles of Success*. Ames, IA: Iowa State University Press. 270 p. ISBN: 0813806682. A biographical directory focused on twenty U.S. women veterinarians who with determination and perseverance have made a difference and expanded opportunities within their profession.

Schrader, G.W. 1863. *Biographisch-Literarisches Lexicon der Thierarzte aller Zeiten und Lander*. Stuttgart: Ebner & Seubert. 490 p.   This biographical-bibliographical dictionary of 2,001 veterinarians of all times and countries is arranged alphabetically by last name of biographee. Thirty-five additional biographies appear in a separate alphabetical sequence at the end of the dictionary. This title was reprinted in 1967 by the Zentral-Antiquariat der DDR.

Smith, F., Sir. 1919–33. *The Early History of Veterinary Literature and its British Development*. London: Bailliere, Tindall and Cox. 4 vols.   This history of early veterinary literature includes numerous biobibliographies of British veterinarians from 1700 to 1860. Each volume is separately indexed with no cumulative index.

Stalheim, O.V. 1996. *Veterinary Conversations with Mid-Twentieth Century Leaders*. Ames, IA: Iowa State University Press. 302 p. ISBN: 081382995X. Twenty-eight prominent American veterinarians, born between 1904 and 1944, and active in government service, education, the pharmaceutical industry, clinical practice, and research, were interviewed. The interviews provide biographies and report the state of veterinary medicine between the 1930s and the 1960s. A concluding chapter provides sources of veterinary biographies.

*Who's Who in Veterinary Science and Medicine, 1991/1992*. 1991. 2nd ed. Van Nuys, CA: Crown Publications. 290 p. ISBN: 0944437002.   This publication includes biographical profiles for more than 5,200 individuals deemed to be of interest by virtue of their positions of responsibility in or by their contributions to the veterinary profession (primarily in the U.S. or Canada). The directory is composed of two sections: individual biographical profiles arranged in alphabetical order according to the last name of the biographee, and an index of biographees listed under their professional specialties. Each specialty listing is arranged geographically by state, province, and country.

## BOOKS OR CORE LITERATURE

A highly selected list of core texts in veterinary medicine is provided to assist in the development of collections relevant to the teaching, research, or practice interests of those served. The list is arranged by basic veterinary science and clinical veterinary medicine areas. Reference works discussed elsewhere in this chapter are not included here. Earlier works by Kerker and Murphy (1973) and Gibb (1990) provide an overview of texts in the field of animal health and veterinary science at the time.

## Basic Veterinary Sciences

### Anatomy

Ashdown, R.R. 1996. *The Horse*. (Color Atlas of Veterinary Anatomy, v. 2). London: Mosby-Wolfe. ISBN: 0723425744.

Ashdown, R.R. 1996. *The Ruminant*. (Color Atlas of Veterinary Anatomy, v. 1). London: Mosby-Wolfe. ISBN: 0723426627 (paperback); 0723424918 (hardback).

Dyce, K.M. 1996. *Textbook of Veterinary Anatomy*. 2nd ed. Philadelphia, PA: Saunders. ISBN: 0721649610.

Evans, H.E., ed. 2000. *Guide to the Dissection of the Dog*. 5th ed. Philadelphia, PA: W.B. Saunders. ISBN: 0721680798.

Popesko, P. 1977. *Atlas of Topographical Anatomy of the Domestic Animals*. 2nd ed. Philadelphia, PA: Saunders.

### Histology

Bacha, W.J. 2000. *Color Atlas of Veterinary Histology*. 2nd ed. Philadelphia, PA: Lippincott Williams & Wilkins. ISBN: 0683306189.

Dellmann, H.D. 1998. *Textbook of Veterinary Histology*. 5th ed. Baltimore, MD: Williams & Wilkins. ISBN: 0683301683.

### Microbiology

Carter, G.R. 1995. *Essentials of Veterinary Microbiology*. 5th ed. Baltimore, MD: Williams & Wilkins. ISBN: 0683014730.

Hirsh, D.C. and Y.C. Zee. 1999. *Veterinary Microbiology*. Malden, MA: Blackwell Science. ISBN: 0865425434.

### Parasitology

Bowman, D.D. 1999. *Georgis' Parasitology for Veterinarians*. 7th ed. Philadelphia, PA: W.B. Saunders. ISBN: 0721670970.

Foreyt, W.J. 2001. *Veterinary Parasitology Reference Manual*. 5th ed. Ames, IA: Iowa State University Press. ISBN: 0813824192.

Urquhart, G.M., ed. 1996. *Veterinary Parasitology*. 2nd ed. Cambridge, MA: Blackwell Science. ISBN: 0632040513.

### Pathology

Jones, T.C. 1997. *Veterinary Pathology*. 6th ed. Baltimore, MD: Williams & Wilkins. ISBN: 0683044818.

Jubb, K.V.F. and P.C. Kennedy, eds. 1993. *Pathology of Domestic Animals*.

4th ed. San Diego, CA: Academic Press. 3 vols. ISBN: 0123916054 (v. 1), 0123916062 (v. 2), 0123916070 (v. 3).

McGavin, M.D., W.W. Carlton, J.F. Zachary. 2001. *Thomson's Special Veterinary Pathology*. 3rd ed. St. Louis, MO: Mosby. ISBN: 0323005608.

## Pharmacology

Adams, H.R., ed. 2001. *Veterinary Pharmacology and Therapeutics*. 8th ed. Ames, IA: Iowa State University Press. ISBN: 0813817439 (alk. paper).

Plumb, D.C. 1999. *Veterinary Drug Handbook*. 3rd ed. White Bear Lake, MN: Pharma Vet Pub.; Ames, IA: Iowa State University Press. ISBN: 0813824443 (desk edition).

## Physiology

Cunningham, J.G., ed. 2002. *Textbook of Veterinary Physiology*. 3rd ed. Philadelphia, PA: W.B. Saunders. ISBN: 0721689949.

Reece, W.O. 1997. *Physiology of Domestic Animals*. 2nd ed. Baltimore, MD: Williams & Wilkins. ISBN: 0683072404.

## Toxicology

Osweiler, G.D. 1996. *Toxicology*. Philadelphia, PA: Williams & Wilkins. ISBN: 0683066641.

## Clinical Veterinary Medicine

### Birds, Poultry

Altman, R.B., et al. 1997. *Avian Medicine and Surgery*. Philadelphia, PA: Saunders. ISBN: 0721654460.

Calnek, B.W., ed. 1997. *Diseases of Poultry*. 10th ed. Ames, IA: Iowa State University Press. ISBN: 0813804272.

Ritchie, B.W., G.J. Harrison, and L.R. Harrison. 1994. *Avian Medicine: Principles and Application*. Lake Worth, FL: Wingers Publishing. ISBN: 0963699601.

Sainsbury, D. 2000. *Poultry Health & Management*. 4th ed. Cambridge, MA: Blackwell Science. ISBN: 0632051728.

Samour, J. 2000. *Avian Medicine*. St. Louis, MO: Mosby. ISBN: 072342960X.

### Large Animals

Adams, O.R. 2000. *Adams' Lameness in Horses*. 5th ed. Baltimore: Williams & Wilkins. ISBN: 0683079816.

Adams, S.B. and J.F. Fessler. 2000. *Atlas of Equine Surgery*. Philadelphia, PA: W.B. Saunders. ISBN: 721646433.

Andrews, A.H. 2000. *Health of Dairy Cattle*. Cambridge, MA: Blackwell Science. ISBN: 063204103X.

Butler, J.A. 2000. *Clinical Radiology of the Horse*. 2nd ed. Cambridge, MA: Blackwell Science. ISBN: 0632052686 (hardbound).

Colahan, P.T., et al. 1999. *Equine Medicine and Surgery*. 5th ed. St. Louis, MO: Mosby. ISBN: 0815117434.

Fraser, A.F. 1992. *Behaviour of the Horse*. Wallingford, UK: CAB International. ISBN: 0851987850 (paperback).

Greenough, P.R., ed. 1997. *Lameness in Cattle*. 3rd ed. Philadelphia, PA: W.B. Saunders. ISBN: 0721652050.

Hafez, E.S.E., and B. Hafez. 2000. *Reproduction in Farm Animals*. 7th ed. Philadelphia, PA: Lippincott Williams & Wilkins. ISBN: 0683305778.

Martin, W.B. 2000. *Diseases of Sheep*. 3rd ed. Cambridge, MA: Blackwell Science. ISBN: 0632051396.

Mayhew, I.G. 1989. *Large Animal Neurology*. Philadelphia, PA: Lea & Febiger. ISBN: 0812111834.

Radostits, O.M., et al. 2000. *Veterinary Medicine: A Textbook of the Diseases of Cattle, Sheep, Pigs, Goats and Horses*. 9th ed. London; New York: W.B. Saunders. ISBN: 0702026042.

Rebhun, W.C. 1995. *Diseases of Dairy Cattle*. Baltimore, MD: Williams & Wilkins. ISBN: 0683071939.

Smith, B.P., ed. 2002. *Large Animal Internal Medicine*. 3rd ed. St. Louis, MO: Mosby. ISBN: 0323009468.

Smith, M.C. 1994. *Goat Medicine*. Philadelphia, PA: Lea & Febiger. ISBN: 0812114787.

Straw, B.E., et al. 1999. *Diseases of Swine*. 8th ed. Ames, IA: Iowa State University Press. ISBN: 0813803381 (alk. paper).

Youngquist, R.S., ed. 1997. *Current Therapy in Large Animal Theriogenology*. Philadelphia, PA: W.B. Saunders Co. ISBN: 0721653960.

## Small Animals

Birchard, S.J. and R.G. Sherding. 2000. *Saunders Manual of Small Animal Practice*. 2nd ed. Philadelphia, PA: W.B. Saunders. ISBN: 0721670784.

Bojrab, M.J., ed. 1998. *Current Techniques in Small Animal Surgery*. 4th ed. Baltimore, MD: Williams & Wilkins. ISBN: 0683008900.

Bonagura, J. 2000. *Kirk's Current Veterinary Therapy XIII: Small Animal Practice*. 13th ed. Philadelphia, PA: W.B. Saunders. ISSN: 0070-2218.

DiBartola, S.P. 2000. *Fluid Therapy in Small Animal Practice*. 2nd ed. Philadelphia, PA: W.B. Saunders. ISBN: 0721677398.

Feldman, E.C. 1996. *Canine and Feline Endocrinology and Reproduction.* 2nd ed. Philadelphia, PA: W.B. Saunders. ISBN: 0721636349.

Fossum, T.W. 2002. *Small Animal Surgery.* 2nd ed. St. Louis, MO: Mosby. ISBN: 0323012388.

Greene, C.E. 1998. Infectious Diseases of the Dog and Cat. 2nd ed. Philadelphia, PA: W.B. Saunders. ISBN: 0721627374.

Hillyer, E.V., ed. 1997. *Ferrets, Rabbits and Rodents: Clinical Medicine and Surgery.* Philadelphia, PA: W.B. Saunders. ISBN: 0721640230.

Mader, D.R., ed. 1996. *Reptile Medicine and Surgery.* Philadelphia, PA: W.B. Saunders. ISBN: 0721652085.

Piermattei, D.L. 1993. *Atlas of Surgical Approaches to the Bones and Joints of the Dog and Cat.* 3rd ed. Philadelphia, PA: W.B. Saunders. ISBN: 0721610129.

Piermattei, D.L. 1997. *Brinker, Piermattei, and Flo's Handbook of Small Animal Orthopedics and Fracture Repair.* 3rd ed. Philadelphia, PA: W.B. Saunders. ISBN: 0721656897.

Morrison, W.B. 1998. *Cancer in Dogs and Cats.* Philadelphia, PA: Lippincott Williams & Wilkins. ISBN: 0683061054.

Muller, G.H. 2000. *Muller & Kirk's Small Animal Dermatology.* 6th ed. Philadelphia, PA: W.B. Saunders. ISBN: 0721676189.

Shaw, D.H. 1997. *Small Animal Internal Medicine.* Baltimore, MD: Williams & Wilkins. ISBN: 0683076701.

Slatter, D.H., ed. 1993. *Textbook of Small Animal Surgery.* 2nd ed. Philadelphia, PA: W.B. Saunders. ISBN: 0721683304 (set). (New edition due in 2003.)

Tilley, L.P. 2000. *5 Minute Veterinary Consult: Canine and Feline.* Philadelphia, PA: Lippincott Williams & Wilkins. ISBN: 0683304615.

Tilley, L.P. 1999. *ECG for the Small Animal Practitioner.* Jackson, WY: Teton NewMedia. ISBN: 1893441008.

Turner, D.C., ed. 2000. *Domestic Cat: The Biology of Its Behaviour.* 2nd ed. Cambridge, UK; New York: Cambridge University Press. ISBN: 0521636485 (paperback).

Wilkinson, G.T. 1994. *Color Atlas of Small Animal Dermatology.* 2nd ed. London: Wolfe. ISBN: 0723418985.

Withrow, S.J. and E.G. MacEwen, eds. 2001. *Small Animal Clinical Oncology.* 3rd ed. Philadelphia, PA: W.B. Saunders. ISBN: 072167755X.

## Veterinary Specialties

Bistner, S.I., R.B. Ford, and M.R. Raffe. 2000. *Kirk and Bistner's Handbook of Veterinary Procedures and Emergency Treatment.* 7th ed. Philadelphia, PA: W.B. Saunders. ISBN: 0721671667.

Boothe, D.M. 2001. *Small Animal Clinical Pharmacology and Therapeutics*. Philadelphia, PA: W.B. Saunders. ISBN: 0721643647 (Paperbook).

Duncan, J.R. 1994. *Veterinary Laboratory Medicine: Clinical Pathology.* 3rd ed. Ames, IA: Iowa State University Press. ISBN: 0813819172 (acid-free paper).

Ettinger, S.J. and E.C. Feldman, eds. 2000. *Textbook of Veterinary Internal Medicine.* 5th ed. Philadelphia, PA: W.B. Saunders. ISBN: 0721672566 (set).

Feldman, B.V., J.G. Zinkl, N.C. Jain, et al. 2000. *Schalm's Veterinary Hematology.* 5th ed. Philadelphia, PA: Lippincott Williams & Wilkins. ISBN: 0683306928.

Gelatt, K.N. 1999. *Veterinary Ophthalmology.* 3rd ed. Philadelphia, PA: Lippincott Williams & Wilkins. ISBN: 0683300768.

Kealy, J.K. 2000. *Diagnostic Radiology and Ultrasonography of the Dog and Cat.* 3rd ed. Philadelphia, PA: W.B. Saunders. ISBN: 0721650902.

Lagoni, L.S. 1994. *Human-Animal Bond and Grief.* Philadelphia, PA: Saunders. ISBN: 0721645771.

*Merck Veterinary Manual.* 1998. 8th ed. Whitehouse Station, NJ: Merck & Co./Merial Limited. ISBN: 0911910298.

Meyer, D.J. 1998. *Veterinary Laboratory Medicine.* 2nd ed. Philadelphia, PA: W.B. Saunders. ISBN: 0721662226.

Morrow, D.A., ed. 1986. *Current Therapy in Theriogenology.* 2nd ed. Philadelphia, PA: W.B. Saunders. ISBN: 0721665802.

Muir, W.W. et al. 2000. *Handbook of Veterinary Anesthesia.* 3rd ed. St. Louis, MO: Mosby. ISBN: 0323008011.

Murphy, F.A., et al. 1999. *Veterinary Virology.* 3rd ed. San Diego, CA: Academic Press. ISBN: 0125113404.

Nyland, T.G., ed. 2002. *Small Animal Diagnostic Ultrasound.* 2nd ed. Philadelphia, PA: W.B. Saunders. ISBN: 0721677886.

Ogilvie, G.K. 1995. *Managing the Veterinary Cancer Patient.* Trenton, NJ: Veterinary Learning Systems. ISBN: 1884254209.

Oliver, J.E. 1997. *Handbook of Veterinary Neurology.* 3rd ed. Philadelphia, PA: W.B. Saunders. ISBN: 0721671403.

Reagan, W.J. 1998. *Veterinary Hematology.* Ames, IA: Iowa State University Press. ISBN: 0813826640.

Severin, G.A. 2000. *Severin's Veterinary Ophthalmology Notes.* 3rd ed. rev. Fort Collins, CO: Veterinary Ophthalmology Notes. ISBN: 0683306928.

Sloss, M.W. 1994. *Veterinary Clinical Parasitology.* 6th ed. Ames, IA: Iowa State University Press. ISBN: 0813817331.

Thrall, D.E., ed. 2002. *Textbook of Veterinary Diagnostic Radiology.* 4th ed. Philadelphia, PA: W.B. Saunders. ISBN: 0721688209.

Thurmon, J.C., ed. 1996. *Lumb & Jones' Veterinary Anesthesia*. 3<sup>rd</sup> ed. Baltimore, MD: Williams & Wilkins. ISBN: 0683082388. ISBN: 0721682189.

Tizard, I.R. 2000. *Veterinary Immunology*. 6<sup>th</sup> ed. Philadelphia, PA: W.B. Saunders. ISBN: 0721682189.

## Zoo and Wildlife Diseases

Fowler, M.E., ed. 1999. *Zoo & Wild Animal Medicine: Current Therapy*. 4<sup>th</sup> ed. Philadelphia, PA: W.B. Saunders. ISBN: 0721686648.

Williams, E.S., ed. 2001. *Infectious Diseases of Wild Mammals*. 3<sup>rd</sup> ed. Ames, IA: Iowa State University Press. ISBN: 0813825563.

## Zoonoses and Communicable Diseases

Hugh-Jones, M.E., W.T. Hubbert, and H.V. Hagstad. 2000. *Zoonoses: Recognition, Control, and Prevention*. Ames, IA: Iowa State University Press. ISBN: 0813825423.

Palmer, S.R., Lord Soulsby, and D.I.H. Simpson, eds. 1998. *Zoonoses: Biology, Clinical Practice, and Public Health Control*. Oxford; New York: Oxford University Press. ISBN: 019262380X.


## DICTIONARIES

Dictionaries, including thesauri and nomenclatures, common to veterinary medicine are the focus of this section. The dictionaries begin with those general works that identify or define veterinary-related terms or synonyms in the same language and those that include equivalent terms in other languages. These works are grouped under the English Language and Polyglot sections. Foreign-language veterinary dictionaries, such as Bakulov's *Solvar Veterinarnykh Terminov* (Moskva: Edelveis, 1995), Villemin's *Dictionnaire des Termes Veterinaires et Zootechniques* 3<sup>rd</sup> edition, revised (Paris: Vigot Freres Editeurs, 1984), or Wiesner's *Worterbuch der Veterinarmedizin* 3<sup>rd</sup> edition, neu bearbeitete Aufl. (Jena: G. Fischer, 1991) were not included here.

In addition to the general dictionaries listed, there are a number of specialized dictionaries dealing with an individual species or a specific veterinary discipline or group of disciplines. A selection of special subject dictionaries in the English language is included. Certain special subject dictionaries, such as P.D. Rossdale's *A Horse's Health from A to Z*, revised and updated edition, (Newton Abbot, Devon, UK: David & Charles, 1998), which were written by or in consultation with veterinarians for the animal owner and/or the general public, are not listed nor are any polyglot special-subject dictionaries, such as the *Dictionary of Animal Production Terminology: in English, French, Spanish, German and Latin*, 2<sup>nd</sup> edition (Amsterdam: Elsevier, 1993).

Currently there are no individual works covering abbreviations or eponyms in veterinary medicine. However, the dictionaries listed usually provide some information of this nature as do more general medical works, such as B. G. Firkin's *Dictionary of Medical Eponyms*, 2ⁿᵈ edition (Park Ridge, NJ: Parthenon Publishing, 1996) or S.B. Sloane's *Medical Abbreviations and Eponyms*, 2ⁿᵈ edition (Philadelphia, PA: W.B. Saunders, 1997). A Web-based list of veterinary abbreviations and acronyms, compiled by Mitsuko Williams, the University of Illinois at Urbana-Champaign Veterinary Medicine Library, can be found at http://www.library.uiuc.edu/vex/vetdocs/abbreviation.htm (cited June 26, 2001).

Medical dictionaries frequently will list terms of both medical and veterinary significance, as well as words of purely veterinary usage. Although *Stedman's Medical Dictionary Illustrated* (27ᵗʰ ed., 2000) appears to define more uniquely veterinary terms, the choice between *Dorland's Illustrated Medical Dictionary* (29ᵗʰ ed., 2000) or *Stedman's* is a matter of personal preference.

The need for thesauri or controlled vocabularies suitable for indexing and retrieving veterinary literature arose with the creation of printed indexes, but became even more critical with the advent of online databases. Building on earlier veterinary thesauri, such as the Commission of the European Communities' five-volume *Veterinary Multilingual Thesaurus* (K.G. Saur, 1979), the Commonwealth Bureau of Animal Health's *Controlled Vocabulary* (1985), and the Veterinary Services' *Animal Disease Thesaurus* (12th rev., APHIS, 1986), the *CAB Thesaurus*, described in the following text, reflects the development and progressive standardization of veterinary terminology and its close relationship to both medical and agricultural terminology.

Veterinary nomenclature continues to evolve as seen in the selected works described. Although the *Standard Nomenclature of Veterinary Diseases and Operations* (SNVDO) (2ⁿᵈ ed., 1975 and 1977 coding supplement) were widely accepted and utilized in veterinary medical records systems from 1966 through 1988, it was supplanted by the *Systematized Nomenclature of Medicine: Microglossary for Veterinary Medicine (SNOVET)* (1984) that was published in late 1988. Today, the standard is *SNOMED: Systematized Nomenclature of Human and Veterinary Medicine* (College of American Pathologists, 1993).

## English Language

Blood, D.C., and V.P. Studdert. 1999. *Saunders Comprehensive Veterinary Dictionary*. 2ⁿᵈ ed. London; New York: W.B. Saunders. 1,380 p. ISBN: 0702024422. When initially published in 1988 as *Balliere's Comprehensive Veterinary Dictionary*, this authoritative title, written by veterinarians, filled a long-standing need for an English-language authoritative source covering veterinary technical and scientific terminology. It now includes over 65,000 entries and subentries and extensive appendices of anatomical information as well as laboratory and clinical data.

Boden, E., ed. 1998. *Black's Veterinary Dictionary*. 19th ed. Lanham, MD: Barnes and Noble Books. 595 p. ISBN: 038921017X. First published in 1928 under the title, *Black's Veterinary Cyclopedia*, and edited for the last fourteen editions by Geoffrey West, this one-volume British dictionary alphabetically lists anatomical terms, drugs, diseases, treatments, and other veterinary terminology relating to various types of animals. Intended for professionals and laymen who deal with animals, it is a standard reference work whose entries frequently include citations to key articles as well as scientific names of animals and relevant cross-references. Earlier editions have been translated into Spanish by F. Perez y Perez (Barcelona: Iatros).

Brown, C.M., D.A. Hogg, and D.F. Kelly, eds. 1988. *Concise Veterinary Dictionary*. Oxford: Oxford University Press. 890 p. ISBN: 0198542089. Covering all major fields in veterinary science, this title is intended for farmers, agricultural workers, veterinary students and assistants, and researchers in related fields, as well as veterinarians and animal owners. Entries avoid the use of jargon, and any scientific or alternate terms used in a definition are starred and defined elsewhere in the work.

## Polyglot Dictionaries

Blaha, T. and G. Ilchmann. 1993. *Fachworterbuch, Veterinarmedizin (Dictionary, Veterinary Medicine)*. Berlin: Verlag Alexander Hatier. 414 p. ISBN: 386117037X. This work is a German, French, Russian, and English polyglot dictionary of veterinary medicine.

Mack, R. 1972. *Russian-English Veterinary Dictionary (Russko-Angliiskii Veterinarnyi Slovar')*. Farnham Royal, UK: Commonwealth Agricultural Bureaux. 104 p. ISBN: 0851982557. Covering 6,000 specialist terms normally found in the Russian literature on veterinary science and relevant biological sciences, this dictionary also includes terms for drugs and chemicals. The entries consist of Russian terms and English equivalents. Abbreviations and acronyms are also included, but pronunciation is not indicated.

Mack, R., ed. 1992. *Dictionary of Animal Health Terminology: In English, French, Spanish, German, and Latin*. Amsterdam; New York: Elsevier. 426 p. ISBN: 0444880852. Compiled by the Office International des Épizooties, Paris, France, this is a revised edition of the veterinary medicine portion of the *Dictionary of Animal Production Terminology* (2nd ed., 1979).

Mack, R. 1996. *Dictionary of Veterinary Sciences and Biosciences [Worterbuch fur Veterinarmedizin und Biowissenschaften]*. 2nd ed. Berlin; Boston: Blackwell

Wissenschafts-Verlag. 823 p. ISBN: 3826330552. Initially published in 1988, this second edition supplements standard German-English, English-German dictionaries by focusing on some 20,000 technical terms in the fields of anatomy, microbiology, physiology, parasitology, pathology, pharmacology, toxicology, and animal husbandry, with special reference to domestic animals and their diseases. Anatomical terms are listed in accordance with the nomenclature of *Nomina Anatomica Veterinaria* (4th ed.) and *Nomina Histologica* (2nd ed.). In Part 3 of the work, 6,500 Latin terms are listed separately with their German, English, and French equivalents.

Mack, R., and E. Meissonnier. 1991. *Dictionnaire des Termes Vétérinaires et Animaliers: Français-Anglais, Anglais-Français = Veterinary and Animal Science Dictionary: French-English, English-French*. Maisons-Alfort: Editions du Point Vétérinaire. 575 p. ISBN: 2863260863. This French-English, English-French dictionary of veterinary and animal science has a Latin-English-French comparison in Part 3.

Schulz, H.E. 1963. *Vocabularium Veterinarium Polyglotte*. 3. Aufl. Halle (Westf.): Deutsches Archiv fur Veterinarmedizinische Nomenklatur. 1v. (various paging). Arranged in two parts, this dictionary first lists internationally recognized Greek and Latin terms relevant to veterinary medicine, in alphabetical order, followed by the equivalent English, German, French, Italian, Spanish, and Russian terms. The second part of this dictionary is comprised of six individual language indexes that refer one from a known term in one of the languages to the equivalent terms as listed in the general index.

## Special-Subject Dictionaries

Handy-Marchello, B. 1984. *The Veterinary Technician's Guide to Medical Terminology*. Reston, VA: Reston Pub. 286 p. ISBN: 0835983137. Written specifically for student and practicing veterinary technicians, this is a dictionary of medical terms used in veterinary situations. It includes animal diseases with names common to human disease, with names derived directly from the name of a parasite, or with common, well-known names, such as rabies. An entry may provide the word, its pronunciation, any alternative spelling, any abbreviation and/or synonyms, and the definition. It has an appendix of colloquial terms that are sometimes used in veterinary situations but are not necessarily medical in origin, such as broken wind.

Harre, R., and R. Lamb. 1986. *The Dictionary of Ethology and Animal Learning*. Cambridge, MA: The MIT Press. 171 p. ISBN: 0262580764. Based on the *Encyclopedic Dictionary of Psychology* (MIT Press, 1983), this dictionary includes updated articles on the discoveries and theories relevant to the behavior of nonhuman animals, particularly ethological ideas and studies of animals in their natural habitat. Covering such issues as conditioning, reinforcement, and imprinting, this publication discusses basic organic processes as well as the more social topics. Most entries also include a current bibliography.

Herren, R.V., and J.A. Romich. 2000. *Delmar's Veterinary Technician Pocket Dictionary.* Albany, NY: Delmar Thomson Learning. 265 p. ISBN: 0766814211. This dictionary includes over 6,000 terms frequently used by practicing veterinary technicians and nurses. It is cross-referenced to point out key related concepts with pronunciation keys for uncommon terms. Appendices provide weight-conversion tables, common veterinary abbreviations, and clinical lab values for large and small animals.

Hurnik, J.F., A.B. Webster, and P.B. Siegel. 1995. *Dictionary of Farm Animal Behaviour.* 2nd ed. Guelph, Ontario, Canada; University of Guelph. 200 p. ISBN: 0585072078. This dictionary contains concise definitions of behavioral terms, as well as terms from disciplines relating to animal behavior, such as genetics, disease, and animal husbandry. However, the selection of peripheral terms appears to be somewhat arbitrary and at times unnecessary (i.e., horse dressage terms are included).

Hurov, L. 1978. *Handbook of Veterinary Surgical Instruments and Glossary of Surgical Terms.* Philadelphia, PA: Saunders. 214 p. ISBN: 0721648487. Intended for veterinary practitioners and others who perform surgery on animals, the first section of this volume covers surgical instruments, including instrument cleaning and care, with illustrations. Arranged by type of use (ophthalmic, orthopedic, etc.), general hand instruments appear in alphabetical order with product information. A glossary of surgical diseases, related conditions, and the methods of correction appear in section two.

Lane, D.R. and S. Guthrie. 1999. *Dictionary of Veterinary Nursing.* Oxford; Boston: Butterworth-Heinemann. 271 p. ISBN: 0750636157. Written for veterinary nurses and animal technicians, this English-language dictionary provides concise definitions for more than 3,500 words relative to the nursing care of companion and pet animals including exotics. Included are line drawings, and a variety of appendices on such topics as temperature, pulse, respiratory rates for various species, along with biochemistry, normal urine, and hematology parameters for dogs and cats.

Toma, B., M. Pascal, and J.P. Vaillancourt, et al. 1999. *Dictionary of Veterinary Epidemiology.* Ames, IA: Iowa State University Press. 284 p. ISBN: 081382639X. Incorporating a translation of *Glossaire d'Épidémiologie Animale* (Editions du Point Vétérinaire, 1991), this dictionary defines epidemiological terms and those from related fields, such as economics and biostatistics. It presents both French and North American perspectives on the definitions of the terms. Currently there is no consensus within the veterinary epidemiology community on the meaning of some expressions included. The debates will continue with hopefully a greater focus for discussion as a result of this work. Entries include examples of a term's use, comments from colleagues, and cross-references.

## Thesauri

*CAB Thesaurus*. 1999. 5th ed. 2 vols. Wallingford, UK: CAB International. ISBN: 0851993664. Since 1984, this thesaurus has been used to index the subject content of CAB Abstracts and other products of CAB International, as well as the AGRICOLA database prepared by the NAL. As a controlled vocabulary, this fifth edition contains some 59,000 terms relevant to all aspects of the agricultural sciences, including animal health and veterinary science. Arranged first in a single alphabetical display, descriptors are listed in a hierarchy showing terms, which are more general (BT) or more specific (NT) than the descriptor listed, as well as any cross-references or related terms (rt) relevant to the descriptor but not part of its hierarchy.

SAMPLE ENTRY

**veterinary parasitology**
PD000028

| | |
|---|---|
| uf | parasitology, veterinary |
| BT | parasitology |
| | veterinary science |
| NT | veterinary helminthology |
| rt | veterinary entomology |
| | veterinary medicine |

With this edition, a new classified section has been added that lists preferred terms, their reference numbers and any equivalent American spellings (AF), for example:

| | |
|---|---|
| NK00101 | Flaviviridae |
| NK00124 | Pestivirus |
| NK00125 | border disease virus |
| NK00126 | bovine diarrhoea virus |
| | AF bovine diarrhea virus |
| NK00127 | swine fever virus |

The author suggests that any search of the CAB Abstracts database and its derivations should include both British and American spellings of terms to achieve maximum retrieval. Further discussion of this title is provided elsewhere in this book.

## Nomenclature

*Nomina Anatomica Veterinaria*. 1994. 4th ed.; *Nomina Histologica*, 2nd rev. ed.; and *Nomina Embryologica Veterinaria*. Zürich; Ithaca, NY: The Committees; Ithaca, NY: Distributed by the Department of Veterinary Anatomy, Cornell University. 1v. (various pagings). ISBN: 0960044477. This fourth edition of *Nomina Anatomica Veterinaria* was revised by the International Committee on Veterinary Gross Anatomical Nomenclature and authorized by the Twelfth General Assembly of the World Association of Veterinary Anatomists held in Ghent in 1992. In the introduction, the history of standardizing the nomenclature of human and veterinary anatomy is outlined together with that of the committees that produced this and earlier editions. The main section of this work is arranged by broad structures. With few exceptions, each anatomical concept is designated by a single term in Latin that is simple yet descriptive. Structures that are closely related topographically have similar names, and eponyms are not used. Terms of direction are expressed as related to parts of the body, not the originally standard anatomical position of a standing human with arms at his side (a position that is impossible in most animals). Substantial footnotes explain usage and variant terms. Also included in this volume is the second edition of *Nomina Histologica* as revised by the International Committee on Veterinary Histological Nomenclature and authorized at Ghent in 1992. As the third edition of *Nomina Histologica*, prepared by the Subcommittee on Histology of the International Anatomical Nomenclature Committee in 1989, deleted many terms and footnotes of importance to veterinary histologists, the World Association of Veterinary Anatomists at Ghent adopted the revisions and corrections made as a revised second (not a third) edition. *Nomina Embryologica Veterinaria* as prepared by the International Committee on Veterinary Embryological Nomenclature and authorized at Ghent in 1992 is also included in this volume. A separate history and index are provided for each *Nomina*.

Schaller, O., ed. 1992. *Illustrated Veterinary Anatomical Nomenclature*. Stuttgart: Ferdinand Enke Verlag. 614 p. ISBN: 3432995911. Building on the *Nomina Anatomica Veterinaria* published by the World Association of Veterinary Anatomists as a list of terms without illustrations or definitions, this work defines and depicts more than 6,500 structures (terms) to aid the user in understanding anatomy and its concepts. Adhering strictly to the arrangement of the Nomina Anatomica Veterinaria, gross anatomical structures of the cat, dog, pig, ox, sheep, goat, and horse are the primary focus.

*SNOMED*: *Systematized Nomenclature of Human and Veterinary Medicine*. 1993. 4 vols. Northfield, IL: College of American Pathologists. ISBN: 0930304489. Edited by Drs. Roger A. Cote, David Rothwell, Ronald Beckett, and James Palotay, DVM, this four-volume title provides a detailed, integrated, and structured nomenclature for all aspects of diagnosis and treatment in human and veterinary medicine. This work is far more than an update of the second edition of *SNOMED* (1979) or Palotay and Rothwell's *SNOVET* (1984) because now more than 180,000 terms, and their interpretive code numbers, are available to use within a data structure suitable for describing and indexing virtually all events found in the medical record. Consisting of 11 modules (topography; morphology; function; living organisms; chemical drugs and biological products; physical agents, forces, and activities; occupations; social context; diseases/diagnoses; procedures; and general linkage/modifiers), this controlled vocabulary is a tool to manage massive amounts of information in an expanding electronic environment. Cross-referencing for 28,600 terms corresponding to ICD-9-CM (International Classification of Diseases) codes is provided, as are cross-references to anatomic site; morphologic and functional changes; and etiologic agents. It has an alphabetic index combining terms from the diseases/diagnoses, morphology, and function modules as well as a procedures index. For up-to-date information on veterinary developments related to *SNOMED*, see the AVMA SNOMED Secretariat Web page found at http://snomed.vetmed.vt.edu/ (cited June 14, 2001).

## DIRECTORIES

Several types of directories provide useful information relevant to animal health and veterinary science. First, the primary veterinary directory for any country, if available, is the one that best lists the veterinarians and veterinary-related organizations within that nation's boundaries. In the United States, this directory is the *AVMA Membership Directory*, frequently subtitled *Caring for Animals*. In addition, directories to the veterinary profession may cover the world such as the *World Veterinary Directory*; a group of countries such as *Eurovet*; or veterinarians in a given location (e.g., state, region, or city) or specialty (e.g., equine practitioners, veterinary immunologists). Although this type of directory is useful, the frequency of publication may vary greatly from association to association, and, many times, a directory is only distributed to members of the particular association. Some specialty groups also make their membership directory available through their Web site (see URL list under Proceedings).

A broader category of directories addressing a diversity of veterinary-related subjects, such as veterinary schools or zoos also are available. Although many of these publications were written with the veterinarian or biological scientist in mind, several titles were prepared primarily for the general public.

American Association of Veterinary Clinicians. 2000. *Directory of Internships and Residencies: Matching Programs for 2001–2002*. Columbus, OH: AAVC. 672 p.   As an annual directory conceived by the American Association of Veterinary Clinicians (AAVC), this title is designed to disseminate internship and residency information to senior students and recent graduates interested in post-DVM training and to assist institutions in matching ranked applicants with preferred programs. Each program entry gives the name of the institution offering training, its address, telephone number, authorized administrative official, descriptive title and inclusive dates of the program, salary, and number of positions available. Also included are the average number of cases and faculty in direct support of the program; other prerequisites and requirements for application; a program disclosure form; and a description of the program. Supplements listing additional or withdrawn programs and changes are provided. Usually published in August or September of each year.

American Veterinary Medical Association. 2000. *AVMA Membership Directory & Resource Manual*. 49th ed. Schaumburg, IL: American Veterinary Medical Association. 1,022 p. ISSN: 1095-3884. This annual publication begins with resource information on AVMA officers; committees; bylaws; and related statistics, procedures, and policies. It is followed by information on American and international veterinary organizations; government agencies; veterinary schools and libraries; specialty boards as well as the veterinary practice acts in force in each state; and audiovisuals available from the AVMA. Next, AVMA members in the U.S., Canada, and other countries and those nonmember veterinarians for whom data is available in AVMA records are listed first *alphabetically*, arranged by the veterinarian's name and giving city, state, province, or country. This is followed by a *geographical* listing by state and town (or country) and giving the veterinarian's name, full address (frequently including telephone number and/or e-mail), professional specialty (species, medical discipline), type of employment, and employment function as well as the institution granting his/her degree and the year graduated. The "veterinarian's oath" can be found on the inside front cover and a brief index appears at the end of the work.

Association of American Veterinary Medical Colleges. 2000. *Veterinary Medical School Admission Requirements in the United States and Canada. 2000 edition for 2001 matriculation*. West Lafayette, IN: Purdue University Press. 166 p. ISBN: 1557532192. Prepared by the Association of American Veterinary Medical Colleges, this 16th edition contains current information from the 31 veterinary medical schools in the U.S. and Canada. Entries give the name of the veterinary school; information on the admissions office contact (address; telephone number; and frequently fax, e-mail and URL); prerequisite courses, application deadlines; standardized test requirements; residency implications; timetables; cost of education and other admission policies; and procedures. It also includes information on special programs, statistical data on the national applicant pool, and minority student opportunities. New to the edition is a segment on "Financing Your Veterinary Medical Education."

*Directory of Animal Disease Diagnostic Laboratories, July 1992*. 1992. Washington, DC: USDA, Animal & Plant Health Inspection Service, Veterinary Ser-

vices. 239 p. ISSN: 0146-1621.   Prepared in cooperation with the American
Association of Veterinary Laboratory Diagnosticians (AAVLD), this title lists
the animal disease diagnostic laboratories in the United States by state and city
in which they are located, giving laboratory name, name of director, address,
phone, affiliation, who may submit specimens, major species accepted for examination, and the services offered. Also lists domestic and foreign animal disease
reference centers. Indexed by type of laboratory. Includes a list of the AAVLD
accredited laboratories. AAVLD Web site at http://www.aavld.org (cited February 20, 2002) has a current list of accredited laboratories available to members.

*Eurovet Guide: A Guide to Veterinary Europe, 1998–1999.* 1998. 2nd ed.
Maisons-Alfort, France: Les Éditions du Point Vétérinaire. 679 p. ISBN:
286326141X.   Beginning with a list of suppliers to the veterinary profession
within Europe, this directory then provides a calendar of upcoming veterinary
conferences and programs and a list of European specialist referral clinics arranged by country, followed by a section on the history of veterinary medicine
including images, museums, and relevant associations. Next a description of the
European Union (EU) and related veterinary committees and directorates-general,
along with the Erasmus program for veterinary students is provided. The section
following lists European and international veterinary associations, giving the
name; address; telephone; and fax with names of officers (address, telephone,
and fax); and member associations, if any. The directory continues with two
sections covering European countries: EU member countries each have an entry
giving a map, statistics (general, animals, and veterinary), summary of the veterinary situation and registration procedures, veterinary services, principle research
institutes and veterinary schools (and faculty), official veterinary representative
bodies, other professional associations, and veterinary journals. Included for each
non-EU country are lists of veterinary schools (and faculty), professional associations, and journals. A section on specialist associations provides an overview
of veterinary specialization in Europe followed by two lists: specialist species
associations and associations of specialist disciplines with each giving the name
of the association; membership; address with telephone and fax; and officers
(name, address, telephone, and fax). The directory concludes with a glossary of
organizational abbreviations and index.

*2001 AZA Membership Directory.* 2001. Silver Spring, MD: American Zoo and
Aquarium Association.   Published as *Zoological Parks and Aquariums in the
Americas* (1979–97) and as *Directory of Zoological Parks and Aquariums*
(1998–99) by the American Association of Zoological Parks and Aquariums
(AAZPA), this annual directory includes an alphabetical institutional listing for
the U.S. and for Bermuda and Canada. Each entry gives the name of the zoo or
aquarium; its address, telephone number, fax, and e-mail; ownership information;
hours; admission cost, along with statistics on annual attendance, budget, and
number of employees; and the names of the administrative staff, including veteri-

narian(s). Tables show institution-related statistics as well as species and specimen statistics. Also provided is an alphabetical listing of commercial members, and lists of relevant government agencies and conservation organizations, both U.S. and Bermuda/Canadian. The American Zoo and Aquarium Association (AZA) charter and bylaws, code of professional ethics conservation programs, and more conclude the work. The AZA Web site at http://www.aza.org/members/zoo/ (cited June 13, 2001) also provides alphabetical and state searching for zoos and aquariums, giving name, address, telephone number, fax number, and e-mail as well as a Web page link if available.

Wijgergangs, A. and I. Kati'c 1997. *Guide to Veterinary Museums of the World.* Copenhagen: Kandrup; Viby Sjælland, Denmark: Historia Medicinae Veterinariae. 77 p. ISBN: 8788682188.   This directory to veterinary museums in 26 countries features short descriptions of each collection as well its address (and curator if known). Illustrated with black-and-white photographs of the museums or their displays as taken by the authors, this title appeared originally in *Historia Medicinae Veterinariae*, volumes 21:1, 21:2, and 21:3 & 4, 1996.

*World Veterinary Directory, 1991.* 1991. Madrid: World Veterinary Association. 430 p.   Arranged alphabetically by country, this directory provides information on over 380 of the 400 centers of university-level veterinary education worldwide, listing name of the center and year founded; address; phone; telex; fax; type of center; clinical resources; division of academic year and length of studies in years; as well as number of teachers and students admitted; postgraduate degrees; and areas of study. Each country's entry begins with number of veterinarians, veterinary centers, and veterinarians graduated each year, as well as their animal populations. It concludes with a list of veterinary associations, research institutes, and veterinary journals. More than 3,000 audio-visual aids of veterinary interest are also listed with an index of titles by format. Published in cooperation with the FAO, the World Health Organization (WHO), and the International Office of Epizootics (Office International des Épizooties), this title updates the *World Directory of Veterinary Schools* (Geneva: World Health Organization, 1973).

## ENCYCLOPEDIAS

Although George Henry Wooldridge's two-volume *Encyclopaedia of Veterinary Medicine, Surgery and Obstetrics* (2nd ed., London: Oxford University Press, H. Milford, 1934); Thomas Dalling's five-volume *International Encyclopedia of Veterinary Medicine* (Edinburgh: W. Green and Sons, Ltd.; London: Sweet & Maxwell, 1966); and Kjeld Wamberg's four-volume *Veterinary Encyclopedia: Diagnosis and Treatment* (English edition edited by E.A. McPherson, Copenhagen: Medical Book Co., 1968) are excellent examples of animal health and veterinary science encyclopedias of their time, no general English-language veterinary encyclopedias have been published since the mid-1960s. Foreign-language veterinary encyclopedias, such as F.M. Konrad's *Bl-Lexikon Haustierkrankheiten* (Leipzig: VEB Bibliographisches Institut Leipzig, 1987), exist but are not included here. Because few special-subject encyclopedias exist that were not primarily written for a lay audience, only two classic works on animals are listed because of their usefulness in a veterinary collection. Other animal or species encyclopedias should be collected as the need arises.

## Special-Subject Encyclopedias

Grzimek, B., ed. 1972–75. *Grzimek's Animal Life Encyclopedia*. New York: Van Nostrand Reinholt. 13 vols. In thirteen volumes, this extensively illustrated work covers lower animals to mammals. Each volume of this translation of *Tierleben* is arranged by animal orders and families, giving information such as an animal's length and weight, dentition, geographical range, mating habits, size of litter, or rearing of young. It has a systematic classification index and an animal dictionary in English, German, French, and Russian.

Grzimek, B., ed. 1990. *Grzimek's Encyclopedia of Mammals*. 5 vols. New York: McGraw-Hill Publishers. ISBN: 0079095089. A translation of a 1988 German edition, this title is not an update of the four volumes on mammals found in *Grzimek's Animal Life Encyclopedia*. Although still arranged by classification, this title features new chapter authors describing animals in their natural environments with the latest information on their evolution, biochemistry, and genetics. It includes some 3,500 illustrations and photographs, range maps, and scientific nomenclature as well as English, French, and German common names.

## GUIDES TO THE LITERATURE

From 1973 through 1989, the premier guide to the literature of veterinary medicine was Ann E. Kerker and Henry T. Murphy's *Comparative and Veterinary Medicine: A Guide to the Resource Literature* (1973). Two chapters, Marjan Merala's "Veterinary Medicine" chapter section in Blanchard and Farrell's *Guide to Sources for Agricultural and Biological Research* (1981) and D.E. Gray's "Veterinary Science" chapter in Lilley's *Information Sources in Agriculture and Food Science* (1981), appeared in the early 1980s to supplement but not replace Kerker and Murphy's work. Not until 1990 did the next comprehensive guide to the literature appear—Mike Gibb's *Keyguide to Information Sources in Veterinary Medicine*. Regardless of its British slant, selective nature, and the omission of some expected titles, this excellent work is the most recent guide to the veterinary literature available. In 1993, Jo Anne Boorkman's "Veterinary Medicine" chapter section in Olsen's *The Literature of Animal Science and Health* identified the reference works that had appeared since Merala's work in 1981.

Guides to agricultural or medical literatures frequently mention relevant veterinary medicine resources, and more specialized guides, such as those listed at the end of this section, may include references to key veterinary works.

Boorkman, J. 1993. "Veterinary Medicine." In: *The Literature of Animal Science and Health*, by W.C. Olsen, 266–83. Ithaca, NY: Cornell University Press. ISBN: 0801428866. As section C of the "Reference Update" chapter by Jo Anne Boorkman and Judith Levitt, this segment highlights the numerous reference or working tools in veterinary medicine that had appeared since Marjan Merala's chapter segment in Blanchard and Farrell's *Guide to Sources for Agricultural and Biological Research* (1981). Featured, with limited annotations and publisher's addresses (where needed), are recent literature guides and lists of periodicals; abstracts and major indexes; bibliographies and catalogs; dictionaries; nomenclature; directories; congresses/conferences/symposia; literature and course reviews; manuals and handbooks; reference texts; and laws and disease reporting.

Gibb, M. 1990. *Keyguide to Information Sources in Veterinary Medicine*. London; New York: Mansell Pub. 459 p. ISBN: 0720120187. This work is the first extensive guide to the major information sources in veterinary medicine published since 1973. Beginning with a chapter surveying the history and scope of veterinary medicine and another discussing the origins and use of veterinary information, the following chapters in Part one are essays on the sources of veterinary information available. Part two begins with an extensive bibliography of the reference sources cited in Part one. It concludes with chapters on selected, annotated lists of monographs, conferences, and journals by species (large and small animal) and by specialty. The emphasis in all chapters is on English-language monographs and major journals in all languages. Part three of the work is an international directory of some 747 associations and societies, with a selected list of libraries, online systems and databases, and publishers. References that appear in parts two and three of the work are comprehensively indexed, but no indexing is provided to the bibliographic essays in Part one. Although reviewers have identified some omissions and errors, the consensus is that this work is an extremely useful addition to the veterinary literature.

Gray, D.E. 1981. "Veterinary Science." In: *Information Sources in Agriculture and Food Science*, by G.P. Lilley, 418–39. London; Boston: Butterworths. ISBN: 0408106123. Utilizing a bibliographic essay format, D.E. Gray's "Veterinary Science" chapter summarizes the major reference resources in veterinary science, especially those of British or American origin, and then identifies key texts by species and by type of disease. A unique section of Gray's chapter lists British government publications of veterinary interest, and a concluding segment briefly notes veterinary works for lay readers.

Kerker, A.E. and Murphy, H.T. 1973. *Comparative & Veterinary Medicine: A Guide to the Resource Literature*. Madison, WI: University of Wisconsin Press.

308 p. ISBN: 0299063305.   Although selective in nature, this guide is very comprehensive for those titles published prior to 1973. It covers the literature of comparative and veterinary medicine and those related biomedical disciplines that utilize animals as subjects. Beginning with an annotated list of some 100 indexing and abstracting services, it also lists bibliographies, reviews, periodicals, and reference books relevant to the field. Stressing English-language publications, a major portion of this work is comprised of some 2,300 veterinary texts listed by specific discipline and over 1,500 texts arranged by species. A special section on laboratory animals is also included. Access is provided through author and subject indexes.

Merala, M. 1981. "Veterinary Medicine." In *Guide to Sources for Agricultural and Biological Research*, by J.R. Blanchard and L. Farrell, eds., 287–307. Berkeley, CA: University of California Press. ISBN: 0520032268.   Appearing in a greatly enlarged and updated version of J.R. Blanchard and H. Ostvold's *Literature of Agricultural Research* (1958), Marjan Merala's excellent chapter on the literature of the animal sciences includes sections on the resources in veterinary medicine, animal husbandry, poultry husbandry, wildlife and wildlife management, commercial fishing and fisheries, and entomology and nematology. The veterinary medical section provides annotated references to key resources such as abstracts and indexes, dictionaries, and reference texts subdivided by basic science (e.g., anatomy, pathology) and by clinical specialty (e.g., large animal, small animal, birds and poultry, wildlife diseases). Also listed is information on laboratory animals, laws including disease reporting, and the veterinary profession. An appendix giving relevant acronyms and abbreviations and author, title, and subject indexes concludes Blanchard's work.

### Other

> Cregier, S.E. 1989. *Farm Animal Ethology: A Guide to Sources*. North York, Ontario, Canada: Captus University Publications. 213 p. ISBN: 0921801408.
> Magel, C.R. 1989. *Keyguide to Information Sources in Animal Rights*. London: Mansell; Jefferson, NC: McFarland. 267 p. ISBN: 0720119847.
> Wexler, P. 2000. *Information Resources in Toxicology*. 3rd ed. New York: Elsevier. 921 p. ISBN: 0127447709.

## HANDBOOKS AND MANUALS

A variety of handbooks and manuals are published in animal health and veterinary science, and published titles frequently use one of these appellations. A selection of general and drug handbooks and manuals are included here. Various formularies are now available for large, small, and exotic animals, but they are not included here.

Aiello, S., ed. 1998. *The Merck Veterinary Manual*. 8ᵗʰ ed. Whitehouse Station, NJ: Merck & Co./Merial Limited. 2,305 p. ISBN: 0911910298. Considered a "bible" for veterinary students, this manual, which is subtitled "a handbook of diagnosis, therapy and disease prevention and control for the veterinarian," concisely covers the topic. It includes information on diseases seen worldwide in all the common domestic species as well as physical examination and procedures for those species. Arranged by anatomic systems with specific conditions appearing in the system primarily affected, it provides additional sections on behavior, clinical pathology and procedures, emergency medicine and critical care, exotic and laboratory animals, management and nutrition, pharmacology, poultry, toxicology, and zoonoses. It has an extensive alphabetic index, but provides no references.

Allen, D.G., J.K. Pringle, D.A. Smith, et al. 1998. *Handbook of Veterinary Drugs*. 2ⁿᵈ ed. Philadelphia, PA: Lippincott-Raven. 886 p. ISBN: 0397584350. This pharmaceutical handbook for the practicing veterinarian is divided by speciality (small animal, large animal, exotic, and avian) with each section providing both a summary of common dosages and a detailed description of drugs used. These drug monographs, organized alphabetically, include indications, routes of administration, adverse and common side effects, drug interactions, source of supply and, in many cases, both the U.S. and Canadian generic names.

Bistner, S.I. and R.B. Ford. 1995. *Kirk and Bistner's Handbook of Veterinary Procedures & Emergency Treatment*. 6ᵗʰ ed. Philadelphia, PA: W.B. Saunders Co. 1,006 p. ISBN: 0721649726. This condensed, yet comprehensive manual for small animal practitioners addresses emergency care, patient evaluation and organ system examination, clinical signs, clinical procedures, and interpretation of laboratory tests. It contains an extensive number of charts and tables on such topics as optimum body weights by breed, breed predilection for disease, and normal laboratory values for dogs and cats.

*Compendium of Veterinary Products*. 1999. 5ᵗʰ ed. Port Huron, MI: North American Compendiums, Inc. 2429 p. ISBN: 1889750107. Beginning with a section providing information on veterinary drug-related organizations and guidelines, this handbook provides an alphabetical index of manufacturers and their products; a brand name/ingredient index; a therapeutic index; biological, anthelminthic, and parasiticide charts; withdrawal charts; and a new formulary of systemic antimicrobials. Over 45,000 monographs on pharmaceutical, biological, diagnostic, and pesticide products are arranged by trade name. Each monograph gives the active ingredients, indications, pharmacology, dosage and administration, contraindications, precautions, cautions, antidotes, warnings, toxicology, adverse reactions, references, discussion, and presentation (quantities) for each product as relevant. Handy reference and antimicrobial dosage tables are also provided. It concludes with an alphabetical index of products.

Fenner, W.R., ed. 2000. *Quick Reference to Veterinary Medicine*. 3rd ed. Philadel-phia, PA: Lippincott Williams & Wilkins. 731 p. ISBN: 0397516088. Utilizing an outline format to provide quick access to the facts, this handbook addresses clinical signs and client complaints, laboratory abnormalities, principles of fluid, osmotic and electrolyte balance, systems disturbances, and special topics such as physical injuries or intoxications. Its goal is to assist in diagnosis and treatment of common problems in small animals. References provided at the end of each chapter.

*Foreign Animal Diseases*. 1998. 6th ed. Richmond, VA: United States Animal Health Association. 462 p. Presenting the latest information on forty foreign animal diseases that are the greatest threat to the U.S. livestock and poultry indus-tries, this handbook provides a definition of each disease (with variant names) followed by its etiology, host range, geographic distribution, transmission, patho-genesis, clinical signs, gross lesions, diagnosis, control and eradication, public health aspects, and a guide to the literature. Its appendices include information on preparation and submission of specimens, a glossary, and colored photographs to assist in disease recognition.

Gfeller, R.W. and S. Messonnier. 1998. *Handbook of Small Animal Toxicology & Poisonings*. St. Louis, MO: Mosby. 405 p. ISBN: 0815164548. This handbook provides comprehensive information on toxicoses and poisonings of dogs and cats covering over 1,000 toxic agents and 500 common poisonous plants. Follow-ing an overview of patient evaluation and symptomatic treatment of the poisoned pet, there is an alphabetic list of toxins giving mechanism of action, clinical signs, treatment, and emergency treatment, as well as suggested readings. Appendices provide a formulary, daily fluid requirements, recipes for dialysate and continu-ous drug infusions, and a poison treatment flow chart.

Plumb, D.C. 1999. *Veterinary Drug Handbook*. 3rd ed. White Bear Lake, MN: Pharma Vet Pub.; Ames, IA: Iowa State University Press. 750 p. ISBN: 0813802350. First published in 1991, this handbook provides drug monographs for drugs approved for use in veterinary species as well as those nonapproved drugs that are routinely used in veterinary practices. Each monograph gives the chemistry, storage/stability/compatibility, pharmacology, uses/indications, phar-macokinetics, contraindications/precautions/reproductive safety, adverse effects/ warnings, overdosage/acute toxicity, drug interactions, doses (by species), moni-toring parameters, client information, dosage forms/preparations/FDA approval status/withholding times, and human-approved products. This edition has an ap-pendix that includes information on ophthalmic products, therapeutic diets for small animals, chemotherapy protocols, various conversion tables, reference lab-oratory values for domestic species, as well as a new index listing drugs by their therapeutic classification or major indication. All references appear at the end of the volume.

*Veterinary Pharmaceuticals & Biologicals (VPB), 1999–2000.* 1998. 11th ed. Veterinary Medicine Publishing Group. 1,212 p. ISSN: 0272-4669, ISBN: 0935078738.   This compilation of veterinary product inserts opens with indexes for manufacturers, brand and generic names, and product category/therapeutic classification. Product listings cover pharmaceuticals and other approved products; biologicals, parasiticides, and insecticides; therapeutic and maintenance diets; nutritional supplements; diagnostic aids and supplies; and other chemicals used for animals. It concludes with a section covering withholding, antigenic, and other drug information including withdrawal times in food animals; vaccine charts; veterinary drug regulation policy and guidelines; animal health-product approval process; food animal residue avoidance information; and convenient conversion aids and reporting forms.

*Veterinary Values.* 1998. 5th ed. Lenexa, KS: Veterinary Medicine Publishing Group. 342 p. ISBN: 0935078665.   As a clinical handbook that complements the *Veterinary Pharmaceuticals and Biologicals*, this title includes reference range values such as hematology and body fluid values; drug interactions, incompatibilities, and effects on test result; small, large, and exotic animal formularies; food animal drug-withdrawal times; poison management guidelines; drug compounding and prescribing; rabies control; and small animal vaccination schedules. Additional resource information on veterinary hotlines, interstate shipping regulations, Web site addresses, conversion aids, and generic and trade name drug indexes is provided.

## HISTORIES

Beginning with general English-language historical works in animal health and veterinary science, in alphabetical order by author, this section lists veterinary histories for the United States. Annotations are not provided in this section due to the clarity of most titles and the desire to include as many titles as available. General foreign-language histories and works that cover the historical development of veterinary medicine in other countries are excluded, as are dissertations on narrower aspects of veterinary history. Also omitted are the numerous veterinary school or state professional association "anniversary" histories that have been published in the United States over the past thirty years. These titles usually can be found by searching online or print library catalogs or bibliographic databases, such as WorldCat (OCLC) using the terms "veterinary medicine AND history."

Those interested in extensive bibliographies on early veterinary medicine should also consult Smith's *The Early History of Veterinary Literature and its British Development* (Smith, 1976), Smithcors' *Evolution of the Veterinary Art*, and Dunlop and Williams' *Veterinary Medicine: An Illustrated History*. Although not included in this publication, numerous journal articles on veterinary medical history have been published and can be identified using the National Library of Medicine's *Bibliography of the History of Medicine, Index Veterinarius*, or CAB Abstracts, and other appropriate sources. Additional resources also can be found through Smithcors and Smithcors' *Five Centuries of Veterinary Medicine: A Short-Title Catalog of the Washington State University Veterinary History Collection* (1997) and Stalheim's *Guide to Collections of Papers Pertaining to American Veterinary History* (1996).

## General Histories

Dunlop, R.H. and D.J. Williams. 1996. *Veterinary Medicine: An Illustrated History*. St. Louis, MO: Mosby. 692 p. ISBN: 0801632099.

Karasszon, D. 1988. *A Concise History of Veterinary Medicine*. Budapest: Akadémiai Kiadó. 458 p. ISBN: 9630546108.

McCoy, J.J. [1964]. *The World of the Veterinarian*. New York: Lothrop, Lee & Shepard. 223 p.

McCullough, L.D. and J.P. Morris III, eds. 1978. *Implications of History and Ethics to Medicine: Veterinary and Human*. College Station, TX: Texas A&M University. 158 p.

Michell, A.R., ed. 1993. *History of the Healing Professions: Parallels between Veterinary and Medical History*. (The Advancement of Veterinary Science: the Bicentenary Symposium Series, vol. 3). Wallingford, UK: CAB International. 137 p. ISBN: 0851987613.

Pugh, L.P. [1962]. *From Farriery to Veterinary Medicine, 1785–1795*. Cambridge: Published for the Royal College of Veterinary Surgeons by W. Heffer. 198 p. (Reprinted: West Orange, NJ: Saifer, 1970.)

Schwabe, C.W. 1978. *Cattle, Priests, and Progress in Medicine*. Minneapolis, MN: University of Minnesota Press. 277 p. ISBN: 0816608253.

Smithcors, J.F. 1957. *Evolution of the Veterinary Art: A Narrative Account to 1850*. Kansas City, MO: Veterinary Medicine Pub. Co. 408 p.

Swabe, J. 1999. *Animals, Disease, and Human Society: Human-Animal Relations and the Rise of Veterinary Medicine*. London; New York: Routledge. 243 p. ISBN: 0415181933.

Toynbee, J.M. 1973. *Animals in Roman Life and Art*. Ithaca, NY: Cornell University Press. 431 p. ISBN: 0801407850. (Includes section on Roman Veterinary Medicine by R.E. Walker.)

Walker, R.E. [1991]. *Ars Veterinaria: the Veterinary Art from Antiquity to the End of the XIXth Century: Historical Essay*. Kenilworth, NJ: Schering-Plough Animal Health. 99 p.

Wilkinson, L. 1992. *Animals and Disease: An Introduction to the History of Comparative Medicine*. Cambridge; New York: Cambridge University Press. 272 p. ISBN: 0521375738.

## United States Histories

Bierer, B.W. 1939. *History of Animal Plagues of North America: With an Occasional Reference to Other Diseases and Diseased Conditions.* 5v. in 1. Baltimore, MD: [B. Bierer]. (Reprint: Washington, DC: U.S. Department of Agriculture, 1974. 97 p.)

————. 1940–41. *American Veterinary History (on the Duty and Advantage of Studying the Diseases of Domestic Animals).* 7 parts in 1 vol. Baltimore, MD: [B. Bierer]. (Reprint: Madison, WI: Carl Olson. 222 leaves.)

————. 1955. *A Short History of Veterinary Medicine in America.* East Lansing, MI: Michigan State University Press. 113 p.

Clark, W.H.H. [1991?]. *The History of the United States Army Veterinary Corps in Viet Nam 1962–1973.* Ringgold, GA: W.H.H. Clark. 230 p.

Larsen, P.H. 1997. *Our History of Women in Veterinary Medicine: Gumption, Grace, Grit, and Good Humor.* Littleton, CO: The Association for Women Veterinarians. 115 p.

Merillat, L.A. and D.M. Campbell. 1935. *Veterinary Military History of the United States: with a Brief Record of the Development of Veterinary Education, Practice, Organization and Legislation.* 2 vols. Chicago: Veterinary Magazine Corp.; Kansas City, MO: Haver-Glover Laboratories.

Miller, E.B. 1961. *United States Army Veterinary Service in World War II.* Washington, DC: Office of the Surgeon General, U.S. Department of the Army. 779 p.

————. 1966. *A Veterinarian's Notes on the Civil War.* Schaumburg, IL: American Veterinary Medical Association. 22 p.

Miller, R.M. 1991. *RMM, the Second Oldest Profession: The History of Veterinary Medicine: An Anthology.* Goleta, CA: American Veterinary Publications. 678 p. ISBN: 0939674351. (Veterinary history through caricatures and cartoons.)

Smithcors, J.F. 1963. *The American Veterinary Profession: Its Background and Development.* Ames, IA: Iowa State University. 704 p.

————. 1975. *The Veterinarian in America, 1625–1975.* Santa Barbara, CA: American Veterinary Pub. 160 p. (Most of text abridged from *American Veterinary Profession*, with revisions.)

Stalheim, O.H.V. 1988. *Veterinary Medicine in the West.* Manhattan, KA: Sunflower University Press. 82 p. ISBN: 0897451090.

————. 1994. *The Winning of Animal Health: 100 Years of Veterinary Medicine.* Ames, IA: Iowa State University Press. 251 p. ISBN: 081382429X.

United States Army. 7th Service Command. 1946. *History of the Veterinary*

*Service, Seventh Service Command*. Chicago, IL: [Seventh Service Command]. 71 p.

Waddell, W.H. [1982]. *The Black Man in Veterinary Medicine: Afro-American, Negro, Colored.* Rev. ed. [Honolulu, HI: W.H. Waddell]. 176 p.

———. [between 1984 and 1986]. *Indigenous Historical Veterinary Medical Facts, the Pioneer Black Veterinarian.* [s.l.: s.n.], 6, [22] leaves.

Williamson, S., ed. [1983]. *50 Years of Educational Excellence and Practice Improvement: American Animal Hospital Association.* [Mishawaka, IN]: American Animal Hospital Association. 64 p.

Wiser, V.D. 1987. *100 Years of Animal Health, 1884–1984.* [Series: *Journal of NAL Associates* New series; 11, nos. 1/4. ISSN: 0277-2841]. Beltsville, MD: Associates of the National Agricultural Library. 230 p.

## JOURNALS

In 1976, the Veterinary Medical Libraries Group of the Medical Library Association established an ad hoc committee to prepare a list of veterinary journals essential to the operation of a veterinary medical library. Chaired by Atha Louise Henley, the committee completed a ''Basic List of Veterinary Serials'' which was approved in 1977 and published in 1978 (Henley, 1978). In 1978, the Veterinary Serials Committee, chaired by Trenton Boyd, began its work to maintain, update, and publish the ''Basic List of Veterinary Medical Serials'' on a regular basis. The renamed Veterinary Medical Libraries Section approved a second edition of the ''Basic List'' in 1980. Given the ever-changing world of serials, the committee continued to make revisions until 1986 when they published ''Basic List of Veterinary Medical Serials, 2ⁿᵈ ed., 1981, with Revisions to April 1, 1986'' (Boyd et al., 1986). A third edition of the ''Basic List'' is in preparation by the committee at this time.

Beginning in 1977, an ad hoc committee, chaired by Kathrine MacNeil, was charged with compiling a list of foreign (published outside of North America) veterinary serials noting owning libraries and their holdings for use as an interlibrary loan tool. In 1980, this committee produced and distributed a microfiche publication, *Veterinary Serials: A Union List of Selected Titles Not Indexed by Index Medicus and Held by Veterinary Collections in the U.S. and Canada* (1980). By 1990, the committee, chaired by David Anderson, had compiled a union list of veterinary serials, which was published in two volumes as *Veterinary Serials: A Union List of Serials Held in Veterinary Collections in Canada, Europe, and the U.S.A. 2d, 1987–1988* (1990).

Other publications on veterinary serials include Trenton Boyd's recurring chapter on ''Veterinary Science'' in *Magazines for Libraries* from 1989 to 2000 (Boyd, 2000), and Norma Bruce's review of new, redesigned, or retitled veterinary journals (Bruce, 1990). Quantitative studies including veterinary serials appear in Wallace C. Olsen's chapter on "Primary Journals and the Core List" (Olsen, 1993) and W. Houston's article on "The Application of Bibliometrics to Veterinary Science Primary Literature" (Houston, 1983).

The following list of animal health and veterinary science journals was selected by the author and does not necessarily reflect those that will appear in the forthcoming edition of "Basic List of Veterinary Serials." Selected veterinary journals of national or international interest have been included, as have veterinary news magazines. Recent title changes are noted. Due to space restrictions, veterinary journals from all countries or practice specialties have not been covered, and laboratory animal journals have been excluded.

The number of veterinary journals electronically available is increasing. The *Journal of the American Veterinary Medical Association* and the *American Journal of Veterinary Research* are coming online to AVMA members this year. URLs, each cited February 28, 2002, are provided for those journals that have current issues available through the Internet, usually for an additional subscription cost or fee.

> *Acta Veterinaria Scandinavica* w/supplements. 1960– . Quarterly. ISSN: 0044-605X.
> *Advances in Veterinary Medicine.* 1997– . Irregular. ISSN: 1093-975X.
> *American Journal of Veterinary Research.* 1940– . Monthly. ISSN: 0002-9645. (http://www.electronicipc.com/JournalEZ/toc.cfm?code=429001)
> *Anatomia Histologia Embryologia—Journal of Veterinary Medicine/Zentralblatt fur Veterinarmedizin.* Series/Reihe C. 1973– . Bimonthly. ISSN: 0340-2096. (http://www.blackwell-synergy.com/rd.asp?goto=journal&code=ahe)
> *Animal Behaviour.* 1958– . Monthly. ISSN: 0003-3472. (http://www.academicpress.com/anbehav)
> *Animal Biotechnology.* 1990– . Semiannual. ISSN: 1049-5398.
> *Animal Genetics.* 1986– . Bimonthly. ISSN: 0268-9146. (http://www.blackwell-synergy.com/rd.asp?goto=journal&code=age)
> *Animal Health Research Reviews.* 2000– . Semiannual. ISSN: 1466-2523. (www.ingenta.com/journals/browse/cabi/ahr)
> *Animal Reproduction Science.* 1978– . Monthly. ISSN: 0378-4320. (http://www.elsevier.com/locate/anireprosci)
> *Animal Welfare.* 1992– . Quarterly. ISSN: 0962-7286.
> *Annales de Medecine Veterinaire.* 1852– . Bimonthly. ISSN: 0003-4118.
> *Anthrozoos.* 1987– . Quarterly. ISSN: 0892-7936.
> *Applied Animal Behavior Science.* 1984– . Semimonthly. ISSN: 0168-1591. (http://www.elsevier.com/locate/applanim)
> *Australian Veterinary Journal.* 1927– . Monthly. ISSN: 0005-0423.
> *Australian Veterinary Practitioner.* 1971– . Quarterly. ISSN: 0310-138X.

*Avian and Poultry Biology Reviews* (formerly *Poultry Science Reviews*). 2000– . Quarterly. ISSN: 1470-2061.

*Avian Diseases.* 1957– . Quarterly. ISSN: 0005-2086.

*Avian Pathology.* 1972– . Quarterly. ISSN: 0307-9457. (http://bioline.bdt. org.br/ap)

*Berliner und Munchener Tierarztliche Wochenschrift.* 1938– . Monthly. ISSN: 0005-9366.

*Bovine Practitioner.* 1967– . Annual. ISSN: 0524-1685.

*Canadian Journal of Veterinary Research—Revue Canadienne de Recherche Veterinaire.* 1986– . Quarterly. ISSN: 0830-9000.

*Canadian Veterinary Journal—Revue Veterinaire Canadienne.* 1960– . Monthly. ISSN: 0008-5286.

*Cattle Practice.* 1993– . Quarterly. ISSN: 0969-1251.

*Clinical Techniques in Small Animal Practice.* 1998– . Quarterly. ISSN: 1096-2867.

*Comparative Immunology, Microbiology and Infectious Diseases.* 1978– . Quarterly. ISSN: 0147-9571. (http://www.elsevier.com/locate/ cimid)

*Compendium on Continuing Education for the Practicing Veterinarian.* 1979– . Monthly. ISSN: 0193-1903.

*Diseases of Aquatic Organisms.* 1985– . Monthly. ISSN: 0177-5103. (http://www.int-res.com/journals/dao/)

*Domestic Animal Endocrinology.* 1984– . Quarterly. ISSN: 0739-7240. (http://www.elsevier.nl/locate/domaniend)

*DTW. Deutsche Tierarztliche Wochenschrift.* 1971– . Monthly. ISSN: 0341-6593.

*DVM.* 1975– . Monthly. [News] ISSN: 0012-7337. (http://www. dvmnewsmagazine.com/dvm/)

*Equine Athlete.* 1988– . Bimonthly. ISSN: 1047-8620.

*Equine Veterinary Education.* 1989– . Bimonthly. ISSN: 0957-7734.

*Equine Veterinary Journal with supplements.* 1968– . Bimonthly. ISSN: 0425-1644.

*European Journal of Veterinary Pathology.* 1994– . Triennial. ISSN: 1124-5352.

*Exotic DVM.* 1999– . Bimonthly. ISSN: 1521-1363.

*Exotic Pet Practice.* 1996– . Monthly. ISSN: 1086-4288.

*Fish Pathology.* 1994– . Quarterly. ISSN: 0388-788X.

*Fish & Shellfish Immunology.* 1991– . Bimonthly. ISSN: 1050-4648. (http://www.academicpress.com/fsi)

*Folia Veterinaria.* 1956– . Semiannual. ISSN: 0015-5748.

*Historia Medicinae Veterinariae.* 1976– . Quarterly. ISSN: 0105-1423.

*In Practice.* 1979– . Monthly. ISSN: 0263-841X.

*Indian Veterinary Journal.* 1924– . Monthly. ISSN: 0019-6479.

*Irish Veterinary Journal*. 1946– . Monthly. ISSN: 0368-0762.

*Israel Journal of Veterinary Medicine*. 1986– . Quarterly. ISSN: 0334-9152. (http://www.isrvma.org/journal.htm)

*Japanese Journal of Veterinary Research*. 1953– . Quarterly. ISSN: 0047-1917.

*Journal of Animal Science*. 1942– . Monthly. ISSN: 0021-8812. (http://www.asas.org/jas/)

*Journal of Applied Animal Welfare Science*. 1998– . Quarterly. ISSN: 1088-8705. (http://www.catchword.com/erlbaum/10888705/contp1-1.htm)

*Journal of Aquatic Animal Health*. 1989– . Quarterly. ISSN: 0899-7659.

*Journal of Avian Medicine and Surgery*. 1995– . Quarterly. ISSN: 1082-6742. (http://www.bioone.org/bioone/?request=get-journals-list&issn=1082-6742)

*Journal of Comparative Pathology*. 1965– . Bimonthly. ISSN: 0021-9975.

*Journal of Dairy Science*. 1917– . Monthly. ISSN: 0022-0302. (http://www.adsa.org/jds)

*Journal of Equine Veterinary Science*. 1981– . Monthly. ISSN: 0737-0806.

*Journal of Fish Diseases*. 1978– . Bimonthly. ISSN: 0140-7775. (http://www.blackwell-synergy.com/rd.asp?goto=journal&code=jfd)

*Journal of Herpetological Medicine and Surgery* (formerly *Bulletin of the Association of Reptilian & Amphibian Veterinarians*). 1992–99. Quarterly. ISSN: 1076-3139.

*Journal of Reproduction and Fertility* [esp. supplements]. 1962–2000. Bimonthly. ISSN: 0022-4251.

*Journal of Small Animal Practice*. 1960– . Monthly. ISSN: 0022-4510.

*Journal of the American Animal Hospital Association*. 1968– . Bimonthly. ISSN: 0587-2871. (http://www.iknowledgenow.com/byissue.cfm)

*Journal of the American Veterinary Medical Association*. 1915– . Semimonthly. ISSN: 0003-1488.

*Journal of the South African Veterinary Association—Tydskrif van die Suid Afrikaanse Veterinere Vereniging*. 1927– . Quarterly. ISSN: 0038-2809.

*Journal of Veterinary Cardiology*. 1999– . Semiannual. ISBN: NA.

*Journal of Veterinary Dentistry*. 1984– . Quarterly. ISSN: 0898-7564.

*Journal of Veterinary Diagnostic Investigation*. 1989– . Bimonthly. ISSN: 1040-6387.

*Journal of Veterinary Emergency and Critical Care*. 1985/1991– . Quarterly. ISSN: 1056-6392.

*Journal of Veterinary Internal Medicine*. 1987– . Bimonthly. ISSN: 0891-6640.

*Journal of Veterinary Medical Education*. 1974– . Semiannual. ISSN: 0748-321X.

*Journal of Veterinary Medical Science.* 1991– . Bimonthly. ISSN: 0916-7250. (http://jvms.jstage.jst.go.jp/en/)

*Journal of Veterinary Medicine. Series A*: *Physiology, Pathology, Clinical Medicine.* [*Zentralblatt fur Veterinarmedizin. Reihe A*]. 1986– . Monthly. ISSN: 0931-184X. (http://www.blackwell-synergy.com/member/institutions/issuelist.asp?journal=jva)

*Journal of Veterinary Medicine. Series B: Infectious Diseases and Veterinary Public Health.* [*Zentralblatt fur Veterinarmedizin. Reihe B*]. 1986– . Monthly. ISSN: 0931-1793. (http://www.blackwell-synergy.com/member/institutions/issuelist.asp?journal=jvb)

*Journal of Veterinary Pharmacology and Therapeutics.* 1978– . Bimonthly. ISSN: 0140-7783. (http://www.blackwell-synergy.com/rd.asp?goto=journal&code=jvp)

*Journal of Wildlife Diseases.* 1970– . Quarterly. ISSN: 0090-3558.

*Journal of Zoo and Wildlife Medicine.* 1989– . Quarterly. ISSN: 1042-7260. (http://www.bioone.org/bioone/?request=get-journals-list&issn=1042-7260)

*Kleintier-Praxis.* 1956– . Monthly. ISSN: 0023-2076.

*Le Point Veterinaire.* 1973– . Bimonthly. ISSN: 0335-4997.

*Medical and Veterinary Entomology.* 1987– . Quarterly. ISSN: 0269-283X. (http://www.blackwell-synergy.com/rd.asp?goto=journal&code=mve)

*New Zealand Veterinary Journal.* 1956– . Bimonthly. ISSN: 0048-0169.

*Onderstepoort Journal of Veterinary Research.* 1951– . Quarterly. ISSN: 0030-2465.

*Pig Journal.* 1994– . Semiannual. ISSN: 1352-9749.

*Poultry Science.* 1921– . Monthly. ISSN: 0032-5791.

*Praktische Tierarzt.* 1951– . Monthly. ISSN: 0032-681X.

*Pratique Medicale et Chirurgicale de l'Animal de Compagnie.* 1983– . Bimonthly. ISSN: 0758-1882.

*Preventive Veterinary Medicine.* 1982– . Semimonthly. ISSN: 0167-5877. (http://www.elsevier.com/locate/prevetmed)

*Recueil de Medecine Veterinaire.* 1824– . Monthly. ISSN: 0034-1843.

*Reproduction in Domestic Animals.* 1990– . Bimonthly. ISSN: 0936-6768. (http://www.blackwell-synergy.com/rd.asp?goto=journal&code=rda)

*Research in Veterinary Science.* 1960– . Bimonthly. ISSN: 0034-5288. (http://www.harcourt-international.com/journals/rvsc/)

*Revue de Medecine Veterinaire.* 1937– . Monthly. ISSN: 0035-1555.

*Revue Scientifique et Technique de l'Office International des Épizooties.* 1982– . Triennial. ISSN: 0253-1933.

*Schweizer Archiv fur Tierheilkunde.* 1859– . Monthly. ISSN: 0036-7281.

*Seminars in Avian and Exotic Pet Medicine.* 1992– . Quarterly. ISSN: 1055-937X.

*Small Ruminant Research.* 1988– . Bimonthly. ISSN: 0921-4488. (http://www.elsevier.com/locate/smallrumres)

*Society & Animals.* 1993– . Semiannual. ISSN: 1063-1119.

*Swine Health and Production.* 1993– . Bimonthly. ISSN: 1066-4963.

*Theriogenology.* 1974– . Monthly. ISSN: 0093-691X. (http://www.elsevier.nl/locate/theriogenology)

*Tierarztliche Praxis. Ausgabe Grosstiere Nutztiere.* 1997– . Bimonthly. ISSN: 1434-1220.

*Tierarztliche Praxis. Ausgabe Kleintiere Heimtiere.* 1997– . Bimonthly. ISSN: 1434-1239.

*Tierarztliche Umschau.* 1946– . Monthly. ISSN: 0049-3864.

*Tijdschrift voor Diergeneeskunde.* 1916– . Semimonthly. ISSN: 0040-7453.

*Trends* [AAHA]. 1988– . Bimonthly [News]. ISSN: 0883-1696.

*Tropical Animal Health and Production.* 1969– . Bimonthly. ISSN: 0049-4747. (http://www.kluweronline.com/issn/0049-4707)

*Veterinary Anaesthesia and Analgesia* (formerly *Journal of Veterinary Anaesthesia*). 2000– . Quarterly. ISSN: 1467-2987. (http://www.blackwell-synergy.com/member/institutions/issuelist.asp?journal=vaa)

*Veterinary and Comparative Orthopaedics and Traumatology.* 1988– . Quarterly. ISSN: 0932-0814.

*Veterinary and Human Toxicology.* 1977– . Bimonthly. ISSN: 0145-6296.

*Veterinary Clinical Pathology.* 1977– . Quarterly. ISSN: 0275-6382.

*Veterinary Clinics of North America. Equine Practice.* 1985– . Triennial. ISSN: 0749-0739.

*Veterinary Clinics of North America. Exotic Animal Practice.* 1998– . Triennial. ISSN: 1094-9194.

*Veterinary Clinics of North America. Food Animal Practice.* 1985– . Triennial. ISSN: 0749-0720.

*Veterinary Clinics of North America. Small Animal Practice.* 1979– . Bimonthly. ISSN: 0195-5616.

*Veterinary Dermatology.* 1989– . Quarterly. ISSN: 0959-4493. (http://www.blackwell-synergy.com/rd.asp?goto=journal&code=vde)

*Veterinary Economics.* 1960– . Monthly. ISSN: 0042-4862.

*Veterinary Heritage.* 1982– . Semiannual. ISSN: 1096-5904.

*Veterinary History.* 1973– . Semiannual. ISSN: 0301-6943.

*Veterinary Immunology and Immunopathology.* 1979– . Semimonthly. ISSN: 0165-2427. (http://www.elsevier.nl/locate/vetimm)

*Veterinary Journal.* 1997– . Bimonthly. ISSN: 1090-0233. (http://www.harcourt-international.com/journals/tvjl/)

*Veterinary Medicine.* 1985– . Monthly. ISSN: 8750-7943. (Proquest access)

*Veterinary Microbiology.* 1976– . Semimonthly. ISSN: 0378-1135. (http://www.elsevier.nl/locate/vetmic)

*Veterinary Ophthalmology.* 1998– . Quarterly. ISSN: 1463-5216. (http://

www.blackwell-synergy.com/Journals/member/institutions/
issuelist.asp?journal=vop)

*Veterinary Parasitology.* 1975– . Semimonthly. ISSN: 0304-4017. (http://
www.elsevier.nl/locate/vetpar)

*Veterinary Pathology.* 1971– . Bimonthly. ISSN: 0300-9858.

*Veterinary Practice News.* 2000– . Monthly. ISSN: 1528-6398.

*Veterinary Quarterly.* 1979– . Quarterly. ISSN: 0165-2176.

*Veterinary Radiology and Ultrasound.* 1992– . Bimonthly. ISSN: 1058-
8183.

*Veterinary Record.* 1888– . Weekly. ISSN: 0042-4900.

*Veterinary Research.* 1993– . Bimonthly. ISSN: 0928-4249. (http://
www.edpsciences.org/docinfos/INRA-VET/)

*Veterinary Research Communications.* 1989– . Bimonthly. ISSN: 0165-
7380.

*Veterinary Surgery.* 1978– . Bimonthly. ISSN: 0161-3499.

*Veterinary Technician.* 1984– . Monthly. ISSN: 8750-8990.

*Veterinary Therapeutics.* 2000– . Quarterly. ISSN: 1528-3593.

*Veterinary Times.* 1984– . Monthly [News]. ISSN: 1352-9374.

*Vlaams Diergeneeskundig Tijdschrift.* 1931– . Bimonthly. ISSN: 0303-
9021.

*Wiener Tierarztliche Monatsschrift.* 1914– . Monthly. ISSN: 0043-535X.

*Zoo Biology.* 1982– . Bimonthly. ISSN: 0733-3188. (http://www.
interscience.wiley.com/jpages/0733-3188/)

## PROCEEDINGS

Many veterinary organizations, on national, geographical, or specialty levels, sponsor congresses, conferences, or symposia for their members. Such conferences frequently publish the papers presented, either in a separate proceedings volume, as a volume of the organization's journal, or as articles within their journal. This list of national, regional, and specialty veterinary organizations gives the abbreviation and Web address for each organization/association/society. Conference or symposia announcements and those publications produced by the organization can usually be found through these Web sites.

Although state or provincial veterinary associations and veterinary schools frequently sponsor conferences that generate proceedings or other publications as well, those organizations are not listed here. Other sources for tracking veterinary conference proceedings would be CAB Abstracts, WorldCat (OCLC), or the Veterinary Conference Proceedings Web page at http://www.medvet.umontreal.ca/biblio/gopher/bases/ (cited June 15, 2001).

Academy of Feline Medicine (AFM) (http://www.aafponline.org/afm/
afm.htm) (cited June 17, 2001).

Academy of Veterinary Allergy and Clinical Immunology (AVACI) (http://www.avaci.org/) (cited June 17, 2001).

Academy of Veterinary Consultants (AVC) (http://gpvec.unl.edu/avc/default.htm) (cited June 17, 2001).

Academy of Veterinary Emergency & Critical Care Technicians (AVECCT) (http://www.veccs.org/technicians/) (cited June 17, 2001).

Academy of Veterinary Homeopathy (AVH) (http://www.acad vethom.org/) (cited June 17, 2001).

Alliance of Veterinarians for the Environment (AVE) (http://www.aveweb.org/) (cited June 17, 2001).

American Academy of Veterinary Acupuncture (AAVA) (http://aava.org/) (cited June 17, 2001).

American Academy of Veterinary Disaster Medicine (AAVDM) (http://www.cvmbs.colostate.edu/clinsci/wing/aavdm/aavdm.htm) (cited June 17, 2001).

American Academy of Veterinary Informatics (AAVI) (http://netvet.wustl.edu/aavi.htm) (cited June 17, 2001).

American Academy of Veterinary Nutrition (AAVN) (http://acvn.vetmed.vt.edu/) (cited June 17, 2001).

American Academy of Veterinary Pharmacology and Therapeutics (AAVPT) (http://www.vet.purdue.edu/bms/aavpt/) (cited June 17, 2001).

American Animal Hospital Association (AAHA) (http://www.aahanet.org/) (cited June 17, 2001).

American Association for Laboratory Animal Science (AALAS) (http://www.aalas.org/) (cited June 17, 2001).

American Association of Avian Pathologists (AAAP) (http://cahp www.nbc.upenn.edu/~aaap/) (cited June 17, 2001).

American Association of Bovine Practitioners (AABP) (http://www.aabp.org/) (cited June 17, 2001).

American Association of Equine Practitioners (AAEP) (http://www.aaep.org/) (cited June 17, 2001).

American Association of Feline Practitioners (AAFP) (www.aafponline.org/) (cited June 17, 2001).

American Association of Food Hygiene Veterinarians (AAFHV) (http://www.avma.org/aafhv/default.htm) (cited June 17, 2001).

American Association of Human-Animal Bond Veterinarians (AAHABV) (http://members.aol.com/guyh7/aahabv.htm) (cited June 17, 2001).

American Association of Public Health Veterinarians (AAPHV) (http://www.avma.org/aaphv/) (cited June 17, 2001).

American Association of Small Ruminant Practitioners (AASRP) (http://www.aasrp.org/) (cited June 17, 2001).

American Association of Swine Veterinarians (AASV) (http://www. aasp.org/) (cited June 17, 2001). Formerly the American Association of Swine Practitioners (AASP).

American Association of Veterinary Anatomists (AAVA) (http:// civic.bev.net/aava/) (cited June 17, 2001).

American Association of Veterinary Clinicians (AAVC) (http:// cvm.msu.edu:80/~judy/aavcl.htm) (cited June 17, 2001).

American Association of Veterinary Immunologists (AAVI) (http:// www.cvm.missouri.edu/aavi/) (cited June 17, 2001).

American Association of Veterinary Laboratory Diagnosticians (AAVLD) (http://www.aavld.org/) (cited June 17, 2001).

American Association of Veterinary Medical Colleges (AAVMC) (http:// www.aavmc.org/) (cited June 17, 2001).

American Association of Veterinary Parasitologists (AAVP) (http:// www.vetmed.ufl.edu/aavp/) (cited June 17, 2001).

American Association of Veterinary State Boards (AAVSB) (http:// www.aavsb.org/) (cited June 17, 2001).

American Association of Wildlife Veterinarians (AAWV) (http:// www.aawv.net/) (cited June 17, 2001).

American Association of Zoo Veterinarians (AAZV) (http://www. aazv.org/) (cited June 17, 2001).

American Board of Veterinary Practitioners (ABVP) (http://www. abvp.com/) (cited June 17, 2001).

American Board of Veterinary Toxicology (ABVT) (http://www.abvt. org/) (cited June 17, 2001).

American Canine Sports Medicine Association (ACSMA) (http:// www.acsma.com/) (cited June 17, 2001).

American College of Laboratory Animal Medicine (ACLAM) (http:// www.aclam.org/) (cited June 17, 2001).

American College of Theriogenologists (ACT) (http://www.therio genology.org/) (cited June 17, 2001).

American College of Veterinary Anesthesiologists (ACVA) (http:// www.acva.org/) (cited June 17, 2001).

American College of Veterinary Behaviorists (ACVB) (http://www. var.vet.uga.edu/behavior/html/ACVB.htm) (cited June 17, 2001).

American College of Veterinary Clinical Pharmacology (ACVCP) (http:// www.acvcp.org/) (cited June 17, 2001).

American College of Veterinary Dermatology (ACVD) (http://www. dermvet.com/page3.html) (cited June 17, 2001).

American College of Veterinary Emergency and Critical Care (ACVECC) (http://www.acvecc.com/) (cited June 17, 2001).

American College of Veterinary Internal Medicine (ACVIM) (http:// acvim.org/) (cited June 17, 2001).

American College of Veterinary Microbiologists (ACVM) (http://cem.vet.utk.edu/acvm.html) (cited June 17, 2001).

American College of Veterinary Nutrition (ACVN) (http://acvn.vetmed.vt.edu/) (cited June 17, 2001).

American College of Veterinary Ophthalmologists (ACVO) (http://www.acvo.com/) (cited June 17, 2001).

American College of Veterinary Pathologists (ACVP) (http://www.afip.org/acvp/) (cited June 17, 2001).

American College of Veterinary Preventive Medicine (ACVPM) (http://www.acvpm.org/) (cited June 17, 2001).

American College of Veterinary Radiology (ACVR) (http://www.acvr.ucdavis.edu/) (cited June 17, 2001).

American College of Veterinary Surgeons (ACVS) (http://www.acvs.org/) (cited June 17, 2001).

American College of Zoological Medicine (ACZM) (http://www.isis.org/aczm/aczmindex.htm) (cited June 17, 2001).

American Heartworm Association (AHA) (http://www.heartwormsociety.org/) (cited June 17, 2001).

American Holistic Veterinary Medical Association (AHVMA) (http://www.altvetmed.com/AHVMA_brochure.html) (cited June 17, 2001).

American Pre-Veterinary Medical Association (APVMA) (http://www.vetsci.sdstate.edu/~apvma/) (cited June 17, 2001).

American Society of Laboratory Animal Practitioners (ASLAP) (http://www.aslap.org/) (cited June 17, 2001).

American Society of Veterinary Medical Association Executives (ASVMAE) (http://www.asvmae.org/) (cited June 17, 2001).

American Society of Veterinary Ophthalmology (ASVO) (http://www.asvo.com/) (cited June 17, 2001).

American Veterinary Assistants Association (AVAA) (http://www.avaa.bigstep.com) (cited June 17, 2001).

American Veterinary Chiropractic Association (AVCA) (http://www.animalchiropractic.org/) (cited June 17, 2001).

American Veterinary Dental College (AVDC) (http://members.aol.com/AMVETDENT/) (cited June 17, 2001).

American Veterinary Dental Society (AVDS) see Journal of Veterinary Dentistry (http://www.jvdonline.org//)

American Veterinary Medical Association (AVMA) (http://www.avma.org/) (cited June 17, 2001).

American Veterinary Medical Foundation (AVMF) (http://www.avmf.org/) (cited June 17, 2001).

American Veterinary Medical History Society (AVMHS) (http://www.cvm.missouri.edu/avmhs/) (cited June 17, 2001).

American Veterinary Medical Law Association (AVMLA) (http:// www.avmla.org/) (cited June 17, 2001).

American Veterinary Society of Animal Behavior (AVSAB) (http:// www.avma.org/avsab/) (cited June 17, 2001).

Association for Equine Sports Medicine (AESM) (http://www.aesm.org/) (cited June 17, 2001).

Association for Veterinary Informatics (AVI) (http://netvet.wustl.edu/ avi.htm) (cited June 17, 2001).

Association for Women Veterinarians (AWV) (http://www.awv-women-veterinarians.org/) (cited June 17, 2001).

Association of American Veterinary Medical Colleges (AAVMC) (http:// www.aavmc.org/) (cited June 17, 2001).

Association of Avian Veterinarians (AAV) (http://www.aav.org) (cited June 17, 2001).

Association of Reptilian and Amphibian Veterinarians (ARAV) (http:// www.arav.org/) (cited June 17, 2001).

Association of Veterinarians for Animal Rights (AVAR) (http:// www.avar.org/) (cited June 17, 2001).

Association of Veterinary Clinical Pharmacology and Therapeutics (AVCPT) (http://www.avcpt.org/) (cited June 17, 2001).

Association of Veterinary Technician Educators (AVTE) (http://www. br.cc.va.us/avte) (cited June 17, 2001).

Associations of Teachers of Veterinary Public Health and Preventative Medicine (ATVPHPM) (http://www.cvm.uiuc.edu/atvphpm/) (cited June 17, 2001).

Christian Veterinary Mission (CVM) (http://www.vetmission.org/) (cited June 17, 2001).

Consortium of North American Veterinary Interactive New Concept Education (CONVINCE) (http://www.convince.org/) (cited June 17, 2001).

Lesbian and Gay Veterinary Medical Association (LGVMA) (http:// www.lgvma.org/) (cited June 17, 2001).

Mid-Atlantic States Association of Avian Veterinarians (MASAAV) (http://masaav.org/) (cited June 17, 2001).

National Association of Federal Veterinarians (NAFV) (http://users. erols.com/nafv/) (cited June 17, 2001).

National Board Examination Committee for Veterinary Medicine (NBEC) (http://www.nbec.org) (cited June 17, 2001).

North American Veterinary College Administrators (NAVCA) (http:// navca.cvm.tamu.edu/) (cited June 17, 2001).

North American Veterinary Technician Association (NAVTA) (http:// www.avma.org/navta) (cited June 17, 2001).

Society for Theriogenology (SFT) (http://www.therio.org/) (cited June 17, 2001).

Society for Tropical Veterinary Medicine (STVM) (http://forest.bio.
ic.ac.uk/stvm/Welcome.htm) (cited June 17, 2001).
Society of Aquatic Veterinary Medicine (SAVM) (http://www.savm.org/)
(cited June 17, 2001).
Society of Veterinary Nuclear Medicine (SVNM) (http://www.acvr.
ucdavis.edu/societ/nm/nm.html) (cited June 17, 2001).
United States Public Health Service Veterinarians (USPHSV) (http://
www.fda.gov/cvm/links/vcc/default.htm) (cited June 17, 2001).
Veterinary Cancer Society (VCS) (http://www.vetcancersociety.org/)
(cited June 17, 2001).
Veterinary Emergency & Critical Care Society (VECCS) (http://www.
veccs.org) (cited June 17, 2001).
Veterinary Hospital Managers Association (VHMA) (http://www.
vhma.org/) (cited June 17, 2001).
Veterinary Orthopedic Society (VOS) (http://www.vet-ortho-soc.org/)
(cited June 17, 2001).
Western Veterinary Conference (WVC) (http://www.wvc.org/) (cited June
17, 2001).
Wildlife Disease Association (WDA) (http://www.vet.uga.edu/wda/)
(cited June 17, 2001).

## STATISTICS

Statistical resources for animal health and veterinary science vary from country
to country. In the United States, statistics on livestock may be found in the *Agricultural Statistics* (USDA, 2000) or more selectively in the *Statistical Abstract
of the United States* (United States Department of Commerce, 2000). Selected
pet population and veterinary market statistics can be found on the AVMA Web
site at http://www.avma.org/cim/default.htm (cited June 14, 2001) with detailed
data available in the AVMA publications described in the following text. Additional market information can be found in titles from the American Animal Hospital Association and the American Pet Products Manufacturers Association. Animal disease statistics can be found through the Office International des Épizooties
(International Office of Epizooties) or the FAO as described.

*Animal Health Yearbook, 1995*. 1997. [FAO animal production and health series;
no. 36.] [s.l.] Food and Agricultural Organization. 279 p. ISBN: 9250039611
(paperback).   First published as the *World Livestock Disease Report* (1956) and
then as the *FAO/OIE Animal Health Yearbook* (1957–60), this title reports on
142 animal diseases for more than 15 animal species worldwide. It includes tables
on selected zoonotic diseases; on animal populations (cattle, dairy cows, buffa-
loes, horses, mules and asses, camels, sheep, goats, pigs, chickens, and other

poultry) per country; and on the number of veterinarians (government, research and university, private, and other) and animal health auxiliary personnel per country.

*APPMA National Pet Owners Survey, 1999–2000*. 1999. Greenwich, CT: American Pet Products Manufacturer Association. 359 p. ISBN: 0963255215. Since 1988, the APPMA has gathered data on pet ownership, pet care practices, purchasing behavior of pet owners for pet-related products, and sources of pet-related goods and services. Data from this sixth survey (1998) is summarized by dog, cat, fish, bird, reptile, and other small animal ownership including rabbits, hamsters, guinea pigs, gerbils, mice/rats, or ferrets. Of particular interest is the data on veterinarians visits, veterinary expenses, and services obtained from veterinarians, as well as reasons for changing veterinarians.

*Economic Report on Veterinarians & Veterinary Practices*. 2001. [Schaumburg, IL]: American Veterinary Medical Association. 227 p. This authoritative compilation of veterinary economic statistics and trends includes more than 100 tables and charts providing median and mean values for total professional income; hours worked; and income per hour for all employment categories such as small (animal) predominant or equine practice, for practice owners, and for practice associates (employed veterinarians). Key practice operating expenses, personnel statistics and salaries, and breakdowns of client transactions per species are also provided.

*Financial and Productivity Pulsepoints*: *A Comprehensive Survey and Analysis of Performance Benchmarks*. 1998. Lakewood, CO: AAHA Press. 121 p. ISBN: 0941451682. Conducted by the American Animal Hospital Association, this survey of randomly selected veterinary practices in the U.S. and Canada seeks to assist individual small animal veterinary practices to identify their strengths and weaknesses and maximize their productivity; generate more revenue; and control expenses. Data provided is segmented by region, number of veterinarians, gross income, and cost-of-living indices. It includes total income, dental income, laboratory income, drugs and medical expenses, and nonveterinarian staff expenses.

*U.S. Livestock Market for Veterinary Medical Services and Products*. 1995. [Schaumburg, IL]: Center for Information Management, American Veterinary Medical Association. 103 p. Updating J. Karl Wise's *U.S. Market for Food Animal Veterinary Medical Services*. [Schaumburg, IL: American Veterinary Medical Association, 1987. 185 p.], this study describes trends in food-animal agriculture and producer characteristics, and reports the results of national surveys of livestock enterprises and veterinarians. Authors of this study hope to identify areas for potential increased demand by assessing the current market for veterinary medical services in the United States.

*U.S. Pet Ownership & Demographics Sourcebook.* 1997. [Schaumburg, IL]: American Veterinary Medical Association. 135 p. ISBN: 1882691024. Updating the AVMA's Center for Information Management. *Veterinary Services Market for Companion Animals* [(Schaumburg, IL): American Veterinary Medical Association, 1992. 120 p.], this study of over 60,000 U.S. households evaluates trends in pet ownership, use of veterinary services, and reasons for selecting a veterinarian. Includes statistics on pet populations for dogs, cats, horses, fish, ferrets, rabbits, rodents, reptiles and other specialty pets, as well as dog and cat breed registrations.

*World Animal Health in 1998.* 1999. 2 vols. Paris, France: Office International des Épizooties. ISBN: 929044469X. This annual compilation of animal health data from over 180 countries is published in two parts. Part 1 reports the year's significant epidemiological events for farmed mammals and birds with emphasis given to the most contagious and economically significant diseases (OIE List A). In addition, an overview of emerging wildlife diseases and national details on disease control methods are provided. Part 2 consists of tables showing the impact of each of the List A and List B diseases (less contagious but still a threat to economy or public health), with tables presenting the number of outbreaks, cases, deaths, and number of animals slaughtered or vaccinated.

## THESES

Dissertations and theses in animal health and veterinary science from colleges and universities within the United States can be identified through use of *Dissertation Abstracts International* (1969 to date) or the Dissertations Express Web page at http://wwwlib.umi.com/dxweb/gateway (cited June 27, 2001). Some relevant dissertations are also indexed in CAB Abstracts or can be identified through WorldCat (OCLC).

European veterinary dissertations can be found through the European Veterinary Dissertations (EVD) database at http://campert.library.uu.nl/evd/ (cited June 27, 2001). Updated biomonthly, the EVD project is carried out by Euroscience, in co-operation with the Library of the Faculty of Veterinary Medicine, Utrecht University, The Netherlands, and under the auspices of the European Association of Establishments for Veterinary Education (EAEVE). The first edition of this database, made available in February of 2001, contains the bibliographic data for 10,165 dissertations of the period 1991–2001 and defended at 32 European veterinary faculties.

## WORLD WIDE WEB SITES

The leading guides to Internet resources for animal health and veterinary science professionals currently is Dr. Ken Boschert's *NETVET: Mosby's Veterinary Guide to the Internet* (1998) and his premier Web site, NETVET & Electronic

Zoo, found at http://netvet.wustl.edu (cited June 28, 2001). In addition, Dr. Richard L. Crawford's bibliography of Selected Web Sites for Biomedical, Pharmaceutical, Veterinary, and Animal Sciences is available at http://www. nal.usda.gov/awic/pubs/awic9802.htm (cited June 28, 2001) (Crawford, 1999).

The following are Web sites relevant to animal health and veterinary science selected by the author from those identified in the preceding publications and Web sites, as well as from the author's personal bookmarks. Sites discussed elsewhere in this chapter are not repeated in this section.

American Veterinary Medical Association (AVMA) (http://www.avma.org/) (cited June 26, 2001). As the primary veterinary organization for the U.S., the AVMA's Web site provides information for the member veterinarian, as well as information for the public. For the public, it has information on emergency preparedness, educational resources for dog bite prevention, AVMA position statements on subjects such as compounding drugs, and public health information on topics such as bovine spongiform encephalopathy, foot-and-mouth disease, psittacosis (avian chlamydiosis), and rabies. Lists of U.S. veterinary medical colleges and veterinary technology programs are also available.

Animal Health/Emerging Animal Diseases (AHEAD) (http://www.fas.org/ ahead/) (cited June 26, 2001). Sponsored by the Federation of American Scientists, this Web page addresses outbreaks of animal and zoonotic diseases and related global security issues.

Animal Poison Control Center (ASPCA) (http://www.napcc.aspca.org/) (cited June 26, 2001). The National Animal Poison Control Center, a division of the American Society for the Prevention of Cruelty to Animals (ASPCA), provides a fee-based service with veterinary toxicologists manning the center 24 hours a day. The Web site provides basic information about the center, emergency phone numbers, what to do in the event of an animal poisoning, and how to prevent poisoning in your pet.

Animal Welfare Information Center (AWIC) (http://www.nal.usda.gov/awic) (cited June 26, 2001). This Web site provides information for improved animal care and use in research, teaching, and testing.

Animals and Emergencies (FEMA) (http://www.fema.gov/fema/anemer.htm) (cited June 26, 2001). This U.S. Federal Emergency Management Agency Web site has information on dealing with animals in time of disaster including preparedness measures.

APHIS Web (http://www.aphis.usda.gov/oa/pubs.html#third) (cited June 26, 2001). This Web page provides access to all publications currently available from the U.S. Animal and Plant Health Inspection Service (APHIS), including fact sheets, popular publications, and scientific and technical reports. Whenever possible, the items are available in both English and Spanish.

Care for Pets (AVMA) (http://www.avma.org/care4pets/) (cited June 26, 2001). This site, created by the AVMA, covers the spectrum of animal health with pet health news, pet loss information, what you need to know when buying a pet, animal safety, pet stories, how to select a veterinarian, veterinary career information, and a "Kid's Korner."

Center for Emerging Issues (http://www.aphis.usda.gov/vs/ceah/cei/) (cited June 26, 2001). Prepared by the U.S. Centers for Epidemiology and Animal Health, this Web page identifies and analyzes both emerging animal health issues and emerging market conditions for animal products.

Consultant (http://www.vet.cornell.edu/consultant/consult.asp) (cited June 26, 2001). This veterinary diagnostic support program, created by Drs. M. White and J. Lewkowicz of Cornell University, allows users to search for diagnoses based on one or more symptoms. Supported by a database of approximately 500 signs/symptoms, about 4,000 diagnoses, and over 10,000 literature references, the program also links to additional sources of information on the disease described.

Delta Society (http://www.deltasociety.org/) (cited June 26, 2001). The not-for-profit Delta Society Web site provides information on pet loss and bereavement, animal-assisted therapy, and dogs for people with disabilities.

DVM News Magazine (http://www.dvmnewsmagazine.com/) (cited June 26, 2001). This is the online version of a monthly print publication for veterinarians in private practice.

Endangered Species Information (http://endangered.fws.gov/) (cited June 26, 2001). From the U.S. Fish and Wildlife Service, this Web page provides information on the animals and plants listed as threatened or endangered within the United States.

FDA Approved Animal Drug Database (Green Book) (http://www.fda.gov/cvm/greenbook/greenbook.html) (cited June 26, 2001). From the U.S. Food and Drug Administration, this is a list of approved animal drugs on the Web and other important information.

Food Animal Residue Avoidance Databank (FARAD) (http://www.farad.org/) (cited June 26, 2001). A computer-based decision support system designed to provide livestock producers, extension specialists, and veterinarians with practical information about drugs used to treat animal diseases in the U.S., as well as corresponding safe withdrawal times for meat and milk production. Access to veterinary level information requires password, but producer level information is publicly available.

Healthy Pet (http://www.healthypet.com/) (cited June 26, 2001). Created by the American Animal Hospital Association, the Web site has a Veterinary Hospital Locator; newsletter; pet care tips including a library of information on behavior, pet health problems, and nutrition; and other topics, as well as a kid's coloring page.

International Veterinary Information Service (IVIS) (http://www.ivis.org/) (cited July 1, 2001). This not-for-profit Web site provides information to veterinarians, veterinary students, and animal health professionals worldwide. It has free access to original, up-to-date publications organized in electronic books, each edited by highly qualified editors; proceedings of veterinary meetings; short courses; continuing education (lecture notes, manuals, autotutorials, and interactive Web sites); an international calendar of veterinary events, image collections; and much more.

National Board Examination Committee for Veterinary Medicine (NBEC) (http://www.nbec.org/) (cited June 26, 2001). This site provides information on the NBEC standardized licensing examinations used by state and provincial licensing boards as part of their licensure procedure for veterinarians.

National Wildlife Health Center (NWHC) (http://www.nwhc.usgs.gov/) (cited June 26, 2001). This Web site has information on wildlife diseases and mortality reports, diagnostics, and other helpful links to NWHC information.

NetVet (http://netvet.wustl.edu) (cited June 26, 2001). The most comprehensive Web site for everything veterinary and animal related, this provides a path for both public (Electronic Zoo) and professional (NetVet) information.

Office International des Épizooties—International Office of Epizooties) (OIE) (http://www.oie.int/) (cited June 26, 2001). As the world organization for animal health, the OIE through its Web site provides a wealth of related animal health and disease information (in English, French, and Spanish).

Online Mendelian Inheritance in Animals (OMIA) (http://www.angis.su.oz.au/Databases/BIRX/omia/) (cited June 26, 2001). This is a database of gene and phene (familial trait or phenotype) information in a wide range of animal species by the Department of Animal Science, University of Sydney, Australia. Entries consist of references, sorted by year, and within year by authors.

Orthopedic Foundation for Animals (http://www.offa.org) (cited June 26, 2001). This site includes breed-specific genetic databases for hip dysplasia, elbow dysplasia, autoimmune thyroiditis, congenital cardiac disease, and patella luxation.

Pets and Pet Health (MEDLINEPlus) (http://www.nlm.nih.gov/medlineplus/petsandpethealth.html) (cited June 26, 2001). Prepared as part of MEDLINE Plus, a consumer-orientated Web site by the NLM, this Web page provides links to quality resources on the topic.

Pig Health (http://www.pighealth.com) (cited June 26, 2001). From the Pig Disease Information Center (UK), this Web site focuses on swine health issues.

Poisonous Plants. Various Web sites exist on plants poisonous to pets and livestock including: Plants Poisonous to Livestock and Pets (Purdue University ADDL) (http://www.vet.purdue.edu/depts/addl/toxic/cover1.htm) (cited June 26, 2001). Plant Poisonings in Livestock and Pets (Mid Atlantic Region) (http://education.vetmed.vt.edu/Curriculum/VM8424/toxicplants/index.html) (cited June 26, 2001). Plants Toxic to Animals (UIUC) (http://www.library.uiuc.edu/vex/vetdocs/toxic.htm) (cited June 26, 2001).

Preparing for a Career in Veterinary Medicine (AAVMC) (http://www.aavmc.org/prevet/prevet.htm) (cited June 26, 2001).

USAHA Information by Species (USAHA) (http://www.usaha.org/species.html) (cited June 26, 2001). Maintained by the U.S. Animal Health Association, this Web page provides information on livestock diseases arranged by species (cattle, fish, horses, poultry, sheep, swine, and wildlife).

Vetbase (http://www.vetinfo.demon.nl) (cited June 26, 2001). A large electronic veterinary formulary prepared by Dr. Hans Kuiper, a Dutch veterinarian, with 13,000 veterinary drug dosages, 75 pharmaceutical classes, 170 animal species, more than 800 drugs, dose extrapolation, and literature references. It includes a new Gerbil Formulary with 180 nonantibiotic drug dosages.

Veterinarian "Members-Only" Web Sites. Network Of Animal Health (NOAH)—AVMA members–only access (http://www.avma.org/network.html) (cited June 26, 2001). E-VET—A free profession veterinary site for members only (http://www.e-vet.com) (cited June 26, 2001). Veterinary Information Network (VIN)—Veterinarians/members-only access on America Online (AOL) (http://www.vin.com/) (cited June 26, 2001).

Veterinary Abbreviations and Acronyms (UIUC) (http://www.library.uiuc.edu/vex/vetdocs/abbreviation.htm) (cited June 26, 2001).

Veterinary Conference Proceedings (http://www.medvet.umontreal.ca/biblio/gopher/bases/default.htm) (cited June 26, 2001). Prepared by Jean-Paul Jetté (Université de Montréal), this is a database of tables of contents for a variety of veterinary conference proceedings.

Veterinary Emergency Drug Calculator (http://www.cvmbs.colostate.edu/clinsci/wing/emdrughp.html) (cited June 26, 2001). Created by Wayne E. Wingfield, DVM, Colorado State University, this Web tool features canine and feline versions that provide a variety of emergency drug doses based on the animal's current weight.

Veterinary Glossary: Definitions and Abbreviations of Veterinary Terms (http://www.vetmed.wsu.edu/glossary/glossary.asp) (cited June 26, 2001).

Veterinary Medicine Libraries (http://www.usask.ca/~ladd/vet_libraries.html) (cited June 26, 2001). Maintained by Ken Ladd, this is an international list of veterinary libraries, librarians, home pages, and other resources.

Veterinary Oncology/Cancers (Oncolink/U of Penn) (http://oncolink.upenn.edu/specialty/vet_onc/) (cited June 26, 2001).

VetGate (http://vetgate.ac.uk/) (cited June 26, 2001). Entitled "the UK's guide to high quality Internet resources on animal health," this excellent resource is one of five of the BIOME gateways to biomedical information.

Vetinfo (http://www.vetinfo.com/) (cited June 26, 2001). Prepared by Michael Richards, DVM (a.k.a. Dr. Mike) and his staff, this site provides information on veterinary medicine, especially canine and feline health and disease.

The World Wide Web Virtual Library: Veterinary Medicine (http://netvet.wustl.edu/vetmed.htm) (cited June 26, 2001). This is a selected collection of veterinary medical Internet resources now maintained by Dr. Ken Boschert, DVM.

Zoonosis Web Page (http://medicine.bu.edu/dshapiro/ zool.htm) (cited June 26, 2001). Prepared by Dr. David Shapiro, MD, Boston Medical Center, this Web page allows one to search for zoonotic diseases that humans may acquire from a given animal.

Zoonotic Diseases (CDC) (http://www.cdc.gov/ncidod/dpd/parasiticpathways/animals.htm) (cited June 26, 2001). Prepared by the Centers for Disease Control (CDC), this Web page provides links to information on diseases spread from animals to people.

## YEARBOOKS

Yearbooks, as a publishing format for animal health and veterinary science, appear to be fading. The *Veterinary Annual* ceased in 1996 and the last volume of *Advances in Veterinary Medicine* was published in 1999 after a two-year delay.

*Advances in Veterinary Medicine*. 1997– . vol. 40– . San Diego, CA: Academic Press. ISSN: 1093-975X. With the latest volume (vol. 41) appearing in 1999, this annual has focused individual volumes on diverse topics such as veterinary vaccines and diagnostics, veterinary medical specialization, and comparative vertebrate exercise physiology. This title continues *Advances in Veterinary Science and Comparative Medicine* (vols. 13–39, 1969–95) and *Advances in Veterinary Science* (vols. 1–12, 1953–68).

*The Veterinary Annual.* 1959–96. vols. 1–36. Bristol, UK: Wright-Scientechnica. ISSN: 1093-975X. For thirty-six years, this annual presented highly topical articles of interest to practicing veterinarians, written by international authors.

## REFERENCES

Boschert K. NETVET: Mosby's Veterinary Guide to the Internet. St. Louis, MO: Mosby, 1998, pp. 1–285 and disk. [Updates are available quarterly by registering with Mosby at http://www.harcourthealth.com/MERLIN/netvet/ (June 15, 2001)].

Boyd T. Veterinary Science. Magazines for Libraries. 10th ed. New York: Bowker, 2000, pp. 1408–1415.

Boyd CT, et al. Basic List of Veterinary Medical Serials, 2nd ed., 1981, with Revisions to April 1, 1986. Serials Librarian. 11(2):5–39, 1986.

Bruce N. Veterinary Medicine and Animal Health Journals—The New, the Redesigned, and the Retitled. Serials Review. 16(3):39–46, 1990.

Crawford RL compiler. Selected Web Sites for Biomedical, Pharmaceutical, Veterinary, and Animal Sciences. (AWIC Series 98-02) Bethesda, MD: Animal Welfare Information Center, USDA, 1999. http://www.nal.usda.gov/awic/pubs/awic9802.htm (June 15, 2001).

Dunlop, RH, Williams DJ. Veterinary Medicine: An Illustrated History. St. Louis: Mosby, 1996, pp. 1–692.

Future Trends in Veterinary Public Health. Weekly Epidemiological Record 19:154–156, May 14, 1999.

Henley AL. Organization and compilation of a basic list of veterinary medical serials. Journal of Veterinary Medical Education. 5(3):136–138, 1978.

Houston W. The Application of Bibliometrics to Veterinary Science Primary Literature. IAALD Quarterly Bulletin. 1983, 28(1):6–13.

Loew FM. Foreword. In: Dunlop and Williams. Veterinary Medicine: An Illustrated History. St. Louis: Mosby, 1996. pp. vii–viii.

Olsen WC. Primary Journals and the Core List. In: Olsen, WC, ed. The Literature of Animal Science and Health. Ithaca, NY: Cornell University Press, 1993, pp. 204–222.

Smith F. The Early History of Veterinary Literature and its British Development. London: Bailliere, Tindall & Cox, 1919–33. 4 vols. [Reprint: London: J.A. Allen, 1976.]

Smithcors JF, Smithcors A. Five Centuries of Veterinary Medicine: A Short-Title Catalog of the Washington State University Veterinary History Collection. Pullman, WA: Washington State University Press, 1997, pp. 1–145.

Stalheim OHV. Guide to Collections of Papers Pertaining to American Veterinary History. Ames, IA: OHVEE, Inc., [1996], pp. 1–39. (Describes the collections maintained in libraries and museums in the United States and Canada.)

Veterinary Serials: A Union List of Selected Titles Not Indexed by Index Medicus and Held by Veterinary Collections in the U.S. and Canada. [Guelph, Ontario: University of Guelph Library Systems and Data Processing Division], (microfiche) 1980.

Veterinary Serials: A Union List of Serials Held in Veterinary Collections in Canada, Europe, and the U.S.A. 2d, 1987–1988. 2 vols. [Davis, Calif.: Veterinary Medical Libraries Section, Medical Library Association.], 1990.

# 5

## Animal Science and Livestock Production

**Jodee Kawasaki**
Montana State University, Bozeman, Montana, USA

The emphasis of this chapter is livestock production of animal science. All other aspects of animal science can be found in Chapter 4 ''Animal Health and Veterinary Sciences.'' Livestock production is the applied aspect of the combination of many disciplines, including breeding, physiology, nutrition, farm management and economics, agricultural engineering for animal housing and ergonomics, and veterinary medicine and health. Within the last two decades biotechnology also has been incorporated into the advancement of livestock production.

During the late twentieth century, livestock breeders became more involved in feeding and breeding programs to increase production. This growing interest contributed to the change from an earlier focus on animal husbandry to the broader study of animal science. However, the initial concept of selective breeding reached agricultural producers as early as the late 1800s and early 1900s. Exhibitions, such as the International Live Stock Exhibition held in Chicago, brought awareness of livestock breeding programs to the forefront and led to the experimentation of crossbreeding of livestock species throughout the past century. Through the initiation of such practices, meat products were produced to meet societal needs and preferences for consumption. In addition, improved livestock nutrition and feeding practices were developed as scientists came to recognize the importance of vitamins and minerals to animal diets. Currently, biotechnology applications are being incorporated to enhance nutrition and breeding programs. Thus, it can be seen that the field of livestock production has been greatly influenced by advances provided through new research and technologies. This has led to improved breeding, health, nutrition, management, and feeding programs, and to a movement from small purebred farms to commercial operations.

Much of this research has been conducted at universities and within government agencies throughout the world. Within the U.S. Department of Agriculture (USDA), experiment stations have tested research findings within particular geographic and climatic areas, and against other regional influences and practices. Other entities and countries have similar research centers where field trials of production practices are performed.

This chapter is intended to be international in scope and to cover a wide variety of livestock. However, due to the extent of U.S. livestock production research and publications, there is an emphasis on materials from the United States. The layout of the chapter begins with general materials covering livestock or multiple species, including sections on biotechnology and history. Following this are sections dedicated to aquaculture, beef cattle, dairy, horses, poultry (chickens and turkeys), sheep and goats, and swine. The final section is "Other Livestock," covering species for which there is little research material available, including mink, buffalo, musk oxen, alligators, ducks, and rabbits. An important aspect of this chapter is that materials are not segregated by format, such as electronic journals or Web sites. If a Web site appears to be a directory, it is listed in the directory section. If a journal is published electronically, it will appear in the journal list, not in a separate category. Finally, as a companion to Chapter 4, this chapter defers to individual items or search strategies thoroughly covered in Chapter 4 whenever appropriate.

## GENERAL LIVESTOCK LITERATURE

### Abstracts, Indexes, and Databases

The basic agricultural indexes [CAB Abstracts, AGRIS, and AGRICOLA] cover livestock production, especially for cattle, horses, swine, poultry, sheep, and goats. A general science database covering the core agricultural journals is Science Citation Index. These four databases are described at the beginning of this work.

See also Chapter 4 for details on searching CAB Abstracts and a complete list of appropriate CABI CODES to use for refining searches. These codes are meant to be general and a starting point for retrieving appropriate articles. For instance, to search the topic of swine production, use the subject *pigs* with the general code, LL120 Meat Producing Animals, to retrieve all relevant articles.

If you search by *swine*, the result is a severely reduced set of articles. The *CABI Thesaurus* is very helpful in this regard. If you look up *swine*, the thesaurus refers you to *pigs*. The word *swine* is only used within the name of diseases, such as *swine-fever virus*.

## Atlases

Several atlases are also listed Chapter 4 covering livestock anatomy. Searching a library catalog nets many of these publications by combining the keyword *atlas* with the keyword *livestock*. An example of this is:

Sambraus, Hans Hinrich. *Color Atlas of Livestock Breeds*. 1992. Saint Louis, MO: Mosby.

## Bibliographies

Bibliographies, selected or annotated, are useful for beginning a literature search on a particular subject. Major works can be found by searching library catalogs where a keyword can be combined with the subject heading *bibliography* to obtain relevant records. Examples of keywords include *animal housing*, *swine*, *goats*, *beef cattle*, *horses*, *animal genetics*, *livestock* and so forth. Also, the section, Guides to the Literature, explains how to search indexes for bibliographies, most often using keywords *literature review* or *reviews*. See also Chapter 4 for an extensive list of bibliographies and for a description of the published series of bibliographies complied by the U.S. National Agricultural Library (NAL) such as *Quick Bibliography* and *Special Reference Briefs*.

Edsall, M.E. and A.T. Young. 1994. *Fish Farming*: *January 1989–April 1994*. Quick bibliography series; 94-44. Beltsville, MD: National Agricultural Library.

Science & Life Consultants Association Staff. 1996. *Animal Husbandry Directory of Authors of New Medical & Scientific Reviews with Subject Index*. Annandale, MD: ABBE Publishers Association of Washington, D.C. A bibliography indexing literature reviews of livestock and animal production, plus many other scientific subjects.

Young, A.T. and M.E. Edsall. 1994. *Aquaculture*: *Salmon and Trout Farming*: *January 1979–March 1994*. Quick bibliography series 94-37. Beltsville, MD: National Agricultural Library.

## Biographies

Animal scientists have few biographies dedicated to their discipline. However, these researchers can be found in general agriculture biographies or manuscript collections. Searching the agricultural databases or library catalogs can retrieve desired records. As with the previous section on bibliographies, search library catalogs by subject heading, in this case, *biography* for the most accurate retrieval. Use the appropriate designation for the person, that is, etiologists, animal scientists, cattle breeders, farmers, cowboys, or agriculturists. A sample search might look like this: subject = *agriculturists* and subject = *biography* or keyword = *livestock* and subject = *biography*. Additionally, searching in CAB Abstracts or AGRICOLA by *biographies* will retrieve relevant citations from various journals and monographs. Note the difference between searching a library catalog (biography) and indexes/abstracts (biographies). The correct variation, singular or plural, will significantly improve the number of records or citations found.

From 1987 through 1999, the *Journal of Animal Science* included biographies in its first section. Browsing these sections provides names of animal scientists who have been active in the profession. However, the biography section does not appear in every monthly issue as only four to seven sketches were provided per year; check the cumulative index in the December issue for a listing of the sketches. As of 2000, the sketches are published only on the American Society of Animal Scientists Web page at http://www.asas.org.

DeBaca, R.C. 1990. *Courageous Cattlemen*. Huxley, IA: R.C. de Baca; Champaign, IL: KOWA Graphics. This book gives a history of the growth of the beef industry in the U.S. Contributions to the advancement of animal husbandry by animal scientists, together with cattle breeders, are described.

*Who's Who in Science in Europe: a Biographical Guide to Science*, *Technology*, *Agriculture, and Medicine*. 1995. 9th ed. London: Cartermill. Also co-published in the United States and Canada by Stockton Press, this is a two-volume set with indexes by country and subject.

See also Chapter 4 for standard titles such as *Agricultural and Veterinary Sciences International Who's Who*.

## BIOTECHNOLOGY ISSUES

The following is a sampling of articles and monographs on biotechnology within the animal science discipline. There are many resources covering the topic. Again, searching the indexes or catalogs will retrieve many more citations. A combination search will net the best results, using the subject *biotechnology* and subjects, such as *livestock* or *sheep* or *dairy*. If *biotechnology* as a subject term returns a small number of records, expand the search by changing the search field from subject to the keyword or word anywhere field.

Bonneau, M.,B. Laarveld, and A.L. Aumaitre. 1999. ''Biotechnology in Animal Nutrition, Physiology and Health.'' *Livestock Production Science* 59(2–3): 223–

241. A review covering how biotechnology is already widely used in animal production, along with numerous other potential applications. Nutrients, enzymes, prebiotics and probiotics or immune supplements, plant biotechnology, and transgenic manipulation are discussed.

Cunningham, E.P. 1999. "The Application of Biotechnologies to Enhance Animal Production in Different Farming Systems." *Livestock Production Science* 58(1): 1–24. This review covers the economic and ethical issues on such topics as artificial intelligence, embryo transfer, in vitro fertilization, sex diagnosis and control, embryo cryopreservation, cloning, marker-assisted selection, gene mapping, identification of important single-gene traits, identification of quantitative trait loci, transgenesis, the use of recombinant hormones, immunomodulation of hormones, and the production of recombinant proteins.

Geldermann, H. and F. Ellendorff, eds. 1990. *Genome Analysis in Domestic Animals*. Weinheim, Germany; New York: VCH. Proceedings of a symposium held November 2–3, 1988, in Hanover, Germany. Papers cover the following topics: genetic engineering, chromosome mapping, gene mapping, and animal breeding.

Hafs, H.D., ed. 1993. *Genetically Modified Livestock: Progress, Prospects, and Issues*. Champaign, IL: American Society of Animal Science.

Murray, J.D. 1999. *Transgenic Animals in Agriculture*. New York: Oxford University Press. A representation of the issues and ethics of using genetically modified livestock in animal production.

Wallace, R.J. and A. Chesson. 1995. *Biotechnology in Animal Feeds and Animal Feeding*. Weinheim, Germany; New York: VCH. One of the first works discussing the influence of biotechnology in animal feeds and feeding.

## Directories

As mentioned in the introduction, items listed here include both print and electronic directories. These directories are selected based on their overall usefulness. In library catalogs, search the subject *directories* with another keyword, such as *livestock* or *meat industry*. Many associations and companies also have directories of members or employees listed on the Web, some available only for a fee.

AgNIC (http://www.agnic.org). Beltsville, MD: National Agricultural Library. The Agriculture Network Information Center (AgNIC) is an excellent Web resource for all types of agricultural materials, including animal science and livestock production. On the home page choose the subject Animal and Veterinary Sciences. Another excellent link from the top page is Calendars, which provides information about conferences, seminars, and meetings.

American Meat Institute (www.meatami.com/). This Web site is an information resource for the meat and poultry packing industry including statistics, meat safety regulations, information, and guidelines. It also has links to associated associations, members, and legislative bodies.

American Sheep Industry Association (www.sheepusa.org/). Provides information on the sheep, wool, and lamb industry and production; a directory of sheep breeds; links to sheep breed associations and sheep experts; and sheep production resources for sale. Also, this site provides an extensive list of links to related sheep Web sites, including associations, universities, electronic journals, and newsletters.

Livestock, Dairy & Poultry Science Journals (http://www.sciencekomm.at/journals/farm.html). Sciencekomm, Inc. Directory of journals; the links provide information on accessing these electronically but does not supply any contents or articles.

Livestock Library (http://www.ansi.okstate.edu/library/). Department of Animal Science, Oklahoma State University. This is a comprehensive directory covering many livestock species including ''other species'' and stock production Web site links.

*Meat Trades Journal Directory*. 1995– . Croydon, UK: n.p. Directory of the meat industry and trade in Great Britain.

Sustainable Agriculture Directory of Expertise (http://www.sare.org/expertise/). This is made available through the Sustainable Agriculture Network for the Sustainable Agriculture Research and Education Program funded through the USDA. Searchers may select the term for animal production or individual species.

**Encyclopedias and Dictionaries**

Searching for encyclopedias and dictionaries is a straightforward process. Within library catalogs, use either term as a subject in plural form.

Breeds of Livestock (http://www.ansi.okstste.edu/breeds/). Stillwater, OK: Oklahoma State University, Department of Animal Science. This site allows users to search for information on livestock by world region or by species name. Data is available on cattle, horses, swine, goats, and sheep and organized as encyclopedia entries.

Christman, C.J., D.P. Sponenberg, and D.E. Bixby. 1997. *A Rare Breeds Album of American Livestock*. Pittsboro, NC: American Livestock Breeds Conservancy.

Gispert, C. 1999. *Enciclopedia Práctica de la Agricultura y la Ganadería*. Barcelona: Océano/Centrum. A one-volume Spanish encyclopedia on livestock and agriculture topics.

Hurnik, J.F., A.B. Webster, and P.B. Siegel. 1995. *Dictionary of Farm Animal Behavior*. See Chapter 4.

Mason, I.L. 1988. *A World Dictionary of Livestock Breeds, Types and Varieties*. Wallingford, UK: CAB International.

Swatland, H.J. 2000. *Meat Cuts and Muscle Foods: an International Glossary*. Nottingham: Nottingham University Press.

Wallis, D. 1986. *The Rare Breeds Handbook*. Poole, UK: Blandford. This encyclopedia lists rare and historical livestock and animals found in Great Britain.

## Government Documents

See also other sections, such as statistics, handbooks and manuals, or journals. There are many world agencies that publish a wealth of animal science literature. The more common ones are listed. Search these as a corporate author, combined with a subject in a library catalog for specific materials, for example: author *International Livestock Centre for Africa* with subject *cattle* or *statistics*. Searching by acronyms also can be helpful, especially if searched as a keyword instead of an author search term.

> United States Department of Agriculture (USDA)
> Food and Agriculture Organization of the United Nations (FAO)
> International Livestock Centre for Africa (ILCA)
> Technical Centre for Agricultural and Rural Cooperation (CTA)
> Consultative Groups on International Agricultural Research (CGIAR)

## Guides to the Literature

Literature guides are useful, especially for anyone new to the subject or who are coming from a different discipline, as they often explain the literature and content more than bibliographies. Searching catalogs for literature guides is similar to searching for bibliographies. More often than not, a guide is given the subject term *bibliography*. Other subject terms used in library catalogs include *information services*, *literature*, *indexes*, and *catalogs*. A combination subject search of *livestock* and *information services* also can retrieve appropriate literature guides.

When searching indexes and databases for literature guides, different strategies are needed to produce relevant records. Specify the subject words and combine them with *literature review*. The *CAB Thesaurus* uses the term *reviews* as the appropriate term. This search produces review articles explaining historical and current research with an extensive bibliography covering the subject. Several of these review articles are listed under the history section of this chapter. Chapter 4 also includes additional literature guides.

Olsen, W.C. ed. 1993. *The Literature of Animal Science and Health*. Ithaca, NY: Cornell University Press. This comprehensive work examines the literature and covers monographs, journals, and reference works for livestock production.

## Handbooks and Manuals

Handbooks are excellent reference tools covering practical and condensed information on a topic for a particular audience. Listed here are general livestock handbooks and manuals. Handbooks for a specific livestock, such as sheep, are listed under that particular species heading later within this chapter. Searching approaches to locate handbooks and manuals can include library catalogs, indexes, and databases. Monographs or chapters from monographs are usually given a subject heading of *handbooks*. Combining this term with a particular livestock species, such as, *sheep* or *livestock* will retrieve relevant references. Works by Church are standard classics in feeds for and feeding of livestock. M. Eugene Ensminger also has authored many handbooks on general livestock raising and on individual livestock. His works have endured in popularity through many decades and are included here. This list is not comprehensive, but a representation of works dealing with different aspects of livestock production.

Battaglia, R.A. 2001. *Handbook of Livestock Management*. 3rd ed. Upper Saddle River, NJ: Prentice Hall. This resource is current with management practices and techniques of the more common livestock.

Church, D.C. 1991. *Livestock Feeds and Feeding*. 3rd ed. Englewood Cliffs, NJ: Prentice-Hall.

Church, D.C. and W.G. Pond. 1988. *Basic Animal Nutrition and Feeding*. 3rd ed. New York: John Wiley. The 2 handbooks by Church cover the basics on feeds and nutrition and are useful as introductory works.

Ensminger, M.E., J.E. Oldfield, and W.W. Heinemann. 1990. *Feeds & Nutrition*. 2nd ed. Clovis, CA: Ensminger Pub. Co. Formerly *Feeds and Nutrition, Complete*. This book is a more comprehensive work on feeds and nutrition than the two mentioned above by Church et al.

Ensminger, M.E. 1991. *Animal Science*. 9th ed. Danville, IL: Interstate Publishers. Considered by some to be a classic, this is a comprehensive handbook covering all aspects of animal science.

*Grazing Reference Materials Manual*. 1994. Madison, WI: University of Wisconsin, Cooperative Extension Division and College of Agricultural and Life

Sciences. A grazing handbook covering various aspects of pastures and forage management, types of forage, and nutrition of forage for swine, sheep, and cattle.

Aumaitre, A.L., A.J. van der Zijpp, and In K. Han. (eds.). 1999. Animals, Animal Products and their Contribution to the Quality of Human Life. *Livestock Production Science* 59:(2–3, 95–206). The nine articles covering livestock production systems review current practices, regulations, human and animal nutrition, and food needs. These articles provide insight into worldwide differences in the quality and safety of animal products.

*Handbook of Australian Meat.* 1998. 6th ed. South Brisbane, Australia: AUS-MEAT. A thorough work about the Australian meat industry, covering standards, cuts, and terminology. Meats discussed include beef, veal, sheep meat, fancy meats, processed meats, goat, and buffalo.

Putnam, P.A. 1991. *Handbook of Animal Science.* San Diego, CA: Academic Press. The book starts with an overview of the animal industry, while other chapters cover managing livestock based on the specific products produced by these livestock.

Ranken, M.D. 2000. *Handbook of Meat Product Technology.* Oxford: Blackwell. Information about the latest technology and issues for slaughterhouses, meat handling, and safety.

Swallow, B.M. 1990. *Strategies and Tenure in African Livestock Development.* Madison, WI: University of Wisconsin-Madison, Land Tenure Center. This important monograph on African livestock production outlines relevant studies conducted in various African nations.

Theodorou, M.K. and France, J. 1999. *Feeding Systems and Feed Evaluation Models.* Wallingford, UK: CAB International. The main themes of the book are methods of feed evaluation, current feeding systems, and mechanistic mathematical modeling.

## Histories

As mentioned earlier, the history of livestock production may appear as literature reviews in an article format, as bibliographies within classical works, or as compilations of scientists' works. Examples of each of these are listed.

Harris, D.L. 1998. ''Livestock Improvement: Art, Science or Industry?'' *Journal of Animal Science* 76(9): 2294–2302. This article provides a discussion of the shifts in perspective on the selection for livestock improvement that have taken place over the last 70 years.

Cunningham, M. and D. Acker. 2000. *Animal Science & Industry*. 6th ed. Paramus, NJ: Prentice Hall PTR. A thorough discussion of the latest developments and research in the animal industry.

Davies, A. and R. Board, eds. 1998. *The Microbiology of Meat and Poultry*. London: Blackie Academic & Professional. The microbiology of the food industry, especially related to meats and food safety and handling issues.

Gillespie, J.R. 1998. *Animal Science*. Albany, NY: Delmar Thomson Learning. Comprehensive text discussing all areas of livestock production and more; especially good for novices or new students of animal science.

Kinsman, D.M., A. Kotula, and B.C. Breidenstein, eds. 1994. *Muscle Foods*: *Meat, Poultry, and Seafood Technology*. New York: Chapman & Hall.

Lawrie, R.A. 1998. *Lawrie's Meat Science*. 6th ed. Cambridge: Woodhead. This work is a classic, with previous editions titled *Meat Science*. Lawrie has authored numerous books covering various aspects of meat science.

Lee, J.S. 1996. *Introduction to Livestock and Poultry Production*: *Science and Technology*. Danville, IL: Interstate Pub.

Payne, W. JA and R.T. Wilson. 1999. *An Introduction to Animal Husbandry in the Tropics*. 5th ed. Oxford: Blackwell.

Romans, J.R. 2001. *The Meat We Eat*. 14th ed. Danville, IL: Interstate Publishers. A comprehensive work on meat processing, curing, marketing, storing, packing, equipment used, preparing, and consuming; covers the major livestock: cattle, sheep, swine, and poultry.

## Proceedings

Associations, societies, agencies, and universities are the main producers of proceedings resulting from conferences or seminars. Full proceedings by these entities can be located by searching library catalogs. Individual papers are only found through abstracts and indexes. In addition, association names can be searched in library catalogs by specifying the names as corporate authors of the proceedings. Many more organizations are listed in Chapter 4.

Searching can be accomplished using many different combinations of subjects and keywords. In a library catalog such as OCLC's *WorldCat*, different terms net different results. Examples of combinations from *WorldCat* with number of retrieved citations follow:

> title: proceedings and [(title: livestock and title: production)] 87
> title: proceedings and [(title: animal and title: science)] 84
> title: proceedings and [(author: animal and author: science)] 174

title: proceedings and [(author: livestock and author: production)] 37
title: proceedings and author: livestock 321
title: proceedings and subject: livestock 687
title: proceedings and [(subject: animal and subject: science)] 9

Searching abstracts and indexes is slightly different from searching library catalogs, mainly because indexes will produce records for both the entire proceedings and individual papers from the proceedings. Examples of search results are listed.

AgNIC (http://www.agnic.org/mtg/abs.html). Agricultural Conferences, Meetings, Seminars Calendar: Abstracts or Proceedings. A directory-type list of abstracts and proceedings available in chronological order by meeting dates. For the years 2000 and 2001 there is a subcategory: Animals. For previous years, scroll down the listings until an appropriate meeting is located.

American Society on Animal Science. Proceedings. Champaign, IL: American Society on Animal Science. 1ˢᵗ– . 1953– . Also published as the *Journal of Animal Science Supplement*.

Bottcher, R.W. and S.J. Hoff, eds. 1997. *Livestock Environment V: Proceedings of the Fifth International Symposium*. St. Joseph, MI: American Society of Agricultural Engineers. Proceedings from the American Society of Agricultural Engineers; earlier ones are cataloged beginning with the same two words in the title (*livestock environment*). Covers all aspects of environmental needs of livestock, such as housing, ecology, and hygiene.

## Statistics

Locating statistics on livestock production can be straightforward as national and international governments regularly publish statistics on animal production. Examples of major government references are listed. Searching in library catalogs for statistics can be simply a combination of the term *statistics* with the other topics of interest, for example, *cattle* or *livestock* or *meat industry* or *wool industry*, and so forth. The Agricultural Economics chapter also includes an excellent section on statistics.

Agricultural Market Information Virtual Library (http://www.aec.msu.edu/agecon/fs2/market/Welcome.htm). Lansing, MI: Michigan State University, Department of Agricultural Economics. Market information sources available through the Internet: daily to yearly market and outlook reports, prices, commodities, and quotes.

Canada. Agriculture and Agri-Food. Market and Industry Services Branch Online Information (http://ACEIS.AGR.CA/misb.html). Agricultural production and food safety and consumption statistics from Canada.

Canada. Dominion Bureau of Statistics. 1966. *Handbook of Agricultural Statistics Pt. 6: Livestock and Animal Products.* v. 1– . 1966– .

International Livestock Research Institute. 2000. *Handbook of Livestock Statistics for Developing Countries.* Nairobi, Kenya: International Livestock Research Institute.

United Nations. Food and Agriculture Organization. FAOSTAT (http://apps.fao.org/). ''FAO Statistical databases are online and multilingual currently containing over 1 million time-series records covering international statistics in the following areas: Production; Trade; Food Balance Sheets; Fertilizer and Pesticides; Land Use and Irrigation; Forest Products; Fishery Products; Population; Agricultural Machinery; Food Aid Shipments.''

USDA Economics and Statistics System. List by Subject: Livestock, Dairy, and Poultry (http://usda.mannlib.cornell.edu/reports/nassr/livestock/). The datasets and statistics for livestock production in the United States. Covers all traditional livestock and products plus others such as mink or mohair.

USDA National Agricultural Statistics Service. Agricultural Statistics (http://www.usda.gov/nass/pubs/agstats.htm). Annual compilations of agriculture statistics including livestock production in the United States.

U.S. Foreign Agricultural Service and U.S. World Agricultural Outlook Board (http://purl.access.gpo.gov/GPO/LPS1770). *Dairy, World Markets and Trade.* Washington, DC: U.S. Government Printing Office (Semiannual, 1994– ). For current issue, scroll down and click on **Dairy**: **World Markets and Trade**. Also available are archive issues: click on **Archives**, then select a title from the list, and finally click on an issue to view. This title is also a part of the Circular Series from the USDA Foreign Agricultural Service.

U.S. Foreign Agricultural Service and U.S. World Agricultural Outlook Board (http://ffas.usda.gov/livestock%5Farc.html). Livestock and Poultry, World Markets and Trade. Washington, DC: U.S. G.P.O. (Semiannual, 1994– ). This title is also a part of the Circular Series from the USDA Foreign Agricultural Service.

## Theses

Search indexes and databases, such as Dissertation Abstracts or the general agricultural databases mentioned at the beginning of this chapter, for theses and dissertations. Use the terms *thesis* or *dissertation* within the search strategy as a keyword or subject. Also, many university library catalogs include records for these works published at their institutions. In the case of this literature format, indexing and subject terms seem to be more specific than terms given to a handbook or dictionary. Searching for theses or dissertations on livestock in CAB

Abstracts might start with the broad CABI CODE "LL120 Meat Producing Animals," and include a search string "pt = thesis" to cover thesis or dissertation publication types.

## Trade Literature

Trade literature is oriented toward real-life situations, be it marketing a product, adding value to a raw commodity, or identifying new trends in the care of a product. Searching the business index Prompt (electronic form of *F&S Index*) for trade literature will save time and effort. Some publications indexed by Prompt include *AgExporter, Agra Europe, AgraFood East Europe, Feedstuffs, Biotech Business, Food Chemical News, Food Processing (USA), Food Processing (UK), Food Technology, Genetic Engineering News, Meat Processing* and many more. Industry organizations are a good source for trade literature; many have Web sites with the latest news for the industry. Searching library catalogs using subject terms such as *wool industry, meat industry,* and *poultry industry* will retrieve monographs with historical trade information.

American Meat Institute (http://www.meatami.com/). This institute is the oldest in the United States and the largest meat and poultry trade association. The American Meat Institute (AMI) exists to increase the efficiency, profitability, and safety of meat and poultry trade worldwide.

Gracey, J.F., D.S. Collins, and R.J. Huey. 1999. *Meat Hygiene*. 10th ed. London: W.B. Saunders. Meat-industry hygiene and health requirements and restrictions for handling meat.

Matthews, D.D. 1999. *Food Safety Sourcebook: Basic Consumer Health Information About the Safe Handling of Meat, Poultry, Seafood, Eggs, etc*. Detroit, MI: Omnigraphics. An introduction to food safety, guidelines for handling all types of food, food-borne illnesses, and listings of resources for more information on food industry, trade, inspection, and handling of food.

*World Poultry*. v. 13, no. 10– . 1997– . Doetinchem, The Netherlands: Elsevier International Business Information. A monthly trade journal with production, processing, and marketing information for the poultry industry worldwide. Formerly called *Misset World Poultry*.

## SUBJECT-RELATED LITERATURE
### Aquaculture

Canter, L.M. and A.T. Young. 2001. *Directory of Aquaculture Related Associations and Trade Organizations*. Beltsville, MD: U.S. National Agricultural Library. Available from the World Wide Web at http://www.mda.state.md.us/aqua/org.htm.

FAO. 2001. *Fisheries*. Rome: FAO. Available from the World Wide Web at http://www.fao.org/fi/default.asp.

Lovell, T. 1989. *Nutrition and Feeding of Fish*. New York: Van Nostrand Reinhold.

Stickney, R.R., ed. 2000. *Encyclopedia of Aquaculture*. New York: John Wiley. A comprehensive one-volume encyclopedia covering all aspects of aquaculture.

Stickney, R.R. 1994. *Principles of Aquaculture*. New York: John Wiley. An excellent handbook covering world aquaculture including feeding, breeding, water quality, economics and management issues, harvesting, meat quality, diseases, and laws and regulations.

## Beef Cattle

Ensminger, M.E. and R.C. Perry. 1997. *Beef Cattle Science*. 7th ed. Danville, IL: Interstate Publishers. Ensminger, M.E. 1992. *The Stockman's Handbook*. 7th ed. Danville, IL: Interstate Publishers. These two classics from Ensminger are comprehensive beef-production handbooks.

Kaus, R., P. Carroll, and J.W. Lapworth. 1997. *The Stockman's Handbook*. 6th ed. Townsville, Australia: Department of Primary Industries. A modern and thorough handbook on Australian stock raising practices.

National Cattlemen's Beef Association (U.S.). 1999. *Cattle and Beef Handbook*: *Facts, Figures and Information*. Englewood, CO: National Cattlemen's Beef Association. Available from the World Wide Web at http://www.beef.org/library/handbook/index.htm. Chapters cover nutrition, welfare, environment, food safety, and economics.

National Research Council (U.S.). Subcommittee on Beef Cattle Nutrition. 2000. *Nutrient Requirements of Beef Cattle*. 7th rev. ed. Washington, DC: National Academy Press.

## Dairy

American Dairy Science Association. 2000. *Dairy Management Practices, Housing, and Cattle Health*: *a Scientific Reader*. Savoy, IL: American Dairy Science Association.

Bath, D.L., et al. 1985. *Dairy Cattle*: *Principles, Practices, Problems, Profits*. 3rd ed. Philadelphia, PA: Lea & Febiger.

Ensminger, M.E. 1993. *Dairy Cattle Science*. 3rd ed. Danville, IL: Interstate Publishers. An excellent resource is a classic dairy handbook for all farmers.

Gietema, B. 1999. *Modern Dairy Farming in Tropical and Subtropical Regions*. Wageningen, The Netherlands: STOAS. A comprehensive three-volume set about all aspects of dairy production in the tropics.

Kurmann, J.A., J.L. Rašic, and M. Kroger. 1992. *Encyclopedia of Fermented Fresh Milk Products: an International Inventory of Fermented Milk, Cream, Buttermilk, Whey, and Related Products*. New York: Van Nostrand Reinhold. A comprehensive work about milk and products derived from milk.

National Research Council (U.S.). Subcommittee on Dairy Cattle Nutrition. 2001. *Nutrient Requirements of Dairy Cattle*. 7th rev. ed. Washington, DC: National Academy of Sciences.

Short, S.D. 2000. *Structure, Management, and Performance Characteristics of Specialized Dairy Farm Businesses in the United States*. Washington, DC: USDA Economic Research Service. Covering the economic aspects of dairy farming in the United States. This title is from the series *Agricultural Handbook*, (USDA) no. 720.

U.S. Cooperative State Research, Education, and Extension Service, and Wisconsin Milk Marketing Board. 2000. *Dairy Infobase Putting a World of Dairy Information at Your Fingertips*. Verona, WI: ADDS Center. "Includes over 1,100 extension articles and research papers from 35 universities and private industry groups from across the U.S." CD-ROM data file of full text articles and papers. Also, sample version available at http://www.adds.org/, click on **InfoBases**.

## Horses (Equine)

Boulet, J.C. 1997. *Multilingual Dictionary of the Horse = Dictionnaire multilingue du cheval: English, français, español, Deutsch, Latin*. 2nd ed. Cap-Rouge, Quebec, Canada: J.-C. Boulet.

Cross, P.L. and J.H. Brown. 1988. *Horse Business Management Reference Handbook*. 2nd ed. Warwickshire, UK: Warwickshire College of Agriculture. Covers the management and economics of raising and breeding horses.

Ensminger, M.E. 1999. *Horses and Horsemanship*. 9th ed. Danville, IL: Interstate Publishers. Covering all aspects of horses and horsemanship, including training and handling, making an excellent comprehensive handbook.

Ensminger, M.E. 1991. *Horses and Tack*. rev. ed. Boston: Houghton.

Hendricks, B. 1995. *International Encyclopedia of Horse Breeds*. Norman, OK: University of Oklahoma Press.

National Research Council (U.S.). Subcommittee on Horse Nutrition. 1989. *Nutrient Requirements of Horses*. 5th rev. ed. Washington, DC: National Academy of Sciences.

Price, S.D. 2000. *The Horseman's Illustrated Dictionary*. New York: Lyons Press.

Sponenberg, D.P. 1996. *Equine Color Genetics*. Ames, IA: Iowa State University Press. A work about genetics and breeding for hair color and skin pigmentation.

## Poultry

Austic, R.E. and M.C. Nesheim. 1990. *Poultry Production*. 13th ed. Philadelphia, PA: Lea & Febiger.

Damerow, G. 2000. *Storey's Guide to Raising Chickens*. Pittsboro, NC: American Livestock Breeds Conservancy.

Ensminger, M.E. 1992. *Poultry Science*. 3rd ed. Danville, IL: Interstate Publishers. A classic handbook covering all poultry.

Mercia, L. 2000. *Storey's Guide to Raising Poultry*. Pittsboro, NC: American Livestock Breeds Conservancy.

Mountney, G.J. and C.R. Parkhurst. 1995. *Poultry Products Technology*. 3rd ed. Binghamton, NY: Haworth Press.

National Research Council (U.S.). Subcommittee on Poultry Nutrition. 1994. *Nutrient Requirements of Poultry*. 9th rev. ed. Washington, DC: National Academy of Sciences.

Richardson, R.I. and G.C. Mead, eds. 1999. *Poultry Meat Science*. Poultry Science Symposium Series, v. 25. Papers from the 25th Poultry Science Symposium, held at the University of Bristol, September 17–19, 1997. Wallingford, UK: CABI Pub.

## Sheep and Goats

Belanger, J. 2000. *Storey's Guide to Raising Dairy Goats*. Pittsboro, NC: American Livestock Breeds Conservancy.

Dey, D.H. 2000. *Meat Goat Profit$: Profit Planning Tools in an Alberta Startup Meat Goat Enterprise*. Edmonton, Canada: Alberta Agriculture, Food, and Rural Development. Bulletin about raising and marketing goat meat.

Gatenby, R.M. 1986. *Sheep Production in The Tropics and Subtropics*. London and New York: Longman.

Haenlein, G.F.W. 1996. "Status and Prospects of The Dairy Goat Industry in the United States." *Journal of Animal Science*. 74(5): 1173–1181. Covers statistical dimensions, infrastructure, national programs, milk quality standards, prospects, and implications.

*Meat Goat Industry*. 1996. Edmonton, Canada. Alberta Agriculture, Food and Rural Development. Another good resource on goat meat.

National Research Council (U.S.). Subcommittee on Goat Nutrition. 1981. *Nutrient Requirements of Goats: Angora, Dairy, and Meat Goats in Temperate and Tropical Countries*. Washington, DC: National Academy Press.

National Research Council (U.S.). Subcommittee on Sheep Nutrition. 1985. *Nutrient Requirements of Sheep*. 6th rev. ed. Washington, DC: National Academy of Sciences.

Sadler, R.D. 1994. *Wool: a Glossary of Wool Terms Used From the Grower to the Spinner*. Newton, MA: R.D. Sadler.

Simmons, P. 2000. *Storey's Guide to Raising Sheep*. Pittsboro, NC: American Livestock Breeds Conservancy.

## Swine

Andresen, N. 2000. *The Foraging Pig: Resource Utilization, Interaction, Performance and Behaviour of Pigs in Cropping Systems*. Uppsala, Sweden: Swedish University of Agricultural Sciences. Discusses alternative forage system for raising pigs.

Ensminger, M.E. and R.O. Parker. 1997. *Swine Science*. 6th ed. Danville, IL: Interstate Publishers. A comprehensive handbook on swine.

Hansen, J.A. *Swine Nutrition Guide*. Raleigh, NC: North Carolina State University. Available from the World Wide Web at http://mark.asci.ncsu.edu:80/NUTRIT~1/NUTRIT~1/NUTRGUID.HTM. Excellent resource for feeds and feeding of swine covering basic nutritional needs and including sample diets. Some information was taken from the Kansas State University Extension publication *Swine Nutrition* published in 1998.

Harris, D.L. 2000. *Multi-Site Pig Production*. Ames, IA: Iowa State University Press.

Klober, K. 2000. *Storey's Guide to Raising Pigs*. Pittsboro, NC: American Livestock Breeds Conservancy.

Midwest Plan Service. 1983. *Swine Housing and Equipment Handbook*. 4th ed. Ames, IA: Midwest Plan Service.

National Pork Producers Council. 2000. Ag & Swine Links (http://www.nppc.org/PROD/other.html). Provides a starting point for swine information on the Web.

National Research Council (U.S.). Subcommittee on Swine Nutrition. 1998. *Nutrient Requirements of Swine*. 10th rev. ed. Washington, DC: National Academy of Sciences.

*Recent Advances in Swine Production and Health*. v. 1– . 1991– . St. Paul, MN: University of Minnesota, University of Minnesota Swine Center.

U.S. Cooperative State Research, Education, and Extension Service, and Wisconsin Milk Marketing Board. 1999. *The National Pig Information Database*. Verona, WI: ADDS Center. Available from the World Wide Web at http://www.adds.org/.

Whittemore, C.T. 1987. *Elements of Pig Science*. Harlow, UK: Longman.

## Other Livestock

Alligator Bob's Gourmet Alligator (http://www.gatorbob.com/). Thonotosassa, Florida. Web site covering alligator meat and production.

Bison and Other Alternative Livestock (http://www.lib.ndsu.nodak.edu/subjects/ag/bison.htm). North Dakota State University Libraries.

Central Australian Camel Industry Association. 1997. *Camel: Selected Meat Cuts and Information*. Alice Springs, Australia: Central Australian Camel Industry Association. Bulletin listing meat cuts and nutritional value as a food source.

Elsey, R.M., T. Joanen, and L. McNease. 1994. *Louisiana's Alligator Research and Management Program*. Results of a conference held in Pattaya, Thailand. Includes breeding and farming alligators in the state of Louisiana.

Groves, P. 1992. *Muskox Husbandry: A Guide for the Care, Feeding, and Breeding of Captive Muskoxen*. Biological Papers of the University of Alaska. Special Report, no. 5. Institute of Arctic Biology. Fairbanks, AK: University of Alaska Fairbanks.

Hill, D.H. 1988. *Cattle and Buffalo Meat Production in the Tropics*. Harlow, UK: Longman.

Holderread, D. 2000. *Storey's Guide to Raising Ducks*. Pittsboro, NC: American Livestock Breeds Conservancy.

Ligda, D.J. 1996. The Water Buffalo. (http://ww2.netnitco.net/users/djligda/waterbuf.htm).

McKay, J. 1989. *The Ferret and Ferreting Handbook*. Crowood. Because no location of publisher is mentioned within the book, the ISBN (1852231807) might be helpful for acquiring or borrowing this book. Comprehensive text on raising ferrets for meat and pelt production.

Minnaar, M. 1998. *The Emu Farmer's Handbook Commercial Farming, Methods for Emus, Ostriches & Rheas*. Groveton, Australia: Nyoni Publishing Company. Management and production of domestic birds, covering care, breeding, hygiene, facilities, and health.

National Research Council (U.S.). Subcommittee on Furbearer Nutrition. 1982. *Nutrient Requirements of Mink and Foxes*. 2nd rev. ed. Washington, DC: National Academy of Sciences.

National Research Council (U.S.). Subcommittee on Rabbit Nutrition. 1977. *Nutrient Requirements of Rabbits*. 2nd rev. ed. Washington, DC: National Academy of Sciences.

Queensland. Department of Primary Industries, Rural Industries Research and Development Corporation; and Northern Territory. Department of Primary Industry and Fisheries. 1996. *Handbook of Australian Crocodile Meat*. Brisbane, Australia: Queensland Department of Primary Industries.

Sandford, J.C. 1986. *The Domestic Rabbit*. 4th ed. London; Dobbs Ferry, NY: Collins.

Tulloh, N.M. and J.H.G. Holmes, eds. 1992. *Buffalo Production*. Amsterdam: Elsevier. A world treatise on buffalo production, distribution, ecology, adaptation, genetics and breeding, nutrition, growth and development, and management of varieties of buffalo worldwide.

## BIBLIOGRAPHY

Bird JE, Smith JC. A Selected Annotated Bibliography of Agricultural Information. *Library Trends* 1990, 38(winter 1990): 517–541.
Crawford RL. 1999. Selected Web Sites for Biomedical, Pharmaceutical, Veterinary, and Animal Science (http://www.nal.usda.gov/awic/pubs/awic9802.htm) cited [July 17, 2001]. Beltsville, MD: National Agricultural Library.
Hurt RD, Hurt ME. Animal Sciences. In: The History of Agricultural Science and Technology: an International Annotated Bibliography. New York: Garland Publishing. 1994, pp. 241–271.
Jensen RD, Lamb C, Smith NM (compilers). Agricultural and Animal Science Journals and Serials: an Analytical Guide. Westport, CT: Greenwood Press, 1986.
Lean IJ, Campling RC. Animal Production. In: Lilley GP, ed. Information Sources in Agriculture and Food Science. London: Butterworths, 1981, pp. 407–417.

Merala M, Kimball J, and Blanchard JR. Animal Sciences. In: Blanchard, JR, Farrell L, eds. Guide to Sources for Agriculture and Biological Research. Berkeley, CA: University of California, 1981, pp. 283–378.

Olsen WC, ed. The Literature of Animal Science and Health. Ithaca, NY: Cornell University Press, 1993.

# 6

# Environmental Sciences and Natural Resources

**Robert S. Allen**
University of Illinois at Urbana-Champaign, Urbana, Illinois, USA

Environmental sciences and natural resources are broad and multidisciplinary fields. Every agricultural subdiscipline has an environmental component. There are environmental concerns in farm management, agricultural economics, crop sciences, animal science, agricultural engineering, horticulture, food science, and so on. There are also branches of environmental science that are beyond agriculture and unique to other fields. The focus of this chapter will be to introduce the most significant current resources of agricultural information that relate to environmental science and natural resources.

More precisely defined, environmental science is the multidisciplinary study of the environment. The inquiries may be grounded in the natural, biological, or social sciences and often draw on the literature of multiple disciplines. Related to this field of study are natural resources, those naturally occurring resources used to support human life. Agricultural natural resources are usually considered to be forests, soil, water, fish, wildlife, and the atmosphere. Natural resources can be classified as animate or inanimate, plant or animal, and renewable or nonrenewable. Environmental sciences typically involve an interplay of the actions of man and the effect of these actions on natural resources.

It is important to realize that there is nothing completely unique about the type of literature that reflects environmental sciences and natural resources. As with other disciplines, there are journals, books, reference sources, and so forth. The unique feature of these subjects is their breadth and extensive array of interrelated topics. As a great deal of environmental science literature is contained within other disciplines, it maintains the characteristics of the literature of those disciplines. For the purposes of this chapter, the scope will include broad coverage of environmental science in agriculture along with natural resource management.

## SEARCHING THE LITERATURE

Environmental sciences and natural resources in agriculture are composed of an extremely broad and multidisciplinary range of subjects. To effectively search the literature, many available resources must be examined. It is interesting to consider some of the potential topics that might be researched on agriculture-related environmental science subdisciplines.

Agricultural economics is a useful example as it is a discipline unto itself. To successfully research this area requires checking both economic and agricultural resources. In addition, there are many environmental aspects of agricultural economics. A typical subject scenario could be: "what are the economic considerations of environmentally friendly farm management?" To research this topic you might begin with determining examples of environmentally friendly techniques employed in farm management. However, it might also be appropriate to follow this with an assessment of those that are not environmentally friendly. Economic literature would need to be checked, as well as agricultural literature, public policy literature, and general literature.

Another typical topic of environmental science in agriculture is the application of fertilizer or pesticides in farming operations. These potentially toxic and hazardous chemicals are often spread over large areas, making them a form of nonpoint source pollution. This type of pollution is different from point source pollution, which is a release of pollutants from a single source, such as the emission stack of an incinerator. Fertilizers and pesticides are important for modern agriculture, but there is potential for these harmful chemicals to contaminate groundwater and wetland resources from the diffused runoff that occurs during rainfall. This subject, then, may mix the literature of chemistry, civil engineering, geology, biology, ecology, public policy, social sciences, and agriculture.

Most people doing environmental research in agricultural fields will need to consult basic reference works from time to time. This is especially important, as an expert in agricultural research may be a novice in environmental science research. Of course, the tables may be turned and an expert in environmental research may well be a novice in agricultural information sources. Combining two multidisciplinary fields means that a great many new terms and information resources must be examined. It is encouraging that a large number of environmental science reference sources are now available to aid in this quest. It is the intent of this chapter to introduce some of the best and most recent of these.

Scholarly research of an advanced nature will require checking appropriate indexes to journal literature. There are two possible paths to take in this case. The first is to use a single multidisciplinary index, such as Web of Science. This will adequately cover many of the desired disciplines, but may not be as exhaustive as needed. The second path is to use separate subject-specific indexes. This will cover the discrete subjects, but will probably miss a great deal of the multidisciplinary material that is of interest, but tangential to the index's subject. The best method for researching journal literature will require following both paths. Books also may be of interest to these advanced researchers, but are generally less important than journal literature.

The multidisciplinary nature of environmental sciences makes traditional controlled vocabulary searching difficult in many instances. The large number of potentially appropriate databases also makes controlled vocabulary searching difficult to manage, as there is no single controlled vocabulary used by all databases. If an index does have a thesaurus available, then it is useful to research the available terms that define the topic under investigation. Multidisciplinary subjects often require examining a topic from a variety of perspectives and with a great deal of flexibility in the terminology selected.

Literature available on the Web is also important for environmental science research. This is especially true for publications released by government agencies and nonprofit organizations. It is common for these two groups to do the bulk of their publishing on the Web, where the cost of publication and dissemination is much less than traditional paper publishing. An entity such as the U.S. Environmental Protection Agency has a large number of publications available on their many Web sites. It is not always easy to locate exactly the desired information, but it is expected that the Web will become even more important in future years for disseminating information of all types. To facilitate identification of relevant electronic resources, there is a listing of related governmental and nongovernmental organization Web sites included in this chapter.

## ABSTRACTS, INDEXES, AND DATABASES

Scholarly communication in agriculture-related environmental science and natural resource fields is manifested in a fashion similar to other academic disciplines. Journal literature is particularly important to the subdiscipline, and is the primary vehicle for scientific communication. Most disciplines have a unique index that covers their specific area in complete detail. However, effective use of journal literature in an interdisciplinary field requires searching a number of periodical indices to identify relevant information. Environmental science, in particular, will require a researcher to query a number of sources to completely cover the topic. The intent of this section is to introduce a number of bibliographic databases that can be used for researching subjects in agricultural/environmental topics. It is difficult to describe the actual online access to these resources, as it varies depending on how the database is being accessed. Factors influencing access are the proprietary nature of the search software being used, the extent of backfiles purchased, and institutional restrictions on use of licensed products. These resources typically have an extensive paper counterpart that is available for many years prior to online availability.

Databases with broad subject coverage such as AGRICOLA, BIOSIS, CAB Abstracts, and Web of Science contain a wealth of information in the areas of environmental science and natural resources. As mentioned, these topics can be challenging to search because they are general concepts and have no concrete vocabulary to base a search on. A great deal of time can be saved in searching these or other databases by consulting with an information professional to help with the vocabulary and the subject codes. Consulting the *CAB Thesaurus* yields the following Concept Codes that can be useful starting points for conducting literature searches in CAB Abstracts and AGRICOLA.

> Biomass
> Conservation
> Environment
> Environmental Degradation
> Environmental Factors
> Environmental Impact
> Environmental Legislation
> Environmental Management
> Environmental Policy
> Environmental Protection
> Fishery Resources
> Forest Management
> Forest Policy
> Forest Resources
> Forestry
> Forest Trees
> Groundwater
> Groundwater Pollution
> Natural Resources
> Nonrenewable Resources
> Pollutants
> Pollution

Pollution Control
Renewable Resources
Resources
Soil
Soil Conservation
Soil Management
Water
Water Conservation
Water Management
Water Pollution
Water Quality
Water Resources
Wildlife
Wildlife Conservation
Wildlife Management

*Applied Science and Technology Index*. 1913– . New York: H.W. Wilson.   Good general undergraduate-level coverage of physical science and engineering type information on environmental science. This index covers many commonly available journals and is a useful resource for looking at the effects of toxic chemicals on the environment and methods used to remediate pollution. Coverage includes groundwater pollution and hydrologic modeling. Available online through various services.

*Biological and Agricultural Index*. 1916– . New York: H.W. Wilson.   General undergraduate level coverage of life science and agricultural-type information on environmental science. This is especially appropriate for aspects having to do with living organisms, animal waste management, pesticides, fertilizers, and agricultural pollution in general. Available online through various services.

*Biological Abstracts*. Philadelphia, PA: BIOSIS. 1926– .   Comprehensive index to biological literature, with good coverage of ecology and environmental science and natural resources. Includes subjects having to do with biodiversity, entomology, and the influences of environmental changes on the ecosystem. Available in electronic format from 1969 through various vendors.

*Chemical Abstracts*. 1907– . Columbus, OH: American Chemical Society, Chemical Abstracts Service.   A comprehensive index to the chemical literature, this resource includes publications on the chemical aspects of environmental science, such as pesticides and other chemical toxins. *Chemical Abstracts* is especially important for its use of CAS registry numbers. These numbers are unique identifiers representing chemicals that may be important for their environmental influence. A number of other databases also employ these numbers to identify chemical products. Available electronically through the publisher.

*EI Compendex (Engineering Index)*. 1919– . Hoboken, NJ: Engineering Information, Inc. A comprehensive engineering index that contains excellent coverage of environmental engineering and related literature, including groundwater dynamics and hydrology. It is also a resource for locating information about agricultural engineering and soil sciences. The topic of civil engineering is particularly relevant to environmental concerns. Available electronically through the Engineering Village 2.

*Environment Abstracts*. 1971– . Bethesda, MD: Congressional Information Service; Lexis-Nexis. Interdisciplinary coverage of environmental literature. Includes scientific, economic, social, political, economic, and legal issues. International in scope, it covers more than 5,000 primary and secondary source publications including journals, technical reports, conference proceedings, and government documents. The companion Envirofiche collection provides the full text of the majority of indexed materials.

*Environmental Knowledgebase*. 1972– . Santa Barbara, CA: International Academy at Santa Barbara. This is the online version of *Environmental Periodicals Bibliography*. Covers approximately 400 journals in many disciplines of environmental sciences.

*Environmental Sciences and Pollution Management*. Bethesda, CA: Cambridge Scientific Abstracts. A subfile of Cambridge Scientific Abstracts, this includes citations from more than 4,000 scientific journals and other sources including conference proceedings, technical reports, books, and government publications. This multidisciplinary database is composed of the following subsets:

> *Agricultural and Environmental Biotechnology Abstracts*, 1993–
> *ASFA 3*: *Aquatic Pollution and Environmental Quality*, 1990–
> *Bacteriology Abstracts* (Microbiology B), 1982–
> *Ecology Abstracts*, 1982–
> *EIS*: *Digests of Environmental Impact Statements*, 1985–
> *Environmental Engineering Abstracts*, 1990–
> *Health and Safety Science Abstracts*, 1981–
> *Industrial and Applied Microbiology Abstracts* (Microbiology A), 1982–
> *Pollution Abstracts*, 1981–
> *Risk Abstracts*, 1990–
> *Toxicology Abstracts*, 1981–
> *Water Resources Abstracts*, 1967–

GEOREF. 1785–. Alexandria, VA: American Geological Institute. Database covering geological literature. Important for coverage of soils and groundwater environmental information. Covers over 3,000 journals and many international sources. Earliest years cover information only on North America, with international coverage added in 1933. Available electronically from various sources.

Groundwater and Soil Contamination Database. 1975– . Alexandria, VA: American Geological Institute. Includes over 60,000 references to documents in journals, conference proceedings, theses, dissertations, and government publications.

*PAIS—Public Affairs Information Service.* 1915– . Dublin, OH: OCLC Public Affairs Information Service. Index of political science information with some coverage of environmental issues. Especially good for policy and governmental environmental concerns from a social perspective. Available online through OCLC First Search.

Wildlife Worldwide. 1935– . Baltimore, MD: National Information Service Corporation. Online database that contains over 500,000 bibliographic records, some with abstracts. This superfile is composed of the following subfiles:

> *Biodoc*, 1970–99
> *IUCN—The World Conservation Union*, 1946–
> *Swiss Wildlife Information Service*, 1974–
> *Waterfowl and Wetlands Bibliography (Duckdata)*, 1838–2000
> *Wildlife Database*, 1960–
> *Wildlife Review Abstracts*, 1935–

## BOOKS

Books are considered by some to be less important to scholarly research than journal literature, especially in scientific and technical fields. The information contained in books is often less timely than that of journal articles. However, they can be of great importance for presenting overviews of research and are often assembled from many years of scholarly research. Scholarly books in environmental science and natural resources may be of the review sort mentioned or the published proceedings of conferences. (See Proceedings section in this chapter.)

Rather than referencing the many books that have been published in environmental sciences with an agricultural interest, a short list of appropriate Library of Congress Subject Headings and some of the major book series are presented. The series mentioned here are not an exhaustive profile of all possible books of interest on the subject. There are many valuable titles that either are not part of a series or are a part of a series that does not have a major environmental or agricultural focus. Serious researchers will need to consult a broad ranging catalog of books, either from a major research library such as the U.S. National Agriculture Library (NAL), a land-grant university library, or WorldCat from OCLC. Those interested in developing a book collection in this area will also benefit from a broad interdisciplinary book-selling agent with a Web-accessible search engine such as Amazon.com. Many interesting environmental science books are published by small publishing houses or organizations, and these publications can often prove elusive for information seekers and information professionals.

A great deal of information regarding books published in environmental sciences can be found in the Guides section of this chapter. Most of the available guides are somewhat dated, but some rather extensive lists of books published on environmental sciences can be found by looking through these guides. Using these guides along with the appropriate subjects and keywords in an online catalog will help identify the most desirable books on this subject.

To aid in finding books in traditional library catalogs, some of the most beneficial Library of Congress subject headings are presented in the following text. The use of Library of Congress Subject Headings is an imperfect method of finding books on an interdisciplinary and newer subject, but it is hoped that using these and similar headings can aid in successful discovery. Using well-chosen keywords can also aid in discovering books, and these records can then be examined to choose subject headings for similar books. The following list contains some suggested headings to use:

Agricultural Conservation
Agricultural Ecology
Agricultural Pollution
Agricultural Wastes—environmental aspects
Agriculture—environmental aspects
Bioremediation
Conservation of Natural Resources
Deforestation
Ecology
Environmental Chemistry
Environmental Degradation
Environmental Economics
Environmental Engineering
Environmental Ethics
Environmental Geochemistry
Environmental Law
Environmental Management
Environmental Monitoring
Environmental Policy
Environmental Protection
Environmental Sciences
Forest Management

Forests and Forestry
Man—influence on environment
Natural Resources
Nature Conservation
Sustainable Agriculture
Sustainable Forestry
Wildlife Conservation

As mentioned, the number of books published on the subject of environmental and natural resource aspects of agriculture is too substantial to cover in this short chapter. Some of the major book series and their publishers are given in the following text. This is by no means an exhaustive listing, and many books of great value are not part of any of these series. This is offered simply as a starting point to examine the books published on the subject. A representative sample of major book series is as follows:

Advances in Agroecology (Lewis)
Advances in Soil Science (CRC)
Agriculture and Environment Series (CRC/Lewis)
Air Quality Monographs (Elsevier)
ASAE Publication Series (American Society for Agricultural Engineering)
Biopesticides Series (CABI)
Books in Soils, Plants, and the Environment (Marcel Dekker)
Conservation Biology Series (Kluwer)
Developments in Agricultural and Managed-Forest Ecology (Elsevier)
Developments in Environmental Economics (Elsevier)
Developments in Environmental Monitoring (Elsevier)
Developments in Landscape Management and Urban Planning (Elsevier)
Ecosystems of the World (Elsevier)
Environment and the Human Condition (University of Illinois Press)
Environment and Policy (Kluwer)
Environmentally and Socially Sustainable Environment Series (World Bank)
Environmental Philosophies Series (Routledge)
Environmental Research Series (World Conservation Union)
Environmental Science and Technology Library (Kluwer)
Forestry Sciences (Kluwer)
Green Chemistry Series (Oxford University Press)
Man and Biosphere Series (CRC Press/Parthenon Publishers)
NATO Asi Series, Partnership Sub-Series 2, Environmental Security (Kluwer)
NATO Asi Series, Series I, Global Environmental Change (Kluwer)
Natural Resource Management and Policy (Kluwer)

Natural Resources and Environmental Issues (Utah State University)
New Horizons in Environmental Economics (Edward Elgar)
Our Sustainable Future Series (University of Nebraska Press)
Pesticides in the Hydrologic System (Ann Arbor Press)
Society of Environment Toxicology and Chemistry (Elsevier)
Soil and Environment (Kluwer)
Springer Series on Environmental Management (Springer)
Studies in Environmental Science (Elsevier)
Studies in Environmental and History (Cambridge University Press)
Sustainable Agricultural Network Handbook Series (Sustainable Agricultural Network)
Sustainable Rural Development Series (CABI)
System Approaches for Sustainable Agricultural Development (Kluwer)
Topics in Sustainable Agronomy (Oxford University Press)
Trace Metals in the Enviroment (Elsevier)
Waste Management Series (Elsevier)
Wiley Encyclopedia Series in Environmental Science (Wiley)
Wiley Series in Agrochemicals and Plant Protection (Wiley)
World Bank Environmental Papers (World Bank)
World Forests (Kluwer)

## DIRECTORIES

The following resources provide an entry into finding people and organizations involved in environmental studies and natural resources. These directories are typical of those found in library reference collections and can be used in conjunction with similar listings found on the Web.

Brackley, P., ed. 1990. *World Guide to Environmental Issues and Organizations.* Harlow, UK: Longman Current Affairs; Detroit, MI: Gale Research Co. 386 p. ISBN: 0582062705 (Longman) 0810383535 (Gale). Divided into four sections: issues; politics; conventions, reports, directives, and agreements; and organizations, the directory is divided into international, regional, and national segments. Distributed exclusively in the United States and Canada by Gale Research Company.

Buckley-Ess, J., ed. 1988. *A Directory of Natural Resource Management Organizations in Latin America and the Caribbean.* Washington, DC: Partners of the Americas; New York: Tinker Foundation, Inc. 205 p. Contains detailed presentations on countries within larger regions of the area. Each entry is divided into nongovernmental, governmental, and educational organizations. Prepared by Partners of the Americas with a grant from the Tinker Foundation.

Ekström, G., ed. 1994. *World Directory of Pesticide Control Organizations*. 2nd ed. Surrey, UK: British Crop Protection Council; Cambridge, Royal Society of Chemistry, Information Services. 423 p. ISBN: 0948404787. This is a directory of international and national pesticide control organizations, which contains chapters on pesticides as environmental pollutants and international efforts to promote chemical safety.

*The Environment Encyclopedia and Directory 1998*. 2nd ed. London: Europa Publications Limited. 560 p. ISBN: 185743028X. Divided into five sections, this directory includes a series of maps relevant to the environment, a listing of nearly 3,000 environmental organizations, and 1,000 environmental periodicals, a who's who directory of important environmentalists, and a lengthy list of definitions.

Gordon S. and D.B. Tunstall, eds. 1996. *World Directory of Country Environmental Studies: An Annotated Bibliography of Natural Resource Profiles, Plans, and Strategies*. Washington, DC: World Resources Institute. 272 p. ISBN: 1569730954. This annotated bibliography of natural resource profiles, plans, and strategies was produced by the International Environmental and Natural Resource Assessment Information Service. It is divided into international regions: Africa, Americas, Asia and Oceania, Europe and others, and is subdivided by country within that scheme.

Katz, L.S., S. Orrick, and R. Honig. 1993. *Environmental Profiles: A Global Guide to Projects and People*. New York: Garland. 1,083 p. ISBN: 0815300638. Published as part of the Garland Reference Library of Social Science series (v. 736), this alphabetical country-based directory of environmental contact also has an index based on broad subjects.

National Wildlife Federation. Annual. *Conservation Directory*. Washington, DC: National Wildlife Federation. ISSN:0069911X. Former title was *Directory of Organizations and Officials Concerned with the Protection of Wildlife and Other Natural Resources*. Provides a list of organizations, agencies, and officials concerned with natural resource use and management.

Shipp, S. 1997. *Rainforest Organizations: A Worldwide Directory of Private and Governmental Entities*. Jefferson, NC: McFarland & Company. 184 p. ISBN: 0786403810. There are over 120 entries in the directory providing contact information, organizational goals, activities, and publications. The four appendices include specific information on: U.S. tropical forests, international tropical timber agreement, history of a Puerto Rican forest, and the Hawaii Tropical Forest Recovery Act.

*Sustainable Agriculture Directory of Expertise*. 1996. Burlington, VT: Sustainable Agriculture Network. 3rd ed. ISBN: 1888626003. Has indexes based on geography, individuals, organizations, crop and livestock enterprises, areas of expertise, product and services, and management methods. There is an online fourth edition, though it has a much smaller number of entries than the paper version. Available from the World Wide Web at http://www.sare.org/expertise/.

Thiele, C.S., ed. 2000. *Environmental Grantmaking Foundations 2000.* 8th ed. Cary, NC: Resources for Global Sustainability, 2000. 1,061 p. ISBN: 0963194372. Profiles 889 grant programs relevant to environmental issues, providing detailed information about each funding source, sample past grants from the source, application process, special emphases, and restrictions. Gives listings in alphabetical order, but includes a number of indexes based on specific criteria.

United Nations. 1988. *ACCIS Guide to United Nations Information Sources on the Environment.* New York: United Nations. 141 pages. ISBN: 9211003393. Prepared by the Advisory Committee for the Co-ordination of Information Systems (ACCIS), in collaboration with the Programme Activity Centre of the International Environmental Information System (INFOTERRA PAC) of the United Nations Environment Programme (UNEP). This guide functions mainly as a directory to suppliers of environmental information and is divided into four sections: sources of environmental information within the UN system, categorized descriptions of the environmental information sources, other sources of UN information, and select producers of non-UN environmental information. Issued as part of the series, ACCIS Guides to United Nations Information Sources.

*World Directory of Environmental Organizations.* Irregular serial. Claremont, CA: California Institute of Public Affairs. ISSN: 00920908. Published by the International Center for the Environment and Public Policy, California Institute of Public Affairs, in cooperation with IUCN—The World Conservation Union, and initiated and endorsed by the Sierra Club. Latest irregular edition is 2001. Listings include addresses, telephone, fax, and Web sites. Entries for major organizations have detailed descriptions of programs and projects and are often several hundred words long. Contains a guide that identifies organizations working on specific topics. Also contains an index, a glossary, and a list of landmark world events in environmental protection.

## ENCYCLOPEDIAS

Encyclopedias provide a concise introduction into a subject as well as providing the basic vocabulary in a field. An encyclopedia is an excellent place to begin research in an interdisciplinary field as it provides the researcher with a broad overview of the subject. This makes encyclopedias an excellent resource for less scholarly users as well.

Allin, C.W. and R. McClenaghan, eds. 2000. *Encyclopedia of Environmental Issues*. 3 vols. Pasadena, CA: Salem Press. ISBN: 0893569941. Contains 475 alphabetically arranged authored entries. Most entries are broad in coverage and a few paragraphs in length.

Calow, P. 1998. *The Encyclopedia of Ecology & Environmental Management*. Oxford; Malden, MA: Blackwell Science. 805 p. ISBN: 0865428387. Contains nearly 3,000 entries in alphabetical order, 250 of which are substantial in length. Each entry is cross-referenced extensively and is attributed to a specific contributor with the use of initials.

Coyne, M.S., M. Adams, and C.W. Allin. 1998. *Natural Resources*. 3 vols. Pasadena, CA: Salem Press. ISBN: 0893569127. Broad encyclopedia covering entire spectrum of natural resources. Includes some coverage of agricultural issues, but focuses a great deal on mineral, energy, and nonrenewable resources.

Cunningham, W.P., ed. 1998. *Environmental Encyclopedia*. 2nd ed. Detroit; London: Gale. 1,196 p. ISBN: 081039314X. Brief entries are arranged alphabetically and cross-referenced using bold print. Includes two appendices: environmental chronology and summary of environment-related legislation.

Franck, I.M. and D. Brownstone. 1992. *The Green Encyclopedia*. New York: Prentice Hall General Reference. 485 p. ISBN: 0133656772. Ready-reference encyclopedia with over 1,000 short entries in alphabetical order. Contains a list of environmental acronyms and a list of related publications.

Herschy, R.W. and R.W. Fairbridge. 1998. *Encyclopedia of Hydrology and Water Resources*. Dordrecht; Boston: Kluwer Academic. 803 p. Includes maps. ISBN: 0412740605. This is a technical encyclopedia, which includes many involved mathematical formulas and detailed explanations. Contains an author and subject index, and a select list of international journals in hydrology and water resources, and a treatment of units, symbols, and conversion.

Johnson, N.C., R.C. Szaro, and W.T. Sexton, eds. 1999. *Ecological Stewardship: A Common Reference for Ecosystem Management*. Oxford: Elsevier Science. 3 vols. ISBN: 0080432069 (set). This work includes key findings (v. 1), biological and ecological dimensions including humans as agents of ecological change (v. 2), and social, cultural, legal, and economic dimensions as well as information and data management (v. 3). Based on the papers resulting from a two-week workshop in Tucson, Arizona, in December of 1995. Includes computer laser optical disc.

Meyers, R.A., ed. 1998. *Encyclopedia of Environmental Analysis and Remediation*. New York: Wiley. 8 vols. ISBN: 0471117080. Part of the series, Wiley Encyclopedia Series in Environmental Science, this work contains approximately

280 in-depth articles arranged in alphabetical order. Broad topics include air pollution control, environmental law and regulations, environmental sampling and analysis, hazardous waste remediation, pollution in the biosphere, and water reclamation. This is a technical encyclopedia set meant for academia or professional reference. Authors are noted for each article and cross-references are provided.

Mongilo, J.F. and L. Zierdt-Warshaw. 2000. *Encyclopedia of Environmental Science*. Phoenix, AZ: Oryx Press. 450 p. ISBN: 1573561479.   Good stand-alone encyclopedia suitable for undergraduate use. Gives short entries on many topics. Includes a bibliography of approximately 300 classic books on environmental science.

Nierenberg, W.A., ed. 1995. *Encyclopedia of Environmental Biology*. San Diego, CA: Academic Press. 3 vols. Includes maps. ISBN: 0122267303 (set).   Entries contain an outline, glossary, cross-references, and bibliography. Access is through a well-developed table of contents or an extensive subject index and a related article index.

Paehlke, R., ed. 1995. *Conservation and Environmentalism*: *An Encyclopedia*. New York: Garland Pub. 771 p. ISBN: 0824061012.   Part of the series, Garland Reference Library of Social Science, v. 645.

Schneider, S.H., ed. 1996. *Encyclopedia of Climate and Weather*. New York: Oxford University Press. 2 vols. 929 p. ISBN: 0195094859.   Contains entries of varying length in alphabetical order attributed to a particular author. Includes a glossary, an index, a directory of contributors, and a list of abbreviations and symbols.

## GUIDES TO THE LITERATURE

The guides listed here serve as introductions to the literature and information sources for the field of environmental sciences. Although some may be dated and lacking an agricultural focus, they are included here as potential resources for the beginning researcher. It is hoped that this chapter will serve as a useful addition to this short list.

Balachandran, S., ed. 1993. *Encyclopedia of Environmental Information Sources*: *A Subject Guide to About 34,000 Print and Other Sources of Information on All Aspects of the Environment*. Detroit, MI: Gale Research. 1,813 p. ISBN:0-8103-8568-89.   This guide is divided into disciplines and subdisciplines, with much duplication of resources between disciplines. It includes a section on agriculture.

Eagle, S. and J. Deschamps, eds. 1997. *Information Sources in Environmental Protection*. London; New Providence, NJ: Bowker-Saur. 280 p. ISBN: 1857390628.   Written in expository rather than outline style, this book is divided into four sections: official and unofficial; effects and the affected; practicalities; and controls and public awareness. Part of the series, Guides to Information Sources, this item is somewhat more difficult to use than many other comparable guides.

Lees, N. and H. Woolston. 1997. *Environmental Information: A Guide to Sources*. 2nd ed. London: British Library. 271 p. ISBN: 0712308253.   The book contains a specific chapter on each of the following subjects focused on the environmental information available for each: general, business, legal, air pollution, water pollution, solid waste/waste disposal, contaminated land, chemicals, energy, transport, recycling, noise pollution, ecology, and agriculture and food. It is written primarily from a European and international perspective and organized in an easy-to-use outline format. The agricultural section is particularly limited in scope.

Merideth, R. 1993. *The Environmentalist's Bookshelf*. 296 p. New York: G.K. Hall. ISBN: 0816173591.   This is an interesting guide to the most popular books published on environmental topics. The author sent out questionnaires to leading environmentalists and received 296 responses. Respondents were asked to list their choices for the best books on environmental topics. The entries are divided into the top 40 books, core books, strongly recommended books, and other books. There are 500 books annotated in this volume. Includes author, subject, and title indexes.

Miller, J.A., et al. 1993. *The Island Press Bibliography of Environmental Literature*. 396 p. Washington, D.C.: Island Press. ISBN: 1559631899.   Annotated bibliography divided into broad subjects and subdivided into narrower subjects. Contains short annotations to 3,084 items, many on agriculture, forestry, and wildlife. All entries are not necessarily environmental in nature, though most are. Includes books, government publications, and journals.

Moseley, C.J. 1993. *Beacham's Guide to Environmental Issues and Sources*. 5 vols. Washington, DC: Beacham Pub. ISBN: 0933833318 (v. 1).   An extensive guide to older environmental information sources. Contains both annotated and unannotated sources, and is primarily intended as a reference source for under-graduate-level term papers.

## HANDBOOKS AND MANUALS

The handbooks and manuals discussed here are important as desk references for locating quick information on agricultural chemicals and mathematical formulas that are often used in environmental studies. These handbooks are excellent resources for academic library reference collections, especially for institutions serving schools of engineering and hard science environmental studies. Much information regarding agrochemicals and hazardous substances also can be found from browsing the Web, but these print products can serve as important archival sources of information important for future researchers.

ASTM Committee E-47 on Biological Effects and Environmental Fate. 1999. *ASTM Standards on Biological Effects and Environmental Fate*. 2nd ed. West Conshohocken, PA: ASTM. 1,025 p. ISBN: 0803127227. Revised and expanded version of the original 1993 compilation of standards from the American Society for Testing and Materials. This is a classic publication for standards that are used in testing for environmental toxicology. Largely composed of aquatic-related methods, there is also attention paid to terrestrial and physiological testing methods.

Berry, J.F. and M.S. Dennison. 2000. *The Environmental Law and Compliance Handbook*. New York: McGraw-Hill. 807 p. ISBN: 0071340947. Intended to help environmental practitioners understand, implement, and adhere to complex environmental regulations. Though largely based on U.S. laws, there is some coverage of international guidelines.

Bitton, G. 1998. *Formula Handbook for Environmental Engineers and Scientists*. New York: Wiley. 290 p. ISBN: 047113905X. This handbook deals mainly with formulas in biological or biochemical processes in natural or engineered systems. Formulas are presented in alphabetical order with some cross-referencing, and with definitions and references to the appropriate literature. This book contains a number of useful appendices. Published as part of the series Environmental Science and Technology.

Bregman, J.I., C. Kelley, and J.R. Melchor. 1996. *Environmental Compliance Handbook*. Boca Raton, FL: CRC Press. ISBN: 1566701465. Discusses state and federal environmental law. Used as a guide to business to aid in complying with legal guidelines.

Canadian Council of Ministers of the Environment. 1999. *Canadian Environmental Quality Guidelines*. Winnipeg, Canada: Canadian Council of Ministers of the Environment. 600 p. (approximately). This extensive and important work is a continuation of the Canadian government's *Canadian Water Quality Guidelines* (1987). These science-based guidelines are contained in separately numbered chapters integrating current Canadian environmental quality guidelines for water, soil, sediment, tissue residue, and air. Loose-leaf with a compact disk.

Cooper, A.R. 1996. *Cooper's Comprehensive Environmental Desk Reference*. New York: Van Nostrand Reinhold. 1,039 p. ISBN: 0442021593. Includes computer disk. Intended to help in the understanding of environmental jargon. Contains 762 pages of dictionary entries and 164 pages of expanded acronyms. Also includes an extensive subject based jargon finder index.

Copping, L.G. 1998. *The Biopesticide Manual: A World Compendium*. Farnham, Surrey, UK: British Crop Protection Council. 333 p. ISBN: 1901396266.   Divided into five sections, this book focuses on biopesticides that have limited negative environmental effects. The sections include natural products, pheromones, living systems, insect predators, and genes. Also contains a reference section with a Latin-English and English-Latin glossary, a directory of companies, abbreviations, and codes as well as an index of CAS registry numbers and an index of approved names, common names, code numbers, and trade names.

Finucane, E.W. 1999. *Concise Guide to Environmental Definitions, Conversions, and Formulae*. Boca Raton, FL: Lewis Publishers. 206 p. ISBN: 1566703158.   This book is more focused on industrial hygiene than environmental science, but may still serve as a useful reference to agricultural and food engineering professionals. Chapters are concerned with standards and calibration, workplace ambient air, ventilation, thermal stress, sound and noise, ionizing and nonionizing radiation, and statistics and probability.

Kamrin, M.A., ed. 1997. *Pesticide Profiles: Toxicity, Environmental Impact, and Fate*. Boca Raton, FL: CRC/Lewis Publishers. 676 p. ISBN: 1566701902.   Contains information about 137 specific pesticides. Also includes a trade-name index, conversion tables, and a substantial glossary of terms. Each entry gives a chemical figure, trade or other names, regulatory status, introductory discussion, toxicological effects, ecological effects, environmental fate, physical properties, references to exposure guidelines, and the basic manufacturer. This book is well referenced, and contains concept papers that explain in detail human health risks and environmental concerns.

Lehr, J.H., ed. 2000. *Standard Handbook of Environmental Science, Health, and Technology*. New York: McGraw-Hill. 1650 p. 007038309X.   Contains entries from a large number of contributing authors and is divided into three parts. The first part is "the interaction of basic scientific disciplines on contaminant fate and transport in the environment," the second is "site based environmental science, health and technology," and the third is "place based environmental science, health and technology." This is a well-organized and referenced book with many numerical equations, photos, and graphical representations.

Liu, D.H.F. and B.G. Lipták, eds. 1997. *Environmental Engineers' Handbook*. Boca Raton, FL: Lewis Publishers. 1,431 p. ISBN: 0849399718.   A thorough and interdisciplinary treatment of the topic in 11 chapters. Chapters cover laws and regulation, impact assessment, pollution prevention in chemical manufacturing, standards, air pollution, noise pollution, wastewater treatment, contaminant removal, ground and surface water pollution, solid waste, and hazardous waste. The book is extensively illustrated and referenced, as well as being well organized.

Montgomery, J.H. 1996. *Groundwater Chemicals Desk Reference*. 2nd ed. Boca Raton, FL: CRC Lewis Publishers. 1,345 p. ISBN: 1566701651. This book is very similar to the previous entry under the same author, but it also contains an appendix with bulk density values for various soils and rocks and another appendix for porosity ranges.

Montgomery, J.H. 1993. *Agrochemicals Desk Reference*: *Environmental Data*. Boca Raton, FL: Lewis Publishers. 625 p. ISBN: 0873717384. This book is extremely well referenced. It also includes a list of abbreviations and symbols, and an abbreviated introduction to many definitions and formulae of use in field-work. Each entry gives synonyms, structure, designations, properties, transformation products, exposure limits, exposure symptoms, formulation types, toxicity, and use. Includes substantial appendices on conversion factors, approved U.S. EPA test methods, CAS index, empirical formula index, synonym index, and a cumulative index.

*MSDS Reference for Crop Protection Chemicals*. Annual. New York: C&P Press. ISBN: 1-57009-081-5. Provides important value-added information derived from Material Safety Data Sheets (MSDS) and other safety data for common pesticides. Includes a brand name, company name, and common name index.

Olsen, S.S. 1999. *International Environmental Standards Handbook*. Boca Raton, FL: CRC Press. 416 p. ISBN: 1566702704. Includes copies of available treaties, laws, and standards, as well as recommendations for businesses. Contains glossary, appendices, and a guide to information on the Internet.

Waxman, M.F. 1998. *Agrochemical and Pesticide Safety Handbook*. Boca Raton, FL: Lewis Publishers. 616 p. ISBN: 1566702968. This book is divided into two sections. Section one includes an introduction and chapters on the pesticide market, regulations, pests, pesticide formulation, toxicology, handling, first aid, application equipment, and environmental protection. Section two includes response resources and a discussion of pesticide and chemical tables. Also includes a glossary.

## JOURNALS

Journals are the primary means for scientific and scholarly communication today. The following is a listing of the top journals for environmental science (in agriculture) and natural resource management. Each entry includes the number of articles published in calendar year 1999, and the Institute of Scientific Information (ISI) "impact factor" rating. This impact factor is calculated by dividing the number of current citations to articles published in the two previous years by the total number of articles published in those two years. The impact factor is one measure of a journal's merit and relative importance to a particular field. Each journal is listed under its current title, with title changes and mergers noted where appropriate. The historical information for the current title incarnation is also given with the ISSN number.

*Agricultural and Forest Meteorology.* v. 31– . 1984– . Amsterdam: Elsevier. ISSN: 0168-1923. Impact Factor: 1.466. Continues *Agricultural Meteorology.* Text in English, with some French or German; abstracts in English.

*Agricultural Systems.* v. 1– . 1976– . Barking, UK: Applied Science Publishing. ISSN: 0308-521X. Impact Factor: 0.630.

*Agricultural Water Management.* v. 1– . 1976– . Amsterdam: Elsevier. ISSN: 0378-3774. Impact Factor: 0.333.

*Agriculture, Ecosystems & Environment.* v. 9– . 1983– . Amsterdam; New York: Elsevier. ISSN: 0167-8809. Impact Factor: 0.975. Formed by the union of *Agriculture and Environment* and *Agro-ecosystems.*

*Agroforestry Systems.* The Hague; Boston: M. Nijhoff/Dr. W. Junk. ISSN: 0167-4366. Impact Factor: 0.500.

*American Journal of Alternative Agriculture.* v. 1– . 1986– . Greenbelt, MD: Institute for Alternative Agriculture. ISSN: 0889-1893. No impact factor.

*Annals of Forest Science/INRA (ANSFAS).* v. 56– . 1999– . Paris: Éditions Elsevier. ISSN: 1286-4560. Impact Factor: 0.681. Continues *Annales des Sciences forestières.* Text, summaries, tables of contents in English and French.

*Atmospheric Environment.* v. 28– . 1994– . Oxford; New York: Pergamon. ISSN: 1352-2310. Impact Factor: 2.003. Some special issues have distinctive titles. Formed by the union of *Atmospheric Environment Part A, General Topics,* and *Atmospheric Environment Part B, Urban Atmosphere.*

*Biogeochemistry.* v. 1– . 1984– . Dordrecht; Boston: Kluwer Academic Publishers. ISSN: 0168-2563. Impact Factor: 2.039. Available online, subscription to online journal required for access to abstracts and full text.

*Biological Conservation.* v. 1– . 1968– . Barking, UK: Applied Science Publishers. ISSN: 0006-3207. Impact Factor: 1.579.

*Biological Control: Theory and Applications in Pest Management.* San Diego, CA: Academic Press. ISSN: 1049-9644. Impact Factor: 1.272.

*Bioresource Technology: biomass, bioenergy, biowastes, conversion technologies, biotransformations, production technologies.* v. 35– . 1991– . New York: Elsevier Applied Science. ISSN: 0960-8524. Impact Factor: 0.700. Formed by merger of *Biomass* and *Biological Wastes.*

*Canadian Journal of Forest Research. Journal Canadien de la Recherche Forestière.* v. 1– . 1971– . Ottawa, Canada: National Research Council of Canada. ISSN: 0045-5067. Impact Factor: 1.058. English or French with summaries in both languages.

*Climatic Change.* v. 1– . 1977– . Dordrecht-Holland, Boston: Reidel Publishing Co. ISSN: 0165-0009. Impact Factor: 1.871.

*Conservation Biology: The Journal of the Society for Conservation Biology.* v. 1– . 1987– . Boston: Blackwell Scientific Publications. ISSN: 0888-8892. Impact Factor: 3.240.

*Environmental Entomology.* v. 1– . 1972– . College Park, MD: Entomological Society of America. ISSN: 0046-225X. Impact Factor: 0.889.

*Environmental Health Perspectives: EHP.* no. 1– . 1972– . Research Triangle Park, NC: U.S. Dept. of Health, Education, and Welfare, Public Health Service, National Institutes of Health, National Institute of Environmental Health Sciences. Distributed by U.S. Government Printing Office. ISSN: 0091-6765. Impact Factor: 2.469. Also known as *EHP.* Some issues also available from the World Wide Web at http://ehpnet1.niehs.nih.gov/docs/publications.html.

*Environmental and Molecular Mutagenesis.* v. 10– . 1987– . New York: Alan R. Liss. ISSN: 0893-6692. Impact Factor: 1.990. Continues *Environmental Mutagenesis.* Official journal of the Environmental Mutagen Society.

*Environmental Research.* v. 1– . 1967– . New York: Academic Press. ISSN: 0013-9351. Impact Factor: 1.617. Also available online by subscription.

*Environmental Science & Technology.* v. 1– . 1967– . Easton, PA: American Chemical Society. ISSN: 0013-936X. Impact Factor: 3.751. Also known as *ES&T.* Available online by subscription.

*Environmental Toxicology and Chemistry.* v. 1– . 1982– . Elmsford, NY; Oxford: Pergamon Press. ISSN: 0730-7268. Impact Factor: 2.462. Sponsored by the Society of Environmental Toxicology and Chemistry (SETAC).

*Forest Ecology and Management.* v. 1– . 1976/77– . Amsterdam: Elsevier. ISSN: 0378-1127. Impact Factor: 0.962.

*Forest Pathology/Journal de Pathologie Forestière/Zeitschrift für Forstpathologie.* v. 30– . 2000– . Berlin: Blackwell. ISSN: 1437-4781. Continues *European Journal of Forest Pathology.* Text in English, French, and German.

*Forest Products Journal.* No beginning date available. Madison, WI: Forest Products Research Society. ISSN: 0015-7473. Impact Factor: 0.336. Continues *Jour-*

*nal of the Forest Products Research Society.* Available on microfilm from University Microfilms International.

*The Forestry Chronicle.* v. 1– . 1925– . Sainte-Anne-de-Bellevue, Quebec: Canadian Institute of Forestry. ISSN: 0015-7546. Impact Factor: 0.687. Continues *Canadian Institute of Forestry. Annual Report of the Canadian Institute of Forestry.* Supplements accompany some issues. Includes some text in French.

*Forestry: The Journal of the Society of Foresters of Great Britain.* 1927– . London: Oxford University Press. ISSN: 0015-752X. Impact Factor: 0.493.

*Forest Science.* v. 1– . 1955– . Bethesda, MD: Society of American Foresters. ISSN: 0015-749X. Impact Factor: 1.034. Available on microfilm from University Microfilms International.

*Global Biogeochemical Cycles.* v. 1– . 1987– . Washington, DC: American Geophysical Union. ISSN: 0886-6236. Impact Factor: 4.309. Issued also by the publisher on microfiche and on microfilm at the end of the volume year.

*Global Change Biology.* v. 1– . 1995– . Oxford: Blackwell Science. ISSN: 1354-1013. Impact Factor: 3.014.

*Holzforschung: Mitteilungen zur Chemie, Physik, Biologie, und Technologie des Holzes.* 1947– . Berlin: Technischer Verlag Herbert Cram. ISSN: 0018-3830. Impact Factor: 0.995.

*IAWA Journal/International Association of Wood Anatomists.* v. 14– . 1993– . Leiden, The Netherlands: Rijksherbarium/Hortus Botanicus. ISSN: 0928-1541. Impact Factor: 0.722. Continues *International Association of Wood Anatomists. IAWA.*

*Journal of Aerosol Science.* 1970– . Oxford; New York: Pergamon Press. ISSN: 0021-8502. Impact Factor: 1.887. Available on microfilm from Microforms International Marketing Corp. Text in English, French, and German. Published in association with the Gesellschaft für Aerosolforschung. Also called *Aerosol Science.*

*Journal of Agricultural, Biological, and Environmental Statistics.* 1996– . Alexandria, VA: American Statistical Association and the International Biometric Society. ISSN: 1085-7117.

*Journal of Agricultural & Environmental Ethics.* v. 4– . 1991– . Guelph, Ontario, Canada: University of Guelph. ISSN: 0893-4282. Impact Factor: 0.053. Continues *Journal of Agricultural Ethics.*

*Journal of Agriculture and Environment for International Development*. v. 1– . 1907– . Florence, Italy: Istituto Agronomico per l'Oltremare. ISSN: 0035-6026. No impact factor. Continues *Rivista di Agricoltura Subtropicale e Tropicale*.

*Journal of Atmospheric Chemistry*. Dordrecht, Boston: D. Reidel. 1983- ISSN: 0167-7764. Impact Factor: 2.010.

*Journal of Economic Entomology*. v. 1– . 1908– . College Park, MD: Entomological Society of America. ISSN: 0022-0493. Impact Factor: 1.096.

*Journal of Environmental Economics and Management*. v. 1– . 1974– . New York: Academic Press. ISSN: 0095-0696. Impact Factor: 1.216. Related to the Association of Environmental and Resource Economists. Available online by subscription.

*Journal of Environmental Quality*. v. 1– . 1972– . Madison, WI: Published cooperatively by American Society of Agronomy, Crop Science Society of America, and Soil Science Society of America. ISSN: 0047-2425. Impact Factor: 2.357. Also called *Environmental Quality*.

*Journal of Environmental Science and Health*: *Part B*, *Pesticides*, *Food Contaminants*, *and Agricultural Wastes*. *Pesticides*, *Food Contaminants*, *and Agricultural Wastes*. v. B11– . 1976– . New York: Marcel Dekker. ISSN: 0360-1234. Impact Factor: 0.582. Continues *Environmental Letters*. Related titles: *Journal of Environmental Science and Health*: *Part A*, *Environmental Science and Engineering* (ISSN 0360-1226) and *Journal of Environmental Science and Health*: *Part C*, *Environmental Health Sciences* (ISSN 0360-1242).

*Journal of Forestry*. v. 15– . 1917– . Washington, DC: Society of American Foresters. ISSN: 0022-1201. Impact Factor: 0.735. Formed by the union of *Forestry Quarterly*, and *Proceedings of the Society of American Foresters*. Official journal of the Society of American Foresters.

*Journal of Range Management*. v. 1– . 1948– . Denver, CO: Society for Range Management. ISSN: 0022-409X. Impact Factor: 0.577.

*Journal of Sustainable Agriculture*. v. 1– . 1990– . Binghamton, NY: Food Products Press. ISSN: 1044-0046. Impact Factor: 0.169.

*Journal of Sustainable Forestry*. v. 1– . 1993– . Binghamton, NY: Haworth Press. ISSN: 1054-9811. No impact factor available.

*Journal of Vegetation Science*: *Official Organ of the International Association for Vegetation Science*. Knivsta, Sweden: OPULUS Press. ISSN: 1100-9233. Impact Factor: 1.957. Also called *JVS*.

*Natural Areas Journal, A Quarterly Publication of the Natural Areas Association.* v. 2– . 1982– . Rockford, IL: The Association. ISSN: 0885-8608. Impact Factor: 0.736. Continues *Journal of the Natural Areas Association.*

*Organic Gardening.* v. 35– . 1988– . Emmaus, PA: Rodale Press. ISSN: 0897-3792. No impact factor available. Continues *Rodale's Organic Gardening.*

*Pest Management Science.* v. 56– . no. 1– . West Sussex, UK: Wiley for the Society of Chemical Industry. ISSN: 1526-498X. Impact Factor: 1.102. Continues *Pesticide Science.* Available online by subscription.

*Plant Ecology.* v. 128– . 1997– . Dordrecht; Boston: Kluwer. ISSN: 1385-0237. Impact Factor: 1.339. Continues *Vegetatio.*

*Remote Sensing of Environment.* v. 1– . 1969– . New York: Elsevier. ISSN: 0034-4257. Impact Factor: 1.868.

*Silvae Genetica.* Bd. 6– . 1957– . Frankfurt am Main, Germany: J. D. Sauerländer. ISSN: 0037-5349. Impact Factor: 0.513. Continues *Zeitschrift für Forstgenetik und Forstpflanzenzüchtung.*

*Tree Physiology.* v. 1– . 1986– . Victoria, BC: Heron Publishing. ISSN: 0829-318X. Impact Factor: 2.042.

*Trees: Structure and Function.* v. 1– . 1986– . Berlin: Springer-Verlag. ISSN: 0931-1890. Impact Factor: 1.278. Other title: *Trees.*

*Water Research.* v. 1– . 1967– . Oxford; New York: Pergamon Press. ISSN: 0043-1354. Impact Factor: 1.748. Continues *Air and Water Pollution.* Official journal of the International Association on Water Pollution Research. Available on microform from Micro Mark.

*Water Resources Research.* v. 1– . 1965– . Washington, DC: American Geophysical Union. ISSN: 0043-1397. Impact Factor: 2.061.

*Wood and Fiber Science: Journal of the Society of Wood Science and Technology.* v. 15– . 1983– . Lawrence, KS: The Society. ISSN: 0735-6161. Impact Factor: 0.446. Published in cooperation with the Forest Products Research Society. Formed by the merger of *Wood and Fiber* and *Wood Science.*

*Wood Science and Technology.* v. 1– . 1967– . New York: Springer-Verlag. ISSN: 0043-7719. Impact Factor: 0.373.

## PROCEEDINGS

Conference proceedings are an important component in scholarly communication. They often present difficulties for literature researchers, both in finding and referencing them. Conference proceedings can be published as monographs, issues of journals, or part of a series. Some proceedings include the entire group of papers presented at the conference, while others are a subset. Some conference proceedings contain only abstracts of papers, while others are a mix of entire papers and abstracts. It is common for a paper presented at a conference also to be published in a peer-reviewed journal at a later date. Some professional societies and associations that publish literature in environmental sciences and natural resources are given in the following text.

American Chemical Society
American Fisheries Society
American Meteorological Society
American Society of Agricultural Engineers
American Society of Civil Engineers
American Society for Photogrammetry and Remote Sensing
Association for Environmental Health and Sciences
Global Warming International Center
International Association for Landscape Ecology
International Erosion Control Association
International Union of Forest Research Organizations
International Union of Pure and Applied Chemistry
International Water Association
IUCN—International Union for Conservation of Nature and Natural Resources (World Conservation Union)
National Coalition against the Misuse of Pesticides
National Wildlife Federation
Royal Chemistry Society
Society for Conservation Biology
Society for Range Management
Soil and Water Conservation Society

Some recent conference or symposium proceedings, published as monographs, are given in the following text.

Conference on University Education in Natural Resources. 1998. *Proceedings of the Second Biennial Conference on University Education in Natural Resources: March 7–10. 1998, Utah State University, Logan, Utah*. C.G. Heister, compiler. Logan, UT: Utah State University, S.J. and Jessie E. Quinney Natural Resources Research Library. Covers the study and teaching of natural resources, wildlife, fisheries, and forest management. Volume seven of the series Natural Resources and Environmental Issues.

E.C. Conference (8th: 1994: Vienna, Austria). 1995. *Biomass for Energy, Environment, Agriculture, and Industry: Proceedings of the 8th European Bio-*

*mass Conference, Vienna, Austria, 3–5 October 1994.* Ph. Chartier, A.A.C.M. Beenackers, and G. Grassi, eds. 3 vols. Oxford; New York: Pergamon. 2,426 p. ISBN: 0080421350.

International Fertilizer Development Center. 1998. *Environmental Challenges of Fertilizer Production: An Examination of Progress and Pitfalls: Proceedings of An International Workshop, September 17–19. 1997, Atlanta, Georgia, U.S.* J.J. Schultz and E.N. (Beth) Roth, eds. Muscle Shoals, AL: International Fertilizer Development Center. 272 p. Includes maps. ISBN: 0880901160. Part of the series Special Publication IFDC; SP-25. Contains approximately 33 papers divided into 5 technical sessions. Each technical session contains a question-and-answer chapter. Includes directory of workshop presenters and delegates.

International Symposium on Agricultural and Food Processing Wastes. 1995. *Seventh International Symposium on Agricultural and Food Processing Wastes (ISAFPW95): Proceedings of the 7th International Symposium, June 18–20, 1995, Hyatt Regency Chicago, Chicago, Illinois.* C.C. Ross, ed. St. Joseph, MI: American Society of Agricultural Engineers. 636 p. ISBN: 0929355660. Part of the ASAE publication series. Contains 68 papers.

International Symposium of the FORESEA Miyazaki. 1999. *Global Concerns for Forest Resource Utilization: Sustainable Use and Management: Selected Papers from the International Symposium of the FORESEA Miyazaki 1998.* A. Yoshimoto and K. Yukutake, eds. Dordrecht; Boston: Kluwer. 361 p. Includes maps. ISBN: 0792359682. Part of the series Forestry Sciences, v. 62. Contains 29 papers in 3 sections: understanding global forest sector, modeling efforts in forest sector analysis, and the role of Japanese forest policy in a global context.

International Symposium on Nuclear and Related Techniques in Soil-Plant Studies on Sustainable Agriculture and Environmental Preservation. 1995. *Nuclear Techniques in Soil-Plant Studies for Sustainable Agriculture and Environmental Preservation: Proceedings of an International Symposium on Nuclear and Related Techniques in Soil-Plant Studies on Sustainable Agriculture and Environmental Preservation, held in Vienna, 17–21 October 1994.* Vienna: International Atomic Energy Agency. 735 p. ISBN: 9201008953. Jointly organized by the International Atomic Energy Agency and the Food and Agriculture Organization (FAO) of the United Nations, this symposium is dedicated to the 30th anniversary of the joint FAO/IAEA Division of Nuclear Techniques in Food and Agriculture.

International Workshop on Natural Resource Management in Rice Systems: Technology Adaptation for Efficient Nutrient Use. 1999. *Resource Management in Rice Systems: Nutrients: Papers Presented at the International Workshop on Natural Resource Management in Rice Systems: Technology Adaptation for Efficient Nutrient Use, Bogor, Indonesia, 2–5 December 1996.* V. Balasubramanian,

J.K. Ladha, and G.L. Denning, eds. Dordrecht; Boston: Kluwer in cooperation with the International Rice Research Institute. 355 p. ISBN: 0792351991.   Partly reprinted from *Nutrient Cycling in Agroecosystems*, v. 53, no. 1 (January 1999). Part of the series Developments in Plant and Soil Sciences, v. 81.

Lockeretz, William, ed. 1996. *Environmental Enhancement through Agriculture*: *Proceedings of a Conference Held in Boston, Massachusetts, November 15–17. 1995*. Medford, MA: Tufts University Center for Agriculture, Food and Environment, School of Nutrition Science and Policy. 334 p.   Thirty-six papers presented at the named conference are included. Sections include watershed protection; wildlife conservation and biodiversity; livestock systems; waste recycling and nutrient management; energy from agricultural biomass; metropolitan agriculture and farmland preservation; national policies on agriculture and the environment; and agricultural development and the environment. Discusses the need to focus on production optimization rather than production maximization. Organized by Tufts University School of Nutrition Science and Policy, the American Farmland Trust, and the Henry A. Wallace Institute for Alternative Agriculture.

Loseby, Margaret, ed. 1996. *European Association of Agricultural Economists. Seminar (39th: 1995: Viterbo, Italy) Agroforestry and Its Impacts on the Environment: Proceedings of the 39th seminar of the European Association of Agricultural Economists (EAAE) Viterbo, May 19th–21st. 1995*. Rome: Instituto Nazionale di Economia Agraria. 223 p.   This collection of papers is divided into three sections: policies of supranational organizations toward agroforestry, technical aspects of agroforestry, and a number of case studies.

Napier, T.L., S.M. Napier, and J. Tvrdon, 2000. *Soil and Water Conservation Policies and Programs: Successes and Failures*. Boca Raton, FL: CRC Press. 640 p. ISBN: 0849300053.   Proceedings of an international conference convened at the Czech Agriculture University in Prague, Czech Republic in September of 1996. Contains 38 papers, international in scope.

Patterson, P.H. and J.P. Blake, eds. 1996. *Proceedings. 1996 National Poultry Waste Management Symposium, Harrisburg, PA, October 21–23, 1996*. n.p.: National Poultry Waste Management Symposium Committee. 354 p. ISBN: 0962768265.   Contains 64 papers from a symposium held at the Marriott-Harrisburg Hotel, Harrisburg, PA, October 21–23, 1996. Sections include general session (air quality), production (mortality and waste management), processing (waste reduction), nutrient management, and processing (wastewater treatment). Covers recycling of manure, dead birds, and water.

Rosen, D., ed. 1997. *Modern Agriculture and the Environment: Proceedings of an International Conference, held in Rehovot, Israel, 2–6 October 1994: Under the Auspices of the Faculty of Agriculture, the Hebrew University of Jerusalem.*

Dordrecht; Boston: Kluwer. 646 p. ISBN: 079234295X. Volume 71 of the series Developments in Plant and Soil Sciences.

Rodriguez-Barrueco, C., ed. 1996. *International CIEC Symposium (8th: 1994: Salamanca, Spain). Fertilizers and Environment: Proceedings of the International Symposium "Fertilizers and Environment", held in Salamanca, Spain, 26–29, September. 1994.* Dordrecht; Boston: Kluwer. 581 p. ISBN: 0792337298. Part of the series Developments in Plant and Soil Sciences, v. 66 and partly reprinted from *Fertilizer Research*, v. 43, no. 1–3 (1995/1996). Contains 99 papers in 6 sections, with an author and subject index. Sections are food, forest, cycling and sustainability, mineral management, manure management, environmental impact, and code for agricultural practices.

USDA Working Group on Water Quality. 1995. *Clean Water, Clean Environment, 21st Century: Team Agriculture, Working to Protect Water Resources: Conference Proceedings, March 5–8, 1995, Kansas City, Missouri.* St. Joseph, MI: ASAE. 186 pages (v. 1), 254 pages (v. 2), 318 p. (v. 3). ISBN: 0929355601. This three-volume set is composed of the brief descriptions of posters presented. Volume one covers pesticides, volume two covers nutrients, and volume three covers practices, systems, and adoption.

## WORLD WIDE WEB SITES

There has been a tremendous proliferation of informational Web sites in recent years. This is especially true for governmental organizations and not-for-profit groups. The Web has introduced a method for disseminating information quickly and inexpensively to the entire international community. This short list is representative of some of the more authoritative information providers at the present time.

Canadian Council of Ministers of the Environment (http://www.ccme.ca/index.html). This Web site was established as a discussion tool for environmental issues of Canadian and international concern. Contains information for ordering publications.

Commonwealth of Australia, Department of Environment and Heritage. Environment Australia Online (http://www.environment.gov.au/index.html). Searchable Web site for environmental information in Australia. Contains databases, links to government sites, and publications.

Environmental Defense. Environmental Defense Scorecard (http://www.score card.org/). Contains various rankings of environmental statistics on a geographic basis for the United States including a notable collection of information and data on animal waste from factory farms. Contains an interactive map of the United States with point-and-click access to smaller maps of environmental information. Includes a feature for entering a ZIP code and finding pollution information for communities in that location.

Environmental Protection Agency (U.S.). Environmental Protection Agency (http://www.epa.gov/). Contains an extensive collection of information on the environment including the representative sublinks described in the following text.

> EPA Envirofacts Warehouse (http://www.epa.gov/enviro/). Aggregate link of environmental data, much of it geographically referenced.
> EPA Pesticide Topic Pages (http://www.epa.gov/ebtpages/pesticides. html).
> National Environmental Publications Internet Site (NEPIS) (http:// www.epa.gov/ncepihom/nepishom/). A database of over 7,000 full-text online documents.
> National Service Center for Environmental Publications (NSCEP) (http:// www.epa.gov/ncepihom/). A searchable catalog of hardcopy and multimedia documents available free of charge.

Florida State University, Florida Center for Public Management. Environmental Indicator Technical Assistance Series (http://mailer.fsu.edu/~cpm/segip/catalog/index.html). Part of a four-volume series in Adobe Acrobat format. Contains numerous links. Volume one: Catalog of Environmental Indicators; volume two: Catalog of Data Sources; volume three: State Indicators of National Scope; and volume four: Directory of Environmental Indicator Practitioners

George Washington University, Green University Initiative. Environmental Information Resources (http://www.hfni.gsehd.gwu.edu/~greenu/index2. html). Contains a useful list of links for specific environmental information sites organized by country.

EnviroOne. EnviroOne.com (http://www.enviroone.com). Serves as an organizer of environmental information on the Internet. This site has a separate category for agricultural related environmental information and includes links to many news releases and environmental sites.

European Environmental Bureau. European Environmental Bureau (http:// www.eeb.org/). The European Environmental Bureau (EEB) is a nongovernmental environmental group, representing 130 nongovernmental organizations (NGOs) from 24 countries, including all 15 European Union Member States. Member organizations represent some 14,000 member organizations, 500 regional branches, 800 local branches, 260 associated organizations, and more than 11 million individual members. This site contains numerous press releases, links, organizational information, and a significant number of full-text publications available online.

European Union, European Environment Agency (http://www.eea.eu.int/). European Environment Information and Observation Network (EIONET). Serves as an important gateway to environmental information for Europe. Contains numerous press releases, full-text documents, and maintains an interactive information center. Provides an excellent search interface to European Environment Agency (EEA) documents as well as documents and sites for member countries in 13 languages. Provides information based on themes, specific countries, or data sets intended to be used with GIS-type software.

Institute for European Environmental Policy (http://www.ieep.org.uk). The Institute for European Environmental Policy (IEEP) conducts ''research on the European dimension of environmental protection, with a major focus on the development, implementation and evaluation of the environmental policy of the European Union, international bodies and national governments'' (quoted from Web site). Contains a listing of agricultural and nature conservation publications available from the IEEP.

Institute of Science and Public Affairs. Program for Environmental Policy and Planning Systems (http://www.pepps.fsu.edu/about.htm). Includes environmental management and environmental indicator links, as well as links to a large number of electronic publications, technical documents, and subject bibliographies.

National Council for Science and the Environment (U.S.). National Library for the Environment (http://www.cnie.org/nle/). Intended to be an aggregate of environmental information for the United States.

Organization for Economic Co-Operation and Development. Environmental Indicators: A Review of Selected Central and Eastern European Countries 1996 (http://www.oecd.org/sge/ccnm/pubs/gd/cd156201/present.htm) Provides full text of this data report in Adobe Acrobat format.

Pacific Research Institute. Index of Leading Environmental Indicators (http://www.pacificresearch.org/issues/enviro/99eindex/main.html). Selective report that focuses on providing a summary view of those aspects of the environment that are most important to ordinary citizens.

Sustainable Agriculture Network. Sustainable Agriculture Research and Education (SARE) Program (http://www.sare.org). Contains ordering information for publications. Also contains some free online documentation.

University of Minnesota. 2001. Forestry AgNIC (http://forestry.lib.umn.edu/agnic/). Excellent online guide to forestry information is part of the Agriculture Network Information Center (AgNIC).

University of Nebraska-Lincoln. 2001. Water Quality AgNIC (http://deal.unl.edu/agnic/). Guide to agricultural water-quality information on the Internet, made available through AgNIC.

USDA. Alternative Farming Systems Information Center (http://http://www.nal.usda.gov/afsic/). Contains extensive links to full-text online publications. This site has a search interface and is part of the Agriculture Network Information Center (AgNIC) system.

USDA. USDA Forest Service (http://www.fs.fed.us/). Site for the agency overseeing forests in the United States. Includes a library with full-text publications and images of forest maps. Also includes a search function for all USDA Forest Service Web sites.

# 7

## Farming and Farming Systems

**Irwin Weintraub**
Brooklyn College, Brooklyn, New York, USA

Farming systems vary widely around the world. They are dependent upon a variety of factors such as climate, soil types, vegetation, cultural practices, regional preferences, environment, economics, and personal preferences. As world population increases and the number of people on farms decreases, a smaller number of farmers have the responsibility of producing a greater amount of food on a diminishing land base. The interrelationship and complex interactions among the components of farming systems are well illustrated in this definition. A farming system is a ''complicated, interwoven mesh of soils, plants, animals, implements, workers, other inputs and environmental influences with the strands held and manipulated by a person called the farmer who, given his or her preferences and aspirations, attempts to produce output from the available inputs and technology. It is the farmer's unique understanding of the immediate environment, both natural and socio-economic, that results in his or her farming system.'' A farming systems approach to agricultural research and training is based on recognition of these complicated relationships. The aim is to first understand, then add options to existing systems by gathering needed information and generating and testing needed technologies (Consultative Group on International Agricultural Research, 1978). Duckham and Masefield (1969) point out that decision making in farming is based upon experience, tradition, expected profit, personal preferences, available resources, feasibility, operational constraints, and social and political pressures. The successes and failures of farming systems are heavily dependent upon the decisions made by individuals, communities, and governments in planning and designing farming systems and the impacts of those decisions on the community being served.

In the latter part of the twentieth century, widespread use of chemical inputs in agriculture gave rise to public concern about the consequences of conventional farming systems on air, soil, water, and natural resources, and the potential health risks to farmers and consumers. This led to a clamor for alternative farming systems, usually referred to as ''sustainable agriculture'' that will leave a legacy of clean land, air, and water for future generations and a clean and safe food supply that does not damage the health and well being of farmers and consumers. ''A sustainable food and agriculture system is one which is environmentally sound, economically viable, socially responsible, nonexploitative, and which serves as the foundation for future generations. It must be approached through an interdisciplinary focus which addresses the many interrelated parts of the entire food and agricultural system at local, regional, national, and international levels.''(Allen, 1992) Other terms that have become a regular feature of the lexicon of sustainable agriculture are *alternative*, *regenerative*, *organic*, *agroecology*, *low input*, *permaculture*, *biodynamic*, *biological*, *ecological*, *holistic*, and *precision agriculture*. Although some of these terms are sometimes used interchangeably, they comprise specific characteristics and philosophies that distinguish each farming system. The driving force behind the growing importance of sustainable agriculture is the need for a long-term commitment to stewardship of land, humane treatment of animals, viable rural communities, and a safe and high-quality food supply. The common theme among these alternative systems is opposition to the so-called ''industrial'' or ''conventional'' agriculture and a dedication to farming systems that are decentralized, community based, maintain harmony with nature, enhance diversity, and exercise restraint in use of resources (Diver, 1996).

Sustainability involves economic, social, and political dimensions that meet the needs of the present without compromising the ability of future generations to meet their own needs. It gives farmers an opportunity to innovate, widens the knowledge base as they share new information, and improves relationships between government, industry, and nations. The growing public awareness of the significance of ecologically sound agriculture and genetic engineering prompted a growing interest in new ways to apply farming systems to the production of new crops for pharmaceutical, industrial, and edible uses that will serve the needs of burgeoning world populations.

This chapter describes information resources in alternative farming systems from 1980 to the present that have changed attitudes about agricultural practices. The section on databases briefly discusses the abstracting and indexing sources that cite the literature on this important topic and some perspectives on their scope and limitations. The remainder of the chapter cites and describes conferences and symposia; dictionaries and glossaries; directories and bibliographies; encyclopedias and handbooks; monographs; periodicals; reports; and theses and dissertations on a broad range of farming systems. Although space constraints made it impossible to list all the resources on this important topic, the sources selected for inclusion reflect a wide coverage of the topic that provide readers an overview of farming systems literature and its characteristics.

## ABSTRACTS AND DATABASES

Databases, also referred to as abstracting or indexing sources, provide users with bibliographic citations and abstracts (summaries) of world literature in a topic of interest. They offer a range of search options in specified fields or text searching in the full record or entry. In print editions, entries are arranged in alphabetical, numerical, chronological, or geographical order. Citations usually give complete bibliographic information and notes indicating illustrations, maps, or other features of the cited work. In abstracting sources, citations are accompanied by a summary consisting of a few sentences or a short paragraph describing the work. Farming systems are perhaps the most important aspect of a farm business because without a successful system of producing crops, the farmer will not remain in business. Consequently the volume of resources on farming systems is enormous and cited heavily in databases that cover agricultural literature. For comprehensive coverage of the vast array of literature on farming systems, there are five databases that readers should consult for maximum exposure to the subject. Because these sources have been described in detail in earlier chapters, this chapter will focus on coverage of the farming systems literature.

AGRICOLA cites journals, monographs, theses, patents, audiovisual resources, and documents from government agencies and organizations. This database focuses upon literature from North and South America, Canada, and Western Europe. In addition to the material on plant and animal production, references are included to works on the various components of farming systems that involve economics, rural sociology, agricultural engineering, new crops, industrial crops, and soil conservation. Users should search AGRICOLA for articles, books, and miscellaneous materials to get a good overall introduction to this topic.

A search in CAB Abstracts database will expand the search for coverage of farming systems in 140 countries including developing countries in Asia, Africa, and Latin America. CAB Abstracts cites materials from over 9,000 journals and technical reports and 2,500 books and conference proceedings. It contains over 59,000 terms covering agriculture, veterinary science, forestry, life sciences, biotechnology, natural resources, and health. The *CAB Thesaurus* is the official thesaurus for CAB Abstracts and AGRICOLA users. The thesaurus lists descriptor terms such as *alternative farming*, *sustainability*, and *organic farming* as most relevant to retrieve citations for resources in sustainable agriculture. Because farming systems can fall under innumerable categories, the users should not rely upon descriptors alone for large searches. For maximum efficiency in searching AGRICOLA and CAB Abstracts, use keywords or combinations of keywords and descriptors.

*Biological and Agricultural Index.* 1964– . New York: H.W. Wilson.   H.W. Wilson's *Biological and Agricultural Index (BAI)* covers a core of about 335 popular and professional journals in biology and agriculture. The works cited in this database help searchers to find sources that mesh the farming systems knowledge with the biological aspects of the topic such as pest control, plant pathology, weed management, soil science, plant breeding, and animal and livestock management. Citations in BAI also cite items that may not be covered in other databases and are often missed even in a very thorough search. This includes book reviews, biographical sketches, reports of symposia and conferences, review articles, selected letters to the editor, special issues, and monographic supplements are cited in BAI.

AGRIS is a cooperative database providing references to international information in scientific, technical, social, and economic literature relating to agriculture, food, nutrition, and rural development produced by the Food and Agriculture Organization of the United Nations (FAO). AGRIS covers articles, books, theses, reports, and other resources that are usually not cited in other indexing sources. There are approximately 200 national, international, and intergovernmental participating centers that submit nearly 2 million records per year to AGRIS. It is particularly strong on resources from developing countries and is the main database for finding materials issued by agencies in the respective countries and international agencies such as the FAO, the World Bank, and others. The *AGROVOC Thesaurus* is a standard vocabulary for subject classification of citations in AGRIS. *AGROVOC* recommends a variety of descriptor terms when searching the database for literature on farming systems. Use descriptors such as *alternative agriculture*, *biodynamic agriculture*, *cropping systems*, *farming systems*, *low-input agriculture*, and *organic agriculture* to retrieve good coverage of in-country resources. Keyword searches and keyword descriptor combinations are also recommended for maximum hits when searching AGRIS.

*Alternative Press Index.* 1969– . Baltimore, MD: Alternative Press Center.   The *Alternative Press Index* published by the Alternative Press Center in Baltimore, Maryland cites articles from about 380 progressive journals and newspapers issued by universities, organizations, and independent presses in the United States and selected publications from Canada, Australia, and the United Kingdom that report and analyze the practices and theories of cultural, economic, political, and social change. *Alternative Press Index* offers readers a chance to see the topics from the vantage point of writers who discuss farming systems from the social, political, and economic perspectives that are an integral part of the design of sustainable farming systems. The cited items focus on public attitudes regarding farming and farming systems and public perceptions about the philosophies and values behind agricultural systems. Thus, this database offers the opportunity to complement the existing knowledge with some perspectives from writers who have definite viewpoints about the goals and practices of the agricultural industry.

Those seeking information on farming systems should consult all of these databases in order to gain a wide range of insights and strategies for the implementation of farming systems. Users should not restrict their searches to just one database and assume this is adequate. Conducting searches in two or more databases is the most effective way to gain a broad perspective on farming systems and their impact on world agriculture.

## CONFERENCES AND SYMPOSIA

Conferences and symposia are important venues of communication for scientists, scholars, and others to share knowledge of new developments in their fields. At professional meetings they present papers, workshops, and poster sessions detailing the results of their research and project activities. The importance of farming systems to the worldwide agricultural community has resulted in a wide range of such meetings over the past two decades. Following are details of some of the significant conferences in which significant discoveries and research results have been reported.

*Automated Agriculture for the 21st Century: Proceedings of the 1991 Symposium.* 1991. ASAE Publication 11–91. St. Joseph, MI: American Society of Agricultural Engineers. 551 p. ISBN: 0929355210.   Over 60 papers from a symposium on machinery applications and automation in the production and processing of food and fiber. Papers are from four topic areas: sensor applications, control system applications, in-field site-specific crop production, and engineering for plant culture systems. The intent of the symposium was to update engineers and scientists on the state of the art in these areas and introduce new technical skills.

Buck, M. and J. Allen, eds. 2000. *Sustainable Agriculture—Continuing to Grow: A Proceedings of the Farming and Ranching for Profits. Stewardship & Community Conference, March 7–9, 2000.* Logan, UT: Western Region Sustainable Agriculture Research and Education. (SARE). 144 p. Available from the World Wide Web at http://wsare.usu.edu.   Abstracts of papers and interviews with sustainable agriculture researchers, producers, educators, and advocates who attended the conference. Attenders relate their experience and expertise in sustainable farming systems, soil, diseases, pests and weeds, sustainable ranching, marketing, and sustainable agriculture in the western United States. The goals of this publication are to record the technical information and successful farming models presented at the conference and to extend the knowledge to elected officials, opinion leaders, farmers, ranchers, consumers, and the sustainable agriculture community.

*Building a Better Agriculture*: *Proceedings of the Philadelphia Society for Promoting Agriculture 1990–1991*. 1991. Philadelphia, PA: The Philadelphia Society for Promoting Agriculture. 98 p.   Collection of papers on sustainable agriculture addressing requirements for sustainable agriculture, management practices, integrated pest management, biological control, energy conservation, utilization of animal wastes, water quality, and responsible use of pharmaceutical products in livestock production.

Campbell, K.L., W.D. Graham, and A.B. Del'Bottcher, eds. 1994. *Environmentally Sound Agriculture*: *Proceedings of the Second Conference, 20–22 April, 1994, Orlando, Florida*. ASAE Publication 04-94. St. Joseph, MI: American Society of Agricultural Engineers. 583 p. ISBN: 0929355474.   Contains summaries of over 80 research projects and programs presented at the conference to facilitate communication regarding the application of environmental concerns to farm management, education, regulations, technology, waste utilization, soil and water systems, and wetlands. Addresses a broad range of issues and technology regarding environmentally sound agriculture in the United States.

Chou, C.-H. and K.-T. Shao, eds. 1998. *Frontiers in Biology: The Challenges of Biodiversity, Biotechnology and Sustainable Agriculture*. Taipei: Academica Sinica. 298 p. ISBN: 9576715997.   Thirty-four papers from a three-part symposium dealing with questions and issues in biodiversity, biotechnology, and sustainability theories and practices. The papers cover terrestrial and marine ecosystems and the problems, challenges, and future directions for these three areas of ecosystem management. They provide highlights and key elements in the applications of genetics, molecular biology, agroecology, ecotechnology, biodiversity, allelochemicals, crop engineering, and the role of farmers in sustainable farming systems and predictions regarding new innovations in twenty-first century agriculture.

deGroot, J.P. and R. Ruben. 1997. *Sustainable Agriculture in Central America*. New York: St. Martin's. 242 p. ISBN: 0312175558.   Paper from the 1995 annual conference of the Association for European Research on Central American and the Caribbean. A comprehensive assessment of the current prospects for sustainable agriculture under state policies and local action. Part one covers macroeconomic conditions for sustainable agriculture. Part two emphasizes the variety of production systems and their respective climate and soil differences. Part three explores land-use patterns and conflicts with natural resource conservation and protected areas. Part four addresses agrarian policies and implications for sustainable land use regarding conservation and economics. Main focus is on situations in Nicaragua and Honduras.

Edwards, C.A. et al., eds. 1990. *Sustainable Agricultural Systems.* Ankeny, IA: Soil and Water Conservation Society. 712 p. ISBN: 903573421X.   Proceedings of the International Conference on Sustainable Agricultural Systems held at Ohio State University, Columbus, Ohio in September 1988. An extensive and scholarly examination of sustainable agricultural systems and their impact in all areas of the world. Contains 40 chapters written by over 70 contributors covering history, goals, and components of sustainable agriculture, integrated sustainable farming systems, sustainable agriculture in the tropics, policy development, and ecological impacts. Stresses the need to create and maintain economically viable sustainable farming systems.

Glen, D.M., M.P. Greaves, and H.M. Anderson, eds. 1995. *Ecology and Integrated Farming Systems: Proceedings of the 13th Long Ashton International Symposium.* Chichester, UK: Wiley. 343 p. ISBN: 0471955345.   Papers presented at a conference on integrated farming systems that emphasize ecological technologies and sustainable production. Topics include plant diversity, weed control, pests and natural enemies, soil management, and other areas relevant to a holistic approach to farming systems.

Lake, J.V., G.R. Bock, and J.A. Goode, eds. 1997. *Precision Agriculture: Spatial and Temporal Variability of Environmental Quality.* Ciba Foundation Symposium 210. Chichester, UK: New York: Wiley. 259 p. ISBN: 0471974552.   Papers and discussion from a symposium covering a range of spatial and temporal aspects of precision agriculture. Chapters deal with sapling techniques, crop-growth models, remote sensing, geographical information, weed management, and statistical methods.

Lockeretz, W., ed. 1996. *Environmental Enhancement Through Agriculture: Proceedings of a Conference Held in Boston, Massachusetts, November 15–17, 1995.* Medford, MA: Tufts University, Center for Agriculture, Food and Environment, School of Nutrition Science and Policy. 346 p.   Thirty-six papers devoted to improving the environmental situation of U.S. agriculture. Offers a range of viewpoints regarding wateshed protection; wildlife conservation and biodiversity; livestock systems; waste recycling and nutrient management; energy from agricultural biomass; farmland preservation; national policies; and agricultural development.

McDougall, E.A., ed. 1990. *Sustainable Agriculture in Africa*: *Proceedings of the Agricultural Systems and Research Workshop and Selected Papers from the Canadian Association of African Studies Meeting, University of Alberta, Edmonton, May 1987*. Trenton, NJ: Africa World Press. 340 p. ISBN: 0865431477.  A collection of papers from the conference addressing issues regarding sustainable agricultural practices in Africa. The introductory chapter contains comments, evaluations and conclusions from a workshop on farming systems research and development. The papers discus development of sustainable agriculture and the political and economic implications throughout Africa. Presents examples from Kenya, Nigeria, Sierra Leone, Ghana, Botswana, Senegal, Ivory Coast, and Tanazania. Includes a 30-page annotated bibliography on agriculture in Africa.

Meyer, G.E. and J.A. DeShazer, eds. 1999. *Precision Agriculture and Biological Quality*. SPIE Proceedings Series, v. 3543. Bellingham, WA: SPIE—the International Society for Optical Engineering. 399 p. ISBN: 0819431559.  Scholarly papers from a conference on optical and machine systems for agriculture and biological systems. Papers focus on spectrometry, sensing, imaging, identification and detection of crops, weeds, seeds, and agricultural products.

National Research Council, Board on Agriculture. 1991. *Sustainable Agriculture*: *Research and Education in the Field*: *A Proceedings*. Washington, DC: National Academy Press. 446 p. ISBN: 0309045789.  Research reports from a workshop on the scientific and technological basis of sustainable and economically viable agricultural production systems. The workshop explored regional differences and challenges facing American farmers. Gaps in research strategies, common themes and approaches in successful research, and areas needing new approaches were emphasized. Consists of 22 scholarly papers arranged in 6 parts. Available from the World Wide Web at http://books.nap.edu/books/0309045789/html/index.html.

Ogilvie. J.R., J. Smithers, and S.E. Wall, eds. 2000. *Sustaining Agriculture in the 21st Century*: *Proceedings of the 4th Biennial Meeting, North American Chapter, International Farming Systems Association, October 20–24, 1999*. Guelph, Canada: University of Guelph. 350 p. ISBN: 0889555001.  This conference addressed a wide range of issues regarding the strategies for the development and expansion of sustainable farming systems around the world in the twenty-first century. Over 70 contributors presented papers on research and applications to sustainable agriculture focusing on environment, crop diversification, soil conservation, water quality, nutrient management, dairy and livestock industries, greenhouse pest control, computer applications, food safety, international standards, rural development, landscaping, and so forth. Includes research and case studies from Brazil, Canada, Colombia, Ecuador, Guatemala, Peru, the United States, and Zimbabwe.

Olson, R.K., C.A. Francis, and S. Kaffka. 1995. *Exploring the Role of Diversity in Sustainable Agriculture*: *Proceedings of a Symposium*. Madison, WI: American Society of Agronomy; Crop Science Society of America; Soil Science Society of America. 262 p. ISBN: 089118288.   Nine papers explore the relationships among diversity, function, and sustainability in agricultural systems, present a scale of diversity and illustrate how each level in the hierarchy reflects different issues and measures of diversity. The book's purpose is to show that a greater understanding of the concepts of diversity and biodiversity can be applied to developing more viable and profitable agricultural systems. Addresses diversity in soil microbes and microbiology; spatial and temporal diversity in production; diversification in farmscapes and landscapes; diversity within human, community, food, and agricultural systems; and diversity in regional ecosystems.

Persley, G.J. and M.M. Lantin. 2000. *Agricultural Biotechnology and the Poor*: *Proceedings of an International Conference on Biotechnology*. Washington, DC: Consultative Group on International Agricultural Research (CGIAR). 242 p.   A scientific conference devoted to exploring the risks and benefits of biotechnology and its applications to issues of environment and health worldwide. Papers present viewpoints on the opportunities, challenges, constraints, achievements, ethics, and public-policy issues regarding genetic engineering in agriculture and food security. Private- and public-sector research and experiences from selected countries are presented including China, India, Philippines, Thailand, Brazil, Costa Rica, Mexico, Egypt, Iran, Jordan, Kenya, South Africa, and Zimbabwe. Available from the World Wide Web at http://www.cgiar.org/biotech/rep0100/contents.htm.

Regland, J. and R. Lal, eds. 1993. *Technologies for Sustainable Agriculture in the Tropics*. Special Publication no. 56. Madison, WI: American Society of Agronomy, Crop Science Society of America, Soil Science Society of America. 332 p. ISBN: 0891181180.   Papers from two symposia explore seven aspects of tropical agriculture from sub-Saharan Africa: basic concepts, technological options, agroforestry and nutrient cycling, vegetative hedges for erosion management, computer models, socioeconomic considerations, and case studies. This book serves as a valuable source of information on sustaining the valuable agricultural resources of the tropics.

Stolton, S., B. Geier, and J.A. McNeely. 2000. *The Relationship Between Nature Conservation, Biodiversity and Organic Agriculture. Proceedings of an International Workshop Held in Vignola, Italy 1999*. Tholey-Theley, Germany: IFOAM. 224 p. ISBN: 3934055052.   Proceedings of a workshop sponsored jointly by International Federation of Organic Agriculture Movements (IFOAM), World Conservation Union (IUCN) and World Wildlife Fund (WWF) on the theme of the relationship of organic agriculture toward biodiversity and nature conservation. Presents a wide range of papers and discussions on genetic diversity, diversity of agroecosystems and rural landscapes, protected areas for sustainable development, and policies regarding agricultural biodiversity worldwide.

## Pre-1990 Sources of Potential Interest

Bezdicek, D., ed. 1984. *Organic Farming: Current Technology and its Role in a Sustainable Agriculture: Proceedings of a Symposium*. Madison, WI: ASA, CSSA, and SSSA. (ASA Special Publication Number 46). 199 p. ISBN: 0891180761. Compares organic and conventional agriculture in the United States and Europe.

Edens, T.C., C. Fridgen, and S.L. Battenfield. 1985. *Sustainable Agriculture and Integrated Farming Systems: 1984 Conference Proceedings*. East Lansing, MI: Michigan State University Press. 349 p. ISBN: 0870132385. Defines sustainable agricultural systems and the structure, management, economics, ecological impacts, ethics and values, and the integration of these elements into farming systems.

Hutchinson, F.E. 1989. "Opportunities and Implications for U.S. Universities in International Sustainable Agriculture Programs." In: *Proceedings of the Annual Meeting. National Association of State Universities and Land-Grant Colleges, November 19–21, 1989*. Washington, DC: NASULGC, pp. 85–91. Identifies the major constraints, necessary policy changes, and procedures for building upon existing programs and projects in sustainable agriculture.

Thomas, G.W. 1989. "Sustainable Agriculture: Timely Thrust for International Development." In: *Proceedings of the Annual Meeting. National Association of State Universities and Land Grant Colleges, November 19–21, 1989*. Washington, DC: NASULGC, pp. 75–84. Covers the concept of sustainable agriculture, complexities of environmental issues, importance of site-specific approaches, mechanisms for change, and the research and extension imperatives for universities.

Wickens, G.E., N. Haq, and P. Day. 1989. *New Crops for Food and Industry*. New York: Chapman and Hall. 444 p. ISBN: 0412315009. Discusses the potential of new crops as profitable enterprises for developing areas with adaptability to a wide range of climates and topography.

## Conferences and Symposia in Series

Symposia of the Farming Systems Research and Extension (FSR/E) Association

Thirteen symposia were held between 1981 and 1992. Some were published in their entirety, while others were described only in abstracts.

1. *Farming Systems Symposium, 13th (International Sponsorship) Recherches-systeme en Agriculture et Developpement Rural: Symposium International, Montpellier, France—21–25 Novembre 1994, Communications.* M. Sebillotte, ed. Montpellier: CIRAD-SAR. 3 vols.

2. *Toward a New Paradigm for Farming Systems Research/Extension. 12th Annual Symposium, Association for Farming Systems Research and Extension. (1992: Michigan State University). East Lansing, MI, Michigan State University.* 635 p.

3. *Farming Systems Research and Extension in the 1990s: Critical Issues and Future Directions. 11th Annual Symposium, Association for Farming Systems Research and Extension. (1991: Michigan State University).* Not published.

4. *The Role of Farmers in FSR/E and Sustainable Agriculture. 10th Annual Symposium, Association for Farming Systems Research and Extension. (1990: Michigan State University).* Not published.

5. *Impacts of Farming Systems Research/Extension on Sustainable Agriculture. Farming Systems Research/Extension Symposium (1989: University of Arkansas).* 1999. Manhattan, KS: Kansas State University. 94 p.

6. *Contributions of FSR/E Towards Sustainable Agricultural Systems. Farming Systems Research/Extension Symposium (1988: University of Arkansas).* 1988. Fayetteville, AR: University of Arkansas, Winrock International Institute for Agricultural Development. 411 p.

7. *How Systems Work: Farming Systems Research Symposium, 1987. Farming Systems Research/Extension Symposium (1987: University of Arkansas).* 1987. Fayetteville, AR: University of Arkansas, Winrock International Institute for Agricultural Development. 509 p.

8. *Farming Systems Research and Extension: Food and Feed. Farming Systems Research Symposium (1986: Kansas State University).* C.B. Flora and M. Tomecek, eds. 797 p.

9. *Farming Systems Research and Extension: Management and Methodology. Farming Systems Research Symposium (1985: Kansas State University).* C.B. Flora and M. Tomecek, eds. 1 vol.

10. *Farming Systems Research and Extension: Implementation and Monitoring. Farming Systems Research Symposium (1984: Kansas State University).* 1986. C.B. Flora and M. Tomecek, eds. Manhattan, KS: Kansas State University. 588 p. 1 vol. Topics: Institutionalizing FSR/

E: the Asian experience; domestic FSR/E experiences; case studies; technology and farming systems; on-farm trials.

11. *Animals in the Farming System. Farming Systems Research Symposium (1983: Kansas State University).* 1984. C.B. Flora and P.P. Nichols, eds. Manhattan, KS: Kansas State University. 924 columns.

12. *Farming Systems in the Field. Farming Systems Research Symposium (1982: Kansas State University).* 1982: Manhattan, KS: Office of International Agricultural Programs, Kansas State University. 311 p.

13. *Small Farms in a Changing World: Prospects for the Eighties. Farming Systems Research Symposium (1981: Kansas State University).* 1982. W.J. Sheppard, ed. Manhattan, KS: Kansas State University. 190 p.

## National Symposium on New Crops

A symposium dedicated to new, specialty, neglected, and underutilized crops. The participants presented papers on all aspects of new crop development, policy, economics, breeding, marketing, and information systems. Covers new crops for use as vegetables, fruits and nuts, forages, grains, legumes, cereals, fiber, oilseed, industrial, floral, aromatic, herbal, medicinal, bioactive, and space uses. Four meetings have been held and published as follows:

Janick, J., ed. 1999. *Perspectives on New Crops and New Uses: Proceedings of the Fourth National Symposium, New Crops and New Uses: Biodiversity and Agricultural Sustainability, Phoenix, Arizona, November 8–11, 1998.* Alexandria, VA: ASHS Press. 546 p. ISBN: 0961502703.

Janick, J., ed. 1996. *Progress in New Crops: Proceedings of the Third National Symposium NEW CROPS—New Opportunities, New Technologies, Indianapolis, Indiana, October 22–25, 1996.* Alexandria, VA: ASHS Press. 679 p. ISBN: 0961502738.

Janick, J. and J.E. Simon, eds. 1993. *New Crops: Proceedings of the Second National Symposium NEW CROPS—Exploration, Research, and Commercialization, Indianapolis, Indiana, October 6–9, 1991.* New York: Wiley. 731 p. ISBN: 0471593745.

Janick, J. and J.E. Simon, eds. 1990. *Advances in New Crops: Proceedings of the First National Symposium NEW CROPS: Research, Development, Economics.* Portland, OR: Timber Press. 582 p. ISBN: 0881921661.

Also available on CD-ROM as The New Crop Compendium available from Purdue University Center for New Crops and Plant Products, which includes the entire text and figures from the second, third, and fourth national symposiums (http://www.hort.purdue.edu/newcrop/compendium/order.html).

## Special Conferences and Workshops on Precision Agriculture Sponsored by the American Society of Agronomy, Crop Science Society of America, and Soil Science Society of America

Papers and poster sessions presented at these conferences covered various aspects of precision agriculture and its role in farming systems management. Topics included soil resources, managing variability, technology, profitability, environment, technology transfer, development needs, and a range of other issues. A selection of the conferences are listed in the following text.

Pierce, F.J., and E.J. Sadler, eds. 1997. *The State of Site Specific Management for Agriculture*. Madison, WI: American Society of Agronomy; Crop Science Society of America; Soil Science Society of America. 442 p. ISBN: 0891181342. Discussions of precision agriculture research and sharing of new knowledge on the current state of site-specific management to improve crop performance and environmental quality.

*Proceedings of the First Workshop on Soil Specific Crop Management*: a Workshop on Research and Development Issues. 1993. Madison, WI: American Society of Agronomy; Crop Science Society of America; Soil Science Society of America. 406 p. ISBN: 0891181164.

*Proceedings of the Second International Conference on Site-Specific Management for Agricultural Systems*. 1994. Madison, WI: American Society of Agronomy; Crop Science Society of America; Soil Science Society of America. 1,010 p. ISBN: 089118127X.

*Proceedings of the Third International Conference on Precision Agriculture*. 1996. Madison, WI: American Society of Agronomy; Crop Science Society of America; Soil Science Society of America. 1243 p. ISBN: 0891181326.

Robert, P.C., R.H. Rust, and W.E. Larson, eds. *Proceedings of the Fourth International Conference on Precision Agriculture*. 1999. 2 vols. Madison, WI: American Society of Agronomy; Crop Science Society of America; Soil Science Society of America. 1,965 p. ISBN: 0891181407.

## IFOAM Conferences

IFOAM is an international organization that represents and coordinates the organic agriculture movement around the world and provides a platform for global exchange and cooperation through international, continental, and regional conferences. Proceedings of conferences are published or distributed by IFOAM,

Tholey-Theley, Germany. More information on IFOAM can be found on their Web site at http://www.ifoam.org.

*13ᵗʰ International IFOAM Conference*: *The World Grows Organic*. 2000. Amsterdam. T. Alföldi, W. Lockeretz, and U. Niggli, eds. Basel, Switzerland: IOS Press. 790 p. ISBN: 1586030876.

*12ᵗʰ International IFOAM Conference*: *Organic Agriculture, the Credible Solution for the XXIst Century*. 1998. Mar del Plata, Argentina. D. Foguelman and W. Lockeretz, eds. Tholey-Theley: Germany, IFOAM. 275 p. ISBN: 3934055036.

*11ᵗʰ International IFOAM Conference*. Copenhagen, Denmark, August 11–15, 1996. 2 vols. Tholey-Theley: Germany, IFOAM. Volume 1: *Fundamentals of Organic Agriculture*. ISBN: 3930720981. Volume 2: *New Research in Organic Agriculture*. ISBN: 393072099X.

*10ᵗʰ International IFOAM Conference*: *People, Ecology, Agriculture*: *Paper Abstracts and Conference Programme*. 1994. Lincoln University, New Zealand: New Zealand Fruitgrowers Federation. 165 p.

*9ᵗʰ International IFOAM Conference*: *Organic Agriculture, a Key to a Sound Development and a Sustainable Environment*. Sao Paulo, Brazil. 1992. U. Köpke and D.G. Schulz, eds. Tholey-Theley: Germany, IFOAM. 438 p.

*7ᵗʰ International IFOAM Conference*: *Agricultural Alternatives and Nutritional Self-Sufficiency*. Ouagadougou, Burkina Faso. 1990. A. Djigma, ed. Witzenhausen, Germany: Verlagsgruppe Ekopan. 429 p. ISBN: 3927080136

*5ᵗʰ International IFOAM Conference*: *The Importance of Biological Agriculture in a World of Diminishing Resources, Proceedings from Scientific Conference*. University of Cassel, Germany. 1984. H. Vogtmann, E. Boehncke, and I. Fricke, eds. Witzenhausen, Germany: Verlagsgruppe Witzenhausen. 448 p. ISBN: 392605901X.

*4th International IFOAM Conference on Trade in Organic Products*. Frankfurt, Germany. 1995. M. Haccius, A. Bernd, and B. Geier, eds. Tholey-Theley: Germany, IFOAM. 131 p.

*2ⁿᵈ International IFOAM Conference on Organic Textiles*. Bingen, Germany. 1997. Konstanz, Germany: Institute für Marktökologie. 205 p.

### Other IFOAM Conferences

*The Future Agenda for Organic Trade*: *Conference Proceedings, the 5th IFOAM International Conference on Trade in Organic Products*. Christ Church College, Oxford: Tholey-Theley, Germany. 24–27 September. 1997. T. Maxted-Frost, IFOAM, and The Soil Association, eds. 58 p. ISBN: 0905200632.

*1990 Australian Organic Agriculture Conference*: *A Regional Conference of the International Federation of Organic Agriculture Movements (IFOAM)*: *Volumes I, II, & III, Conference papers.* University of Adelaide, Adelaide, South Australia, September 23–30, 1990. 3 vols. Norwood, South Australia. ISBN: 0646025295. Volume 1: Keynote Papers; Volume 2: Submitted Papers; Volume 3: Field Visits Guide-Book.

## DICTIONARIES AND GLOSSARIES

The growth of farming systems literature has resulted in the emergence of new terminology that defines and describes the various systems. Although these terms now appear regularly in the literature, they are sometimes unclear or confusing. Following are some dictionaries and glossaries covering terminology of sustainable agriculture and its components.

Agroecology Glossary: A Glossary of Terms Used in Agroecology (http://www.agroecology.org/glossary/). 1999. Brief definitions of terms used in agroecology.

Frey, R.S. 1996. *A Glossary of Agriculture, Environment, and Sustainable Development.* (Agricultural Experiment Station Bulletin 661). Manhattan, KS: Kansas State University, Agricultural Experiment Station. 24 p. ISSN: 0097-0484. Contains definitions of over 500 terms related to agricultural production, the environment, and sustainable development. Terms were chosen from a variety of disciplines to increase public awareness of issues surrounding sustainable agriculture worldwide. Includes a bibliography of recommended readings for further information. Available from the World Wide Web at http://www.oznet.ksu.edu/_library/misc2/samplers/sb661.htm.

Gold, M.V. 1999. *Sustainable Agriculture*: *Definitions and Terms.* Special Reference Briefs no. SRB 99-02. SRB 94-05 (Update). Beltsville, MD: National Agricultural Library, Alternative Farming Systems Information Center. 40 p. ISSN: 1052-5368. This paper discusses the concepts and elements of sustainable agriculture and the ecologic, economic, social, health, and philosophic considerations that comprise sustainable farming systems and brief descriptions of the methodologies and practices currently associated with them. Many of the descriptions are derived from other sources and full credit is given. Includes a bibliography of books, articles, Web documents, and videocassettes for further reading and viewing. Available from the World Wide Web at http://www.nal.usda.gov/afsic/AFSIC_pubs/srb9902.htm#toc1.

*List of Sustainable Agriculture Terms.* 2nd ed. 1995. Lima, Peru: European Network for Low External Input and Sustainable Agriculture (EULEISA). 166 p. A dictionary of terms related to sustainable agriculture in English, French, and

Spanish. Arranged by categories under agriculture, economics/socioeconomics/ rural sociology, resource management/conservation, land-use systems, crop production, agriculture mechanization/equipment/tools, animal production, forestry, and aquaculture/fisheries.

## DIRECTORIES AND BIBLIOGRAPHIES

Directories are useful guides to organizations, publications, research institutions, government agencies, and individuals involved in farming research and projects. The directories cited here provide information on persons, places, businesses, organizations, farmers, educational institutions, and other information centers that are actively sustainable farming systems. The bibliographies focus on the great diversity of alternative farming systems literature. They cite periodical articles, books, reports and other materials that present an introduction and overview of farming systems literature for readers who want to increase their awareness of the topic. The cited works provide an awareness of the philosophies and practices of farming systems, and the range of issues that impact upon their successful implementation.

*AFSIC Volunteer Staff. Sustainable Agriculture in Print: Current Books. 2001.* Special Reference Briefs Series no. SRB 97-05. Beltsville, MD: National Agricultural Library, Alternative Farming Systems Information Center. Paging varies. Revised and updated annually. ISSN: 1052-536X.   Comprehensive annotated bibliography of books, reports, government publications, conference proceedings, and other printed resources on sustainable farming systems. Provides complete bibliographic information and detailed annotations for each resource. Available from the World Wide Web at http://www.nal.usda.gov/afsic/AFSIC_pubs/ srb9705u.htm.

Burton, V. 1997. *Source Book of Sustainable Agriculture for Educators, Producers and Other Agricultural Professionals*. Burlington, VT: Sustainable Agriculture Network. 140 p. ISBN: 1888626038.   A comprehensive annotated guide to 559 free and low-cost books, newsletters, conference proceedings, bulletins, videos, reports, government publications, and Web sites about sustainable agriculture. Cites print, electronic, and video resources from how to market sustainably grown vegetables to the latest sustainable research findings and new information on the World Wide Web. Gives descriptions of cited items and contact and ordering information. Contains subject, organization and author indexes, and appendices with directories of sustainable agriculture organizations and relevant Web sites. Available from the World Wide Web at http://www.sare.org/htdocs/pubs/ resources/index.html#Directory.

DeMuth, S. and M.V. Gold. 2000. *Organizations and Web Sites Related to Community Supported Agriculture*. Beltsville, MD: National Agricultural Library, Alternative Farming Systems Information Center. Paging varies. Updated periodically. A guide to over 45 state, national, and regional organizations in the United States devoted to support and networking on community supported agriculture. All listings give complete address, telephone and fax numbers, e-mail addresses, and URLs. Includes descriptions of Web sites devoted mainly to community supported agriculture resources. Available from the World Wide Web at http://www.nal.usda.gov/afsic/csa/csaorgs.htm.

Diver, S. Find Ag Information on the Web http://www.attra.org/searchAgWeb.html. 2001 (Updated periodically). Fayetteville, AR: Appropriate Technology Transfer for Rural Areas. 2001. An extensive online directory of agriculture databases, directories, library catalogs, and search engines on the Internet. Provides descriptions and links to publications, databases, bibliographies, current research, library catalogs, directories, periodicals and newsletters, publishers, and miscellaneous sites offering information on all aspects of agriculture.

Duke, J.A., compiler. 1983. Handbook of Energy Crops (http://newcrop.hort.purdue.edu/newcrop/duke%5Fenergy/dukeindex.html). West LaFayette, IN: Purdue University. (Last update 1988). Guide to 200 species of plants arranged alphabetically that may be valuable for energy uses in fuel, chemical, or material applications. Each listing provides discussion and presentation of available information such as nomenclature, uses, folk medicine, chemical composition, botanical description, germplasm, distribution, ecology, cultivation, harvesting, yields and economics, energy, biotic factors, and key references.

Gates, J.P. et al. 2000. Educational and Training Opportunities in Sustainable Agriculture. 13th ed. Volunteers and Staff of Alternative Farming Systems Information Center, National Agricultural Library. Beltsville, MD: National Agricultural Library, Alternative Farming Systems Information Center. 74 p. Directory of selected organizations and institutions involved in organic, alternative, or sustainable agriculture that provide education, training, and information. Covers mainly United States and Canada and some overseas listings from Africa, European, and Scandinavian countries. Listings give complete addresses, telephone and fax numbers, and e-mail addresses, URLs, and short descriptions. Available from the World Wide Web at http://www.nal.usda.gov/afsic/AFSIC%5Fpubs/edtr12.htm.

Gates, J.P. 1995. *Sustainable Agriculture in Print*: *Current Periodicals*. Special Reference Briefs 95-08. Beltsville, MD: National Agricultural Library, Alternative Farming Systems Information Center. 140 p. (Revised periodically). Guide to international, national, and regional journals and newsletters devoted to all aspects of sustainable agriculture including farming, environmental protection, food and health, rural development, and social issues. Titles include publications from farmers and gardeners organizations, citizens groups, professional societies, trade organizations, extension service, government agencies, and commercial publishers. Listings give complete bibliographic information, frequency, addresses, costs, and URLs. Available from the World Wide Web at http://www.nal.usda.gov/afsic/AFSIC_pubs/srb95-08.htm.

Gates, J.P. 1988. *Tracing the Evolution of Organic/Sustainable Agriculture*: *A Selected and Annotated Bibliography*. Bibliographies and Literature of Agriculture no. 72. Beltsville, MD: National Agricultural Library, Alternative Farming Systems Information Center, November, 1988. 20 p. Traces the history of organic and sustainable agriculture and its philosophy and scientific concepts from 1580 to 1986. Cites 53 selected reports and monographs arranged chronologically. Each reference contains a brief annotation. Available from the World Wide Web at http://www.nal.usda.gov/afsic/AFSIC_pubs/tracing.htm.

Gold, M. *Sustainable Agricultural Resources for Teachers*, *K–12*. 2000. Beltsville, MD: National Agricultural Library, Alternative Farming Systems Information Center, May, 1998. (AFSIC Notes Series no. 4). 13 p. ISSN: 1063-262X. (Revised periodically). Guide to books, periodicals, posters, and other materials used for teaching sustainable agriculture in elementary and high schools. Provides names and addresses of suppliers or publishers, types of resources, contact persons, and links to the respective Web sites. Available from the World Wide Web at http://www.nal.usda.gov/afsic/AFSIC_pubs/k-12.htm.

Hegyes, G. and C. Francis. 1998. *Future Horizons*: *Recent Literature in Sustainable Agriculture*. Lincoln, NE: University of Nebraska-Lincoln, Center for Sustainable Agricultural Systems, Extension and Education Materials for Sustainable Agriculture: Volume 6. Detailed reviews of over 90 books, agricultural extension publications and papers on sustainable agriculture and related environmental issues. Includes works addressing environment, economics, sociology, farm management, conservation, ethics, and other topics. Available from the World Wide Web at http://www.ianr.unl.edu/ianr/csas/extvol6.htm.

Hemenway, D. 2000. *The Resources of International Permaculture*. Sparr, FL: Yankee Permaculture. 70 p. A listing of 2,070 organizations involved in permaculture, sustainable food systems, and tree-based agriculture. Arranged by location, name of group, and function. Available in print and disk versions. Disk version includes keyword search capability.

Johnson, J.S., R.C. Fisher, and C.B. Robertson, eds. 2000. *Agricultural Information Resource Centers*: *A World Directory 2000*. 3rd ed. Twin Falls, ID: IAALD, 752 p. ISBN: 0962405221. Available in print and CD-ROM versions. See description in General Sources.

Nanda, M. 1990. *Planting the Future: A Resource Guide to Sustainable Agriculture in the Third World.* Minneapolis, MN: International Alliance for Sustainable Agriculture. 335 p. A comprehensive listing of sustainable agriculture resources and groups in the Third World with detailed descriptions of farming practices in the respective countries. Covers 225 groups and 246 resource centers in 48 countries in Africa, Asia, Latin America, and the Caribbean. Each entry includes philosophy, objectives, finances, publications, networking, and volunteer opportunities. The introduction discusses sustainable agriculture and its benefits. Includes indexes by topic, group, and country; glossary; tables; lists of books, periodicals, videos; and other information.

*National Organic Directory: Guide to Organic Information and Resources Worldwide.* 2001. 18th ed. Davis, CA: Community Alliance with Family Farmers (CAFF). 298 p. ISBN: 1891894056. (This will be the last *National Organic Directory* published by CAFF). Contains the latest information on over 500 organic farmers, food wholesalers, manufacturers, farm suppliers, support businesses, certification groups, resource groups, periodicals, and publications relating to organic agriculture. Each listing provides contact names; telephone and fax numbers; terms and services; regions served; e-mail and Web site addresses; and indexes for importers, exporters, manufacturers, and processors. Includes timely information about state laws regarding organic foods, standardization, certification, and other topics.

*Organic Agriculture Worldwide Directory of the Member Organizations and Corporate Associates of IFOAM.* 2001. Tholey-Theley, Germany: IFOAM. Paging varies. Annual. Contains addresses, telephone and fax numbers, key activities, and other relevant information on about 770 organic agriculture contacts in approximately 150 countries. Also includes logos of IFOAM member organizations and corporate associates. Available from the World Wide Web http://www.ifoam.org.

*The Organic Pages: Organic Trade Association's 1999 North American Resource Directory.* 1999. Greenfield, MA: OTA Press. 330 p. ISBN: 1881427919. This is a comprehensive resource for all aspects of the organic food and fiber industry from farmer to consumer. Contains hundreds of entries providing names, descriptions, addresses, telephone and fax numbers, e-mail, contact persons, and other relevant information for companies, brand names, suppliers, growers, manufacturers, importers/exporters, retailers, associations, publishers, etc. All needs for organic produce, products, and services can be found in this directory. Available from the World Wide Web at http://www.ota.com/online%20directory/Directory%20nest.htm.

Sustainable Agriculture Network. 1997. *Source Book of Sustainable Agriculture.* Burlington, VT: Sustainable Agriculture Network. 140 p. ISBN: 1888626038.

(Revised periodically). The source book lists 559 resource materials, from how to market sustainably grown vegetables to locating the latest sustainable research findings on the World Wide Web. Covers print, electronic, and video resources; and contacts and ordering information. Includes indexes organized by subject, organization, and author; a separate listing of videos; and appendices covering sustainable agriculture centers and Web sites. Available from the World Wide Web at http://www.sare.org/sourcebook/.

Sustainable Agriculture Network. 1996. *Sustainable Agriculture Directory of Expertise*. 3rd ed. Burlington, VT: Sustainable Agriculture Publications. 1 vol. Unpaged. (Revised periodically). ISBN: 1888626003. Contains 723 entries that identify and describe nearly 1,000 individuals and more than 200 organizations throughout the United States. Those seeking information on current projects and research on economically profitable and environmentally sound agricultural systems will find this a valuable resource. Arranged by state and territory, with six indexes to lead users directly to the most pertinent information. Available from the World Wide Web at http://www.sare.org/san/htdocs/pubs.

*Sustainable Agriculture Organizations and Publications*. 2000. Fayetteville, AR: Appropriate Technology Transfer for Rural Areas (ATTRA). 38 p. Guide to over 200 organizations, agencies, and farms in the United States and Canada devoted to promoting sustainable agriculture. Arranged by national, regional, and state categories. Listings give names, addresses, telephone numbers, e-mail address, URL, descriptions, and publications. Available from the World Wide Web at http://www.attra.org/attra-rl/susagorg.html.

*Sustainable Farming Internships and Apprenticeships*. 2001. Fayetteville, AR: Appropriate Technology Transfer for Rural Areas (ATTRA). 120 p. This directory has been published by ATTRA since 1989 as a networking tool for farmers and apprentices to connect with each other for learning opportunities in sustainable agriculture. Twelve new listings were added in 2000 and 34 listings in 2001. There are 280 entries arranged by region and states with each listing giving name; address and contact information; and descriptions of farms and institutions seeking interns. Available from the World Wide Web at http://www.attra.org/attra-rl/intern.html.

Weintraub, I. 1993. ''Alternative Agriculture: Selected Information Sources— Part I: Databases, Abstracts and Indexes, Periodicals, and Newsletters and Newspapers.'' *Journal of Agricultural and Food Information* 1(3): 41–100. ''Alternative Agriculture: Selected Information Sources—Part II: Bibliographies, Reports, and Directories.'' *Journal of Agricultural and Food Information* 1(4): 33–96. An annotated bibliography of selected information resources representing the philosophy, practices, and issues involved in alternative agriculture. Sources cited offer readers an introduction to the wide range of alternative farming systems and motivations for switching from conventional to alternative approaches.

## ENCYCLOPEDIAS AND HANDBOOKS

Encyclopedias and handbooks are designed to provide up-to-date information on a particular topic and its related subtopics in an organized manner. They are published in single-volume and/or multivolume editions and generally arranged as a series of articles providing in-depth coverage of an area of interest. This section cites current works in which readers can find significant information on practices relevant to conventional and alternative farming systems.

*Alternative Field Crops Manual.* 1990. Madison, WI: University of Wisconsin, Cooperative Extension. A source of information on 49 field crops adaptable to Midwest farms that may be considered as alternatives to traditional farm commodities. For each crop the manual provides history; uses; growth habits; environment requirements; cultural practices; yield potential and performance results; economics of production and markets; and sources of information for further reading for each crop. The manual is a joint project between the University of Wisconsin Cooperative Extension Service, the University of Minnesota Extension Service, and the Center for Alternative Plant and Animal Products. Available from the World Wide Web at http://www.hort.purdue.edu/newcrop/afcm/index.html.

Arntzen, C.J. and E.M. Ritter, eds. 1994. *Encyclopedia of Agricultural Science.* 4 vols. San Diego, CA: Academic Press. ISBN: 0122266706. Comprehensive coverage of a wide variety of agricultural topics in animal, plant, soil, and range sciences; food science; agricultural economics; agricultural engineering; pest management; rural sociology; and water resources for students, researchers, farmers, and specialists.

Facciola, S. 1998. *Cornucopia II: A Source Book of Edible Plants.* Vista, CA: Kampong Publications. 727 p. ISBN: 0962808725. Guides to over 3,000 species of edible plants ranging from common vegetables to exotic varieties. Arranged in three sections under botanical listings, cultivar listings, and sources. Entries provide scientific and common names, origins, descriptions, distributions, uses, and commercial and noncommercial sources. Includes bibliographies, appendices with listings by usage categories, and name and species indexes.

Facciola, S. 1990. *Cornucopia: A Source Book of Edible Plants.* Vista, CA: Kampong Publications. 686 p. ISBN: 0962808709.

Sexton, W.T., et al. 1999. *Ecological Stewardship: A Common Reference for Ecosystem Management.* 3 vols. Oxford: Elsevier Science. ISBN:

0080432069. Covers 31 separate topics on scientific and management experiences regarding stewardship of the ecosystem. Addresses biological and ecological dimensions; humans as agents of ecological change; public expectations; values and law; social and cultural dimensions; economic dimension; and information and data management. Includes extensive coverage of agriculture, forestry, soil, and land management.

Yuste, M.P. and J. Gostincar. 1999. *Handbook of Agriculture*. New York: Marcel Dekker. 768 p. ISBN: 0824779142. This handbook covers both conventional chemical-based and natural fertilization and pest-control methods. Chapters are devoted to soils, fertilizers, fruit trees, control methods, agricultural techniques, horticulture, and greenhouse cultivation.

## MONOGRAPHS

The growing interest in environmentally sound agricultural production systems has given rise to a prolific body of literature on all aspects of the topic. Books on farming systems have been published for general, scholarly, and research audiences in order to encourage adoption of sustainable production and enhance the quality of life for farmers and consumers. Many of those works have become classical sources that are consulted regularly as guides to farming systems. The titles cited here offer readers an overview of the advancements in farming systems and its place in the future of agricultural production.

Ableman, M. 1993. *From the Good Earth: A Celebration of Growing Food Around the World*. New York: Abrams. ISBN: 0810933187. A photographic essay comprising 170 full-color photographs of farming practices in China, Kenya, Burundi, Italy, Peru, the American Southwest, and other areas with views on the contrast between conventional, chemical-based agriculture and nonchemical approaches still used in most of the world.

Ableman, M. and C. Wisehart. 1998. *On Good Land: The Autobiography of an Urban Farm*. San Francisco: Chronicle Books. 144 p. ISBN: 0811819213. Ableman recounts the 15 years he spent developing Fairview Gardens, which involved learning to operate an organic farm, countering community opposition, and ultimately gaining support of the community.

Allen, P. 1993. *Food For the Future: Conditions and Contradictions of Sustainability*. New York: Wiley-Interscience. 339 p. ISBN: 0471580821. Essays by 15 scholars explore the social, economic, political, and ethical aspects of the transformation to sustainable agricultural systems. Offers new approaches to understanding sustainability, its limitations and future potential, and the need for all stakeholders to participate in the development of sustainable agriculture around the world.

Altieri, M.A. 1995. *Agroecology: The Science of Sustainable Agriculture*. 2nd ed. Boulder, CO: Westview Press; London: IT Publications. 445 p. ISBN: 0813317177. Presents the concepts and practices of sustainable agriculture and management systems that work in the United States and developing countries. Addresses the theoretical basis of agricultural ecology; design of alternative agricultural systems and technologies; alternative production systems; and ecological management of insect pests, pathogens, and weeds. Case studies with sound scientific data from various countries reveal the social and economic benefits of adopting sustainable agriculture systems.

Altieri, M.A. and S.B. Hecht. 1990. *Agroecology and Small Farm Development*. Boca Raton, FL: CRC Press. 265 p. ISBN: 0849348854. A collection of 21 papers concerning research and case studies on applications of sustainable agriculture to small farms in Asia, Africa, and Latin America in mountain, mid-arid, monsoon, and tropical environments.

Bender, J. 1994. *Future Harvest: Pesticide-Free Farming*. Lincoln, NE: University of Nebraska Press. 177 p. ISBN: 080321233X. Offers a step-by-step approach for conversion to land stewardship models of agriculture that are free of synthetic pesticides. Explores chemical-based contemporary agriculture and offers solid reasons why conversion is necessary. Discusses problems and issues in conversion, role of livestock in alternative systems, and responses to critics who defend conventional production methods.

Bird, E.A.R., G.L. Bultena, and J.C. Gardner. 1995. *Planting the Future: Developing an Agriculture that Sustains Land and Community*. Ames, IA: Iowa State University Press. 299 p. ISBN: 0813820723. A socioeconomic comparison of sustainable and conventional farming in Iowa, Minnesota, North Dakota, Montana, and Oregon. The study describes the types farms; economic impact of the different approaches; principles and characteristics of sustainable farming; and research and policy options needed to advance the widespread adoption of sustainable agriculture.

Campidonica, M., J.S. Auburn, and E. O'Farrell. 1997. *How to Find Agricultural Information on the Internet*. UC Division of Agriculture and Natural Resources Publication 3387. Oakland, CA: University of California, Division of Agriculture and Natural Resources. 100 p. ISBN: 1979906317. A practical guide for anybody involved in agriculture from farmers to marketers. Teaches a basic approach to gathering information and using e-mail with realistic examples to illustrate the benefits and limitations of the Internet. Includes samples of Internet sites on crops and livestock, markets, food and nutrition, research, and weather.

Canter, L.W. 1986. *Environmental Impacts of Agricultural Production Activities*. Chelsea, MI: Lewis Publishers. 400 p. ISBN: 0873710665. A summary of environmental impacts from existing and emerging agricultural production technologies. Identifies water and soil, air quality, noise, and solid waste impacts. Compares emerging technologies that will minimize or eliminate environmental damage.

Carroll, R.C., J.H. Vandermeer, and P. Rosset, eds. 1990. *Agroecology*. New York: McGraw-Hill. 645 p. ISBN: 007052923X. A series of papers reflecting the diversity of biological approaches and multidisciplinary problems that are characteristic of agroecosystems. Arranged in four sections covering general background, ecological background, management questions, and research in agricultural systems.

Charles, Prince of Wales and C. Clover. 1993. *Highgrove: An Experiment in Organic Gardening and Farming*. New York: Simon & Schuster. 283 p. ISBN: 061798790. Prince Charles explains his motivation for choosing organic farming and its importance for the future of agriculture in the United Kingdom. Clover describes the history, cultural practices, and management techniques at the Princes' estate, Highgrove, and the lessons learned along the way. Includes over 160 full-color photographs illustrating the success of the Highgrove experiment.

Collins, W.W., ed. 1999. *Biodiversity in Agroecosystems*. Boca Raton, FL: CRC Press 352 p. ISBN: 1566702909. Presents a wide range of papers covering new research on biodiversity in agricultural ecosystems at both micro and macro levels. Covers biodiversity in microfauna, insects, wildlife, agrolandscapes, plants, and animals.

Collinson, M., ed. 2000. *A History of Farming Systems Research*. Wallingford, UK: CABI Publishing. 448 p. ISBN: 0851994059. A detailed history of farming systems research (FSR) focusing on its role with small-scale resource-poor farmers in developing countries. Over 40 papers illustrate the experiences, problems, diversities, and similarities of FSR in 20 countries.

Conway, G.R. and E.R. Barbier. 1990. *After the Green Revolution: Sustainable Agriculture for Development*. London: Earthscan Publications. 205 p. ISBN: 1853830356. An examination of the priorities and conditions for achieving agricultural sustainability in developing countries at the international, national, and local levels. Describes the hierarchical, economic, and policy restraints and the trade-offs and compromises that are needed to ensure agricultural sustainability.

de Souza Filho, H.M. 1998. *The Adoption of Sustainable Agricultural Technologies: A Case Study in the State of Espírito Santo, Brazil*. Aldershot, Hants, UK; Brookfield, VT: Ashgate. 189 p. ISBN: 184014103. An empirical study using duration analysis and econometric techniques to investigate the factors that determine the adoption and diffusion of sustainable practices among 141 farmers in the state of Espírito Santo in Brazil. Examines the role of the economic, environmental, and social aspects of the Green Revolution as a major influence on farmers' decisions to switch to sustainable practices.

Dahlberg, K. 1986. *New Directions for Agriculture and Agricultural Research: Neglected Dimensions and Emerging Alternatives*. Totowa, NJ: Rowman and Allanheld, 447 p. ISBN: 0847674177.   Analyzes alternative agricultural systems and their importance in setting priorities for research in the future.

Dower, R.C., et al. 1997. *Frontiers of Sustainability: Environmentally Sound Agriculture, Forestry, Transportation and Power Production*. Washington, DC: Island Press. 382 p. ISBN: 1559635460.   This book offers a vision of the sustainable future of the United States and the steps needed to get there. Authors examine environmental performance and trends in four key economic sectors: agriculture, electricity generation, transportation, and forestry. They identify potential environmental dangers that may occur if development continues on the current path, and recommendations for reducing, managing, or reversing them.

Dragun, A.K. and C. Tisdell. 1999. *Sustainable Agriculture and Environment: Globalisation and the Impact of Trade Liberalisation*. Cheltenham, UK; Northampton, MA: Edward Elgar. 329 p. ISBN: 184064172X.   Contains 19 essays covering the economic, social, and cultural impacts of sustainable agriculture and liberalized trade polices on the poor and developing countries. The authors point out that globalization and trade liberalization may have negative impacts for developing countries and may differ substantially between poor and rich countries.

D'Souza G.E. and T.G. Gebremedhin, eds. 1998. *Sustainability in Agricultural and Rural Development*. Aldershot, Hants, UK; Brookfield, VT: Ashgate. 262 p. ISBN: 1855219778.   This work explores the issues, concepts, methods of analysis, and empirical results related to the sustainable agriculture and rural communities. It contains 12 chapters contributed by authors who address economic, social, demographic, community, environmental, and human health perspectives.

Dudley, N., J. Madeley, and S. Stolton. 1992. *Land is Life: Land Reform and Sustainable Agriculture*. London: Intermediate Technology Publications in association with Foundation for Development and Peace. 165 p. ISBN: 1853391468.   Analyses of the principles of land reform and the relationships of these principles to sustainable agriculture. The study indicates that when farmers own their land and are free to make management decisions that are suited to their agricultural enterprise, they are likely to produce more food and to farm sustainably. When agrarian reform programs are carried out and farmers are given land, large increases in food output occur.

Flora, C.B., ed. 2001. *Interactions Between Agroecosystems and Rural Communities*. Boca Raton, FL: CRC Press. 289 p. ISBN: 0849309174.   An international team of sociologists, economics, and agriculturalists explore the positive and negative ways that human behavior can effect the sustainability and functioning of agroecosystems. Provides an analysis of agroecosystems and uses case studies to demonstrate how changes in the economy can influence local people to take certain actions regarding agroecosystem sustainability in their communities.

Francis, C.A., C.B. Flora, and L.D. King. 1990. *Sustainable Agriculture in Temperate Zones*. New York: Wiley. 500 p. ISBN: 0471622273. Collection of papers by farming systems specialists offering research data and references to the literature regarding sustainable agriculture in temperate zones. The papers cover research and case studies on breeding, pest management, weed management, soil fertility, crop rotations, soil biology, pasture management, resource efficiency, economics, rural communities, policy issues, and future dimensions. Papers focus on the environmental soundness and economic viability of sustainable agriculture with emphasis on agriculture in the United States.

Fukuoka, M. 1987. *The Natural Way of Farming*: *The Theory and Practice of Green Philosophy*. Translated by Frederic P. Metreaud. rev. ed. Tokyo; New York: Japan Publications. 284 p. ISBN: 0870406132. Indian Edition: Madras: Bookventure, 1997. ISBN: 8185987009. Fukuoka defines all the theories of modern agriculture with a natural approach to farming that shows a great appreciation for the land and respect for nature and natural diversity. He describes his methods for starting and keeping a natural farm and raising and nurturing rice, wheat, barley, fruit trees, and vegetables applying his ideas on disease and insect control, cultivation, and crop rotations.

Gliessman, S.R., ed. 2001. *Agroecosystem Sustainability*: *Developing Practical Strategies*. Boca Raton, FL: CRC Press. 210 p. ISBN: 0849308941. This book attempts to define and assess the concept of agroecology as it relates to sustainability of farming systems. The authors present case studies, practices in different regions, and strategies that help farmers make the change from conventional to sustainable farming.

Gliessman, S.R. 2000. *Field and Laboratory Investigations in Agroecology*. Eric W. Engles, ed. Boca Raton, FL: Lewis Publishers. 340 p. ISBN: 1566704456. Supplements the author's textbook *Agroecology*: *Ecological Processes in Sustainable Agriculture*. The manual contains 24 investigations in 5 sections covering studies of environmental factors, population dynamics in crop systems, interspecific interactions in cropping communities, farm and field systems, and food systems. Stresses hands-on experience involving observation, interpretation, and analysis of findings.

Gliessman, S.R., ed. 1998. *Agroecology*: *Ecological Processes in Sustainable Agriculture*. Chelsea, MI: Ann Arbor Press. 380 p. ISBN: 1575040433. Introduces the concept of agroecology and the importance of sustainable agricultural systems. Detailed coverage of environmental factors such as light, temperature, humidity, wind, and soils, and system level interactions including population, genetics, species, diversity, and disturbances. Discusses the procedures for making the transition to sustainable agriculture and food systems.

Gliessman, S.R., ed. 1990. *Agroecology*: *Researching the Ecological Basis for Sustainable Agriculture*. New York: Springer Verlag, 395 p. ISBN: 0387970822. Includes papers covering case studies and methodologies on production-oriented and systems-oriented approaches. Provides examples of the diversity and complexity of agroecology research and design and management of agroecosystems in temperate and tropical climates.

Hassanein, N.E. *Changing the Way America Farms*: *Knowledge and Community in the Sustainable Agriculture Movement*. 1999. Lincoln, NE: University of Nebraska Press. 228 p. ISBN: 0803273215. This book describes and analyzes two sustainable farming networks in Wisconsin. One group, the Ocooch Grazers Network, consisted of dairy farmers who practice intensive rotational grazing, a technique that represents a major departure from conventional, confinement-based dairying. The other group, the Wisconsin Women's Sustainable Farming Network, was formed to promote the success of women farmers and their various sustainable farming enterprises. The study points out that information exchange among local farming communities has the potential to transform agriculture.

Hatfield, J.L. and D.L. Karlen, eds. 1993. *Sustainable Agriculture Systems*. Boca Raton, FL: CRC Press. 322 p. ISBN: 1566700493. A description of the characteristics and practices of sustainable agriculture systems. Ten chapters cover historical perspective; water relationships; general management strategies; soil, crop, pest, and weed management; economic considerations; social change and sustainability; and challenges for the twenty-first century.

Hildebrand, P.E., ed. 1986. *Perspectives on Farming Systems Research and Extension*. Boulder, CO: Lynne Rienner. 181 p. ISBN: 0861876695. A collection of readings designed to provide background and historical perspective on farming systems research and extension (FSR/E).

Hildebrand, P.E. and F. Poey. 1985. *On-Farm Agronomic-Trials in Farming Systems Research and Extension*. Boulder, CO: Lynne Rienner. 178 p. ISBN: 0931477107. A manual for carrying out on-farm research trials in agricultural technology for production systems. Covers exploratory, site-specific, researcher-managed, and farmer-managed trials.

Hulse, J.H. 1995. *Science, Agriculture, and Food Security*. Ottawa, Canada: National Research Council of Canada. 256 p. ISBN: 0660162105. Comprehensive discussion of worldwide food security; food self-sufficiency; diet and consumption patterns; and the technical, economic, social, and political issues that shape it. Covers the relevance of sustainable agriculture and biological and ecological issues and the critical research priorities necessary for achieving food security.

Ilbery, B., Q. Chiotti, and T. Richard. 1997. *Agricultural Restructuring and Sustainability: A Geographical Perspective*. Wallingford, UK: CAB International, 362 p. ISBN: 0851991653. This book consists of 21 selected and revised papers from a conference held in North Carolina that brought together rural geographers from Canada, the United Kingdom, the United States, and New Zealand. The chapters examine, at various spatial scales, the broad processes and structural changes that are common to all rural systems in developed countries. Different geographical contexts are used to illustrate the uneven development of these processes and the implication for sustainable agriculture and rural systems. Authors provide both literature reviews and original research.

James, H. and K. Estes. 1997. *The Farmer's Guide to the Internet*. 3rd ed. Lexington, KY: TVA Rural Studies. 346 p. ISBN: 0964974630. Explains how to get started on the Internet; essential services and how to use them; detailed operating instructions for Netscape Navigator and Internet Explorer; how to find an Internet service provider; and hundreds of listings of sites covering crops, livestock, markets, prices, weather, organizations, and academic institutions.

Johnson, R.C. 1996. *Target Farming: A Practical Guide to Precision Agriculture*. 2$^{nd}$ ed. Saskatoon, Canada: R.C. Johnson. 144 p. Contains introduction and overview; monitoring systems; global positioning; computers; mapping and geographical information; and soil navigation and sampling. Includes a directory of people, businesses, and organizations involved in precision farming and a glossary.

Koepf, H.H. 1993. *Research in Biodynamic Agriculture: Methods and Results*. Kimberton, PA: Bio-Dynamic Farming and Gardening Association. 84 p. ISBN: 0938250345. Review of research on all aspects of biodynamic agriculture performed in Germany, Sweden, and the United States.

Koepf, H.H. 1989. *The Biodynamic Farm: Agriculture in the Service of the Earth and Humanity*. In cooperation with Roderick Shouldice and Walter Goldstein. Hudson, NY: Anthroposophic Press. 259 p. ISBN: 0880101725. Describes the essential principles of biodynamic agriculture developed in Germany by Rudolf Steiner. It treats the farm as a holistic organism comprised of a variety of components that must be balanced and nurtured. Koepf discusses the place of the farmer and the farm in relation to fertility, manures and composts, crop production, animal feeding, and health and human nutrition.

Lampkin, N.H. and S. Padel, eds. 1994. *The Economics of Organic Farming: An International Perspective*. Wallingford, UK: CAB International. 485 p. ISBN

085198911X. This book provides the first comprehensive international review of the economics of organic farming, from studies in the United Kingdom, the United States, Canada, Australia, Germany, Denmark, and Switzerland. It covers the physical and financial performance of organic farms, the special features of adoption and the transition process, the implications of widespread adoption, and the analysis of policy implications and initiatives in the different countries.

Lockeretz, W. and M.D. Anderson. 1993. *Agricultural Research Alternatives.* Lincoln, NE: University of Nebraska Press. 249 p. ISBN: 080322901. Explores the history of agricultural research in the United States and the forces that determined the research options. Discusses multidisciplinary cooperation in research projects and the rewards and incentives that are available to all stakeholders. Recommends that future research focus on relevance and applicability to the needs of farmers.

Logsdon, G. 2000. *Living at Nature's Pace: Farming and the American Dream.* White River Junction, VT: Chelsea Green. 269 p. ISBN: 189013256X. Logsdon believes that small-scale, market-oriented sustainable farms can be profitable and produce abundant crops to serve the food and fiber needs of the United States Logsdon observed the Amish community and was impressed by their mixture of self-sufficiency, sustainable farming, and family life. He urges a return to the land, to the wonders of farming, and rural life and change in our approach to agriculture.

Loomis, R.S. and D.J. Connor. 1992. *Crop Ecology: Productivity and Management in Agricultural Systems.* Cambridge: Cambridge University Press. 552 p. ISBN: 052138379X. A textbook dealing with a systems approach to crop and pasture production worldwide from an environmental standpoint. Covers farming systems and their biological components; physical and chemical environments; production processes; and resource management.

Madden, J.P. and S.G. Chaplowe, eds. 1997. *For All Generations: Making World Agriculture More Sustainable.* Glendale, CA: OM Publishing. 653 p. ISBN: 096557670. Explains the efforts of governmental and nongovernmental agencies to develop and encourage adoption of sustainable farming methods and some of the barriers that need to be overcome. The authors take issues with the chemical-intensive methods of industrialized farming systems and urge the widespread adoption of sustainable production systems.

Mason, J. 1997. *Sustainable Agriculture.* East Roseville, Australia: Kangaroo Press. 121 p. ISBN: 0864178654. This book contains ideas that contribute to sustainability including permaculture, biodynamics, organic farming, agroforestry, conservation tillage, and integrated hydroculture in order to complete plans for farm management and study the long-term benefits of sustainable systems.

Covers soils; water management; pest and disease control; weed control and cultivation; managing plants and animals; and general business management techniques for the sustainable farm.

Mollison, B. 1990. *Permaculture: A Practical Guide For A Sustainable Future.* Washington, DC: Island Press. 591 p. ISBN: 1559630485. Permaculture design assembles conceptual, material, and strategic components in a pattern that benefits all life forms. Mollison presents the principles of permaculture design and the forces of nature, climate, water, and other elements that are part of the system.

Mollison, B. and R.M. Slay. 1991. *Introduction to Permaculture.* Tyalgum, Australia: Tagari Publications. 216 p. ISBN: 0908228082. An interdisciplinary approach to creating sustainable communities. Covers principles of permaculture, which include energy saving, and efficient use of soil and water. Provides recommendations on elements of permaculture such as site planning and design; house design for temperate, dryland, and tropical regions; planning home gardens; orchards, farm forestry, and grain crops; animal forage systems; aquaculture; and urban permaculture. Includes lists of plants that are suitable for permaculture operations.

Morgan, M. and D. Ess. 1997. *The Precision-Farming Guide for Agriculturists: the Nuts and Bolts Guide to "Getting up to Speed" Fast and Effectively with this Exciting New Management Tool.* Moline, IL: Deere & Company. 123 p. ISBN: 0866912452. Introduces the basic concepts and technologies involved in precision farming, a method that requires the use of some high-tech equipment to assess field conditions and apply chemicals and fertilizers. Provides information on computer use, satellites, global positioning, remote sensing, monitors, and other communication systems in farm-production systems.

Morse, S., N. McNamara, M. Acholo, and B. Okwoli. 2000. *Visions of Sustainability: Stakeholders, Change and Indicators.* Aldershot, UK; Burlington, VT: Ashgate. 264 p. ISBN: 1840148675. This work, based on a detailed 6-year case study of a Nigerian development project in Eroke village, emphasizes the importance of involving farmers in the decision-making process and addresses key issues that are essential to anyone using sustainable practices.

National Research Council (US). Committee on Assessing Crop Yield—Site Specific Farming, Information Systems, and Research Opportunities. 1997. *Precision Agriculture in the 21st Century: Geospatial and Information Technologies in Crop Management:* Washington, DC: National Academy Press. 161 p. ISBN: 058502278X. Covers the components of precision agriculture and its applications, practice, and policy implications. Discusses how precision agriculture can affect decision making in irrigation, crop selection, pest management, environmental quality, and price and market conditions. Also examines geographical dimensions of precision agriculture and the potential impact on American rural communities. Available from the World Wide Web http://books.nap.edu/catalog/5491.html.

National Research Council (US). Committee on Sustainable Agriculture and the Environment in the Humid Tropics. 1993. *Sustainable Agriculture and the Environment in the Humid Tropics*. Washington, DC: National Academy Press. 717 p. ISBN: 0309047498. A study of the environmental, social, and policy issues regarding land conservation and deforestation in the humid tropics. Part two presents profiles by individual authors of sustainability options in Brazil, the Ivory Coast, Indonesia, Malaysia, Mexico, the Philippines, and Zaire. Includes glossary of terms and author biographies. Available from the World Wide Web at http://books.nap.edu/books/0309047498/html/.

Pretty, J.N. 1999. *The Living Land*: *Agriculture, Food and Community Regeneration in the 21$^{st}$ Century*. London: Earthscan. 336 p. ISBN: 185383517X. Originally published in 1998 as *The Living Land*: *Agriculture, Food and Community Regeneration in Rural Europe*. This book examines environmental, health, and social ramifications of chemical-intensive agriculture and strongly supports participation of all farmers in environmentally sound alternative farming practices in Western Europe. Pretty includes local success stories and makes a strong case for changing agricultural policies to establish sustainable rural communities in Western Europe through alternative agriculture, local food systems, and rural community development.

Pretty, J.N. 1995. *Regenerating Agriculture*: *Policies and Practice for Sustainability and Self-Reliance*. Washington, DC: Joseph Henry Press. 328 p. ISBN: 0309052483. (Also published in London: Earthscan. ISBN: 1853831522.) Pretty presents evidence from over 50 projects in 28 countries that illustrate the importance of self-reliance and local initiatives in promoting sustainable practices. Available from the World Wide Web at http://www.nap.edu/books/0309052467/html/index.html.

Reintjes, C., B. Haverkort, and A. Walters-Bayer. 1992. *Farming for the Future*: *An Introduction to Low-External-Input and Sustainable Agriculture*. London: Macmillan. 271 p. ISBN: 0333 570111. This book is a manual for development workers to assist small-scale farmers to establish low-cost local systems that are more productive and do not deplete the ecological resources on which they depend. Provides an overview of sustainable agriculture, and many examples and practical ideas that enable farmers to make sound decisions regarding their specific farm-management situations.

Rodale, R. 1991. *Save Three Lives*: *A Plan for Famine Prevention*. San Francisco, CA: Sierra Club. 271 p. ISBN: 1559633085. Rodale advocates a strategy for returning Africa to sustainable indigenous farming systems, alley cropping with leguminous trees to save the dwindling wood supply, biological pest control, sensible use of available water, population control, and gender equality.

Ruttan, V., ed. 1994. *Agriculture, Environment, and Health: Sustainable Development in the 21st Century*. Minneapolis, MN: University of Minnesota Press. 409 p. ISBN: 0816622914.   A series of papers by agricultural, environmental, and health scholars addressing the biological and technical constraints; resource and environmental constraints; and health constraints on crop and animal production.

Ruttan, V., ed. 1992. *Sustainable Agriculture and the Environment: Perspectives on Growth and Constraints*. Boulder, CO: Westview Press. 205 p. ISBN: 0813385075.   A series of papers by biological and social scientists offering their perspectives on agricultural research priorities and the demands it would place on the economies of developed and developing economies.

Sanders, R. 2000. *Prospects for Sustainable Development in the Chinese Countryside: The Political Economy of Chinese Ecological Agriculture*. Aldershot, UK; Brookfield, VT: Ashgate. 239 p. ISBN: 1840149248.   Sanders investigates the experiences of seven villages and two counties that adopted Chinese Ecological Agriculture (CEA) in the past two decades. He examines the feasibility of sustainable agriculture and the state of the environment in China; the social, political and economic policies toward the countryside since 1945; and the conditions that restrict large-scale adoption of CEA and how to ease them.

Sattler, F. and E.V. Wistinghausen. 1992. *Bio-Dynamic Farming Practice*. Translated by A.R. Meuss. Clent, Stourbridge, UK: Bio-Dynamic Agricultural Association. 333 p. ISBN: 0951897608.   A textbook on biodyamic farming practices describing in detail scientifically proven farming methods and skills gained from many years of practical experience. Covers techniques for proper management of soil, tillage, manures, and compost. Offers a range of techniques for crop and animal husbandry, and landscape management. Includes information on training, labor, marketing, and conversion to biodynamic methods. The text is supplemented with 85 diagrams, 82 tables, and 36 color photographs.

Schilthuis, W. 1994. *Biodynamic Agriculture*. Hudson, NY: Anthroposophic Press. 111 p. ISBN: 0880103825.   First published in Dutch in 1994 by Christofoor Publishers this work is a concise and illustrated introduction to the principles and practices of biodynamic agriculture. Describes how to design farms that are in sound ecological balance. Discusses the role of agriculture in the environment, the earth as a living organism, and a brief history of biodynamics. Tells how to raise crops and livestock using biodynamic techniques, social aspects, and considerations for the future.

Shaner, W.W., P.F. Philipp, and W.R. Schmehl, eds. 1982. *Readings in Farming Systems Research and Development*. Boulder, CO: Westview Press. 189 p. ISBN: 0865315027. Nine scholarly essays that address international farming systems research and development, and its applications.

Solbrig, O.T. and D.J. Solbrig. 1994. *So Shall You Reap: Farming and Crops in Human History*. Washington, DC: Island Press. 304 p. ISBN: 1559633085. Covers the period from ancient times before the adoption of agriculture through the eras of hunter-gatherers, early agriculture, domestication of plants, rise of civilization, spread of agriculture to European societies, the medieval period, and the agricultural revolution to the high-yield market-oriented agriculture of today. Presents details about crop origins and the contacts among societies that brought about global changes in food production and consumption practices.

Soule, J.D. and J.K. Piper. 1992. *Farming in Nature's Image: An Ecological Approach to Agriculture*. Washington, DC: Island Press. 305 p. ISBN: 0933280882. Examination of the ecological damage caused by modern agricultural practices and the economic effects on farmers and rural communities. Offers new perspectives and insights on the importance of establishing permanent sustainable agricultural systems for the future.

Thirsk, J. 1997. *Alternative Agriculture: A History from the Black Death to the Present Day*. Oxford; New York: Oxford University Press. 376 p. ISBN: 0198206623. A chronological survey of alternative agriculture in England from 1350 to the present. Describes the successes; failures and near failures; familiar strategies; diversification and innovation; and current initiatives with a wide range of crop and livestock enterprises.

Thrupp, L.A. 1996. *New Partnerships for Sustainable Agriculture*. Washington, DC: World Resources Institute. 148 p. ISBN: 1569731039. Nine case studies from Bangladesh, the Philippines, Cuba, Nicaragua, Senegal, Kenya, and the United States shows how ecologically oriented integrated pest-and-crop management practices can maintain or increase yields; improve soil quality and resilience; and reduce agrochemical inputs and costs.

Tivy, J. 1990. *Agricultural Ecology*. Harlow, UK: Longman; New York: Wiley. 306. p. ISBN: 0582301637. Analyzes the relationships between crops, livestock, and the environment and the extent to which that system has been managed and modified. Compares managed and unmanaged ecosystems and the effects on organisms, soils, climate, nutrient cycles, crop and livestock production, and land capability. Examines environmental problems where soil and climate conditions impose limitations on agricultural use or where a particular type of agriculture is dominant.

Tripp, R., ed. 1991. *Planned Change in Farming Systems: Progress in On-Farm Research.* Chichester, New York: J. Wiley; Exeter, UK: Sayce Publications. 358 p. ISBN: 0471934178. Focuses on the principles and procedures of agricultural research on farming systems in Africa, Asia, and Latin America with discussions of its accomplishments and weaknesses. Examines the decision making that encompasses changes in on-farm research and the extent of farmers' participation in meeting the research objectives.

Turner, B.L. II and S.B. Brush. 1987. *Comparative Farming Systems.* New York: Guilford Press. 442 p. ISBN: 089862780X. Case studies of 12 farming systems in 11 countries representing a range of climates and topographic settings.

Vasey, D.E. 1992. *An Ecological History of Agriculture: 10,000 B.C.–A.D. 10,000.* Ames, IA: Iowa State University Press. 374 p. ISBN: 0813809096. A detailed historical exploration of present, past, and future agricultural systems with a focus on ecological practices.

Walters, C. and C.J. Fenzau. 1996. *Eco-Farm.* 2nd rev. ed. Raytown, MO: Acres U.S.A. 460 p. ISSN: 0911311503. Covers photosynthesis, light, temperature, water, cells, minerals, soil management, weeds, insects, animal health, silage, and other topics. Contains diagrams, formulas, recommendations, and a wealth of valuable information in a readable, easy-to-understand format.

Young, M.D. 1991. *Towards Sustainable Development.* London, UK: Belhaven Press. 356 p. ISBN: 185293137X. A collection of studies prepared at the request of the Organization for Economic Cooperation and Development (OECD) to examine the direction of agricultural policy and environmental problems associated with agricultural production practices in eight countries: Germany, Sweden, the United States, France, the Netherlands, Portugal, Austria, and the United Kingdom.

Zimmer, G. 2000. *The Biological Farmer: A Complete Guide to the Sustainable and Profitable Biological System of Farming.* Austin, TX: Acres U.S.A. 365 p. ISBN: 0911311629. A primer on biological farming that focuses on ways to reduce the cost of inputs and increase profits through sustainable agriculture. Offers a practical, hands-on approach to balancing the chemical, physical, and biological properties of the soil.

## PERIODICALS

Periodicals are the main channels of communication for presenting the results of scientific research and new knowledge. This section describes periodicals that are devoted to keeping readers informed about the latest developments in farming systems research. In addition to scholarly articles, the periodicals may include book reviews, new publications, and miscellaneous items. They offer articles addressing both theory and applications of farming systems in today's agriculture for scientists and researchers, farmers, students, legislators and other practitioners.

*Acres U.S.A.: The Voice for Eco-Agriculture.* 1971– . Austin, TX: Acres USA. Monthly. ISSN: 1076-4968. Provides practical cutting-edge information on all aspects of ecological agriculture. Includes timely information on fertility management; weed and insect control; organic production; industry news; specialty crops; markets and trends; grazing; composting; soils; natural veterinary care; and human and animal health.

*Agricultural Systems.* 1976– . Kidlington, UK: Elsevier Science. Monthly. ISSN: 0308-521X. A forum of communication for integrating the disciplines that comprise agriculture and contribute to the debate on efficient use of resources to feed the world. Addresses the biological, physical, and chemical components of farming systems from production to marketing and distribution in both conventional and sustainable agriculture. Results of research and experimentation on methods and models applied to systems analysis in agriculture are included.

*Agriculture, Ecosystems & Environment.* v. 9– . 1983– . Amsterdam: Elsevier Science Publishers. 15 issues/yr. ISSN: 0167-8809. Continues *Agriculture and Environment.* Publishes scientific research, review articles and commentary on the relationship of food and agricultural production systems to the biosphere. Topics include methods of production and how those methods affect pollution of soil, water, air, and food; effect of industrial pollutants on agriculture; use of energy and nonrenewable resources; and policy issues involving agriculture and the environment. Includes book reviews.

*Agriculture and Human Values.* 1984– . Dordrecht: Kluwer. Quarterly. ISSN: 0889-048X. Devoted to the economic, legal, political, environmental, and social issues concerning agricultural policies and practices, use of natural resources, and technologies involved in the production of food and fiber. Includes papers reflecting a wide array of philosophical, empirical, theoretical, and conceptual approaches.

*American Farmland*: *The Magazine of the American Farmland Trust.* 1981– . Washington, DC: American Farmland Trust. Quarterly. American Farmland Trust (AFT) is dedicated to the preservation of productive farmland and promoting sustainable agriculture through action-oriented programs of education, technical assistance, and farmland protection projects. The magazine offers the latest information on problems facing today's farmers and approaches for preserving farmland, AFT's activities, general information about preservation efforts, legislative developments, question-and-answer columns, and letters from members.

Available from the World Wide Web at http://www.farmland.org/magazine/index.htm.

*American Journal of Alternative Agriculture.* 1986– . Arlington, VA: Henry A. Wallace Center for Agricultural and Environmental Policy at Winrock International. Quarterly. ISSN: 0889-1893. A forum for interdisciplinary research on biological, physical, economic, and social aspects of alternative and sustainable agriculture and food systems. Contains original research, review articles, policy and opinion pieces, and book reviews.

*American Small Farm.* Canoga Park, CA: Magnet Communications, Inc. 10 issues/yr. ISSN: 1064-7473. No date available. Covers the business and science of agriculture relating to owner-operated farms between 5 and 300 acres. Provides information that will help farmers make their enterprises more efficient, more profitable, and less labor intensive. Serves as a forum for sharing of ideas among small farmers.

*Biodynamics.* 1941– . San Francisco, CA: Bio-Dynamic Farming and Gardening Association Inc. Bimonthly. ISSN: 0006-2863. Dedicated to the principles and practices of biodynamic farming. Articles cover a wide range of practices that promote soil conservation and improvement of soil fertility in order to produce food of high nutritional quality. Based upon the principles of Rudolf Steiner who advocated the importance of composting, green manuring, herbal preparations, and the interaction of natural forces for a productive and healthy agriculture.

*Biological Agriculture & Horticulture*: An International Journal for Sustainable Production Systems. 1982– . Bicester, UK: A B Academic Publishers. Quarterly. ISSN: 0144-8765. A focus of communication for a wide range of articles, research reports, and reviews relating to the development and application of biological husbandry in agricultural systems. Also covers comparative studies and experimental trials on alternative and conventional agriculture. Includes book reviews, news items, and editorials.

*Computers and Electronics in Agriculture*: An International Journal. 1985– . Amsterdam: Elsevier Science. Monthly. ISSN: 0168-1699. Provides international coverage of advances in the application of computer hardware, software, and electronic instrumentation and control systems to agriculture, forestry, and related industries.

*Eco-Farm & Garden.* 1998– . Ottawa, Canada: R.E.A.P (Resource Efficient Agricultural Production). Quarterly. A periodical dedicated to improving farm profits and productivity, while minimizing adverse health and environmental effects. Articles cover organic agriculture for the home garden and the farm.

*Ecological Applications*. 1991– . Washington, DC: Ecological Society of America, Quarterly. ISSN: 1051-0761. Publishes articles dealing with the application of ecological concepts to a broad range of problems. Special emphasis is given to papers on environmental decision making and environmental policy and management. Includes agroecosystems, soils, groundwater, range management, and landscape ecology.

*Ecological Engineering: The Journal of Ecotechnology*. 1992– . New York: Elsevier. 8 issues/yr. ISSN 0925-8574. A journal intended to bridge the gap between ecologists and engineers involved in the design of ecosystems that benefit humans and nature. Topics covered include conservation of renewable and nonrenewable resources in global change, alternative energy, ecological economics, environmental conservation, and global geopolitics. Articles deal with applications of ecological engineering such as wetlands creation and restoration; pollution control in ecosystems; restoration and rehabilitation of forests; grasslands; lakes; and development of sustainable agroecosystems.

*Ecology and Farming*. 1990. Tholey-Theley, Germany: IFOAM. 3 issues/yr. ISSN: 1016-5061. A forum for the exchange of information and experiences with organic/sustainable agriculture around the world. Covers practice, research, agricultural politics, news items, conference reports, and book reviews. Articles appear in English, with summaries in French and Spanish.

*Economic Botany*. New York: Society for Economic Botany/New York Botanical Garden. Quarterly. ISSN: 0013-0001. Documents the relationship among plants and people and the past, present, and future uses of plants. Covers medicinal, industrial, edible, decorative, and special uses. Contains research articles, book reviews, annotated bibliographies, notes, and miscellaneous information.

*Experimental Agriculture*. 1965– . London, New York: Cambridge University Press. Quarterly. ISSN: 0014-4797. Publishes analytical research on the agronomy of crops, with emphasis on the food, forage, and industrial crops of the warmer regions of the world. Experiments conducted in the field and designed to explain agronomic results in biological and environmental terms are of particular concern. Includes new experimental techniques, new methods in experimental crop production, and critical discussions of specific problems.

*Industrial Crops and Products*. 1992– . Amsterdam: Elsevier. 6 issues/yr. ISSN: 0926-6690. An international journal of original research, short communications, and critical reviews on all aspects of production, harvesting, storage, and processing of nonfood crops for industrial uses. Includes applications of pharmaceuticals, lubricants, fuels, fibers, essential oils, biologically active materials, and crop by-products.

*International Permaculture Solutions Journal.* 1990– . Sparr, FL: Yankee Permaculture. Irregular. ISSN: 1046-8366.   Presents the latest international information on ideas, techniques, and strategies for application to permaculture systems. Articles cover permaculture, aquaculture, sustainable agriculture, forests, appropriate technology, and environment. Contains a wide diversity of views from writers in Japan, the United States, Mexico, the Philippines, New Zealand, Germany, the United Kingdom, Australia, and other countries. Includes letters, book reviews, announcements, and news items.

*Journal of Agricultural and Environmental Ethics.* 1991– . Dordrecht: Kluwer. 3 issues/yr. ISSN: 1187-7863.   Contains articles concerning ethical issues in agriculture and related disciplines. Provides a forum for moral issues arising from actual and projected social policies concerning agricultural production, technological changes, farmland utilization, intensive agriculture, animal welfare, ecosystems, biotechnology, and the responsibilities of professionals in these areas.

*Journal of Soil and Water Conservation.* 1946– . Ankeny, IA: Soil and Water Conservation Society. Bimonthly. ISSN: 0022-4561.   Promotes the science and art of conservation of soil, water, and natural resources throughout the world. Articles present results of research and experience that promote the concept of stewardship of natural resources. Includes editorials, news items, upcoming events, and book and video reviews.

*Journal of Sustainable Agriculture.* 1990– . Binghamton, NY: Food Products Press. Quarterly. ISSN: 1044-0046.   Promotes the study and application of sustainable agriculture and the maintenance and enhancement of agricultural resources, production, and environmental quality. Includes original articles on research, innovative practices, new technology, integrated pest management, farming practices, energy use, conservation, and present and future trends. Covers the economic, social, political, and philosophical aspects of sustainable agriculture.

*Journal of Turfgrass Management.* 1995– . Binghamton, NY: Haworth Press. Quarterly. ISSN: 1070-437X.   Gathers and disseminates current advances on all facets of basic and applied turfgrass research for scientists and practicing turfgrass managers. Includes review articles, book reviews, and positions available.

*Organic Farming: Britain's Journal for Organic Food Production.* Bristol, UK: Soil Association. Quarterly. ISSN: 1464-1224. No date available.   A forum on principles, practices, and politics of organic farming for commercial producers in over 30 countries. Covers organic fruit, vegetable, field crop, and livestock production. Included in each issue is a section of gleanings from organic and sustainable farming publications throughout the world. Includes technical information, news, and reader commentary.

*The Permaculture Activist*. Black Mountain, NC: Permaculture Activist. 3 issues/ yr. ISSN: 0897-7348. No date available. Serves as an important networking resource for permaculture in the United States, Canada, and Central America. Articles cover the development and management of rural and urban permaculture systems, edible landscaping, bioregionalism, aquaculture, natural building, earthworks, renewable energy, and other topics. Provides a current listing of upcoming permaculture design courses. Also contains news, permaculture activities, new publications, and advertisements.

*Permaculture International Journal*. South Lismore, Australia: Permaculture International Ltd. Quarterly. ISSN: 1037-8480. Ceased publication in 2000. Contains practical information on all aspects of permaculture such as village and community development, edible landscapes, appropriate technology, earthbank, and ethical enterprises. Includes updates, research findings, practical tips, news, and permaculture activities around the world.

*Resource: Engineering and Technology for a Sustainable World*. 1994– . St. Joseph, MI: American Society of Agricultural Engineers. Monthly. ISSN: 1076-3333. Timely articles on farming systems technology and engineering from an environmental and conservation perspective on agriculture, food biotechnology, aquaculture, forestry, machinery, soil, and water. Also covers news about the society and its members, upcoming events, reader opinions, news briefs business developments, new product releases, and an employment section.

*Resources, Conservation and Recycling*. 1998– . Amsterdam: Elsevier. Monthly. ISSN: 0921-3449. Continues *Resources and Conservation*. Provides analysis and review of the interdisciplinary aspects of renewable and nonrenewable management with emphasis on conservation. Papers emphasize scientific, technological, or institutional methodology and may cover politics; resource management and allocation; and social and legal aspects of air, water, and land resources.

*Small Farm Today*. Clark, MO: Missouri Farm Publishing Inc. Bimonthly. ISSN: 1079-9729. No beginning date available. The original how-to magazine of alternative and traditional crops, livestock rearing, and direct marketing. Dedicated to the preservation and promotion of small farming, rural living, community, sustainability, and ''agripreneurship.'' Offers both traditional and nontraditional farming alternatives adaptable to small farms. Most of the readers are full or part-time family farmers, and many are using alternatives, such as growing high-value crops, raising unusual livestock, and direct marketing.

## REPORTS FROM GOVERNMENTAL AND
## NONGOVERNMENTAL AGENCIES

Reports are issued by organizations, government agencies, universities, and research institutes involved in design, planning, and adoption of alternative farming systems. These agencies represent international, national, regional, or local constituencies who believe that environmentally sound and economically viable agricultural production systems are vital for producing the necessary food and fiber for current and future generations. Reports of government agencies usually present results of research and activities on farming system inputs with the goal of improving agriculture and the quality of life. Nonprofit agencies also carry out agricultural development activities and often issue reports comparing the conventional and alternative farming systems. International agencies such as the World Bank and the FAO provide monetary and staff support and consultants to countrywide programs in agricultural development. The reports cited here reflect a diversity of viewpoints and strategies undertaken by the agencies involved. Although there are major differences in the approaches covered in the reports, they all are working toward the goal of promoting the adoption of sustainable farming systems around the world.

*Agriculture and the Environment: The 1991 Yearbook of Agriculture.* 1991. D.T. Smith, ed. Washington, DC: USDA. 339 p. ISBN: 0160341442. Examines environmental concerns facing agriculture and the programs of the U.S. Department of Agriculture (USDA) that are addressing those concerns. Covers research efforts, technical assistance, education, and programs designed to preserve and enhance the quality of water, land, air, and food. Includes chapters on sustainable agriculture in the United States and abroad; integrated pest management; soil and water conservation; and other topics with environmental impacts. Includes 13 chapters on "what you can do" to conserve the nation's land, water, and air, and promote a productive and ecologically viable agriculture.

*Alternative Agriculture: Federal Incentives and Farmers' Opinions.* (GAO/ PEMD-90-12). 1990. Washington, DC: U.S. General Accounting Office, Program Evaluation and Methodology Division. A study of the effects of Federal agricultural support programs on farmers' decisions to adopt alternative production practices.

*Alternative Agriculture: Scientists' Review.* 1990. Special Publication no. 16. Ames, IA: Council for Agricultural Science and Technology (CAST). 182 p. A report by 44 scientists responding to the National Research Council's (NRC) report, *Alternative Agriculture.* The CAST report presents views of the contributing scientists and their specific opinions regarding coverage of the subject in the NRC report. A 14-page summary at the beginning, outlines the concerns raised in the report. Includes a 20-page bibliography.

Arden-Clarke, C. 1988. *Environmental Effects of Conventional and Organic/ Biological Farming Systems*. Research report 16–17.2 vols. Oxford: The Political Ecology Research Group. 265 p. ISSN: 0142-7199. An extensive review of the international literature comparing the environmental impacts of conventional and organic farming systems. Focuses mainly on Britain and Europe. Cites over 795 references from journals, conference proceedings, reports, books, and government documents. Also published as:

Arden-Clarke, C. and R.D. Hodges. 1987. ''The Environmental Effects of Conventional and Organic/Biological Farming Systems. I. Soil Erosion with Special Reference to Britain.'' *Biological Agriculture and Horticulture* 4(4): 309–357. Arden-Clarke, C. and R.D. Hodges. 1988. ''The Environmental Effects of Conventional and Organic/Biological Farming Systems. II. Soil Ecology, Soil Fertility and Nutrient Cycles.'' *Biological Agriculture and Horticulture* 5(3): 223–287.

Faeth, P. 1995. *Growing Green: Enhancing the Economic and Environmental Performance of U.S. Agriculture*. Washington, DC: World Resources Institute. 120 pages. ISBN: 1569730296. A comprehensive, empirical analysis of environmental and economic considerations regarding agricultural sustainability in the United States.

Hrubovcak, J., U. Vasavada, and J. Aldy. 1999. *Green Technologies for a More Sustainable Agriculture*. Agriculture Information Bulletin (USDA) no. 752. Washington, DC: USDA, Economic Research Service, Resource Economics Division. 42 p. Examines the role of agriculture in the sustainability debate. Available from the World Wide Web at http://www.ers.usda.gov/epubs/pdf/aib752/ index.htm.

Smith, M. et al. 1998. *The Real Dirt: Farmers Tell about Organic and Low-input Practices in the Northeast*. 2nd ed. Burlington, VT: Northeast Organic Farming Association. 264 p. Contains summaries of interviews with more than 60 farmers in 8 Northeastern states on practical methods for ecological soil, pest, disease, crop, greenhouse, and livestock management.

National Research Council (U.S.) Board on Agriculture and Board on Science and Technology for International Development. 1991. *Toward Sustainability: A Plan for Collaborative Research on Agriculture and Natural Resource Management*. Washington, DC: National Research Council, National Academy Press. 147 p. ISBN: 0309045401. Report of the National Research Council's Panel for Collaborative Research Support for the Sustainable Agriculture and Natural Resource Management (SANREM) Program of the U.S. Agency for International Development (USAID). A 144-page appendix contains a series of papers by individual authors on environmental issues, operational concerns, soil research, crop production, integrated pest management, and agroecosystems relating to sustainability. Includes bibliography, list of panel participants, and backgrounds of authors.

*New Crops, New Uses, New Markets: Industrial and Commercial Products from U.S. Agriculture. Yearbook of Agriculture 1992.* Washington, DC: USDA Office of Publishing and Visual Communication. 320 p. ISBN: 0160380294. Report on research and programs by the USDA to find, promote, and encourage new ways to develop and use the abundant output of American agriculture to develop new products for industrial and consumer use.

*New Directions for Agriculture, Forestry and Fisheries: Strategies for Sustainable Agriculture.* Rome, Italy: Food and Agriculture Organization of the United Nations. 66 p. ISBN: 9251035865. Discusses the background of poverty and underdevelopment in participating countries and then considers the challenges, goals, and strategies necessary to achieve favorable policy implementation and wise use of human and natural resources in the agriculture, forestry, and fishery industries.

Norman, W.D. et al. *Sustainable Agriculture: Concepts, Issues and Policies in OECD Countries.* Paris: OECD. 68 p. ISBN: 9264146466. A survey of efforts by OECD member countries to address the issues surrounding sustainable agriculture.

*Organic and Conventional Farming Compared.* 1980. Report no. 84. Ames, IA: Council for Agricultural Science and Technology (CAST). 32 p. Report of a task force of 24 scientists comparing organic and conventional farming systems and the evolution, scientific, economic, and sociological bases of both systems.

Paarlberg, R. 1994. *Countrysides at Risk: The Political Geography of Sustainable Agriculture.* Policy essay no. 16. Washington, DC: Overseas Development Corporation. 112 p. ISBN: 1565170210. This report discusses the social and political abuses in developing countries that threaten the environment and prevent the achievement of environmentally sound agriculture.

*Report and Recommendations on Organic Farming.* 1980. Washington, DC: USDA Study Team on Organic Farming. 94 p. A team of 16 agricultural scientists studied the characteristics, philosophy, motivations, successes, costs, benefits, limitations, and future of organic farming in the United States, Europe, and Japan. The data was derived from case studies on 69 farms in 23 states, a survey of the readership of *New Farm* magazine, literature reviews, interviews, and study tours. They offer 26 specific recommendations for action on research, education, extension, and organization and policy for organic farming.

Serageldin, I. and G.J. Persley. 2000. *Promethean Science: Agricultural Biotechnology, the Environment, and the Poor.* Washington, DC: Consultative Group on International Agricultural Research (CGIAR). 48 p. A companion to the larger volume from the conference proceedings, *Agricultural Biotechnology and the Poor*, published in January 2000. Explores the urgency of shifting research priorities to mobilize the new technologies and innovations in biotechnology and genetics for improvement in the productivity of agroecological systems and crops, livestock, fish, trees, and other agricultural products. Available from the World Wide Web at http://www.worldbank.org/html/cgiar/publications/prometh/pscont.html.

Shah, M., and M. Strong. 1999. *Food in the 21ˢᵗ Century: From Science to Sustainable Agriculture*. Washington, DC: Consultative Group on International Agricultural Research (CGIAR). 72 p. Highlights the roles and accomplishments of the CGIAR. Available from the World Wide Web at http://www.worldbank.org/html/cgiar/publications/shahbook/shahbook.pdf.

*Sustainable Agriculture in the National Research Initiative*. 1991. Walthill, NE: Center for Rural Affairs. 31 p. Report and recommendations of a panel of 32 research administrators, natural and social scientists, research policy analysts, and farmers concerning a national research initiative (NRI) for research on agriculture, food, and environment with emphasis on sustainable agriculture.

*Sustainable Agriculture*: *Task Force Report*. 1997. Washington, DC: President's Council on Sustainable Development. 20 p. The report offered 4 goals and 9 policy recommendations aimed at achieving environmentally sound and economically viable agricultural production; revitalizing rural farming communities; producing a safe and high-quality food supply; encouraging research on integrating productivity, profitability, and environmental stewardship into the U.S. agricultural system; and achieving international cooperation on intellectual property rights.

## FAO Farm Systems Management series

Numbered series published by the FAO, Rome.

1. McConnell, D.J. 1997. *Farm Management for Asia: A Systems Approach*. 374 p. ISBN: 925104077X. (English).
2. De Grandi, J.C. 1996. *El Desarrollo De los Sistemas de Agricultura Campesina en América Latina un Análisis de la Influencia del Contexto Socioeconómico*. (Spanish).
3. De Grandi, J.C. 1996. *L'évolution des Systèmes de Production Agropastorale par Rapport au Développement Rural Durable Dans les Pays D'afrique Soudano-sahélienne*. 173 p. ISBN: 9252038175. (French).

4.  Norman, D.W. 1995. *The Farming Systems Approach to Development and Appropriate Technology Generation.* 241 p. ISBN: 9251036446. (English, French, Spanish).
5.  Upton, M. and J.M. Dixon. 1994. *Methods of Micro-level Analysis for Agricultural Programmes and Policies: A Guideline for Policy Analysts.* 213 p. ISBN: 9251034737. (English, French, Spanish).
6.  Dillon, John L. and J.B. Hardaker. 1993. *Farm Management Research for Small Farmer Development.* 315 p. (English, French, Spanish). ISBN: 9251033056.
7.  Norman, David W. and M. Douglas. 1994. *Farming Systems Development and Soil Conservation.* 181 p. (English, French, Spanish). ISBN: 9251034486.
8.  Dixon, J.M. 1994. *Farm and Community Information Use for Agricultural Programmes and Policies.* 109 p. ISBN: 9251034745. (English).
9.  Norman, D.W. and M. Douglas. 1994. *Farming Systems Development and Soil Conservation.* 181 p. ISBN: 9251034486. (English, French, Spanish).
10. Dillon, J.L. and J.B. Hardaker. 1993. *Farm Management Research for Small Farmer Development.* 315 p. ISBN: 9251033056. (English, French, Spanish).
11. *Institutionalization of a Farming Systems Approach to Development: Proceedings of Technical Discussions.* 1993. 334 p. ISBN: 9251032505. (English).
12. Prabowo, D. and D.J. McConnell. 1992. *Changes and Development in Solo Valley Farming Systems, Indonesia.* 127 p. ISBN: 9251028974. (English).
13. McConnell, D.J. 1992. *The Forest Garden Farms of Kandy, Sri Lanka.* 126 p. ISBN: 9251028982. (English).
14. Anderson, J.R. and J.L. Dillon. 1992. *Risk Analysis in Dryland Farming Systems.* 119 p. ISBN: 9251032041. (English).
15. *Farming Systems Development—Guidelines for the Conduct of a Training Course in Farming Systems Development.* 1991. 266 p. ISBN: 9251032491. (English, French, Spanish).

## World Bank Reports on Sustainable Agriculture and Biodiversity

Pagiola, S., et al. *Mainstreaming Biodiversity in Agricultural Development: Toward Good Practice.* World Bank Environment Paper no. 15. Washington, DC: World Bank. 60 p. ISBN: 0821338846. Discusses the challenges of finding ways to increase agricultural production for growing populations without destroying the many benefits provided by biodiversity. Outlines a conceptual framework for analyzing interactions between agriculture and biodiversity, causes of conflict, policy reforms, and the World Bank's commitment to conserve biodiversity in agricultural development. Available from the World Wide Web at http:// www.worldbank.org/fandd/english/0398/articles/060398.htm.

Simmonds, N.W. 1985. *Farming Systems Research: A Review*. World Bank Technical Paper no. 43. Washington, DC: World Bank. 109 p. ISBN: 0821305662.   A report on the history and scope of farming systems research around the world.

Srivastava, J.P., N.J.H. Smith, and D.A. Forno. 1999. *Integrating Biodiversity in Agricultural Intensification: Toward Sound Practices*. Environmentally Sustainable Development Studies and Monographs. Washington, DC: World Bank. 52 p. ISBN: 0821342630.   This report highlights case studies in Nigeria in which modern and traditional agriculture has successfully enhanced biodiversity without sacrificing yield.

Srivastava, J.P., N.J.H. Smith, and D.A. Forno. 1996. *Biodiversity and Agricultural Intensification: Partners for Development and Conservation*. Environmentally Sustainable Development Studies and Monographs Series no. 11. Washington, DC: World Bank. 140 p. ISBN: 0821337599.   Explores ways to intensify crop and livestock production systems while mitigating adverse environmental impacts.

Srivastava, J.P., N.J.H. Smith, and D.A. Forno. 1996. *Biodiversity and Agriculture: Implications for Conservation and Development*. World Bank Technical Paper no. 321. Washington, DC: World Bank. 40 p. ISBN: 0821336169.   Explores linkages between biodiversity and agriculture and takes steps toward identifying policies and practices that enhance biodiversity for improving rural well-being.

## THESES AND DISSERTATIONS

The theses and dissertations cited here reflect qualitative and quantitative research in alternative farming systems. Some describe empirical research on farming systems, others analyzed quantitative implications and sought answers to some of the sociocultural issues of alternative farming. The following selected list of theses and dissertations reflect a range of hypotheses and research approaches designed to gain broader insights regarding specific variables.

Allen, P.L. 1998. *Sustainability and Sustenance: the Politics of Sustainable Agriculture and Community Food Security*. Ph.D. Diss. University of California, Santa Cruz (U.S.). 243 p.   A study of the resistance to locally based, ecologically sound, and socially just agrifood systems. Allen's approach draws upon political ecology, social movement theory, political economy, and discourse analysis to

examine national and California programs in sustainable agriculture and community food security.

Barham, M.E. 1999. *Sustainable Agriculture in the United States and France: a
Polanyian Perspective.* Ph.D. Diss. Cornell University (U.S.). 313 p. This study
compares the emergence and coalescence of movements in support of sustainable
agriculture in the United States and France during the period following World
War II. These movements are analyzed within a theoretical framework derived
from the work of Karl Polanyi. The author predicts these movements and other
country and region-based sustainable agriculture movements will increasingly
make contact with one another and eventually pursue joint goals at an international or global level.

Belcher, K.W. 1999. *Agroecosystem Sustainability: An Integrated Modeling Approach.* Ph.D. Diss. University of Saskatchewan (Canada). 194 p. The purpose
of this study was to evaluate the sustainability of agroecosystems and the linkages
between system components. Evaluations revealed that decision making for policies of sustainability, requires a weighting of system effects based on societal
preferences, ethical responsibilities, degradation thresholds, and system coevolu-
tion.

Chou, T.-H. 1993. *Energy and Economic Analyses of Comparative Sustainability
in Low-input and Conventional Farming Systems.* Master's Thesis, Michigan
State University (U.S.). The sustainability of two low-input (LIP) cropping systems and one conventional system (CONV) that uses commercial fertilizers and
pesticides, all from the Rodale Farming Systems Trial, are compared from 1981
to 1992 using energy and economic indicators. The results demonstrate that LIP
are more energy sustainable than CONV. Adjustments to social and economic
settings are proposed that could make LIP operations as profitable and economically sustainable as CONV.

Dunbar, T.V. 1997. *An Evaluation of Community-based Integrated Farming Systems Creating Conditions for Sustainability.* Ph.D. Diss. Michigan State University (U.S.). 109 p. Community-based Integrated Farming Systems (IFS) focus
on the unique conditions and interrelationships that exist in small limited-resource
farm communities. Dunbar sets a framework to help assess the extent to which
farming practices are changing, the kinds of changes most commonly made, and
the reasons farmers give to explain why they have or have not made changes in
Brinkley, Arkansas.

Glenda, L.L. 1997. *On Becoming Ecologically Rational: a Social and Environmental Critique of Agriculture.* Ph.D. Diss. University of Missouri—Columbia
(U.S.). 252 p. Constructs and evaluates a model of an agricultural system according to its ability to solve environmental problems within the broad developments of industrial society. It is concluded that critics of the conventional agricul-

ture system should direct their resources toward assisting in the creation and development of ecologically rational forms of agricultural production.

Godsey, L.D. 1996. *Selecting Indicators of Sustainable Farming Systems*. Master's Thesis, University of Missouri—Columbia (U.S.). 178 p. Assesses the sustainability of alternative farming systems in terms of 11 attributes that reflect economic, environmental, and social directions. The study revealed the importance of selecting indicators that are simple, available, reliable, and valid. These criteria can be used by farmers, policy makers, and rural communities to plan their farming systems strategies.

Green, S. 1991. *Facilitating the Transition from Conventional to Sustainable Farming Systems on Six Farms in Southern Quebec*. Master's Thesis, McGill University (Canada). 159 p. ISBN: 0315721138. This study summarizes the discussion pertaining to planning the farm transition from conventional to sustainable farming systems on six Quebec farms in the early stage of transition. The study examines characteristics of farm-level planning and activities, creative visioning and approaches to problem solving, value adjustments, decision-making criteria, and the farmers' perceived restraining force. A comprehensive, practical strategy is constructed and procedures for using the strategy are outlined, along with some requirements for further development.

Grønvold, N.M. 1996. *Adoption of Sustainable Agricultural Practices by Conventional Western Agriculture: Can Practices be Adopted from the Old Order Amish*. Master's Thesis, Washington State University (U.S.). 80 p. A literature research study of sustainable agricultural practices among the Old Order Amish and the adaptability of those practices to conventional agriculture. The author concludes that economics and government policies are barriers that can be overcome but differences in culture such as use of horsepower may be hindrances to complete transfer of sustainable practices. Crop rotation and diversification are suitable for most farms.

Howard, I.S. 1995. *Sustainable Agriculture and Sense of Place*. Ph.D. Diss. University of Guelph (Canada). 301 p. ISBN: 0315974842. An analysis of the economic, ecological, and stewardship perspectives of sustainable agriculture and the perceived connections between agriculture and food. The author offers a new perspective—sense of place—where the matrix of meanings and values that are associated with community and place extend the possibility of sustainable agriculture for the nurturing of human fulfillment in an environmental context.

Javor, R. 1996. *Sustainable Agriculture in Dairy Farming*. Master's Thesis, University of Wisconsin—River Falls (U.S.). 87 p. A study of 560 dairy farmers in Wisconsin to determine the profitability of sustainable operations. Compares conventional and sustainable operations and measures profitability due to reduced operating expenses and the degree of environmental benefits by using less natural resources and decreasing waste disposal. Results showed a wide variation among the groups of farmers studied.

Kelly, T.C. 1995. *A Bioeconomic Systems Approach to Sustainability Analysis at the Farm Level*. Ph.D. Diss. University of Florida (U.S.). 218 p.   Develops an integrated, whole-farm simulation model that provides a framework with which to analyze the effects over time of a farmer's decisions regarding cropping patterns, cultural practices, and farming technologies as they relate to the sustainability of the farming system. The model is applied to a hypothetical North Florida peanut farm and illustrates the richness and flexibility of the approach to analyzing the sustainability of a farming system.

Lorand, A.C. 1996. *Biodynamic Agriculture: a Paradigmatic Analysis*. Ph.D. Diss. Pennsylvania State University (U.S.). 112 p.   The objective of this study was to develop a comprehensive foundation for understanding the paradigm of biodynamic agriculture. Includes a summary, interpretation, extensive recommendations for adoption and diffusion, and Knowledge Map.

Minarovic, R.E. 1995. *North Carolina Cooperative Extension Service Professionals' Attitudes Toward Sustainable Agriculture*. Ed.D. Diss. North Carolina State University (U.S.). 260 p.   This research described the attitudes of North Carolina Cooperative Extension Service (NCCES) professionals and determined that there were no differences in attitudes based on position within the organization, length of time employed by the organization, and discipline in which professionals received their educational degrees. This study suggested the need for NCCES professionals to broaden their perspective of agriculture beyond the focus of their expertise and to acknowledge the complexities and magnitude of agriculture: a global and national perspective; degrees of sustainability such as organic and low-input production; and social impacts of agriculture.

Mothoa, M.G. 1994. *The Design and Development of Sustainable Agricultural Systems for Small Farmers in The Third World: a Review*. Master's Thesis, University of Pretoria (South Africa). 129 p.   A literature review regarding small-scale farm development in the developing world. Emphasizes the role of indigenous knowledge and the preservation of traditional social institutions and farming systems.

Muir, B.A. 1996. *A Comparison of High Input, Low Input and Ultra-low Input Cash Cropping Systems*. Master's Thesis, University of Guelph (Canada). 168 p. ISBN: 0612097153.   Examines whether a lower-cost alternative to conventional cash cropping systems could be sustainable. Results indicate that no differences between high input (HI) and low input (LI) yields for corn, soybean, and wheat, respectively. Ultra-low input (ULI) treatments resulted in 36% lower yields, apparently because of weed competition and lower available nitrogen (N) levels. Differences among ULI manure treatments were minimal.

Nault, J. 1991. *Participatory Extension Strategies for the Implementation of Sustainable Agriculture*. Master's Thesis, McGill University (Canada). 151 p. ISBN: 0315721154. This study proposes an alternative, participatory approach model for agricultural extension to expand the potential of participatory extension strategies to facilitate the development of sustainable agricultural systems. It presents the experiences of a group consisting of 6 farmers and a university coresearch team employed over a 17-month period to develop more ecologically sustainable farm systems.

Olson, R.K. 1998. *Evaluating Alternative Farming Systems: a Case Study for Eastern Nebraska*. Ph.D. Diss. University of Nebraska–Lincoln (U.S.). 319 p. Demonstrates a low-cost procedure for conducting economic, energy, and environmental analyses of farming systems and synthesizing the results into a qualitative assessment of relative sustainability. The results suggest that farming systems can be developed that allow smaller farms to be economically and environmentally competitive with larger conventional farms.

Smalley, S.B. 2000. *The Emergence of Stakeholder Consensus: Examining Issues in Evaluating Sustainable Agriculture Research and Education (SARE)*. Ph.D. Diss. University of Michigan (U.S.). 223 p. ISBN: 0599-918403. This study sought to determine whether it was possible to construct a framework to allow and support stakeholder definition of evaluation issues, problems, and recommendations associated with the North Central Region Sustainable Agriculture Research and Education (NCR SARE) program. Based upon survey responses, Smalley offered recommendations on project identity and expectations, funding, follow-up methods, media utilization, information sharing, investment portfolio, working with and through people, and goal clarity.

Soule, M.J. 1994. *Experimentation and Learning About New Agricultural Technologies: an Application in Sustainable Agriculture*. Ph.D. Diss. University of California–Berkeley (U.S.). 116 p. Farmers are faced with a number of options for maximizing crop production when a new technique is introduced. In this study, a dynamic, expected utility maximization model with Bayesian learning is developed to explain the decision to experiment with a new technology: velvetbean intercropped with maize. The model extends earlier work by employing the Kalman filter to model uncertainty about the profitability of the new technology. Results of the study indicate that velvetbean technology appears to be most appropriate for the poorest, most marginal farmers, and it is unlikely that farmers with good off-farm work opportunities or high labor productivity using their current technology (usually farmers using herbicide and fertilizer) will adopt the velvetbean intercrop.

Udoto, M.O. 1999. *Perceptions of Agricultural Education Teachers Toward Sustainable Agricultural Practices: Implications For Agricultural Education*. Ed.D. Diss. North Carolina State University (U.S.). 98 p. A survey of 94 agricultural education teachers in North Carolina to determine their perceptions of sustainable agricultural practices and the implications of their perceptions on high school agricultural education. Findings support the need for more research into integrating sustainable agricultural practices in agricultural-education teacher-training programs and high school agricultural education.

van Eijk, A.M. 1998. *Farming Systems Research and Spirituality: An Analysis of the Foundations of Professionalism in Developing Sustainable Farming Systems.* Ph.D. Diss. Landbouwuniversiteit, Wageningen (The Netherlands). 321 p. ISBN: 9045859806. (Includes summary in Dutch). A study of the impact of FSR on resource-poor farmers in Mozambique, Kenya, Tanzania, and Zambia and the difficulties encountered in attempting to develop sustainable farming systems. Concludes that most FSR practitioners are not in favor of formal modeling exercises and do not want modeling to become an end in itself that might hinder participation of resource poor farmers.

## WORLD WIDE WEB SITES

The World Wide Web, or the Internet, made it possible to disseminate information to an international audience in a short time. This has been particularly crucial to farmers in all countries who shared information in print format in the past. The printed materials were not distributed to all those who would benefit and were often written in languages they could not understand. The Internet changed the access in two ways—speed and enhanced distribution. It allows the world's farmers to share information about farming practices and find out which methods are beneficial and which are not. In a matter of minutes, and with some basic training in computer use, those who previously had no reliable source of information can now be involved in the communication and interactions that are so vital to improving their farm businesses and improving their standards of living. Following are descriptions of selected Internet sites that are informative, up-to-date, and instrumental in influencing farmers to adopt alternative farming methods. There are, of course, many more sites that could have been included here. However, these were chosen because their scope of coverage is wide and they are sites that users will find helpful, current, and reliable.

Agroecology Home (http://www.agroecology.org). This Web site is developed by the Environmental Studies Department and the Center for Agroecology and Sustainable Food Systems at the University of Santa Cruz mainly for an academic audience. It provides information on developing international sustainable agroecosystems, with emphasis on training, research, and application of agroecological science to solving real-world problems. Also gives descriptions of training courses and intensive workshops, the site provides tools for understanding principles of agroecology using a series of case studies, and a glossary of agroecological terms. Provides links to other agroecology Web sites.

Alternative Farming Systems Information Center—National Agricultural Library (http://www.nalusda.gov/afsic/). The Alternative Farming Systems Information Center (AFSIC) is one of several topic-oriented information centers at the U.S. National Agricultural Library (NAL). AFSIC specializes in locating and accessing information related to alternative cropping systems including sustainable, organic, low-input, biodynamic, and regenerative agriculture. AFSIC also focuses on alternative crops, new uses for traditional crops, and crops grown for industrial production. AFSIC staff create and publish bibliographies, directories, reference guides, and publications covering specific topics. Most AFSIC publications are available in ASCII text through the Web site under *publications* or on computer diskette or hard copy (limited availability).

Appropriate Technology for Rural Areas (http://www.atra.org). Appropriate Technology Transfer for Rural Areas (ATTRA) is the U.S. sustainable farming information center operated by the private nonprofit National Center for Appropriate Technology (NCAT). ATTRA provides technical assistance to farmers, extension agents, market gardeners, agricultural researchers, and other agriculture professionals in all 50 states. ATTRA's areas of interest cover three broad areas: sustainable farming production practices; alternative crop and livestock enterprises; and innovative marketing. Technical assistance, publications, and resources are provided free of charge to appropriate users.

Biodynamics—Biodynamic Farming and Gardening Association (http://www.biodynamics.org). A comprehensive access point for biodynamics information. Includes discussions of biodynamic agriculture, links to other biodynamic organizations, publications, products, jobs, and related resources.

Center for New Crops and Plant Products (http://www.hort.purdue.edu/newcrop). This Web site is located at the Center for New Crops and Plant Products at Purdue University. It offers a wealth of information about everything you need to know regarding new crops and new uses for conventional crops. Users can find lists of crops, information on production and uses of new crops, books, directories, conferences, and other resources. Includes aromatic, medicinal, and culinary crops, unconventional crops (usually consumed only in times of famine) and links to related Web sites, external databases, and libraries.

Ecological Agriculture Projects—Canada's Leading Resource Center for Sustainable Agriculture (http://eap.mcgill.ca). Located on the Macdonald Campus of McGill University, Ecological Agriculture Projects (EAP) has a large resource center, containing one of the world's best collections of materials on sustainable food and agriculture systems. The holdings are cataloged in print and electronic form. Visitors to the EAP site can find descriptions of the books and periodicals in the collection, full text of important articles on issues of ecological farming, guides, directories, and links to other ecological farming sites around the world.

Future Harvest (http://www.futureharvest.org). An organization with 16 research centers around the world that works to promote awareness and educate the general public and decision makers about the importance of international research on sustainable agriculture for a world with global peace, prosperity, environmental renewal, health, and the alleviation of suffering. The site contains information and reports of the Future Harvest programs and links to Web sites of other organizations doing similar work.

Henry A. Wallace Center for Agricultural and Environmental Policy (http://www.winrock.org/what/wallace_center.asp). Henry A. Wallace Center is affiliated with Winrock International, a nonprofit organization that works with people around the world to increase economic opportunity, sustain natural resources, and protect the environment. Wallace Center is concerned primarily with policy analysis, research, and evaluation to further sustainable and equitable agriculture and food systems, promote natural resources management, strengthen rural communities, and shape U.S. agricultural and food policy agendas. Publications of the Wallace Center are available at the Web site or in printed format.

International Federation of Organic Agriculture Movements (http://www.ifoam.org). An international forum for global exchange of information and cooperation on all aspects of the organic agriculture movement around the world. IFOAM advocates a holistic approach to organic farming systems and the importance of maintaining a humane and sustainable environment for future generations. The International Organic Accreditation Services, Inc., (IOAS) runs the IFOAM Accreditation Program to ensure equivalency of certification programs worldwide.

Plants for a Future: A Resource Center for Edible and Other useful Plants (http://www.comp.leeds.ac.uk/pfaf/index.html). A resource center for rare and unusual plants, particularly those that have edible, medicinal, or other uses. Contains a database of over 7,000 plants with information about cultivation and habitats. Includes a series of leaflets on uses of plants, plant portraits, catalogs, and checklists of plant characteristics.

Precision Agriculture (http://www.nal.usda.gov/wqic/preci.html) Maintained by the Water Quality Information Center at the NAL. This site provides links to papers and discussions, bibliographies, universities, and related links covering a wide range of resources on precision agriculture.

Sustainable Agriculture Network/Sustainable Agriculture Resource and Education Network. (SARE) (http://www.sare.org). The Sustainable Agriculture Network (SAN) is a cooperative effort of university, government, farm, business, and nonprofit organizations dedicated to the exchange of scientific and practical information on sustainable agricultural systems. SAN encourages the exchange of information with a variety of printed and electronic communications tools, and produces information on sustainable farming practices, principles and systems in a variety of formats, including print, World Wide Web, and e-books (electronic books). SAN publications are designed to answer practical questions about sustainable agriculture.

Sustain—The Alliance for Better Food and Farming (United Kingdom) (http://www.sustainweb.org). This site advocates food and agriculture policies and practices that enhance the health and welfare of people and animals; improve the working and living environment; enrich society and culture; and promote equity. Sustain represents over 100 national public interest organizations working at international, national, regional, and local levels. Includes links to other organizations involved in similar activities in the United Kingdom and Europe.

## OTHER USEFUL WEB SITES

Acres USA (http://www.acresusa.com).

American Small Farm (http://www.smallfarm.com).

CAST (http://www.cast.science.org).

Permaculture Activist (http://www.permacultureactivist.net).

Resource Efficient Agricultural Production (R.E.A.P) (http://www.reap.ca).

Soil Association United Kingdom (http://www.soilassociation.org).

I would like to thank Mary V. Gold, Librarian/Information Specialist, Alternative Farming Systems Information Center, National Agricultural Library, ARS, USDA for reviewing this chapter and offering valuable recommendations for changes and additions. Her knowledge of the farming systems literature and its characteristics broadened my view of the topic and made it a more comprehensive chapter.

## REFERENCES

Allen P. Sustainability in the Balance: Expanding the Definition of Sustainable Agriculture. Santa Cruz, CA: Agroecology Program, University of California. 1992, p. 8.

Consultative Group on International Agricultural Research. Technical Advisory Committee. Farming Systems Research at the International Agricultural Research Centers. Rome, Food and Agriculture Organization of the United Nations. 1978, p. 8.

Duckham AN, Masefield GB. Farming Systems of the World. New York: Praeger, 1969, pp. 94–101.

Diver S. Towards a Sustainable Agriculture. New Renaissance, 6(2): 1–9, 1996. Available online at http://www.ru.org/artagri.html.

# 8

## Field Crops

**Kathleen Ann Clark***

Purdue University, West Lafayette, Indiana, USA

This chapter is concerned with the specialization of agriculture concerned with the theory and practice of field crop production. It includes the literature of crop production, crop breeding, biotechnology for crop improvement, pest management, and plant pathology. Excluded are resources that focus on general agriculture, farming systems, alternative agriculture, horticulture, precision agriculture, and soil science, which are covered in other chapters.

### ABSTRACTS, INDEXES, AND DATABASES

The primary indexes for crop science literature are AGRICOLA, parts of CAB Abstracts, and AGRIS. Additionally, both Web of Science and BIOSIS/Biological Abstracts offer coverage of the primary journals of interest to field crop scientists.

AGRICOLA has built-in category codes, which are a broad type of subject heading. The most frequently used codes of interest to agronomists include agricultural history and biography; entomology related; pests of plants—animals (or: —insects, —nematodes); physiology and biochemistry of field crops; plant breeding and genetics; plant diseases—bacterial (or: —fungal, —general, —physiological, —viral); plant nutrition; plant physiology and biochemistry; plant protection; plant production—field crops; plant structure and anatomy; and

---

* *Current affiliation*: University of Illinois at Urbana-Champaign, Urbana, Illinois, USA

weeds. These codes appeared on between 10,000 and 170,000 articles (depending on the descriptor) published since 1978. Many category codes define topics of interest to field crop agronomists. To learn how to limit searches to the appropriate category code, consult the documentation for your version of AGRICOLA.

In addition to category codes, AGRICOLA records are assigned a varying number of descriptors (subject headings) from the *CAB Thesaurus*. These are more specific terms than the category codes and provide a wealth of alternative vocabulary terms to enrich searching. Many vendors have built-in access to the *CAB Thesaurus*. A clever way to discover the appropriate subject headings or category codes is to conduct a general search, find several articles that appear to be right on target, and then look to see which subject headings and category codes have been added to the records for these articles. Following this review, go back to the search input screen and reiterate the search. For example, if you are interested in the biological control of insect pests, start by searching for *biological control and insect pests*. Upon perusing the records of several articles, the following descriptors would be found: biological control, entomogenous fungi, insect viruses, biological control agents, microbial pesticides, natural enemies, as well as the specific epithets of numerous control agents. At this point the original search could be modified to: ((biological control or natural enemies) and insects) or insect viruses or entomogenous fungi or microbial pesticides). One caveat must be noted: although AGRICOLA uses the *CAB Thesaurus* for descriptors, many of the British terms in the thesaurus have been Americanized for use in AGRICOLA. For example, soyabeans = soybeans; colour = color; nucleolar organiser = nucleolar organizer; mould = mold; tumour = tumor; behaviour = behavior; and plough = plow. When searching for these terms, be sure to search both forms of the word.

AGRIS records are assigned subject headings from AGROVOC, a multilingual agricultural thesaurus. AGRIS records are also assigned AGRIS categories. Among the AGRIS categories most useful for field agronomists are the plant production and plant protection categories that include crop husbandry, plant propagation, seed production, fertilizing, irrigation, soil cultivation, cropping patterns and systems, plant genetics and breeding, plant ecology, plant structure, plant physiology and biochemistry, plant taxonomy and geography, plant diseases, weeds, and plant pests. AGRIS, AGROVOC, and the AGRIS/CARIS subject categories are freely available on the Internet from the Food and Agriculture Organization of the United Nations (FAO) at http://www.fao.org/agris/ (cited February 21, 2002).

Although AGRICOLA may be better known in some circles, CAB Abstracts actually offers more comprehensive, worldwide coverage of the agronomic literature. Among the print titles included in CAB Abstracts that are of interest to the field agronomist are *AgBiotech News and Information*, *Biocontrol News and Information*, *Crop Physiology Abstracts*, *Field Crop Abstracts*, *Grass-*

*lands and Forage Abstracts, Irrigation and Drainage Abstracts, Maize Abstracts, Plant Breeding Abstracts, Plant Genetic Resources Abstracts, Plant Growth Regulator Abstracts, Postharvest News and Information, Potato Abstracts, Review of Agricultural Entomology, Review of Plant Pathology, Rice Abstracts, Seed Abstracts, Soils and Fertilizers, Soyabean Abstracts, Sugar Industry Abstracts, Weed Abstracts,* and *Wheat, Barley and Triticale Abstracts.* CAB Abstracts is a highly structured database so, as with AGRICOLA, use the *CAB Thesaurus* descriptors and the subject codes (called CABI Codes) to optimize searching.

Cambridge Scientific Abstracts. Agricultural and Environmental Biotechnology Abstracts. 1993–. This rather small database indexes 123 journals that cover the literature of how biotechnological techniques are being used in agriculture. It includes topics such as plant genomics, genetic engineering, and transgenic plants. Agricultural and Environmental Biotechnology Abstracts is available online from Cambridge Scientific Abstracts (CSA) as a subfile of the Environmental Sciences and Pollution Management Database.

## ATLASES

Arkansas Center for Advanced Spatial Technologies. *Census of Agriculture.* Fayetteville, AR: The University of Arkansas. Available from the World Wide Web at http://www.cast.uark.edu/local/catalog/national/html/AgCensus_Main.html (cited February 21, 2002). Using data derived from the 1978–93 *Census of Agriculture,* the Center for Advanced Spatial Technology graphically presents maps of U.S. land used in the production of various major crops. The purpose of this site is to show GIS representation of selected census data. For more complete *Census of Agriculture* data (1987–97), see the Government Information Sharing Project: Census of Agriculture, maintained by Oregon State University (http://sasquatch.kerr.orst.edu/ag-stateis.html) (cited February 21, 2002).

*CMI Distribution Maps of Plant Diseases.* 1942– . Wallingford, UK: CABI Publishing. ISSN: 0012-396X. Published in loose-leaf format, this resource is frequently updated as new occurrences of plant pathogens are reported in the literature (CAB Abstracts), or are accessioned into the UK's International Mycological Institute (IMI) or recognized by the European and Mediterranean Plant Protection Organization. The worldwide "dot maps" are supported by references on the back of each map, together with a code for the extent or status of the pest. Information from this resource is incorporated into CABI's Crop Protection Compendium, and available on the Web with a subscription.

*Crop Protection Compendium.* 1997– . Wallingford, UK: CAB International. ISSN: 1365-9065. An encyclopedic, interactive tool that brings together a wide range of different types of information on all aspects of crop protection. Full data files are available for over 1,560 pests, diseases, weeds, and natural enemies of worldwide or regional importance, each with text, illustrations, and distribution maps. Outline data is available for an additional 10,000 species. Other features include diagnostic keys, links to Web sites, statistical databases, a glossary, and more. Currently available on CD-ROM, a Web version is expected soon.

International Association of Agricultural Economists, Committee for the World Atlas of Agriculture, and Istituto Geografico De Agostini. 1969. *World Atlas of Agriculture*: *Under the Aegis of the International Association of Agricultural Economists*. Novara, Italy: Istituto Geografico De Agostini.    This four-volume set, which is arranged by continent and then by country, graphically presents information about the climatic conditions, land utilization, and principle crops (1950–60s). The text is enhanced with many tables derived from the FAO and other agencies.

*Major World Crop Areas and Climatic Profiles*. 1994. *Agriculture Handbook* (USDA) no. 664, rev. Washington, DC: USDA World Agricultural Outlook Board Joint Agricultural, Weather Facility. ISSN: 0065-4612.    Provides benchmark climate and crop data for the key producing areas of the world. Country maps illustrate the areas where the major crops are produced, and the temperature and precipitation by month. Tables present the historical averages for yield and production. Organized by country, and then by major crop within each country.

Maramorosch, K. and A.J. Dalton. 1977. *The Atlas of Insect and Plant Viruses*: *Including Mycoplasmaviruses and Viroids*. Ultrastructure in Biological Systems, Vol. 8. New York: Academic Press. ISBN: 0124702759.    Includes a description of viruses infecting insects, higher plants, fungi, mycoplasmas, and spiroplasmas. Liberally illustrated with electron micrographs.

Pillsbury, R. and J.W. Florin. 1996. *Atlas of American Agriculture*: *The American Cornucopia*. New York; London: Simon & Schuster, Macmillan, Prentice Hall International. ISBN: 002897333X.    Includes chapters on the major agricultural crops that describe the history and evolution of the crop, the growing conditions required, and the major varieties in use. Distribution maps of the United States show the major production areas during the nineteenth and twentieth centuries.

Van Royen, W. 1952. *Atlas of the World's Resources. Vol. 1, The Agricultural Resources of the World.* New York: Prentice-Hall for the University of Maryland.    Provides an overview of the geographical distribution of the world's agricultural production. A large proportion of the maps are devoted to North America. Chapters for each crop include production dot-maps as well as a short synopsis of the growing conditions, and climate and soil required.

## BIBLIOGRAPHIES

While today's researcher can easily obtain access to the current literature through electronic databases such as AGRICOLA or CAB Abstracts, it is less easy to obtain citations for literature published prior to the 1970s. Hence, bibliographies that pull together this literature are valuable resources.

Farr, D. F. and the American Phytopathological Society. 1989. *Fungi on Plants and Plant Products in the United States*. Contributions from the U.S. National Fungus Collections, no. 5. St. Paul, MN: APS Press. ISBN: 0890540993. Contains information about the fungi involved in vascular plant diseases and destruction. With data on more than 78,000 unique fungus-host combinations, more than 13,000 names of fungi, and citations from over 4,000 publications (articles, books, monographs, and herbarium records), this is an invaluable aid to all mycologists and plant pathologists.

Grainge, M. and S. Ahmed. 1988. *Handbook of Plants with Pest-Control Properties*. New York: Wiley. ISBN: 0471632570. Catalogs 2,400 plant species that have been reported to have pest-control properties. Arranged by species, it has nearly 1,400 references.

Jacobson, M. 1990. *Glossary of Plant-Derived Insect Deterrents*. Boca Raton, FL: CRC Press. ISBN: 0849332788. Compilation of 1,500 species from 175 families that have been reported to have insect-deterrent compounds. Arranged by family and species. Over 1,200 citations.

Rossman, A.Y., M.E. Palm, L.J. Spielman, and the American Phytopathological Society. 1987. *A Literature Guide for the Identification of Plant Pathogenic Fungi: Contributions from the U.S. National Fungus Collections*. St. Paul, MN: APS Press. ISBN: 0890540802. When a plant pathologist is called upon to identify a pathogenic fungus, they may find that they are able to narrow it down to a particular order, or genus. However, to go beyond that they need to consult the literature. This book is a bibliography of the seven orders of plant pathogenic fungi, as well as the individual genera within these orders. Works included span from the mid 1930s to the mid 1980s, and are geographically international. Comprehensive indexes for the authors and genera are included.

## BIOGRAPHIES

There are many interesting biographical books written about agronomists, including many about pioneer farmers. To locate biographies one may search a library catalog for entries with Library of Congress Subject Headings such as agriculturists—[state/country]—biography; farmers—[state/country]—biography; plant breeders—[state/country]—biography; or plant pathologists—[state/country]—

biography. If searching an online catalog using a keyword, one might try something such as ((agricult* or farm* or crop* or plant*) and (biograph*)).

Another source of biographies is the journal literature. For example, the *Annual Review of Phytopathology* usually has a biography or autobiography as its lead article. CAB Abstracts is also a rich source because it has subject headings (descriptors) for "obituaries" and "biographies" as well as a CABI descriptor for "history and biography."

Carver, G.W. and J. Zidar. 1993. *George Washington Carver Papers, Notes and Letters*. Beltsville, MD: NAL National Agricultural Text Digitizing Program. This CD-ROM produced by the U.S. National Agricultural Library is a wonderful resource that provides access to the George Washington Carver papers, notes, letters, and photographs. It was distributed to most of the land-grant institutions, and so should be widely available.

Christensen, C.M. 1984. E.C. *Stakman, Statesman of Science*. St. Paul, MN: APS Press. ISBN: 089054056X. This book chronicles the important events in plant pathology in the twentieth century, including E. C. Stakman's pioneering studies on the epidemiology and nature of genetic variation in *Puccinia graminis*, the stem rust pathogen.

McEwan, B. 1991. *Thomas Jefferson, Farmer*. Jefferson, NC: McFarland. ISBN: 089950633X. Although we usually think of Jefferson as a statesman, he would portray himself as an independent-thinking farmer. Based on Jefferson's notes, letters, and diaries, this small volume recounts activities on the Jefferson farm including the crops he grew, the plants he tried to introduce into Virginia, his innovative cropping systems and farming inventions, discussions on farming he held with other "farmers" (e.g., George Washington and James Madison), and much more.

Perry, J. 1999. *Unshakable Faith: Booker T. Washington & George Washington Carver: A Biography*. Sisters, OR: Multnomah Publishers. ISBN: 1576734935. In their individual and collaborative work, Booker T. Washington and George Washington Carver quietly proved their oppressors wrong and along the way made remarkable discoveries and contributions that have benefited mankind to this day. Their sense of divine appointment guided them through a world of dark prejudice with humility and self-confidence, allowing their lives to transcend racial prejudice and jealousy.

Peterson, P.A. and A. Bianchi. 1999. *Maize Genetics and Breeding in the 20th Century*. Singapore; River Edge, NJ: World Scientific. ISBN: 981022866X. The main section of this book includes the biographies of the preeminent corn breeders of the twentieth century reproduced from the journal *Maydica*. Another section covers various communities or groups of maize geneticists or breeders, such as the Bussey Institute, Emerson's school, and the Italian maize geneticists. The concluding part reviews the stories of those involved in the recent biotechnological developments.

Soyfer, V. 1994. *Lysenko and the Tragedy of Soviet Science*. New Brunswick, NJ: Rutgers University Press. ISBN: 0813520878. Dr. Soyfer, who met Lysenko in the 1930s, chronicles the destructive influence of Lysenko on Soviet science. Although not a well-trained agronomist, Lysenko had the ear of the Soviet press, Stalin, and Khrushchev, so his claims of producing superior cattle and wheat went untested for years. He detested academics, so he was able to effectively eliminate all serious work in the Soviet Union in genetics and agriculture from the 1930s into the 1960s. A fascinating story about how one man's poor science can set back progress for decades.

## BOOKS

### Crop Production

The Agronomy series, published irregularly by the American Society of Agronomy (Madison, WI), is uniformly recommended. Many volumes focus on soil science, which is outside the scope of this chapter, but among the recent field crop titles in this series:

Barker, K.R., G.A. Pederson, G.L. Windham. 1998. *Plant and Nematode Interactions*. Agronomy no. 36. Madison, WI: American Society of Agronomy, Crop Science Society of America, Soil Science Society of America. ISBN: 0891181369.

Hanks, R.J. and J.T. Ritchie. 1991. *Modeling Plant and Soil Systems*. Agronomy no. 31. Madison, WI: American Society of Agronomy, Crop Science Society of America, Soil Science Society of America. ISBN: 0891181067.

Hanson, A.A., D.K. Barnes, and R.R. Hill. 1988. *Alfalfa and Alfalfa Improvement*. Agronomy no. 29. Madison, WI: American Society of Agronomy. ISBN: 089118094X.

Heyne, E.G. 1987. *Wheat and Wheat Improvement*. 2nd ed. Agronomy no. 13. Madison, WI: American Society of Agronomy, Crop Science Society of America, Soil Science Society of America. ISBN: 0891180915.

Marshall, H.G. and M.E. Sorrells. 1992. *Oat Science and Technology*. Agronomy no. 33. Madison, WI: American Society of Agronomy, Crop Science Society of America, Soil Science Society of America. ISBN: 0891181105.

Moser, L.E., D.R. Buxton, and M.D. Casler. 1996. *Cool-Season Forage Grasses*. Agronomy no. 34. Madison, WI: American Society of Agronomy, Crop Science Society of America, Soil Science Society of America. ISBN: 089118130X.

Olson, R.A. and K.J. Frey. 1987. *Nutritional Quality of Cereal Grains*: *Genetic and Agronomic Improvement*. Agronomy no. 28. Madison, WI: American Society of Agronomy. ISBN: 0891180923.

Schneiter, A.A., G.J. Seiler, and J.M. Bartels. 1997. *Sunflower Technology and Production*. Agronomy no. 35. Madison, WI: American Society of Agronomy. ISBN: 0891181350.

Sprague, G. and J.W. Dudley. 1988. *Corn and Corn Improvement*. 3rd ed. Agronomy no. 18. Madison, WI: American Society of Agronomy. ISBN: 0891180990.

Waddington, D.V., R.N. Carrow, and R.C. Shearman. 1992. *Turfgrass*. Agronomy no. 32. Madison, WI: American Society of Agronomy, Crop Science Society of America, Soil Science Society of America. ISBN: 0891181083.

Wilcox, J.R. 1987. *Soybeans: Improvement, Production, and Uses*. 2nd ed. Agronomy no. 16. Madison, WI: American Society of Agronomy. ISBN: 0891180907.

Brouk, B. 1975. *Plants Consumed by Man*. London; New York: Academic Press. ISBN: 012136450x. There are several books on the market that list the thousands of plants of economic importance, but they usually do not provide much information about the plant. This book, on the other hand, includes important details such as the etymology, history, geography, chemistry, morphology, and physiology for 366 plants of economic importance.

Desai, B.B., P.M. Kotecha, and D.K. Salunkhe. 1997. *Seeds Handbook*: *Biology, Production, Processing, and Storage*. New York: Marcel Dekker. ISBN: 0824700422. Reviews new advances in seed science and technology such as techniques for the production of synthetic seeds, loss-reduction biotechnologies, and new developments in the seed production industry. Topics covered range from seed morphology and development to the biochemistry of seed dormancy. The production technique for some 80 agronomic and horticultural crops is covered. Postharvest seed issues include seed drying, cleaning, storage, certification, and so forth. Of use to agronomists, biotechnologists, and geneticists.

Kelly, A.F. and R.A.T. George. 1998. *Encyclopaedia of Seed Production of World Crops*. Chichester; New York: John Wiley. ISBN: 0471982024. The primary focus is on the production of seed for particular vegetable and agricultural species important in world agriculture. Most of the book is devoted to covering the distinguishing traits of cultivars, the genetics of hybrids, environmental conditions required, seed crop establishment, harvesting, and storage. Additional chapters discuss international legislation that governs the development and transport of seed, and the structure of the seed industry.

KingCorn, the Corn Growers' Handbook (http://www.kingcorn.org/) (cited February 21, 2002) 1994– . Maintained by Purdue University Agronomy professor, Dr. R.L. Nielsen, KingCorn seeks to provide all the information a corn grower would need to know. The publications section of KingCorn has categorized these resources into such areas as cropping systems, storage, diseases, irrigation, soil management, and so forth. A search engine is also available on this Web site.

Langer, R.H.M. and G.D. Hill. 1991. *Agricultural Plants*. 2nd ed. Cambridge; New York: Cambridge University Press. ISBN: 0521405459. Serves as an introduction to over 100 agricultural species of worldwide importance, providing information about their structure, botany, cultivation, and use. Arranged by family and then individual crop. Clear line drawings help illustrate the text.

McDonald, M.B. and L.O. Copeland. 1997. *Seed Production*: *Principles and Practices*. New York: Chapman & Hall. ISBN: 0412075512. The first part of this book goes through the basics of seed production, from flowering and seed set, to harvesting, storage, and marketing. The bulk of the book is devoted to a description of seed development, production, harvesting, and storage of particular crop plants. Coverage is broad, ranging from major crops such as corn and soybeans to buckwheat, rapeseed, and the prairie grasses.

Petersen, R.G. 1994. *Agricultural Field Experiments*: *Design and Analysis*. New York: Marcel Dekker. ISBN: 0824789121. The reader is assumed to have some background in statistics. This volume will help the researcher design experiments so they can be analyzed statistically. Topics include variety trials, pasture trials, experiments with perennial crops, the optimal plot size, randomized block design, and others.

Satorre, E.H. and G.A. Slafer, eds. 1999. *Wheat*: *Ecology and Physiology of Yield Determination*. New York: Food Products Press. ISBN: 1560228741. Chapters cover the environmental, physiological, cultural, and breeding aspects that influence yield in wheat. An important section discusses the potential role hybrid wheat and biotechnology will play in the improvement of wheat yield.

## Crop Breeding

Allard, R.W. 1999. *Principles of Plant Breeding*. 2nd ed. New York: Wiley. ISBN: 0471023094. This is a complete revision of the classic 1960 first edition, which is still heavily used. Whereas the first edition was used for many years as a textbook, the second edition seems more information dense, and is suitable for the advanced student. Starting with several introductory chapters that cover Darwinian evolution, the origins of agriculture, and a discussion of the mating systems of plants, the book progresses to chapters that discuss the biological foundations of breeding (e.g., heredity, environmental factors, hybridization). The final half of the book is devoted to modern breeding schemes, such as breeding of self-pollinated plants (wheat and many legumes), outcrossing plants (corn), and clonally propagating plants (potatoes and cassava). Includes a glossary, references, and index.

Carozzi, N. and M. Koziel. 1997. *Advances in Insect Control: the Role of Transgenic Plants*. London; Bristol, PA: Taylor & Francis. ISBN: 0748404171. In fifteen chapters, this book discusses the current technologies that have proven successful for the engineering of specific insect-tolerant crops (corn, rice, potatoes, and beans), as well as an overview of the prospects for the future of the genetic engineering of crops.

Hancock, J.F. 1992. *Plant Evolution and the Origin of Crop Species*. Englewood Cliffs, NJ: Prentice Hall. ISBN: 0136789505. The first half of the book describes the natural forces of plant evolution, such as gene duplication, speciation, and genetic variability. The second part of the book explains the forces that drove the evolution of the major crops (cereals, beans, and starchy crops) over time. An extensive bibliography is provided for each chapter.

Harlan, J.R. 1992. *Crops & Man*. 2nd ed. Madison, WI: American Society of Agronomy; Crop Science Society of America. ISBN: 0891181075. The fundamental thesis of this book is that crops and the human societies they have nourished have coevolved. Professor Harlan cautions that to maintain genetic diversity we must continue to maintain and utilize gene banks for the major and minor crops. He also cautions that society should be wary of the extension of monocultures. One of the chapter titles hints at the scope of this book: "What is a crop?" His answer: ". . . the word "crop" covers all that which is harvested regardless of it's status as a domesticate." In other words, crops include those plants that are cultivated (sown for harvest) as well as those native plants that are domesticated for man's use. Geneticists, plant breeders, cultural anthropologists, and historians will all profit from reading this book.

Hodgkin, T. and International Plant Genetic Resources Institute. 1995. *Core Collections of Plant Genetic Resources*. Chichester; New York: Wiley. ISBN: 0471955450. This volume addresses worldwide conservation of plant genetic resources via gene banks with methods for developing and maintaining core collections and the type of data that must be kept about the core collections.

Murray, D.R. 1991. *Advanced Methods in Plant Breeding and Biotechnology*. Wallingford, UK: CAB International. ISBN: 0851987060. Reviews the recently developed techniques developed in plant breeding and biotechnology, such as gene transfer via *Agrobacterium*, electroporation, microprojectile-mediated gene transfer, somatic embryogenesis, and manipulation of the chloroplast or mitochondrial genomes. Particular emphasis is given to the breeding of crops for disease or pest resistance.

National Research Council (U.S.). Committee on Managing Global Genetic Resources: Agricultural Imperatives. 1991. *The U.S. National Plant Germplasm System (NPGS)*. Washington, DC: National Academy Press. ISBN: 0309043905. Relates the history and background that led to the development of the National Plant Germplasm System (NPGS), whose responsibility it is to preserve the tissues, seeds, and plants that comprise the nation's plant germplasm resources. Describes the challenges of managing the plant germplasm collections, which are spread throughout the United States. Available from World Wide Web at http://www.nap.edu/ (cited February 21, 2002).

Poehlman, J.M. and D.A. Sleper. 1995. *Breeding Field Crops*. 4th ed. Ames, IA: Iowa State University Press. ISBN: 0813824273. Basic textbook on the principles involved in breeding field crops. Covers the genetic basis of modern plant breeding, such as crop reproduction and gene recombination, the tools used by breeders, such as mutations and molecular biology, as well as the methods used in breeding specific crops.

Richards, A.J. 1997. *Plant Breeding Systems*. 2nd ed. London; New York: Chapman & Hall. ISBN: 0412574403. Basic and advanced information about angiosperm breeding systems including such topics as floral diversity, dicliny, self-incompatibility, pollination biology, self-fertilization, vegetative reproduction, and agamospermy. Well illustrated and documented; includes a glossary of terms and index.

Smartt, J. and N.W. Simmonds. 1995. *Evolution of Crop Plants*. 2nd ed. New York: Wiley. ISBN: 0470233729; 0582086434. Organized by crop, this volume covers the cytotaxonomic background, early history, and recent domestication of major crops such as barley and wheat as well as many lesser crops such as cacao and turnips.

Vasil, I.K., ed. 1999. *Molecular Improvement of Cereal Crops*. Advances in Cellular and Molecular Biology of Plants, v. 5. Dordrecht; Boston: Kluwer Academic Publishers. ISBN: 0792354710. Introductory chapters cover the methods used to transform plants, while later chapters detail the specific procedures and approaches that are leading to the successful transformation of particular cereal species such as wheat, rice, corn, oats, and sorghum.

## Crop Protection

Agarwal, V.K. and J.B. Sinclair. 1997. *Principles of Seed Pathology*. 2nd ed. Boca Raton, FL: Lewis Publishers. ISBN: 0873716701. An important text for agronomists, this volume covers topics such as the factors that affect the transmission of pathogens by seed, how seed becomes inoculated with pathogens, how to protect seed from contamination, and the effects of fungi on seed. Contains extensive tables that list seedborne fungi, bacteria, and viruses, and the crops affected. Well illustrated, with an extensive bibliography.

Agrios, G.N. 1997. *Plant Pathology*. 4th ed. San Diego, CA: Academic Press. ISBN: 0120445646. This is a frequently used textbook for plant pathology, covering all the basics including the genetics, epidemiology, and control of plant pathogens, as well as the physiological responses and defenses used by plants. Well over half of the book is comprised of chapters about the major classes of pathogens and specific plant diseases. A final section reviews biotechnological advances that are being employed by plant pathologists.

Copping, L.G. 1996. *Crop Protection Agents From Nature: Natural Products and Analogues*. Critical Reports on Applied Chemistry, v. 35. Cambridge: The Royal Society of Chemistry. ISBN: 0854044140. This book reviews biologically derived compounds that may have potential as crop protection agents. Chapters focus on biologically active metabolites derived from microorganisms, algae, higher plants, and animals that may serve to protect field crops as herbicides, fungicides, or insecticides. The mode of action and biological effects of the biological secondary compounds are documented.

Devine, M., S.O. Duke, and C. Fedtke. 1993. *Physiology of Herbicide Action*. Englewood Cliffs, NJ: Prentice Hall. ISBN: 0133690679; 013679663X; 0131403109. Herbicides kill weeds in a variety of modes. Arranged by mechanism of action, this book explains the physiological function in plants at which the various classes of herbicides work, such as amino acid biosynthesis, photosynthesis, protein synthesis, and so forth. Additionally it explains how herbicides are selectively absorbed and translocated by plants.

Duke, S.O., ed. 1996. *Herbicide-Resistant Crops: Agricultural, Environmental, Economic, Regulatory, and Technical Aspects*. Boca Raton, FL: CRC Lewis Publishers. ISBN: 1566700450. The largest section of this book is on the technology that enables the creation of herbicide-resistant crops. However, there are also important chapters on the regulatory aspects of genetically modified organisms in the United States and abroad, the risk assessment that is required, and the economics of the matter from the perspective of the seed company, the farmer, and the consumer.

Fox, R.T.V. and CAB International. 1993. *Principles of Diagnostic Techniques in Plant Pathology*. Wallingford, UK: CAB International. ISBN: 0851987400. Concisely delineates the methods used by plant pathologists to determine phytopathological agents. Includes immunological, microscopical, biochemical, physiological, and nucleic acid–based methods. Well documented and indexed.

The dichotomous keys in these three volumes, widely distributed to libraries by CIBA-GEIGY, may be used by the novice or expert to identify many of the world's weeds. Descriptions of the underground organs, stems, leaves, flowers, fruits, seedlings, and geographic distribution will aid the user in discovering the identity of a weed. The line drawings of each species clearly illustrate the distinguishing features; additionally, beautiful color plates for many species are also included. While the dichotomous keys are in English, French, German, and Spanish, the body of each volume is in English.

> Häfliger, E. 1982. *Monocot Weeds 3*: *Monocot Weeds Excluding Grasses*. Basel, Switzerland: Documenta CIBA-GEIGY.
> Häfliger, E. and H. Scholz. 1980. *Grass Weeds 2*. Basel, Switzerland: Documenta CIBA-GEIGY.
> Häfliger, T.J. and M. Wolf. 1988. *Dicot Weeds*: *Dicotyledonous Weeds of 13 Families*. Basel, Switzerland: Documenta CIBA-GEIGY.

Hall, F.R. and J.J. Menn, eds. 1999. *Biopesticides*: *Use and Delivery*. Totowa, NJ: Humana Press. ISBN: 0896035158. Until replaced by synthetic pesticides in the 1950s, biopesticides held a place of importance in crop production. This volume summarizes the scientific status of biopesticides in modern agriculture, describes monitoring procedures, and covers the mechanism of action of secondary products derived from microbials, plants, and other biological species that hold potential to protect crops. Includes practical information (such as formulations) as well as a scientific review of the literature.

Holm, L.G. 1997. *World Weeds*: *Natural Histories and Distribution*. New York: Wiley. ISBN: 0471047015. Comprehensive and updated information on 104 of the worlds worst weeds is presented. Each entry has a botanical description that includes the habitat, global distribution, seed production, ecology, physiology, impacted crops, and so forth. Examines how changes in agricultural practices such as herbicide use, tillage methods, and agricultural runoff change the character of weed problems. Each treatment includes excellent line drawings of the plant, with enlarged views of the diagnostic features. Articles range in length from a few pages to 20 or more. Added features include a glossary, extensive bibliography, and list of common weed names.

Lorenzi, H. and L.S. Jeffery. 1987. *Weeds of the United States and Their Control*. New York: Van Nostrand Reinhold. ISBN: 0442258844. Describes and illustrates over 300 species of weeds, including clear color photographs for each plant. A full page for each weed provides a distribution map, description, habitat, and suggested means of control for the home gardener and for the farmer. Tables provide information on methods of herbicide application and herbicides recommended for specific crops, though the main use of this volume would be as a weed identification guide.

Manners, J.G. 1993. *Principles of Plant Pathology*. 2nd ed. Cambridge; New York: Cambridge University Press. ISBN: 0521434025. Rather than serving as a manual for the identification of pathogens, this volume assumes the reader knows how to identify phytopathogens and covers, instead, principles that underlie the discipline of plant pathology. For example, chapters describe the physiology of the host-parasite relationship (chemistry, physiology, genetics, metabolism), epidemiology of disease development, and plant disease control methods. Although this volume is in need of revision, it is still widely used.

Maude, R.B. 1996. *Seedborne Diseases and Their Control: Principles and Practice*. Wallingford, UK: CAB International. ISBN: 0851989225. The primary focus is on how seeds are infected by and become carriers of disease, and on the methods that can be used to control or eradicate the spread of disease by seed.

Pedigo, L.P. 1999. *Entomology and Pest Management*. 3rd ed. Upper Saddle River, NJ: Prentice Hall. ISBN: 013780024X. A popular entomology textbook, this book covers the basics such as how to identify insects, as well as how to manage insect pests using pesticides (natural and chemical), biological agents, resistant plants, and by modifying insect behavior. Includes a glossary, index, and several appendices, including lists of insect pests and insecticides.

*The Plant Viruses*. 1985–96. 5 vols. New York: Plenum Press. ISBN: 0306419580 (v. 1); 0306422581 (v. 2); 0306427052 (v. 3); 0306428458 (v. 4); 0306452251 (v. 5). A five-volume set that is part of a larger series, The Viruses. Each volume is devoted to a different group of plant viruses: v. 1: *Polyhedral Virions With Tripartite Genomes*; v. 2: *Rod-Shaped Plant Viruses*; v. 3: *Polyhedral Virions with Monopartite RNA Genomes*; v. 4: *The Filamentous Plant Virus*; and v. 5: *Polyhedral Virions and Bipartite RNA Genomes*.

Rao, V.S. 2000. *Principles of Weed Science*. 2nd ed. Enfield, NH: Science Publishers. ISBN: 157808069X. Deals with the basic principles and technology of weed science including weed biology, weed management, herbicides (mechanisms of action including absorption, persistence in the environment, interactions, application, and so forth), and bioherbicides. Importantly, this second edition also includes chapters on herbicide resistance and genetic engineering. Suitable for use as a textbook for undergraduates.

Sharma, S.B. 1998. *The Cyst Nematodes*. Dordrecht; Boston: Kluwer Academic Publishers. ISBN: 0412755300. Although initially assumed to be a major problem only in temperate regions, nematodes are now considered to be of global importance across a range of different agro-ecological zones. Among the nematode topics discussed: evolution, genetic diversity, genetic engineering, taxonomy, distribution, biology and physiology, pathogenicity, and control. Includes a dichotomous key to the 105 known species.

Singh, U.S., R.P. Singh, and K. Kohmoto, eds. 1995. *Pathogenesis and Host Specificity in Plant Diseases*: *Histopathological, Biochemical, Genetic, and Molecular Bases*. 3 vols. Oxford; Tarrytown, NY: Pergamon. ISBN: 008042273X (set); 0080425100 (v. 1); 0080425119 (v. 2); 0080425127 (v. 3). The three volumes focus on pathogens that are prokaryotes, eukaryotes, or viruses and viroids, respectively. Included in the eukaryotic volume are the fungi, lichens, nematodes, and parasitic plants. The series is richly illustrated with color photos and electronmicrographs. Each chapter is well documented.

## DICTIONARIES

Ainsworth, G.C., G.R. Bisby, D.L. Hawksworth, and the International Mycological Institute. 1995. *Ainsworth & Bisby's Dictionary of the Fungi*. 8th ed. Wallingford, UK: CAB International. ISBN: 0851988857. Now in its 52nd year, this volume attempts to meet the needs of all mycologists, from those that study rusts on cereals, to the amateur mycologist who is trying to understand the terminology of mycology. The eighth edition includes a key to the families of fungi for the first time.

Bojňanský, V. and A. Fargašová. 1991. *Dictionary of Plant Virology, In Five Languages, English, Russian, German, French, and Spanish*. Amsterdam; New York: Elsevier. ISBN: 0444987401. This dictionary not only covers plant virology, but also virus and viroid diseases and diseases caused by mycoplasma, rickettsia, and viruslike phenomena. The main entry for each term is in English, followed by the equivalent term in Russian, German, French, or Spanish. A brief definition of the term in English is provided. Cross-referenced registers refer the reader from the Russian, German, French, or Spanish back to the English equivalent.

Dermine, P., L. Paradis, and Canada Translation Bureau. Terminology and Linguistic Services Directorate. 1990. *Vocabulaire de l'Agriculture*. Ottawa: Secrétariat d'Etat du Canada. ISBN: 066057446X. The main entries are organized in English, with definitions in English and French. A second index refers French readers to the appropriate English word or phrase. With over 9,000 entries, this was designed to be a useful tool for translators and was derived in part from the first author's personal glossary. Brief definitions are provided.

Herren, R.V. and R.L. Donahue. 1991. *The Agriculture Dictionary*. Albany, NY: Delmar Pub. ISBN: 0827340958. Covering all aspects of agriculture with the definitions of over 15,000 terms and 700 illustrations, this resource is valuable to the researcher, farmer, and novice. Includes references and a table of conversion factors.

Holliday, P. 1998. *A Dictionary of Plant Pathology*. 2nd ed. Cambridge; New York: Cambridge University Press. ISBN: 0521594537. The 11,000 entries in this dictionary provide the names and brief descriptions of plant pathogens; the names and symptoms of plant diseases; pesticides and other control agents; and past prominent plant pathologists.

International Board for Plant Genetic Resources. 1991. *Elsevier's Dictionary of Plant Genetic Resources*. Amsterdam; New York: Elsevier. ISBN: 0444889590. The International Board for Plant Genetic Resources (IBPGR) is an independent international group that works under the aegis of the Consultative Group on International Agricultural Research (CGIAR) to promote and coordinate genetic resources. This is a glossary of terms specific to the area of classical plant genetic resources.

International Society for Horticultural Science, and Ministry of Agriculture, Nature Management and Fisheries (The Netherlands). 1990. *Elsevier's Dictionary of Horticultural and Agricultural Plant Production: In Ten Languages, English, Dutch, French, German, Danish, Swedish, Italian, Spanish, Portuguese, and Latin*. Amsterdam; New York: Elsevier Science. ISBN: 0444880623. Word lists for ten languages covering all aspects of agriculture. The primary list is in English, with references back to it from the other lists.

Lipton, K.L. 1995. *Dictionary of Agriculture: From Abaca to Zoonosis*. Boulder, CO: Lynne Rienner Publishers. ISBN: 1555875238. A practical guide to more than 3,000 terms associated with agriculture. While detailed, the definitions aim to be clear to those unfamiliar with farming, food and agricultural policy, trade, conservation, and other topics. Numerous appendices include useful lists of commonly used acronyms.

MacKay, S.E. 1984. International Programs in Agriculture. Purdue University. *Field Glossary of Agricultural Terms in French and English: Emphasis, West Africa*. West Lafayette, IN: Purdue University. ISBN: 0961410906. This petite volume is handy for agricultural researchers traveling to French-speaking areas, and vice versa. No definitions, just the word equivalents are provided.

Williams, G., W. van der Zweep, and Centrum voor Landbouwpublikaties en Landbouwdocumentatie (The Netherlands). 1990. *Interdisciplinary Dictionary of Weed Science: Dansk, Deutsch, English, Español, Français, Italiano, Nederlands, Portugues*. Wageningen, Netherlands: Pudoc. ISBN: 9022010074. Using this book, one may look up a term in one of eight languages, and be referred to the equivalent word in the other seven languages. No definitions, just terms.

Wrobel, M. and G. Creber. 1998. *Elsevier's Dictionary of Fungi and Fungal Plant Diseases: in Latin, English, German, French, and Italian*. Amsterdam; New York: Elsevier. ISBN: 0444827749. Compiled alphabetical listing of the

Latin names of about 5,000 families, genera, and species. Also includes many of the symptoms of fungal plant diseases along with their synonyms. The English, German, French, and Italian equivalents are listed in the respective indices and keyed to the Basic Table by reference number.

Zimdahl, R.L. 1989. *Weeds and Words*: *the Etymology of the Scientific Names of Weeds and Crops*. Ames, IA: Iowa State University Press. ISBN: 0813801281. The etymology of the scientific names of over 200 weed species and 35 crops is discussed. In many cases, an understanding of the origin of a word can be an aid to identification. Short descriptions of each plant are also provided when they relate to the etymology of the species name.

## DIRECTORIES

Resources of Scholarly Societies: Agricultural and Food Sciences (http://www.lib.uwaterloo.ca/society/agricfood_soc.html) (cited February 21, 2002). Part of the Scholarly Societies Project sponsored by the University of Waterloo in Ontario, Canada, this site provides links to World Wide Web sites for the major scholarly societies such as the American Society of Agronomy, the Crop Science Society of America, the American Crop Protection Association, the American Phytopathological Society, and the Royal Agricultural Society of England.

## ENCYCLOPEDIAS

Arntzen, C.J. and E.M. Ritter, eds. 1994. *Encyclopedia of Agricultural Science*. San Diego, CA: Academic Press. ISBN: 0122266706 (set). This four-volume set contains many authored articles of interest to the agronomist on such topics as production and genetics of the major crops, forages, fungicides, grain storage, herbicides and herbicide resistance, heterosis, pest management, plant biotechnology, plant gene mapping, plant pathology, plant stress, plant virology, turfgrasses, and weed science. Articles are generally eight to ten pages long, are arranged alphabetically by subject, and can be referenced through the subject index. Each chapter begins with a glossary of terms and definitions.

Illustrated Encyclopedia of Forage Crop Disease (http://ss.ngri.affrc.go.jp/disease/detitle.htm) (cited February 21, 2002). Created by the National Grassland Institute (Japan), this site provides full-sized, clear photos of the effects of various diseases on forages, as well as close-up micrographs of the causative organism.

McNeil, I. 1990. *An Encyclopaedia of the History of Technology*. London; New York: Routledge. ISBN: 0415013062. Includes a chapter titled ''Agriculture: the Production and Preservation of Food and Drink.''

## GOVERNMENT DOCUMENTS

E-Answers (http://www.e-answersonline.org/) (cited February 21, 2002). Hosted by the University of Florida and sponsored by a grant from the Cooperative State Research, Education and Extension Service (CSREES)-USDA, E-Answers is a search engine for locating Extension fact sheets and bulletins from over 40 different universities across the United States plus several other resources such as the Alaska Sea Grant and the *California Agriculture Magazine*. Broader in subject scope than PlantFacts (see below), but the maximum number of retrievals has been set to only 25.

National Library for the Environment. National Council for Science and the Environment. Agriculture (http://www.cnie.org/NLE/CRSreports/Agriculture/) (cited February 21, 2002). The National Library for the Environment has gathered together a listing of the Congressional Research Service Reports in the area of agriculture. At the time of this writing it included recent reports titled "Biosafety protocol for genetically modified organisms," "Agriculture: A List of Web sites" and "Federal Crop Insurance." Reports are listed chronologically back to 1994.

PlantFacts (http://plantfacts.osu.edu/) (cited February 21, 2002). Maintained by Dr. Tim Rhodus, Department of Horticulture and Crop Science, Ohio State University, PlantFacts is a search engine for locating plant- and crop-related Web Extension fact sheets and bulletins from over 45 different universities and government institutions across the United States and Canada. It is possible to restrict a search to a particular geographical region. Although PlantFacts is supposedly focused on just plant resources, it actually often finds materials on other topics as well.

USDA/Office of Pest Management Policy (OPMP) Crop Profiles Database (http://pestdata.ncsu.edu/cropprofiles/cropprofiles.cfm) (cited February 21, 2002). Developed by Steve Toth, Department of Entomology, North Carolina State University, and Ron Stinner, National Science Foundation Center for Integrated Pest Management, and the Department of Entomology at North Carolina State University. The usage of several widely used pesticides that are organophosphates, carbamates, or carcinogenic may be severely curtailed by the EPA. In order to better determine their usage and impact, the Crop Profiles Database is being developed, with input from each state about its major crops: their value, cultural practices (such as soil types, irrigation practices, land preparation, planting times, thinning practices), pesticide usage, and current research activities directed at finding replacement strategies for the pesticides of concern. Search by state, and browse for the crop of interest.

## GUIDES TO THE LITERATURE

AgNIC Plant Science Home Page (http://www.unl.edu/agnicpls/agnic.html) (cited February 21, 2002). Hosted and maintained by the University of Nebraska—Lincoln Libraries, this site has gathered together quality World Wide Web sites in agricultural plant sciences. Hierarchically arranged, one will find links for such topics as row crops, plant protection, plant genetics, and forages. In addition links to associations, directories, and research institutions have been gathered.

Demeter's Genomes (http://arsgenome.cornell.edu/) (cited February 21, 2002). Provided as a service from the USDA-Agriculture Research Service Center for Agricultural Bioinformatics, Demeter's Genomes is primarily a site for communication of plant genome data among scientists worldwide. Links to the home pages for 18 grass, dicot, and tree genome projects are provided. Additionally a uniform search engine for searching each genome database is available, which allows one to find information about plant genes, chromosome maps, DNA fingerprints, beneficial traits, and crop varieties that possess those traits.

Kraska, T. Plant Pathology Internet Guide Book (http://www.ifgb.unihannover. de/extern/ppigb/ppigb.htm) (cited February 21, 2002). Hosted by the Institute for Plant Diseases and Plant Protection, Department of Horticulture, University of Hanover. This is a "virtual library" of well-annotated, international links to resources about the major types of plant pests: bacteria, fungi, insects, nematodes, and viruses. A convenient site-wide search engine will find specific resources, such as soybean diseases. Most of the data links are to resources provided by academic departments, government agencies, or societies. In turn, many of these organizations are pointing to the Guide Book, which speaks for the site's credibility. Comprehensive lists of links are also provided to plant pathology organizations, departments, upcoming meetings, jobs, courses, software, online journals, databases, culture collections, and newsgroups.

Variety and Yield Trials (http://www.unl.edu/agnicpls/stats.html) (cited February 21, 2002). Part of the AgNIC Plant Sciences page, this site has pulled together links for the multitude of variety and yield trials that are held throughout the United States for corn, sorghum, small grains, soybeans, forages, and other crops.

## HANDBOOKS AND MANUALS

### Crop Breeding

USDA Agricultural Research Service. National Plant Germplasm System (NPGS) (http://www.ars-grin.gov/npgs/) (cited February 21, 2002). Presents information that aids scientists in creating new plant varieties that are pest and disease resistant. Includes information on germplasm collections and how to request germplasms. Allows users to search the Germplasm Resources Information Network (GRIN), a Web server that provides germplasm information about plants, animals, microbes, and invertebrates. Currently GRIN has information on over 37,000 taxa and 14,000 genera, most of which are major, minor, or potential crops or their wild, weedy relatives. GRIN is fully searchable, allowing the user to query for such data as clones resistant to a particular disease, the date a particular cultivar was registered, whether a particular cultivar is protected by the Plant Variety Protection area, and so on. In addition, links to the various plant genome projects are provided.

## Crop Production

Food and Agriculture Organization of the United Nations (FAO). 1993. *Technical Handbook on Symbiotic Nitrogen Fixation: Legume/Rhizobium.* 2nd ed. Rome: FAO. ISBN: 9251031991.   The purpose of this technical handbook is to enable field workers and researchers to understand the techniques and procedures for increasing nitrogen fixation in the field. It includes laboratory methods, as well as methods for field study of the phenomenon.

## Crop Protection

The Compendium of. . . . Diseases series, published irregularly by the American Phytopathological Society (APS) (Madison, WI) is uniformly recommended. Each volume is generally composed of three sections: infectious agents (the largest section); noninfectious and abiotic disorders; and approaches for management. A prominent feature of each volume that is useful for disease diagnosis is color plates that show plants infected by the various diseases. Many of the compendiums are also available on CD-ROM. Among the recent field crop titles in this series:

> Frederiksen, R.A. and G.N. Odvody. 2000. *Compendium of Sorghum Diseases.* St. Paul, MN: American Phytopathological Society in cooperation with Department of Plant Pathology and Microbiology, Texas A&M University. ISBN: 0890542406.
>
> Hall, R. 1991. *Compendium of Bean Diseases.* St. Paul, MN: APS Press. ISBN: 0890541183.
>
> Hartman, G.L., J.B. Sinclair, and J.C. Rupe. 1999. *Compendium of Soybean Diseases.* 4th ed. St. Paul, MN: APS Press. ISBN: 0890542384.
>
> Jones, J.B. and American Phytopathological Society. 1991. *Compendium of Tomato Diseases.* St. Paul, MN: APS Press. ISBN: 0890541205.
>
> Kirkpatrick, T.L. and C.S. Rothrock. 2001. *Compendium of Cotton Diseases.* St. Paul, MN: APS Press. ISBN: 0890542791.

Kokalis-Burelle, N. 1997. *Compendium of Peanut Diseases*. 2nd ed. St. Paul, MN: APS Press. ISBN: 089054218X.

Mathre, D.E. 1997. *Compendium of Barley Diseases*. 2nd ed. St. Paul, MN: APS Press. ISBN: 0890541809.

Smiley, R.W., P.H. Dernoeden, B.B. Clarke, and American Phytopathological Society. 1992. *Compendium of Turfgrass Diseases*. 2nd ed. St. Paul, MN: APS Press. ISBN: 0890541248.

Stuteville, D.L. and D.C. Erwin. 1990. *Compendium of Alfalfa Diseases*. 2nd ed. St. Paul, MN: APS Press. ISBN: 0890541086.

Webster, R.K. and P.S. Gunnell. 1992. *Compendium of Rice Diseases*. St. Paul, MN: APS Press. ISBN: 0890541264.

White, D.G. and University of Illinois at Urbana-Champaign, Department of Crop Sciences. 1999. *Compendium of Corn Diseases*. 3rd ed. St. Paul, MN: APS Press. ISBN: 0890542341.

Wiese, M.V. and American Phytopathological Society. 1987. *Compendium of Wheat Diseases*. 2nd ed. St. Paul, MN: APS Press. ISBN: 0890540764.

Ahrens, W.H., ed., and Weed Science Society of America Herbicide Handbook Committee. 1994. *Herbicide Handbook*. 7th ed. Champaign, IL: Weed Science Society of America. ISBN: 0911733183. Published at irregular intervals since 1967, this is the "bible" for most weed scientists. For 140 herbicides, herbicide additives, desiccants, and plant growth regulators, it provides comprehensive information including nomenclature (structure, common name, chemical name, manufacturer); chemical and physical properties; use; precautions; behavior in plants or soil; toxicology; synthesis; and references. Numerous tables include a list of chemicals that were in earlier editions, but have been withdrawn from this edition for various reasons; pronunciation guide; several glossaries; conversion factors; company addresses; trademarks; and several cross references.

American Phytopathological Society. Common Names of Plant Diseases (http://www.aspnet.org/online/common/) (cited February 21, 2002). The common names established by the APS Committee on Standardization of Common Names for Plant Diseases are now the official names for use in APS and other plant pathology journals. The committee publishes these names in *Phytopathology News* and *Plant Disease*. This Web-accessible compilation provides an updated, combined version of those lists. You may search the Common Names database by keyword or select one of the plant disease collections listed in the table of contents, which lists the diseases by plant affected.

American Phytopathological Society. 1986– . *Biological and Cultural Tests for Control of Plant Diseases: B & C Tests*. St. Paul, MN: APS Press. ISSN: 0887-2236. Published annually to inform plant pathologists and agronomists about the results of current plant genotype disease-resistance tests. Arranged by plant type, it includes crops, horticulturals, vegetables, fruit trees, and turfgrasses. Indexed by host and disease name or causative agent. Starting with the 2001 edition, only available online.

American Phytopathological Society. 1968– . *Fungicide and Nematicide Tests.* St. Paul, MN: APS Press. ISSN: 0148–9038. Published annually to inform plant pathologists and nematologists about the results of treatment trials to protect plants from diseases or nematodes. Arranged by plant type. Appendices provide information about the formulations tested and the manufacturers. Fully indexed. Starting with the 2001 edition, only available online.

The Association of Applied Biologists, IACR-Rothamsted and the Scottish Crop Research Institute. 1998– . *Descriptions of Plant Viruses.* Wellesbourne, UK: The Association. ISSN: 1463–7227. Published since 1970 in paper, the *Descriptions* recently became an electronic-only CD-ROM product. Currently it includes complete descriptions of 377 of the most important plant viruses and viroids. Topics covered include the disease caused, geographical distribution, host range, strains, transmission vectors, serology, relationships, purification methods, structure, genome properties, and references. Each record is illustrated with electron micrographs as well as photos of infected plants. Updates may be downloaded from http://www.iacr.bbsrc.ac.uk/dpv/CDRom/updates/Default. htm (cited February 21, 2002).

Bradbury, J.F. 1986. *Guide to Plant Pathogenic Bacteria.* Farnham Royal, UK: CAB International, Mycological Institute. ISBN: 0851985572. An alphabetical listing of over 2,000 bacteria for which plant pathogenicity has been claimed on over 2,500 hosts. Includes information on synonymy, bacterial characteristics, symptoms, and geographic distribution of the host and bacteria. Fully referenced. For more current literature, refer to CAB Abstracts.

Brunt, A.A. 1996. *Viruses of Plants: Descriptions and Lists from the VIDE Database.* Wallingford, UK: CAB International. ISBN: 085198794X. The nomenclature, host range, transmission and symptoms, physical and biochemical properties; taxonomy, and relationships of all known plant virus species and genera are presented. The basic organization is alphabetical but a comprehensive index of names, synonyms, and acronyms facilitates locating the desired record. Also included are a list of abbreviations and a character list. The VIDE project was supported by the Australian Centre for International Agricultural Research. This resource is also available online at Plant Viruses Online: Descriptions and Lists from the VIDE Database (http://image.fs.uidaho.edu/vide/). Or, one may consult the Web site, ICTVdb: The Authorized Universal Virus Database (http://www.ncbi.nlm.nih.gov/ICTVdb/) (cited February 21, 2002), limiting the search to plants, for similar descriptive information.

Copping, L.G. 1998. *BioPesticide Manual*. Farnham, UK: British Crop Protection Council. ISBN: 1901396266. A compendium of natural compounds, living systems, and genes commercially available for pest, weed, and disease control. One hundred eighty-eight biocontrol agents are included, covering well over 800 products of the following types: natural products, pheromones, living systems, insect predators, and genes. Each entry includes such information as the nomenclature, target crops, biological activity, toxicity, source, mode of action, and environmental impact.

*Crop Protection Reference*: CPR. 2000. 16ᵗʰ edition. New York: C&P Press. [No ISSN]. Contains the complete text of the product labels for 661 products from 23 companies. Products are indexed by brand name, manufacturer, common name, product category, pest, mode of action, and crop for which they are registered for use in the United States by the EPA. Purchase of the book also includes a CD-ROM, CPR Plus CD 2000, which is the electronic version of the book, plus the *Adjuvant Reference Supplement*.

Department of Botany and Plant Pathology, Purdue University. 1998. Herbicide Injury Symptoms on Corn and Soybeans (http://www.btny.purdue.edu/ Extension/Weeds/HerbInj/InjuryMOA1.html) (cited February 21, 2002). Arranged by mode of action (amino acid synthesis inhibitor, growth regulator, growing point disintegrator, seedling growth inhibitor, photosynthetic inhibitor, cell membrane disruptor, pigment inhibitor), this Web site presents clear photos of the injurious effect of herbicides on corn and soybeans.

Dhingra, O.D. and J.B. Sinclair. 1995. *Basic Plant Pathology Methods*. 2nd ed. Boca Raton, FL: Lewis Publishers. ISBN: 0873716388. This book was designed to be a ready-reference in the classroom or laboratory because every protocol includes enough detail to accomplish the task at hand, without the need to refer to additional resources. The authors indicate that this might be particularly useful in remote areas where it is difficult to have a complete plant pathology library at hand. Topics include the basics, such as aseptic technique, as well as methods for culturing, detecting, and studying various pathogens (primarily fungal, but also soilborne microbes). There is also an extensive chapter on techniques in bright-field microscopy. In cookbook fashion, there are several extensive appendices covering such areas as media recipes, plant nutrient solutions, solutions to maintain constant humidity in a closed atmosphere, drying agents, and pH indicator dyes.

Entomological Society of America. 1994– . *Arthropod Management Tests*. College Park, MD: Entomological Society of America. ISSN: 0276–3656. Published annually to inform entomologists about the results of current screening tests for the management of arthropods that may be beneficial (such as honey bees) or harmful (pests or disease vectors) to plants or animals. The pest control agents reported in this volume may be pesticides, growth regulators, pheromones, or biocontrol agents. The resistance of transgenic plant varieties to pests is also reported. Formerly titled *Insecticide and Acaricide Tests* (1976–93).

Entomology Department, Iowa State University. 2000. Plant Diseases and Damage (http://www.ent.iastate.edu/imagegal/plantpath/) (cited February 21, 2002). Part of the Iowa State University Entomology Image Gallery, this site presents clear color photos of the injurious effect of pests and diseases on a variety of field crops, especially corn, soybeans, and wheat.

Farbenfabriken Bayer Aktiengesellschaft. Geschäftsbereich Pflanzenschutz. 1992. *Important Crops of the World and Their Weeds: Scientific and Common Names, Synonyms, and WSSA/WSSU Approved Computer Codes.* 2nd ed. Leverkusen, Germany: Business Group Crop Protection Bayer AG.   Recommended by the editors of *Weed Science* as the source for determining the standard common and scientific names of crops and weeds. Provides the common names for crops and weeds in up to nine languages. Regional differences are noted as well, such as the difference between what a plant might be commonly called in the Spanish-speaking countries of Argentina, Cuba, or Puerto Rico; or the common name for a plant in Australia, Great Britain, or the United States.

*Farm Chemicals Handbook.* 1951– . Willoughby, OH: Meister Publishing Co. ISSN: 0430-0750.   Published for over 90 years, the bulk of the 2000 edition is a dictionary listing of pesticides (herbicides as well as growth regulators, insecticides, and other pesticides) that is amply cross-referenced. Two smaller sections cover fertilizers and regulatory laws. The main listing for each pesticide gives the chemical structure and properties, trade names, manufacturer(s), action/use, environmental and safety guidelines, and registration notes.

Harvey, L.T. 1998. *A Guide to Agricultural Spray Adjuvants Used in the United States.* 5th ed. Fresno, CA: Thomson Publications. ISBN: 0913702390.   Lists the majority of spray adjuvants used in the United States, including a brief description, the formulation, recommended application rates, and precautions.

*International Code of Nomenclature of Bacteria; and, Statutes of the International Committee on Systematic Bacteriology; and, Statutes of the Bacteriology and Applied Microbiology Section of the International Union of Microbiological Societies: Bacteriological Code. 1990 revision.* 1992. Washington, DC: Published for the International Union of Microbiological Societies by the American Society for Microbiology. ISBN: 155581039X.   Essential guide for naming new bacteriological species whether they are plant pathological, or otherwise. Taxonomy of specific groups is regulated by ICSB-appointed subcommittees that report by minutes published in the *International Journal of Systematic and Evolutionary Microbiology* (ISSN: 1466-5026). Additional information is available on the Web at Bacterial Nomenclature Up-to-date [http://www.dsmz.de/bactnom/bactname.htm (cited February 21, 2002)].

International Committee on Taxonomy of Viruses, M.H.V. van Regenmortel, and the International Union of Microbiological Societies. Virology Division. 2000. *Virus Taxonomy: Classification and Nomenclature of Viruses: Seventh Report of the International Committee on Taxonomy of Viruses*. San Diego, CA: Academic Press. ISBN: 3211825940. Updated every three to five years, this report presents the consensus opinion on the nomenclature of all viruses. The distinguishing features of viral and virion families and genera are presented, together with a list of the species in each genus. Viral nomenclature is contentious and often in a state of flux, so it is important to consult the most recent edition. For online version, see ICTVdb: the Authorized Universal Virus Database. http:// www.ncbi.nlm.nih.gov/ICTVdb/ (cited February 21, 2002).

Klement, Z., K. Rudolph, and D.C. Sands, eds. 1990. *Methods in Phytobacteriology*. Budapest: Akadémiai Kiadó; Distributed by H. Stillman Publishers. ISBN: 9630549557. Conveniently compiled into one volume, this book brings together in a cookbook format the methods for identifying and diagnosing bacterial agents that infect crops. In addition, methods for inducing disease or for studying diseased plants, from the macroscopic to the biochemical level, are presented. Rich in recipes, techniques, and diagrams. Suitable for the novice as well as advanced researcher.

Meade, J. New Jersey Weed Gallery (http://www.rce.rutgers.edu/weeds/) (cited February 21, 2002). Color photos in 3 sizes of over 100 common weeds. Because weeds do not adhere to state boundaries, many living outside New Jersey will find this resource useful. Each photo is accompanied by descriptive text. Other such sites exist for other regions or states, as well. See, for example, for the Midwest: Stevens, R. and C. Coffey. Noble Foundation Plant Image Gallery. Sponsored by the Samuel Roberts Noble Foundation, Ardmore, OK. (http:// www.noble.org/imagegallery/) (cited February 21, 2002).

Montgomery, J.H. 1997. *Agrochemicals Desk Reference*. 2nd ed. Boca Raton, FL: CRC Press. ISBN: 1566701678. The focus of this volume is on the fate and transport of agrochemicals in the air, soil, and groundwater. Of particular interest to agronomists is the data on the biological, chemical, and photolytic degradation of pesticides, as well as the exposure limits, symptoms of exposure, and toxicity. Includes over 1,200 citations to the literature. Numerous appendices cover such areas as conversion factors, solubility data, and synonyms.

Nyvall, R.F. 1999. *Field Crop Diseases*. 3rd ed. Ames, IA: Iowa State University Press. ISBN: 0813820790. Provides basic information in one volume on the diseases of many of the world's important field crops. Each chapter is devoted to a particular crop. Within each chapter, diseases are grouped by causative organism (e.g., bacteria, fungus, nematodes, and viruses) and are then listed alphabetically by the common name of the disease. Includes a glossary, which nonphytopathologists will find handy, and is fully indexed.

Ruberson, J.R. 1999. *Handbook of Pest Management*. New York: Marcel Dekker. ISBN: 0824794338. Includes chapters on the biological and chemical control of plant pathogens, arthropod plant pests, weed control, cultural methods to control plant pests, and plant resistance to pests. One chapter covers issues involved in breeding crops for pest resistance.

Smith, I.M., British Society for Plant Pathology, and Mediterranean Plant Protection Organization. 1988. *European Handbook of Plant Diseases*. Oxford; Boston: Blackwell Scientific Publications. ISBN: 0632012226. The emphasis of this volume is on the biology of plant pathogens that occur in or threaten European crops and trees. The epidemiology of the disease, the economic importance, and the control of the disease are also discussed. Organized by causative group (e.g., viruses, rickettsia, bacteria, and fungal family), it includes a pathogen and host index.

Stubbendieck, J.L., G.Y. Friisoe, and M.R. Bolick. 1994. *Weeds of Nebraska and the Great Plains*. Lincoln, NE: Nebraska Department of Agriculture Bureau of Plant Industry in cooperation with the University of Nebraska-Lincoln. ISBN: 0939870004. One of many books that cover the weeds of a particular region. This book has outstanding, full-page photos of weeds taken in their natural environment, as well as a description of the plant, its habitat, and toxicity. Find such books in library catalogs by searching for the subject heading, Weeds—[geographic region].

Šutić, D.D., R.E. Ford, and M.T. Tošić. 1999. *Handbook of Plant Virus Diseases*. Boca Raton, FL: CRC Press. ISBN: 0849323029. Organized according to the type of plant affected (cereals, forage feed plants, industrial plants), this handbook presents the basic information about viral-caused diseases of crops together with current methods of control. Lists the geographic distribution, host range, symptoms, and pathogenesis for each virus or viruslike agent. Each chapter has an extensive bibliography.

Tomlin, C. 2000. *The Pesticide Manual: A World Compendium: Incorporating the Agrochemicals Handbook*. 12ᵗʰ ed. Cambridge: British Crop Protection Council. ISBN: 1901396126. Provides unbiased information on the active ingredients in thousands of agrochemicals including herbicides, insecticides, other pesticides, plant growth regulators, timber preservatives, and antimicrobial agents. Includes such information about the compounds as their structure, physical and chemical properties, applications, toxicology, and environmental fate. Also included are a glossary of the scientific/common names of pests and a directory of manufacturers. Several indices are useful as well: CAS registry numbers, molecular formulae, manufacturer code numbers, trade names, chemical names, and agrochemical classes. Updated irregularly, every two to three years. The 12ᵗʰ edition provides the option of purchasing a networkable, searchable CD-ROM.

*Weed Control Manual.* 1991– . Willoughby, OH: Meister Publishing Co. ISSN: 0741-9856. Issued biennially, this title provides information about all U.S.-registered herbicides, plus guides for weed identification, worker-protection information, and detailed coverage of herbicide-resistant seed systems. The bulk of the volume is arranged by crop; within each crop are listings of the herbicides that are registered for use. For each herbicide, the recommended time and dosage of application, weeds controlled, state restrictions, and safety concerns are listed. Several tables that list factors such as rainfast intervals, spray additives, and maximum crop size for various herbicides will assist the user in determining which herbicide to use. Indexes for crops, herbicides, weeds, and manufacturers are provided.

Westcott, C. and R.K. Horst. 1990. *Westcott's Plant Disease Handbook.* 5th ed. New York: Chapman & Hall. ISBN: 0412067218. Since it was originally published in 1949, this book has been a mainstay for working plant pathologists. There are two main parts. The first part covers plant diseases and their pathogens, providing a synopsis of the disease that frequently includes photos of affected plants. The second part covers the host plants, providing a list of the diseases that affect them and the host ranges. Also includes a selected bibliography, glossary, and comprehensive index.

## HISTORIES

In addition to books specifically about agricultural history, one can find historical information in a variety of unexpected resources. For example, in the early 1990s the NAL (U.S.) funded a project to create catalog records for the U.S. Experiment Station publications, 1870–1990. The catalog records were input into the OCLC (WorldCat), AGRICOLA, and AGRIS databases. Thus, although AGRICOLA supposedly only covers the literature back to 1970, one will, in fact, find citations for many earlier publications. Other resources are the annual reports of the various state Experiment Stations, U.S. census records, and the *Yearbook of Agriculture* produced by the USDA (1894–1992), which provide a wealth of information on the historical aspects of agronomy, plant protection, and crop production. Of particular interest is the *Census of Agriculture*, which dates back to 1850 and provides an historical record of the crops that were planted and the spread of agriculture across the United States.

*Abridged Agricultural Records; Being a Careful and Judicious Selection, by Practical, Experienced and Expert Writers, from the Most Valuable Discoveries of the Departments of Agriculture and the Various Experiment Stations of Both the United States and Canada.* 1912. Washington, DC: Agricultural Service Co. Volume 4 of this classic covers grass, hay, grains, and vegetables. Although not written as a "history," the age of this seven-volume series provides a glimpse into American agriculture in the early twentieth century. Some print-ings of this widely available series refer to it as *The Farmer's Cyclopedia*.

*Agricultural History*. 1927– . Berkeley, CA: Published for the Agricultural His-tory Society by the University of California Press. ISSN: 0002-1482. Published quarterly, this is the premier journal for agricultural history.

Bidwell, P.W. and J.I. Falconer. 1925. [Reprinted in 1941.] *History of Agriculture in the Northern United States, 1620–1860*. Carnegie Institution of Washington, no, 358. Washington, DC: The Carnegie Institution of Washington. This classic is still highly cited. Many chapters including "Field Husbandry," "Pioneering in the Eighteenth Century," "Crops and Tillage," "Wheat," "Corn," "The Minor Cereals," "Potatoes and Roots," and "Flax and Hemp" cover topics related to field crops.

Campbell, C.L., P.D. Peterson, and C.S. Griffith. 1999. *The Formative Years of Plant Pathology in the United States*. St. Paul, MN: APS Press. ISBN: 0890542333. Covers the early events of plant pathology and related disciplines. Shows how the impact of major figures, organizations, and education led to the successful development of this scientific field in the United States. Major sections include plant diseases and agriculture in early America, the origins of the various branches of U.S. plant pathology as sciences, factors that influenced the rise of plant disease research (such as the Hatch Act and the experiment stations), and the maturation and professionalization of plant pathology due to the formation of the American Phytopathological Society (APS).

Carman, H.J., J. Mitchell, and A. Young. 1939. *American Husbandry*. Columbia University Studies in the History of American Agriculture, no. 6. New York: Columbia University Press. One of the 12-volume Columbia University Studies in the History of American Agriculture series published between 1934 and 1972, this volume is actually an edited reprint of a work published in 1775 that relays the state of agriculture in colonial America. Although largely organized by "colo-nies," it includes several general chapters such as "Importance of the American Colonies to Britain" and "Independency."

Cowan, C.W., P.J. Watson, and N.L. Benco. 1992. *The Origins of Agriculture*: *an International Perspective*. Smithsonian Series in Archaeological Inquiry. Washington, DC: Smithsonian Institution Press. ISBN: 0874749905. A compi-lation of papers originally created for a symposium of the 1985 AAAS meeting, this volume presents the origins of agriculture based on the archaeological evi-dence. Although this is a scholarly volume, it is highly readable.

Harlan, J.R. 1995. *The Living Fields*: *Our Agricultural Heritage*. Cambridge, New York: Cambridge University Press. ISBN: 0521401127. Describes how the modern agricultural systems evolved and explains the danger of depending on a limited number of species for our food.

Harris, D.R. and G.C. Hillman. 1989. *Foraging and Farming: the Evolution of Plant Exploitation*. One World Archaeology Series, no. 13. London; Boston: Unwin Hyman. ISBN: 0044450257. Results of a symposium held at the World Archaeological Congress in 1986, this volume is considered the most scholarly treatment on the subject. It considers the processes of plant domestication by prehistory humans from a worldwide view starting with the earliest record of agriculture, some 9,000 years ago, in Papua, New Guinea.

Heiser, C.B. 1990. *Seed to Civilization: the Story of Food*. 3$^{rd}$ ed. Cambridge, MA: Harvard University Press. ISBN: 0674796810. The origin and uses of food through the ages. Contains chapters on grasses, legumes, starchy staples, oil crops, sugar, and other crops.

Leonard, J.N. and Time-Life Books. 1973. *The First Farmers, Emergence of Man*. New York: Time-Life Books. [No ISBN]. Richly illustrated, this volume provides an easily accessible anthropological overview of the origin of agriculture. Useful introduction for the undergraduate student.

Rasmussen, W.D. 1989. *Taking the University to the People: Seventy-Five Years of Cooperative Extension*. Ames, IA: Iowa State University Press. ISBN: 0813804191. Beginning with the Morrill Act of 1862, Rasmussen delineates the process by which Cooperative Extension provided quality information and education on real concerns to those engaged in agriculture. Rasmussen was the chief of the Agricultural History Branch of the USDA, and this volume celebrated the first 75 years of the Cooperative Extension.

Reed, C.A. 1977. *Origins of Agriculture*. The Hague, The Netherlands: Mouton. ISBN: 9027979197. Over 30 papers from the 9th International Congress of Anthropological and Ethnological Sciences, this book presents a worldwide anthropolitical view on the origins of agriculture, from the Orient to the New World.

Svobida, L. 1986. *Farming the Dust Bowl: a First-Hand Account from Kansas*. Lawrence, KS: University Press of Kansas. ISBN: 0700602895. Originally published as *An Empire of Dust* (1940), this book gives a vivid, heart-rending account of what it was like to live through the Dust Bowl of the mid-1930s in central Kansas. Worth reading today as we seek to cultivate marginal areas that can only support crops when there is enough rain.

Vavilov, N.I. and V.F. Dorofeyev. 1992. *Origin and Geography of Cultivated Plants*. English ed. Cambridge; New York: Cambridge University Press. ISBN: 0521404274. This translation from the 1926 classic paper by Vavilov includes substantive chapters on the origin of cultivated plants. An essential text for plant breeders. The nomenclature was updated by Dorofeyev.

## JOURNALS

The primary journals and book series in crop science are listed in the following text. They have been subdivided into the areas of crop production, crop breeding, and crop protection. The reader is also referred to Wallace C. Olsen's chapter, "Primary Journals and the Core List," in his book *The Literature of Crop Science* for a more exhaustive listing Ithaca, NY: [Cornell University Press, 1995, pp. 299–320 (ISBN: 0801431387)].

The following list was drawn in part from the Institute for Scientific Information's *Journal Citation Reports* (1997) for the subject areas of agriculture, genetics and heredity, plant science, virology, microbiology, entomology, and mycology. Journals with high (greater than 1.0) impact factors, or high citation rates (more than 500 citations/year) have been included. Additional crop science titles from the Purdue University collection that appear to be widely available at land-grant universities have also been included. The titles of auxiliary journals that are not specific to the area of crop science but are widely read by researchers in the area of crop science have also been included.

### Crop Production

*Acta Agronomica Hungarica*. 1986– . Budapest, Hungary: Académia Kiadó. ISSN: 0238-0161. Continues the *Acta Agronomica Academiae Scientiarum Hungaricae*.

*Agricultural and Forest Meteorology*. 1984– . Amsterdam: Elsevier. ISSN: 0168-1923. Continues *Agricultural Meteorology*.

*Agriculture, Ecosystems & Environment*. 1983– . Amsterdam; New York: Elsevier. ISSN: 0167-8809. Continues *Agriculture and Environment*.

*Agronomy Journal*. 1949– . Madison, WI: American Society of Agronomy. ISSN: 0002-1962. Continues the *American Society of Agronomy Journal*.

*American Journal of Potato Research: An Official Publication of the Potato Association of America*. 1998– . Orono, ME: Potato Association of America. ISSN: 1099-209X. Continues the *American Potato Journal*.

*Annals of Applied Biology*. 1914/1915– . London: Association of Applied Biologists. ISSN: 0003-4746.

*Apidologie*. 1970– . New York: Elsevier. ISSN: 0044-8435.

*Australian Journal of Agricultural Research*. 1950– . Melbourne: CSIRO. ISSN: 0004-9409.

*Bee World.* 1919– . Gerrards Cross, UK: International Bee Research Association. ISSN: 0005-772X.

*Cereal Research Communications.* 1973– . Szeged, Hungary: Cereal Research Institute. ISSN: 0133-3720.

*Crop Science.* 1961– . Madison, WI: Crop Science Society of America. ISSN: 0011-183X.

*Economic Botany.* 1947– . New York: New York Botanical Garden. ISSN: 0013-0001.

*European Journal of Agronomy.* 1992– . New York: Elsevier Science. ISSN: 1161-0301.

*Experimental Agriculture.* 1965– . London; New York: Cambridge University Press. ISSN: 0014-4797. Continues *Empire Journal of Experimental Agriculture.*

*Field Crops Research.* 1978– . New York: Elsevier Science. ISSN: 0378-4290.

*Genetic Resources and Crop Evolution.* 1992– . Dordrecht; Boston: Kluwer Academic Publishers. ISSN: 0925-9864.

*Grass and Forage Science*: the journal of the British Grassland Society. 1979– . Oxford: Blackwell Scientific Publications. ISSN: 0142-5242. Continues the *Journal of the British Grassland Society.*

*Japanese Journal of Crop Science* (Nihon Sakumotsu Gakkai Kiji). 1977– . Tokyo: Crop Science Society of Japan. ISSN: 0011-1848. Continues *Proceedings of the Crop Science Society of Japan.* 1976.

*The Journal of Agricultural Science.* 1905– . Cambridge, Cambridge University Press. ISSN: 0021-8596.

*Journal of Natural Resources and Life Sciences Education.* 1992– . Madison, WI: American Society of Agronomy. ISSN: 1059-9053. Updated periodically throughout the year online (subscription access only), with an archival paper version sent to subscribers once a year.

*Journal of Production Agriculture.* Madison, WI: American Society of Agronomy, Crop Science Society of America, Soil Science Society of America. 1987– 94. ISSN: 0890-8524.

*Journal of the Science of Food and Agriculture.* 1950– . London: Blackwell Scientific Publishers. ISSN: 0022-5142.

*The Indian Journal of Agronomy.* 1956– . New Delhi: The Society. ISSN: 0537-197X.

*International Journal of Biometeorology.* 1961– . Amsterdam: Swets & Zeitlinger. ISSN: 0020-7128. Continues the *International Journal of Bioclimatology and Biometeorology.*

*In Vitro Cellular & Developmental Biology: Plant.* 1991– . Columbia, MD: Tissue Culture Association. ISSN: 1054-5476. Continues *In Vitro Cellular & Developmental Biology.*

*Maydica.* 1956– . Bergamo, Italy: Istituto Sperimentale per la Cerealicoltura. ISSN: 0025-6153.

*New Zealand Journal of Agricultural Research.* 1958– . Wellington, NZ: Department of Scientific and Industrial Research. ISSN: 0028-8233.

*New Zealand Journal of Crop and Horticultural Science.* 1989– . Wellington, NZ: DSIR Publishing. ISSN: 0114-0671. Continues the *New Zealand Journal of Experimental Agriculture.*

*Potato Research.* 1970– . Wageningen, The Netherlands: H. Veeman en Zonen. ISSN: 0014-306. Continues the *European Potato Journal.*

*Seed Science Research.* 1991– . Wallingford, UK: CAB International. ISSN: 0960-2585.

*Seed Science and Technology.* 1961– . Wageningen, The Netherlands: International Seed Testing Association. ISSN: 0251-0952.

*Seed Technology.* 1997– . Lawrence, KS: Allen-Press Inc. ISSN: 1096-0724. Continues the *Journal of Seed Technology.*

Auxiliary

Other journals that aren't focused on crop production but would be consulted or read by the crop production agronomist include:

> *American Agriculturist.* 1976– . (ISSN: 0161-8237)
> *Australian Journal of Botany.* 1953– . (ISSN: 0067-1924)
> *Canadian Journal of Botany.* 1951– . (ISSN: 0008-4026)
> *Indian Journal of Agricultural Sciences.* 1951– . (ISSN: 0019-5022)
> *Journal of Apicultural Research.* 1962– . (ISSN: 0021-8839)
> *Journal of Plant Growth Regulation.* 1982– . (ISSN: 0721-7595)
> *Journal of Plant Physiology.* 1984– . (ISSN: 0176-1617)
> *Journal of Plant Research.* 1993– . (ISSN: 0918-9440)
> *Plant and Cell Physiology.* 1959– . (ISSN: 0032-0781)
> *Plant Cell.* 1989– . (ISSN: 1040-4651)
> *Plant Cell Reports.* 1981– . (ISSN: 0721-7714)
> *Plant Cell, Tissue and Organ Culture.* 1981– . (ISSN: 0167-6857)

*Plant Growth Regulation.* 1993– . (ISSN: 0167-6903)
*Plant Journal: For Cell and Molecular Biology.* 1991– . (ISSN: 0960-7412)
*Plant Molecular Biology.* 1981– . (ISSN: 0167-4412)
*Plant Physiology.* 1926– . (ISSN: 0032-0889)
*Plant Physiology and Biochemistry: PPB.* 1987– . (ISSN: 0981-9428)
*Planta.* 1925– . (ISSN: 0032-0935)
*WMO Bulletin [World Meteorological Organization].* 1952– . (ISSN: 0042-9767)

## Crop Breeding

*Euphytica: Netherlands Journal of Plant Breeding.* 1952– . Wageningen, The Netherlands: Centre for Agricultural Publishing and Documentation. ISSN: 0014-2336.

*Genetic Resources and Crop Evolution.* 1992– . Dordrecht; Boston: Kluwer Academic Publishers. ISSN: 0925-9864. Continues *Kulturpflanze*.

*Journal of Genetics & Breeding.* 1989– . Rome: Istituto Sperimentale per la Cerealicoltura. ISSN: 0394-9257.

*Molecular Breeding.* 1995– . Dordrecht: Kluwer Academic Publishers. ISSN: 1380-3743.

*Plant Breeding = Zeitschrift für Pflanzenzüchtung.* 1986– . Berlin: P. Parey. ISSN: 0179-9541. Continues *Zeitschrift für Pflanzenzüchtung*.

### Auxiliary

Other journals that aren't focused on crop breeding but would be read or consulted by crop breeders include:

*Annual Review of Genetics.* 1967– . (ISSN: 0066-4197)
*Cytogenetics and Cell Genetics.* 1973– . (ISSN: 0301-0171)
*Genes & Development.* 1987– . (ISSN: 0890-9369)
*Genetics.* 1916– . (ISSN: 0016-6731)
*Heredity.* 1947– . (ISSN: 0018-067X)
*Hereditas.* 1920– . (ISSN: 0018-0661)
*Journal of Heredity.* 1914– . (ISSN: 0022-1503)
*Molecular and General Genetics.* 1967– . (ISSN: 0026-8925)
*Nature Genetics.* 1992– . (ISSN: 1061-4036)
*Theoretical and Applied Genetics = Theoretische und Angewandte Genetik: TAG.* 1968– . (ISSN: 0040-5752)
*Trends in Genetics.* 1985– . (ISSN: 0168-9525)

## Crop Protection

*Agrochimica*. 1956– . Pisa, Italy: Industrie grafiche V. Lischi & Figli. ISSN: 0002-1857.

*Biological Control*. 1991– . San Diego, CA: Academic Press. ISSN: 1049-9644.

*Canadian Entomologist*. 1868– . Ottawa, Canada: Entomological Society of Canada. ISSN: 0008-347X.

*Canadian Journal of Plant Pathology = Revue Canadienne de phytopathologie*. 1979– . Ottawa, Canada: Canadian Phytopathological Society. ISSN: 0706-0661.

*Canadian Plant Disease Survey*. 1960– . London, Canada: Research Branch Agriculture and Agri-Food Canada. ISSN: 0008-476X. Available from the World Wide Web at http://res2.agr.ca/london/pmrc/report/ (cited February 21, 2002).

*Crop Protection*. 1983– . New York: Elsevier Science. ISSN: 0261-2194.

*Entomologia Experimentalis et Applicata*. 1958– . Dordrecht: Kluwer Academic Publishers. ISSN: 0013-8703.

*Environmental Entomology*. 1972– . College Park, MD: Entomological Society of America. ISSN: 0046-225X.

*European Journal of Plant Pathology*. 1994– . Dordrecht; Boston: Kluwer Academic Publishers. ISSN: 0929-1873.

*Fungal Genetics and Biology: FG & B*. 1996– . San Diego, CA: Academic Press. ISSN: 1087-1845. Continues *Experimental Mycology*.

*International Journal of Pest Management*. 1993– . London: Taylor & Francis. ISSN: 0967-0874. Continues *Tropical Pest Management*.

*International Pest Control*. 1962– . London: Rhodes Industrial Magazines Ltd. ISSN: 0020-8256. Continues *Pest Technology*.

*Journal of Agricultural and Urban Entomology*. 1999– . Clemson, SC: South Carolina Entomological Society. ISSN: 1523-5475.

*Journal of Economic Entomology*. 1905– . College Park, MD: Entomological Society of America. ISSN: 0022-0493.

*Journal of Entomological Science*. 1985– . Tifton, GA: Georgia Entomological Society. ISSN: 0749-8004. Continues *Journal of the Georgia Entomological Society*.

*Journal of Nematology*. 1987– . College Park, MD: Society of Nematologists. ISSN: 0022-300X.

*Molecular Plant-Microbe Interactions*: *MPMI*. 1988– . St. Paul, MN: APS Press. ISSN: 0894-0282.

*Mycologia*. 1906– . Bronx, NY: New York Botanical Garden. ISSN: 0027-5514.

*Pesticide Biochemistry and Physiology*. 1976– . New York: Academic Press. ISSN: 0048-3575.

*Pest Management Science*. 2000– . West Sussex, UK: Wiley. ISSN: 1526-498X. Continues *Pesticide Science*.

*Plant Disease*: *an International Journal of Applied Plant Pathology*. 1980– . St. Paul, MN: APS Press. ISSN: 0191-2917.

*Physiological Entomology*. 1975– . Oxford: Blackwell Scientific Publications. ISSN: 0307-6962. Continues *Journal of Entomology. Series A. General Entomology*.

*Physiological and Molecular Plant Pathology*. 1986– . London: Academic Press. ISSN: 0885-5765. Continues *Physiological Plant Pathology*.

*Phytopathological Papers*. 1956– . Kew, UK: Commonwealth Mycological Institute. ISSN: 0069-7141. Each issue is a monograph on a particular plant disease or disease-causing organism.

*Phytopathologische Zeitschrift* = *Journal of Phytopathology*. 1929– . Berlin: P. Parey. ISSN: 0031-9481. Continues *Forschungen auf dem Gebiet der Pflanzenkrankheiten und der Immunität im Pflanzenreich*.

*Phytopathology*. 1911– . St. Paul, MN: APS Press. ISSN: 0031-949X.

*Plant Pathology*. 1952– . Oxford: Blackwell Scientific Publications. ISSN: 0032-0862.

*Weed Research*. 1961– . Oxford; Boston: Blackwell Scientific Publications. ISSN: 0043-1737.

*Weed Science*. 1968– . Champaign, IL: Weed Science Society of America. ISSN: 0043-1745. Continues *Weeds*.

*Zeitschrift für Pflanzenkrankheiten und Pflanzenschutz* = *Journal of Plant Diseases and Protection*. 1970– . Stuttgart: E. Ulmer. ISSN: 0340-8159. Continues *Zeitschrift für Pflanzenkrankheiten (Pflanzenpathologie) und Pflanzenschutz*.

## Auxiliary

Other journals that aren't focused on crop protection but would be read or consulted by plant pathologists, weed scientists, and plant entomologists include:

*American Entomologist.* 1990– . (ISSN: 1046-2821)
*Annals of the Entomological Society of America.* 1908– . (ISSN: 0013-8746)
*Annual Review of Entomology.* 1956– . (ISSN: 0066-4170)
*Applied Microbiology and Biotechnology.* 1984– . (ISSN: 1432-0614)
*Archives of Insect Biochemistry and Physiology.* 1983– . (ISSN: 0739-4462)
*Canadian Journal of Microbiology.* 1954– . (ISSN: 0008-4166)
*Ecological Entomology.* 1976– . (ISSN: 0307-6946)
*Entomological News.* 1895– . (ISSN: 0013-872X)
*Insect Biochemistry and Molecular Biology.* 1992– . (ISSN: 0965-1748)
*Insect Molecular Biology.* 1992– . (ISSN: 0962-1075)
*International Journal of Insect Morphology and Embryology.* 1971– . (ISSN: 0020-7322)
*Journal of Bacteriology.* 1916– . (ISSN: 0021-9193)
*Journal of General Virology.* 1967– . (ISSN: 0022-1317)
*Journal of Insect Physiology.* 1957– . (ISSN: 0022-1910)
*Journal of Invertebrate Pathology.* 1959– . (ISSN: 0022-2011)
*Journal of the New York Entomological Society.* 1893– . (ISSN: 0028-7199)
*Journal of Virology.* 1967– . (ISSN: 0022-538X)
*Mycological Research.* 1989– . (ISSN: 0953-7562)
*Virology.* 1955– . (ISSN: 0042-6822)

## GENERAL MAGAZINES

Besides professional journals, the practicing agronomist would also read one or more nonscholarly, popular magazines. Additionally, each state in the United States publishes one or more "farm magazines." Many of the national or regional farm magazines have Web sites that provide full or partial contents. The United States Agricultural Information Network (USAIN) has recently created a list of Web-based agricultural magazines, AgZines: A Harvest of Free Agricultural Journals. It can be found at http://usain.org (cited February 21, 2002). The AgNIC Plant Sciences Web site is also maintaining a list of online, full-text newsletters at http://www.unl.edu/agnicpls/ftnews.html (cited February 21, 2002).

Among the more popular farming magazines are the following:

*Farm Chemicals.* 1973– . (ISSN: 0092-0053)
*Golf Course Management.* 1979– . (ISSN: 0192-3048)
*Progressive Farmer.* 1886– . (ISSN: 0033-0760)

*Soybean Digest.* 1940– . (ISSN: 0038-6014)
*Successful Farming.* 1902– . (ISSN: 0039-4432)
*The Farm Journal.* 1956– . (ISSN: 0014-8008)
*The Farmer's Digest.* 1937– . (ISSN: 0046-3337)
*The Farmer's Weekly.* 1934– . (ISSN: 0014-8474)

## PROCEEDINGS

American Society of Agronomy, Crop Science Society of America, and Soil Science Society of America. *Agronomy Abstracts.* Madison, WI: American Society of Agronomy. ISSN: 0375-5495. Combined abstracts from the annual meeting of the "tri-societies," which meet together in October each year.

Nalewaja, J.D., R.G. Goss, R.S. Tann, and the ASTM Committee E-35 on Pesticides. 1998. *Pesticide Formulations and Application Systems: Eighteenth Volume.* ASTM Special Technical Publication, no. 1347. Philadelphia, PA: ASTM. ISBN: 0803124910. Published annually with various editors. The proceedings of an annual symposium held by the American Society for Testing and Materials, Committee E-35 on Pesticides.

*Proceedings of Annual Meeting.* 1944– . Des Moines, IA: North Central Weed Control Conference. ISSN: 0099-6815.

*WSSA Abstracts: Meeting of the Weed Science Society of America.* 1985– . Champaign, IL: The Society. ISSN: 0888-6180. This is the largest professional organization devoted to the study of the control of weeds and meets annually. Sectional divisions of this society also meet and publish their *Proceedings and Progress Reports.* Sections include the Southern Weed Science Society, Northeastern Weed Science Society, Western Society of Weed Science, and North Central Weed Science Society.

## TABLES AND DATA BOOKS

*Climatological Data.* 1890s– . Asheville, NC: National Oceanic and Atmospheric Administration Environmental Data and Information Service National Climatic Center. ISSN: each state has its own ISSN. Agronomists frequently consult this series, issued monthly for each state. For each weather station, this series reports the temperature (daily/average, minimum, maximum, heating days), precipitation (daily and total), soil temperature, evaporation, wind, and snowfall. Recent editions are available on the Web by subscription (free to educational institutions) at http://www5.ncdc.noaa.gov/pubs/publications.html (cited February 21, 2002).

Roberts, T.R., D.H. Hutson, and Royal Society of Chemistry (Great Britain). Information Services. 1998. *Metabolic Pathways of Agrochemicals*. Cambridge Royal Society of Chemistry. ISBN: 0854044949 (pt. 1); 085404499X (pt. 2). This two-part set is a comprehensive treatise on the physio-chemical properties, uses, mode of action, known metabolic pathways (such as degradation), and metabolism of agrochemicals. The first part covers herbicides and plant growth regulators; the second, insecticides and fungicides. Literature references accompany each record.

## THESES

Research agronomists sometimes need to consult a thesis or dissertation to obtain more experimental detail than journal articles, due to their brevity, are able to provide. While the researcher may request a loan of a dissertation through their interlibrary loan service, most universities will only loan copies of dissertations if they have an extra copy in their archive. As an alternative, the researcher may be able to purchase a copy of the dissertation from Bell & Howell Information and Learning, the provider of *Dissertation Abstracts*. The online version of *Dissertation Abstracts* has started to include a digitized version of complete dissertations. Even if the researcher's institution does not provide online access to the full database for *Dissertation Abstracts*, the most recent two years of *Dissertation Abstracts* is currently freely available to all through UMI ProQuest Digital Dissertations, available from the World Wide Web at http://www.lib.umi.com/dissertations (cited February 21, 2002). This free version provides the first 20 or so pages of dissertations in full text, which is often enough to provide a substantial portion of the Literature Review. The researcher may also choose to search *Dissertations Abstracts* by keywords, authors, institutions, or subject headings. Broad subject headings in *Dissertation Abstracts* appropriate to field agronomists would include Agriculture, Agronomy, Plant Culture, and Plant Pathology.

Another resource for locating dissertations is the OCLC database, WorldCat, though this database only provides cataloging records, not abstracts or full text.

## YEARBOOKS AND SERIES

*Advances in Agronomy*. 1949– . New York: Academic Press. ISSN: 0065-2113.

*Advances in Botanical Research, Incorporating Advances in Plant Pathology*. 1963– . London; New York: Academic Press. ISSN: 0065-2296.

*Analytical Methods for Pesticides and Plant Growth Regulators*. 1972–89. New York: Academic Press. ISSN: 0091- 7486.

*Annual Review of Phytopathology.* 1963– . Palo Alto, CA: Annual Reviews Inc. ISSN: 0066-4286.

*Plant Breeding Reviews.* 1983– . Westport, CT: AVI Pub. Co. ISSN: 0730-2207.

*Progress in Plant Breeding.* 1985– . London; Boston: Butterworths. ISBN: 0407007806 (v. 1).

# 9

# Food Marketing

**Sue Wilkinson**
Food Marketing Institute, Washington, DC, USA

How many supermarkets are there in the United States? What is the profit margin for supermarkets? Who are the major food retailers and wholesalers? What is the market share for supermarkets in Atlanta, Georgia? What is the percent of disposable personal income spent on food-at-home versus food-away-from-home over the last 70 years? The answers to these and many other questions about food distribution—food retailing and wholesaling—can be found in a variety of print and electronic resources that include supermarket trade, business, and government resources. The following information represents a selection of principle resources of information on the food distribution industry and is compiled to help users identify relevant sources in this subject area.

## ABSTRACTS AND DATABASES

The literature on the food distribution industry is found in commercial databases either indexed and abstracted or full text. Some representative databases are listed to supplement those described in Chapter 10.

Factiva (http://www.bestofboth.com/). Dow Jones/Rueters Interactive. Accesses a unique collection of nearly 8,000 world-class sources from 118 countries, with content in 22 languages. The database is full text and includes indexing terms for company, industry, region, and subject. Factiva is available for a fee from the publisher either by subscription or pay as you go with a credit card.

Foodline: International Food Market Data (online). Leatherhead, UK: Leatherhead Food Research Association. Has extensive coverage of international markets and includes food and drink products both branded and own label as well as reports on specialty food sectors, alcoholic beverages, pet foods, fresh produce, and food additives. This database provides a detailed analysis of international food and drink markets, as well as identifying key market players, highlighting new product launches, and assessing consumer attitudes and retail trends. Available on Dialog/DataStar as well as in other formats.

Gale Group Trade and Industry Database (online). Foster City, CA: Gale Group. Covers from 1983 to the present and provides in-depth coverage of over 65 major industries, including full-text coverage of management, economic, and other professional journals. Covers international company, industry, product, and market information and includes product evaluations. Searches can be narrowed to one or more groups of industry specific publications. Available on Dialog.

Lexis/Nexis Academic Universe (Business) (online). New York: Reed/Elsevier. Includes business articles from newspapers, magazines, journals, wires, and transcripts; detailed financial data about companies; EDGAR filings, annual and quarterly reports, and proxy statements; news from over 25 countries; and trade show information. The database is full text and allows users to search for companies based on specified criteria.

Reference Point: Food Industry Abstracts (online). Washington, DC: Food Marketing Institute. An information service of the Food Marketing Institute that is updated weekly. This source compiles abstracts of hundred of major articles that appear in trade, business, government, and FMI publications with their cross references classified by subject and company. Access to the Reference Point database is available only through FMI's Web site at http://www.fmi.org/facts_figs/librarydb/. There are no usage fees for FMI members but a fee is charged for nonmembers.

## BOOKS

The following list provides an historical context for the industry.

Charvat, F.J. 1961. *Super Marketing*. New York: Macmillan.

Greer, W., J.A. Logan, and P.S. Willis. 1986. *America the Bountiful: How the Supermarket Came to Main Street*. Food Marketing Institute: Washington, DC.

Hampe, E.C. Jr., and Merle Whittenberg. 1980. *The Food Industry: Lifeline of America.* Ithaca, NY: Lebhar-Friedman.

Lebhar, G.M. 1963. *Chain Stores in America. 1859–1962.* New York: Chain Store Age Publishing Company.

Markin, R.J. 1963. *The Supermarket: An Analysis of Growth, Development, and Change.* Pullman, WA: Washington State University Press.

Peak, H. and E.S. Peak. 1977. *Supermarket Merchandising and Management.* Englewood Cliffs, NJ: Prentice-Hall.

*Super Market: Spectacular Exponent of Mass Distribution.* 1937. New York: Super Market Publishing Co.

Zimmerman, M.M. 1959. *The Supermarket: A Revolution in Distribution.* New York: Mass Distribution Publications, Incorporated.

## DICTIONARIES

The supermarket industry has its own unique vocabulary. To assist users and others in understanding the terms of the trade, the following are helpful:

*Language of the Food Industry: Glossary of Supermarket Terminology.* 1998. Washington, DC: Food Marketing Institute. This desk reference tool contains definitions of over 2,000 terms, acronyms, and jargon commonly used in the food industry.

*The Talk of the Trade: a Glossary of Terminology for Brokers and their Industry Partners.* 1991. Washington, DC: National Food Brokers Association.

## DIRECTORIES

Names and addresses of supermarket companies, food wholesalers, convenience stores, mass merchandisers, and drug stores can be found in the following directories. Also included here are references to annual issues of trade journals that provide basic industry statistics.

*Annual Report of the Grocery Industry.* New York: Bill Communications.   This report is published each April as a supplement to *Progressive Grocer.* A comprehensive report of the grocery industry covering sales, number of stores, and major issues and trends of the year. Also includes sections on prices, competition, consumer relations, and operations of chains and independents.

*Directory of Convenience Stores.* 2000. Wilton, CT: Trade Dimensions.   Profiles the headquarters and divisions of chain convenience-store companies operating four or more stores. Key executives and buyers are listed and companies are organized into 52 domestic market areas. Also profiles Canadian convenience store companies.

*Directory of Drug Store and HBC Chains*. 2000. New York: Business Guides, Inc. Provides complete selling information and company profiles on nearly 1,700 two-or-more store chains operating drug and HBC stores and as well as drug wholesale companies and their divisions.

*Directory of Mass Merchandisers, 2001: The Guide to Fast-Turn, High-Volume Retailers*. 2001. Wilton, CT: Trade Dimensions. Describes the leading mass-merchandisers in the top 52 market areas. Lists consumer demographic information, leading retailers, and company profiles by market area and contains an index of retail operations by state. Includes hypermarkets, wholesale clubs, combination stores, discount department stores, and deep-discount drug stores.

*Directory of Single-Unit Supermarket Operators*. 2001. New York: Business Guides, Inc. Provides a brief company profile and a list of owners, managers, buyers, and other key personnel for over 7,800 single-unit supermarket companies in the United States, arranged geographically.

*Directory of Supermarket, Grocery and Convenience Store Chains*. 2000. New York: Business Guides, Inc. Geographically lists operators of two or more supermarket and grocery stores in the United States and Canada according to the address of their main office, along with the total number of stores operated, market share, the number of stores of differing formats, and the areas in which the chains operate their stores. Includes an estimated sales volume; profile of operations; and the names of officers, merchandising managers, and buyers. For each state there is a map showing Standard Metropolitan Statistical Areas (SMSAs) and counties.

*Directory of Wholesale Grocers*. 2000. New York: Business Guides, Inc. This companion reference to the supermarket directories listed previously geographically profiles grocery wholesalers. U.S. and Canadian co-op chains, voluntary wholesalers, and wholesale grocers who do not sponsor any groups are listed according to the address of their main office. Shows sales volume for all companies where available and lists key company executives and buyers. Also includes a section on service merchandisers and the lines they carry.

*Distribution Study of Grocery Store Sales*. 2000. New York: Business Guides, Inc. Gives the share of market accounted for by specific supermarket companies in over two hundred cities, the principal suppliers, demographic data on the top 50 standard metropolitan areas, grocery store statistical profile, directory of food associations, and a calendar of industry events.

*Food Industry Review*. 2000. Elmwood Park, NJ: Food Institute. This annual review summarizes data from a variety of government, association, and business sources into clear charts and graphs. The book tracks trends in food expenditures, shopper behavior, food store and supermarkets, product and sales analyses, new products, alternative formats, category management, private label, Federal Trade Commission activity, price indices, population, mergers and acquisitions, and food service.

*Marketing Guidebook*. 2000. Wilton, CT: Trade Dimensions. The guidebook is a comprehensive analysis of the retail and wholesale food industry. For each of 52 market areas in the United States, it includes the name and address of top companies, number of stores served, states and counties where the stores are located, major grocery suppliers, three-year financial summary, buying policies, and private label information. In addition, each company has a full list of executive personnel, buyers, and merchandisers. Each market area also provides share-of-market data and county-by-county demographic statistics. A list of brokers, service merchandisers, and specialty food distributors completes each market area.

*Market Scope 2001: The Desktop Guide to Category Sales*: 2001. Wilton, CT: Trade Dimensions. Presents a sales analysis of major food and nonfood categories by market area. Charts also allow users to compare the performance of any given category across all markets.

*Market Scope 2001: The Desktop Guide to Supermarket Share*. 2001. Wilton, CT: Trade Dimensions. Gives a complete analysis of grocery distribution, including retail market share of chains and wholesale companies in top market areas; number of stores supplied in each area; and consumer demographics for each market.

*SN Retailers and Wholesalers Directory*. 2000. Published by Business Guides Inc. Provides an alphabetical listing of U.S. supermarkets and grocery wholesalers. Includes sales statistics, number of stores, brief market share information, and a list of key personnel.

*Thomas Food Industry Register Buying Guide*. 2000/2001. Millerton, NY: Grey House Publishing. Comprehensive listing of suppliers in the food industry. Includes a listing by product category of food product and ingredient suppliers and food equipment manufacturers, along with brief company profiles and a listing, with owners, of food industry brand names and trademarks.

## GOVERNMENT DOCUMENTS

Many U.S. government agencies publish information about the food distribution industry.

trade magazine for the food distribution industry. Available from the World Wide Web at http://www.supermarketnews.com/.

## WORLD WIDE WEB SITES

The popularity of the Internet has made it easier for the end user to find information on the food distribution industry. Major trade journals have their own Web sites and usually a daily news alert service. Trade associations representing the industry offer a wealth of information as well through their Web sites. Government resources are also readily available through the World Wide Web. The following is a selective list of Internet resources.

## U.S. GOVERNMENT WEB SITES

Bureau of Labor Statistics (U.S.) maintains a Web site that provides news releases and data from its research. Surveys and programs, research papers, and regional information on a wide range of employment issues are included. This site also includes the monthly news releases of the Producer Price Index (http://stats.bls.gov/ppihome.htm) and the Consumer Price Index (http://stats.bls.gov/cpihome.htm). Home page available from the World Wide Web at http://stats.bls.gov.

National Agricultural Library (U.S.) is one of the four national libraries in the United States. The library offers extensive resources on agricultural issues, such as trade and marketing, alternative farming, animal welfare, aquaculture, biotechnology, food and nutrition, plant genome, technology transfer, and water quality. Resources available through their Web site include bibliographies, fact sheets, reference briefs, and links to many other university and government agriculture related sites. Available from the World Wide Web at http://www.nalusda.gov.

USDA Economic Research Service is the main source of economic information and research from the USDA. Located in Washington, DC, with approximately 500 employees, the mission of ERS is to inform and enhance public and private decision making on economic and policy issues related to agriculture, food, natural resources, and rural development. To accomplish this mission, highly trained economists and social scientists develop and distribute a broad range of economic and other social science information and analysis. Available from the World Wide Web at http://www.ers.usda.gov/.

## FOOD INDUSTRY ASSOCIATION WEB SITES

Food Distributors International is a trade association comprised of food distribution companies that supply and service independent grocers and food service operations throughout the United States, Canada, and 19 other countries. The association and its food service partner, the International Foodservice Distributors Association (IFDA), have 220 member companies that operate 997 distribution centers. IFDA represents member firms that annually sell food and related products to restaurants, hospitals, and other institutional food service operations. Available from the World Wide Web at http://www.fdi.org/.

Food Marketing Institute (FMI) maintains a Web site offering resources for the food distribution industry. Information on its programs and publications is included. Backgrounders on important food issues are available, as well as a list of minority-owned vendors and suppliers. Key industry statistics are also available. FMI members and subscribers can access the library database and selected newsletters. Available from the World Wide Web at http://www.fmi.org.

FMI also publishes a variety of research reports on an annual basis that report on supermarket operations and consumer trends. Prices vary for members and nonmembers. Examples include:

*Annual Financial Review 1999–2000.* Presents a financial picture of the supermarket industry, including data on profits, return on investment, capital structure, debt, and equity. Also includes an annual balance sheet and income statement for the entire supermarket industry.

*Facts About Store Development.* 2000. Describes characteristics of supermarkets built and remodeled during the previous year by FMI members. Includes costs of constructing and opening new stores, as well as information on leasing arrangements. Data is given for various store formats. Available on CD-ROM.

*The Food Marketing Industry Speaks.* 2001. Consists of the actual detailed tables summarized in *The Food Marketing Industry Speaks (Executive Summary).* Each table is self-documenting and contains base numbers as well as percentages. A copy of the executive summary is included with the detailed tabulations.

*The Food Marketing Industry Speaks.* 2001. Summary report is based on an annual survey of FMI member companies. Includes sales growth, operating averages, expansion, inflation, merchandising practices, human resources, prepared foods, logistics and distribution practices, and other subjects of current interest.

*Trends in the United States: Consumer Attitudes and the Supermarket.* 2001. Compiled from an annual survey on consumer attitudes toward the supermarket industry, and toward business and the economy in general. Includes trend data on attitudes regarding stores, nutrition, food safety, alternative formats, economizing measures, and in-store services and departments, as well as consumer-reported expenditures and shopping behavior. Special current topics are highlighted in each update.

Grocery Manufacturers of America (GMA), founded in 1908, advances the interests of the food, beverage, and consumer packaged-goods industry on key issues that affect the ability of brand manufacturers to market their products profitability and deliver superior value to the consumer. Available from the World Wide Web at http://www.gmabrands.com/.

National Association of Convenience Stores (NACS) (U.S.) founded in 1961, is an international trade association representing 2,300 retail and 1,700 supplier company members. NACS member companies do business in nearly 40 countries around the world, with the majority of members based in the United States. Available from the World Wide Web at http://www.cstorecentral.com/index.html.

The National Grocers Association (NGA) (U.S.) is the national trade association representing retail and wholesale grocers that comprise the independent sector of the food distribution industry. An independent retailer is a privately owned or controlled food retail company operating in a variety of formats. Most independent operators are serviced by wholesale distributors, while others may be partially or fully self distributing. NGA members include retail and wholesale grocers and their state associations in the United States as well as manufacturers and service suppliers. Available from the World Wide Web at http://www.nationalgrocers.org/.

National Restaurant Association (U.S.), founded in 1919, is the leading business association for the restaurant industry. Together with the National Restaurant Association Educational Foundation, the association's mission is to represent, educate, and promote a rapidly growing industry that is comprised restaurant and foods service outlets. Available from the World Wide Web at http://www.restaurant.org.

## BIBLIOGRAPHY

Sources of Facts and Figures. Washington, DC: Food Marketing Institute. 2001.

# 10

## Food Science

**Francine Bernard**
Agriculture and Agri-Food, Quebec, Canada

The literature of food science and food technology has moved from predominately journal articles to a body of literature that also includes books and encyclopedias. Food science covers many aspects of food such as preservation, evaluation, and distribution. It is an applied science, that also covers food chemistry, food analysis, food microbiology, food engineering, and food processing, and is based on a sound knowledge of the pure sciences including chemistry, physics, biology, and mathematics. Food scientists are also involved in establishing international food standards to promote and facilitate world trade. This chapter presents the principal sources of information in food science and technology published after 1985 and has been compiled to assist users in identifying relevant sources in this field. In each section, the aim is to provide brief critical reviews of the most important sources. It is inevitable, however, that some information sources will have been omitted. Emphasis is on English-language sources, although foreign-language materials are included, especially French-language sources.

### ABSTRACTS AND DATABASES

Library catalogs and databases are among the best resources to find information on food science. Most academic and public libraries in the United States use

**291**

Library of Congress Subject Headings to index their collection. These headings cover all disciplines and knowledge. Broader terms should be used to find documents in a catalog such as *Food—analysis*; *Food—composition*; *Food—toxicology*; *Food additives*; and *Food preservatives*.

Most databases differ from catalogs, covering specific disciplines rather than providing access to broad bases of knowledge. They give access to what has been written on a subject through indexes to journal, magazine, and newspaper articles; reports; book chapters; and other publications. One of the best tools to find relevant terms for searching the food science databases is the *FSTA Thesaurus*. The 1996 edition contains 8,274 terms with a structured arrangement of terms, numbered hierarchies for all descriptors, and clear differentiation between descriptors and nondescriptors. History and notes are used where appropriate to clarify usage of terms and British spellings. For most searching, use headings and general descriptors that represent the major topics of the database. Natural language will often retrieve many nonrelevant references. Be sure to use the correct Boolean logical operators as the *and* and the *or* are often easily confused. If you are in doubt, consult an information professional.

Example: To find references on the additives: fumaric acid or tartaric acid

| Library Catalog | Databases |
|---|---|
| Subject Headings | Terms |
| *Food—additives* | *Fumaric acid* or *Tartaric acid* |

Food science and technology is well covered by traditional agriculture and food databases. Related databases such as Chemical Abstracts, BIOSIS, and Engineering Index also contain information useful in finding different aspects of this science.

*Dairy Science Abstracts.* V. 1–. 1939–. Wallingford, UK: CAB International. A monthly publication provides information on all aspects of milk production, processing, and milk products. Over 7,000 references are added from 14,000 abstracts, books, technical reports, theses, patents, and proceedings scanned each year. The print version dates back to 1939 but data from 1973 is available on CD-ROM (as part of the BEASTCD and CAB Abstracts), Internet, and on online hosts as part of the CAB Abstracts database. This is an essential complement to the FSTA or FROSTI databases for anyone seeking exhaustive coverage of the subject.

FOMAD: Food and Drinks Market Intelligence. 1982–. Leatherhead, UK: Leatherhead Food Research Association. This database covers all food and drink product group markets. Updated biweekly, the database includes 120,000 references, with 12,000 records added each year, from 250 market and food-related journals, statistical publications, and market reports. Available in print (*Food*

*Market Abstracts*), over the Internet, on CD-ROM, hard disk, and online hosts. Provides extensive coverage of well-developed food and drink markets across the world with emphasis on UK markets. Copies of articles are available from the Leatherhead Library.

*Foods ADLIBRA*. V. 1– . 1972– . Minneapolis, MN: General Mills Foods Adlibra Publications. Provides information on new food products, markets, and significant research in the area of food processing. Over 250 periodicals are reviewed with an additional 500 periodicals scanned to extract the major references. Contains more than 292,000 records and monthly updates include more than 1,000 records. This is a good source for tracking the introduction of new products but covers mostly the North American market. Available in print and on online hosts.

*Food Science and Technology Abstracts*. V. 1– . 1969– . Shinfield, UK: International Food Information Service. Provides comprehensive coverage on every aspect of food science, food products, and food packaging for all commodities. International coverage of more than 500,000 entries in over 40 languages. More than 20,000 records are added annually from 1,800 journals, patents, books, conference proceedings, reports, pamphlets, and legislation. All abstracts selected for Food Science and Technology Abstracts (FSTA) are translated into English. FSTA is the premier source of information on innovations and developments in the food industry and the latest scientific research for the food sector. It is available in print, over the Internet, on CD-ROM, hard disk, and on online hosts. A thesaurus is available to aid in searching.

FOREGE: Food Additive, Compositions and Labelling Legislation. Current Information Only. Leatherhead, UK: Leatherhead Food Research Association. The only database devoted to laws and regulations governing food additives and those on standards. FOREGE Additives give details on the permitted use of food additives and FOREGE Standards provide detailed information on the compositions and labeling requirements of food. The database has 50,000 citations of current legislation and is updated weekly or quarterly depending on the service provider. FOREGE provides information on the permitted additives in over 120 countries and gives compositional and labeling requirements for foods for 8 countries. The database is available over the Internet, CD-ROM, hard disk, and various online hosts. A key strength of the FOREGE databases is that they interpret complex legal documents, extract the key requirements of the legislation, and present them in an easily understandable form. Each record includes precise legal requirements for particular foods in specific countries.

Leatherhead Food Research Association publishes two other databases: FLAIRS, designed to inform on new food and drink products launched onto the UK retail market and FOSCAN a bibliographic database that covers the press relevant to the food and drink industries.

*FROSTI—Food Science and Technology Abstracts.* 1972– . Leatherhead, UK: Leatherhead Food Research Association.   A database of over 380,000 records, *FROSTI* provides the latest information published worldwide in the fields of food science and technology. The index is updated twice weekly on some systems and adds 24,000 records annually from 550 journals, 500 monographs, standards, statutory instruments, and technical reports. The database is available in print, over the Internet, on CD-ROM, hard disk, and online hosts. A strength of *FROSTI* is its currency, key journals often being indexed within a few weeks after delivery at Leatherhead. Chapter records, books, and conference proceedings are covered. Copies of articles are available from the Leatherhead Library.

IALINE. 1970– . Saint-lô, France: Ialine/Normandie.   The only French-language database in the agri-food sector. Over 330,000 references are taken from 200 serials, books, theses, and proceedings. The database is updated monthly and includes abstracts and keywords in French, from mostly English-language publications. The database is available in print (*Bulletin Industries Agro-Alimentaires-Bibliographie Internationale*), CD-ROM, and over the Internet. A thesaurus is available for searching and photocopies are available from Ialine+.

## BOOKS

Books have been a source of information since the invention of the printing press. They provide access to knowledge at an affordable price and can be considered the core of a professional library where ready access to information is important. Publishers' catalogs are the best source of information on forthcoming publications. A number of journals, such as *Food Technology*, have a book review or new publications column that provides readers with a critical assessment of recent books. The majority of food science books are being produced by a relatively small group of well-known scientific publishers. A selected list of publishers of food science books follows.

### AVI Publishing Company

AVI, a division of Van Nostrand Reinhold now owned by John Wiley & Sons, has specialized in publishing technical books on food processing since 1921. Its catalog has a number of interesting titles, including *Food Science Sourcebook*, which contains definitions, tables, food composition charts, and so forth. Unfortunately, most of their titles are now out of print but locations for the books can be found through WorldCat.

## The Binsted's Booklist for the Food Industry. Food Trade Press

*The Binsted's Booklist for the Food Industry: Food Trade Press* list contains over 2,500 titles of scientific, technical, and reference books on all areas of food processing. The books are published by Food Trade Press or by other publishers such as AVI, Van Nostrand Reinhold, Academic Press, and Campden.
Contact information:
Food Trade Press
Station House, Horton Way
Westerham, Kent TN16 1BZ, United Kingdom
Telephone: +44 959 563944
Fax: +44 959 561285

## Campden Reports

Provides 200 practical publications of guidelines, technical manuals, reviews, and specifications.
Contact information:
Campden & Chorleywood Food Research Association Group
Chipping Campden, Gloucestershire, GL55 6LD, United Kingdom
Telephone: +44 0 1386 842000
Fax: +44 0 1386 842100
URL: http://www.campden.co.uk/

## Leatherhead Reports

This research center is one of the world's leading research centers in food science. It publishes reports covering food science and technology, retail market data, and food regulatory issues.
Contact information:
Leatherhead Food Research Association
Randalls Road
Leatherhead, Surrey KT22 7RY United Kingdom
Telephone: +44 0 13723 76761
Fax: +44 0 13723 86228
URL: http://www.lfra.co.uk

## Tec & Doc Lavoisier

Tec & Doc Lavoisier has been publishing the series Sciences et Techniques Agro-Alimentaires (Agri-food Science and Technology) for over 20 years, and lists more than 300 titles intended for agri-food professionals, and covering all practical and theoretical aspects of food science. With over 30 titles published, it is the most extensive French-language series in the field. Several of the publications are multi-author treatises under the responsibility of co-ordinating editors. The information is presented and analyzed in easily accessible language by authors having a good command of the subject matter. Tec & Doc Lavoisier also distributes publications by organizations such as the Association Pour la Promotion des Industries Agricoles (APRIA) or the Institut National de Recherche Agronomique (INRA).

Contact information:
Tec & Doc Lavoisier
11, rue Lavoisier, 75384 Paris Cedex 08 France
Telephone: +33 1 4265 3995
Fax: +33 1 4740 6702
URL: http://www.lavoisier.fr

## Other Publishers to Consider

Aspen Publisher: http://www.aspenpublishers.com
CRC Press—Food Science and Nutrition: http://www.crcpress.com
CTI Publications Inc.: http://www.ctipubs.com
Elsevier Science—Series Development in Food Science: http://www.elsevier.com
John Wiley & Sons: http://catalog.wiley.com/remsrch.cgi
Marcel Dekker—Series Food Science and Technology: http://www.dekker.com/catalog/catalog-top.htm
Technomic Publishing Company: http://www.techpub.com

## DICTIONARIES

Dictionaries are alphabetical lists of terms, complete with their definition. Although the majority of scientific papers are now published in English, for research work it is also essential to read scientific papers in languages other than English.

Adrian, J., N. Adrian, and K. Harper. 1990. *Dictionnaire Agro-Alimentaire: Anglais-Français, Français-Anglais/Dictionary of Food Science and Industry*. New York: Lavoisier Tec & Doc. 346 p.   Covers the vocabulary of the diverse stages in foodstuff production.

Adrian, J., R. Frangne, and G. Legrand. 1988. *Dictionary of Food and Nutrition/ Dictionnaire de Biochimie Alimentaire et de Nutrition*. New York: VCH. 233 p. Contains data relating to the chemistry, physiology, biochemistry, nutrition, and processing of foods. The publication is based on the original French edition published by Technique et Documentation in 1981.

Frank, H.K. 1992. *Dictionary of Food Microbiology*. Lancaster, PA: Technomic Pub. 298 p. Translation of *Lexikon Lebensmittel-Mikrobiologie*. Includes bibliographical references.

Igoe, R.S. 1989. *Dictionary of Food Ingredients*. 2nd ed. New York: Van Nostrand Reinhold. 225 p. Covers more than 1,000 approved ingredients, detailing the functions, chemical properties, and the applications of each.

International Association for Cereal Science and Technology. Online dictionary/ glossary of science and technology. http://www.icc.or.at/dic.htm. First online dictionary in the field of cereals. English is the basic language for the structure. Search in English, French, German, Spanish, and Russian. It gives the translation and the definition of the term. It is a combination of the revised and completed preprint version of the multilingual ICC Dictionary and the Grain-Glossary prepared by CRC (Quality Wheat CRC Limited).

Luck, E. 1992. *Four Language Dictionary of Food Technology: English, German, Spanish, French/Dictionnaire Quadrilingue de Technologie Alimentaire: Anglais, Allemand, Espagnol, Français*. Hamburg, Germany: Behr's Verlag. 656 p. Includes 5,019 entries in the 4 languages listed in the title.

Luck, E. 1990. *Comprehensive Dictionary of Food: English-German*; *Grobworterbuch des Lebensmittelwesens: Englisch-Deutsch*. Hamburg, Germany: Behr's Verlag. 626 p. First published in 1983, this work includes scientific names in Latin and English.

Runner, E. 1991. *Dictionary of Milk and Dairying*. Munchen, Germany: Volkswirtschaftlicher Verlag. 384 p. English version of *Lexicon der Milch* published in German in 1988. Includes 3,320 entries and 44 tables.

## DIRECTORIES

Directories are alphabetical lists of names and addresses of manufacturers, products, or organizations that present an array of information identifying or locating chosen subjects.

*Food Trades Directory of the UK & Europe*. Annual. London: Hemming Group. Volume one contains details of over 9,000 companies and organizations in the United Kingdom and is arranged by type of activity. Volume two gives details of suppliers of food and ingredients in Continental Europe.
Contact information:
Hemming Information Services
32 Vauxhall Bridge Road

London SW1V 2SS, United Kingdom
Telephone: +44 0 20 7973 640
Fax: +44 0 20 7233 5057

Hodges, K.E. and J. Spring. 1993. *The Directory of Research and Education in Food Science, Technology and Engineering.* 2 vols. Reading, UK: IFIS Publishing; London, UK: EFFoSt. This directory contains over 800 entries from companies, research institutes, universities, and government laboratories throughout the world. The publication was made possible by a collaboration between The International Food Information Service (IFIS) and the European Federation of Food Science and Technology (EFFoST) v. 1: European, v. 2: North American.

Kompass. Current Information Only. Saint Laurent, Cruet, France: Kompass International. One of the best-known directory publishers is Kompass. It originated in Switzerland and is now present in over 70 countries. The Kompass Worldwide database is a good tool to find information on products or companies outside the United States. The database contains 1.6 million companies; 23 million product and service references; 700,000 trade and brand names; and 3.2 million executives' names. Kompass is available on the Internet and through online systems as well as in multi-country, national, and specialized CD-ROMs. Kompass publishes national, regional, sectional, and specialized books in several volumes. All products refer to the +Kompass Classification System.
Contact information:
Kompass International
Saint Laurent-73800 Cruet, France
Telephone: +33 4 79 65 25 08
Fax: +33 4 79 84 13 95
URL: http://www.kompass.com

*Thomas Food Industry Register.* 1898– . New York: Thomas Publishing Company. Thomas's Food Industry Register is the best directory for American food companies. It is available in print, on the Internet, and CD-ROM. The print product, *Thomas Food Industry Register,* comes in three volumes: v. 1: Equipment, Supplies & Services, Distribution (20,000 sources); v. 2: Food Products (25,000 sources); and v. 3: Company Profiles and other interesting lists such as trade associations and industry convention calendars.
Contact information:
Thomas Publishing Company
Five Penn Plaza
New York, NY 10001
Telephone: +1 212 290 7341
Fax: +1 212 290 8749
URL: http://www.tfir.com

Trade journals often publish special issues called *guides and directories*, such as the *Food Processing: Guide and Directory* published annually by Putman Publications, or a buyer's guide, such as the *International Food Ingredients & Analysis Directory*. Generally, these can be found by searching the database.

## ENCYCLOPEDIAS

Good encyclopedias serve as a general text as well as an instant source for details. They offer comprehensive coverage of various branches of knowledge. The most useful ones provide references to further reading.

*Academic Press Encyclopedia of Food Microbiology*. 1999. 3 vols. London: Academic Press. Includes over 400 articles written by the world's leading scientists and covers the field from Acetobacter to Zymomonas.

Adrian, J., J. Potus, and R. Frangne. 1995. *La Science Alimentaire de A à Z*. 2nd ed. Paris: Lavoisier Tec & Doc. 477 p. 3,800 entries with figures, tables, and cross-references.

*Encyclopedia of Food Science, Food Technology, and Nutrition*. 1993. 8 vols. London: Academic Press. This encyclopedia is organized into a number of major entries, each consisting of one or more articles. Targets an audience of academics, students, and professionals from the food industry, and scientific writers seeking a rapid, but thorough review of a given topic. Suggestions for reading at the end of each article allow easy access to the primary literature.

*Encyclopedia of Food and Color Additives*. 1997. G.A. Burdock, ed. 3 vols. Boca Raton, FL: CRC Press. Includes information once scattered among the U.S. Code of Federal Regulations (CFR), other government and technical publications, or only available through the U.S. Government's Freedom of Information Act. This work provides answers to technical, legal, and regulatory questions in clear, nontechnical language.

*Wiley Encyclopedia of Food Science and Technology*. 1999. 2nd ed. 4 vols. New York: John Wiley & Sons. Revised, updated, and expanded from the first edition this work emphasizes emerging areas in the food and nutrition delivery systems, including nutrition monitoring, chemical and microbiological food safety, risk management in food safety, functional foods and nutraceuticals in food formulation, food substitutes, and advances in food biotechnology.

## GOVERNMENT DOCUMENTS

Many worldwide organizations publish information on food, mostly on food safety and on food legislation for national governments. The main sources are the Food and Agriculture Organization of the United Nations (FAO), which has a series called *FAO Food and Nutrition Papers*, and in collaboration with the World Health Organization (WHO), produces many volumes on food quality control; International Atomic Energy Organization (IAEO), which publishes excellent sources of information on food irradiation; and the USDA, which has a large Web site (http://www.foodsafety.gov) covering such aspects as food safety, food and nutrition, and marketing and regulatory programs.

## FLAIR-FLOW

FLAIR-FLOW is a specialized dissemination project of the European Union, which was set up in response to a need for wider dissemination of information and results from European food research and development programs. FLAIR-FLOW uses one-page technical documents as vehicles for disseminating the results of food-related research and development as well as workshops, lectures, and posters at conferences/trade shows. Since its beginning, FLAIR-FLOW has issued more than 260 one-page technical document booklets. Some of these articles can be found at http://exp.hispeed.com/flair/menu.htm.

Contact information:
FAIR-FLOW
Teargas, The National Food Centre
Dunsinea, Castleknock, Dublin 15, Ireland
Telephone: +353 1 805 9500
Fax: +353 1 805 9550

### Food Legislation

Laws and regulations on food change constantly. One of the best sources for updated information is the Web site of the Food Laws and Regulations Division of the Institute of Food Technologists. It has links to various sites such as "Title 21 of the U.S. Code of Federal Regulations relating to Food and Drugs" and covers international organizations and countries other than the United States Available from the World Wide Web at http://www.ift.org/divisions/food_law/jump_gov.htm.

### Leatherhead Food Research Association

Leatherhead Food Research Association has published guides on food regulations in different countries. Guide to food regulations are produced for the Middle East, the United Kingdom, Latin American countries, the Far East, South Asia, and Central Europe. In addition, the Leatherhead also publishes the *E.C. Food Law Manual*. Most of these guides are updated and some of them are also avail-

able on CD-ROM. This association also has publications on topics such as sweet-eners and additives.
  Contact Information:
  Leatherhead Food Research Association
  Randalls Road
  Leatherhead, Surrey KT22 7RY United Kingdom
  Telephone: +44 0 13723 76761
  Fax: +44 0 13723 86228
  URL: http://www.lfra.co.uk

## National Technical Information Service Selected Research in Microfiche (SRIM) Service—Food

The National Technical Information Service (NTIS) delivers complete microfiche copies of selected U.S. government publications. A biweekly set of microfiche publications is provided with a *Food Topic Selection* subscription. The indexes are in the print version but difficult to search. The electronic database is search-able through the Web or through the online services for a fee.
  Contact information
  National Technical Information Service
  Springfield, Virginia 22161 USA
  Telephone: +1 703 605-6000
  URL: http://www.ntis.gov/product/srim.htm

## World Health Organization (WHO)

WHO publishes mostly on food safety issues such as these series on safety evalu-ations: *Evaluation of Certain Food Additives and Contaminants and Evaluation of Certain Veterinary Drug Residues in Food.* It also publishes guidelines for drinking-water quality and has regional offices on all continents. The WHO Web site (http://www.who.int) provides additional information.

## GUIDES

The majority of the bibliographical guides in the field of food science date from the 1970s. However, note in the following text the work of Syd Green, *Keyguide to Information Source in Food Science and Technology*, which was published in 1985. This highly useful guide covers almost exclusively English-language works. The work of Gazeay and Gougeon is recommended for all European sources, and the compilation of Ann Dankbars exclusively covers the United Kingdom.

Bernard, F. 1996. "Importance stratégique de l'information." In: *L'Entreprise Agro-Alimentaire*; *Assurer sa Croissance*. Y. Beaulieu, F. Bernard, and J. Fortin, eds. St-Jean sur Richelieu, Canada: Éditions du Monde Alimentaire. pp. 202–268. This chapter shows the strategic importance of information for a food industry and indicates the main accessible information sources in French. It also presents the concept of competitive intelligence.

Gazeay, M. and C. Gougeon. 1994. *L'Europe de L'Agro-Alimentaire*. Paris: ABDS Editions. 767 p. This guide lists the main information sources related to the food industry field in Europe. It includes European and national sources from seventeen countries, as well as classical and informal sources.

Green, S. 1985. *Keyguide to Information Sources in Food Science and Technology*. London: Mansell Publishing. 231 p. Part one introduces the subject of food science and technology and shows its relationships to other disciplines. Part two is an annotated bibliography of sources of information.

Leatherhead Food Research Association and A. Blakeman. 2000. *Food Information on the Internet*: *An Essential Guide*. 3rd ed. Leatherhead, UK: Leatherhead Publishing. 202 p. Contains references to over 1,250 URLs, including information on newsgroups and e-mail lists. The purchase of the Internet guide gives access to all Web sites listed in the book.

## HANDBOOKS, MANUALS, AND DATA BOOKS

The following monographs have been selected primarily on personal examination of the books as well as favorable reviews appearing in professional journals and textbooks. Handbooks include those reference materials that provide data or information in a compact form conveniently arranged for easy access.

Ash, M. and I. Ash. 1995. *Handbook of Food Additives*: *An International Guide to More Than 7,000 Products By Trade Name, Chemical, Function, and Manufacturer*. Brookfield, VT: Gower. 1,025 p. Information on 5,000 trade name products and 2,500 chemicals that function as food additives. Includes cross-references and is available on CD-ROM.

Bailey, A.E. and Y.H. Hui, eds. 1996. *Bailey's Industrial Oil and Fat Products*. 5th ed. New York: John Wiley & Sons. Comprehensive coverage: v. 1: Edible Oil & Fat Products: General Applications; v. 2: Edible Oil & Fat Products: Oils and Oil Seeds; v. 3: Edible Oils & Fat Products: Products and Application Technology; v. 4: Edible Oil and Fat Products: Processing Technology; and v. 5: Industrial and Consumer Nonedible Products from Oils and Fats.

Baracco, P. and J.C. Frentz. 1988. *L'Encyclopédie de la Charcuterie*: *Dictionnaire Encyclopédique de la Charcuterie*. 2nd ed. Orly, France: Soussana.

790 p. Displays through illustrations and explains every aspect of meat processing in such products as delicatessen meats, sausages, and so forth.

Burdock, G.A. 2001. *Fenaroli's Handbook of Flavor Ingredients*. 4th ed. 1864 p. Boca Raton, FL: CRC Press. Describes natural and synthetic flavor ingredients with detailed characteristics and their applications in food. A fourth edition is due out in the near future.

Bylund, G. 1995. *Dairy Processing Handbook*. Lund, Sweden: Tetra Pak Processing Systems AB. 436 p. This work is excellent for teaching because of its high-quality diagrams and illustrations. A CD-ROM version of the work is also available.

de Roissart, H. and F.M. Luquet. 1994. *Bactéries Lactiques: Aspects Fondamentaux et Technologiques*. 2 vols. Paris: Lorica. Major synthesis on lactic acid bacteria written by 133 authors. Illustrations and tables are included.

Downing, D.L. 1996. *Complete Course in Canning and Related Processes*. 13th ed. Timonium, MD: CTI Publications. 494 p. An illustrated classic that includes fundamental information on canning; microbiology, packaging, HACCP, and ingredients; and processing procedures for canned food products.

Heldman, D.R. and D.B. Lund. 1992. *Handbook of Food Engineering*. New York: Marcel Dekker. 756 p. An illustrated reference of food processing, this work contains 850 literature citations and 1,100 figures in 14 chapters. Properties, rate constants, and related data for food processing are included.

Hui, Y.H. 1991. *Data Sourcebook for Food Scientists and Technologists*. New York: VCH Publishers. 976 p. Part one is a dictionary and Part two covers food composition, properties, and general data.

Hui, Y.H. 1993. *Dairy Science and Technology Handbook*. 3 vols. New York: VCH Publishers. V. 1: Principles and Properties includes chemistry and physics, analyses, sensory evaluation, and proteins; v. 2: Product Manufacturing includes yoghurt, ice cream, cheese, and dry and concentrated dairy products; and v. 3: Applications Science, Technology, and Engineering includes quality assurance, biotechnology, computer applications, equipment and supplies, and processing plant designs. There are company listings included in an appendix.

Jay, J.M. 2000. *Modern Food Microbiology*. 6th ed. Gaithersburg, MD: Aspen Publishers. 679 p. A textbook with illustrations and tables that focuses on the general biology of the microorganisms found in foods.

Jensen, R.G. 1995. *Handbook of Milk Composition*. San Diego, CA: Academic Press. 919 p. Includes illustrations, bibliographical references, and index. Contents include: 1. Introduction; 2. The structure of milk: implications for sampling and storage; 3. Determinants of milk volume and composition; 4. Carbohydrates in milk: analysis, quantities and significance; 5. Nitrogenous components of milk; 6. Milk lipids; 7. Minerals, ions, and trace elements in milk; 8. Vitamins in milk; 9. Defense agents in milk; 10. Comparative analysis of nonhuman milks; and 11. Contaminants in milk.

Karleskind, A. and J.P. Wolff, eds. 1997. *Oils and Fats Manual*. 2 vols. New York: Springer-Verlag. Synthesis of all essential data on oils and fats. Translation of *Manuel des Corps Gras* published in 1992.

Karleskind, A., K.B. Kaylegian, and R.C. Lindsay. 1995. *Handbook of Milkfat Fractionation Technology and Applications*. Champaign, IL: AOCS Press. 662 p. Illustrated with figures and tables, this work summarizes the widely distributed literature on milkfat fractionation technology and applications. Bibliographical references (p. 637–56) and an index are included.

Kosikowski, F.K. and V.V. Mistry. 1997. *Cheese and Fermented Milk Foods*. 3rd ed. 2 vols. Westport, CT: F.K. Kosikowski LLC. This classic published by the author also includes bibliographies and an index. V. 1: Origins and Principles and v. 2: Procedures and Analysis.

Lorenz, K.J. and K. Kulp. 2000. *Handbook of Cereal Science and Technology*. 2nd ed. New York: Marcel Dekker. 808 p. Twenty-four illustrated chapters on major cereals and review of the chemistry and technology of cereals. It also has an extensive bibliography.

Maarse, H. 1997. *Volatile Compounds in Food: Qualitative and Quantitative Data*. 7th ed. Zeist, The Netherlands: TNO-CIVO Food Analysis Institute. 2,600 p. A compilation of more than 30 years of data on volatile compounds naturally occurring in food products has been edited by the TNO Nutrition and Food Research Institute. The work offers access to 560 products and 7,210 volatile compounds, based on 4,000 literature references. Updated by supplements. Disk and network version available.

McCabe, J.T., E.H. Vogel, and Master Brewers Association of the Americas. 1999. *The Practical Brewer: A Manual for the Brewing Industry*. 3rd ed. Wauwatosa, WI: Master Brewers Association of the Americas. 757 p. Illustrated trade standard that includes practical procedures for recent innovative technologies.

Meilgaard, M.M., G.V. Civille, and B.T. Carr. 1999. *Sensory Evaluation Techniques*. 3rd ed. Boca Raton, FL: CRC Press. 416 p. Basic book on sensory evaluation for students and professionals that includes illustrations and figures.

Rahman, S. 1995. *Food Properties Handbook*. Boca Raton, FL: CRC Press. 500 p. Each of the seven chapters deals with specific food properties. Illustrated including tables and figures with an extensive bibliography on pages 463–94.

Ranken, M.D. and R.C. Kill, eds. 1997. *Food Industries Manual*. 24th ed. London:

Chapman & Hall. 650 p. Includes 17 sections with figures and tables on all aspects of food processing including the latest technologies.

Reineccius, G. and H.B. Heath. 1994. *Source Book of Flavors.* 2nd ed. New York: Chapman & Hall. 928 p. Covers all aspects of the flavor industry and includes figures and tables.

Salunkhe, D.K. and S.S. Kadam. 1995. *Handbook of Fruit Science and Technology: Production, Composition, Storage, and Processing.* New York: Marcel Dekker. 611 p. Thirty illustrated chapters on fruits with an emphasis on tropical fruits.

Uhl, S.R. 2000. *Handbook of Spices, Seasonings and Flavorings.* Lancaster, PA: Technomic Publishing. 336 p. Includes 200 individual spices, seasoning blends, and flavoring ingredients along with properties, uses, preparation techniques, and applications.

Valentas, K.J. and R.P. Singh. 1997. *Handbook of Food Engineering Practice.* Boca Raton, FL: CRC Press. 718 p. Includes illustrations, figures, tables, bibliographical references, and an index.

## JOURNALS

Journals constitute a primary resource in the literature of food science. Regular reading of journals is an important step in gathering information as it helps one keep abreast of technical advances, new legislation, recent conferences, and news about competitors. Trade journals are good sources of current, reliable information. The journals listed in the following text are indexed in the major relevant abstracting and indexing services worldwide and are arranged by subject. The listing of URL indicates electronic access to the journal, although often a subscription fee is required for access.

### General Aspects of Food Science and Technology

*Acta Alimentaria.* (An International Journal of Food Science). 1972– . Budapest, Hungary: Central Food Research Institute. Quarterly. ISSN: 0139-3006. Original papers that cover all aspects of food science.

*Critical Reviews in Food Science and Nutrition.* 1970– . Boca Raton, FL: CRC Press. Bimonthly. ISSN: 1040-8398. Authoritative information source that presents critical viewpoints of current technology, food science, and human nutrition. Addresses the acquiring of knowledge and how scientific discoveries are applied as they relate to nutrition.

*Food Chemistry.* 1976– . Amsterdam: Elsevier. Monthly. ISSN: 0308-8146. Available from the World Wide Web at http://www.elsevier.com/locate/

foodchem. Includes research papers dealing with the chemical and biochemical composition of foods and the study of their properties and processing applications. The papers are original peer-reviewed articles.

*Food Research International.* 1968– . Amsterdam: Elsevier. 10 issues/year. ISSN: 0963-9969. Available from the World Wide Web at www.elsevier.com/locate/foodres. Formerly the *Canadian Institute of Food Science and Technology Journal.*

*Food Science and Technology Today.* 1987– . London: Institute of Food Science and Technology. Quarterly. ISSN: 0950-9623. Available from the World Wide Web at http://www.ifst.org/fstt.htm. Journal of the institute founded in 1964. Although based in the United Kingdom, it has members throughout the world.

*Food Technology.* 1947– . Chicago, IL: Institute of Food Technologists. Quarterly. ISSN: 0015-6639. Provides news and analysis of the development, use, quality, safety, and regulation of food sources, products, and process.

*Industries alimentaires et agricoles.* 1883– . Paris: Association des chimistes et ingenieurs de sucrerie, distillerie et industries agricoles de France et des colonies. Monthly. ISSN: 0019-9311. Includes scientific inserts, thematic bibliographies, review of new products and equipment, suppliers guides, and sections on economics.

*Italian Journal of Food Science* (Rivista italiana di scienza degli alimenti). 1989– . Pinerado, Italy: Chiriotti Editori. Quarterly. ISSN: 1120-1770. Main journal on food science in Italy. Articles are in Italian and English.

*Journal of Agricultural and Food Chemistry.* 1953– . Columbus, OH: American Chemical Society. Monthly. ISSN: 0021-8561. Available from the World Wide Web at http://pubs.acs.org/journals/jafcau/index.html. Primary research journal devoted to the application of chemistry in developing more efficient, economical and safe production of foods and agricultural products.

*Journal of Food Science.* 1961– . Chicago, IL: Institute of Food Technologists (IFT). 8 issues/year. ISSN: 0022-1147. Formerly *Food Research*, covers all basic and applied science for food professionals.

*Journal of the Science of Food and Agriculture.* 1950– . New York: John Wiley & Sons. 15 issues/year. ISSN: 0022-5142. Available from the World Wide Web at http://www.wiley.co.uk/sci. Publishes original research and critical reviews in agriculture and food science, with particular emphasis on interdisciplinary studies in agriculture and food.

*Lebensmittel-Wissenchaft und Technologie/Food Science and Technology.* 1968– . New York: Academic Press. Bimonthly. ISSN: 0023-6438. Available from the World Wide Web at http://www.academicpress.com/lwt. Official publication of the International Union of Food Science and Technology. Papers are published in the fields of chemistry, microbiology, biotechnology, food processing, and nutrition, and are written in English, German, and French.

*R.I.A. Revue de l'Industrie Agroalimentaire*. 1953– . Paris: C E P Groupe France Agricole. 11 issues/year, + 4 supplements. ISSN: 0035-4244. Covers industries of meats, milk, grain, drinks, preservation, bread, baking, and packing. Accent is on new technologies and there is good coverage of the French food industry.

*Science des aliments*. (An International Journal of Food Science and Technology). 1981– . Paris: Lavoisier. Bimonthly. ISSN: 0240-8813. Publishes original articles, research notes, and reviews in French and English in the various sectors of the science and technology of food.

*Trends in Food Science & Technology*. 1990– . Amsterdam: Elsevier. Monthly. ISSN: 0924-2244. Available from the World Wide Web at www.elsevier.com/locate/tifs. International peer-reviewed journal that fills the gap between specialized primary journals and general trade magazines by focusing on promising new research developments and their current and potential food industry applications.

## Cereals and Bakery Products

*Cereal Chemistry*. 1924– . St. Paul, MN: American Association of Cereal Chemists. Bimonthly. ISSN: 0009-0352. Available from the World Wide Web at http://www.scisoc.org/aacc/pubs/journ/cc/ccinfo.htm. Covers raw materials, processes, and products utilizing cereal (corn, wheat, oats, rice, rye, and so forth), oilseeds and pulses, along with analytical procedures, technological tests, and basic research in the cereals area.

*Industries des céréales*. 1980– . Paris: Association pour le Progrès des Industries des Céréales (APIC). Bimonthly. ISSN: 0245-4505. Review devoted to research and development of cereal transformation. Chronicles market information and reviews new publications in the field.

*Journal of Cereal Science*. 1983– . New York: Academic Press. Bimonthly. ISSN: 0733-5210. Available from the World Wide Web at http://www.academicpress.com/jcs. Publishes papers of original research covering all aspects of cereal science related to the functional and nutritional quality of cereal grains and their products.

*Technical Bulletin*. 1979– . Manhattan, KS: American Baking Institute. Monthly. Each issue covers a specific topic in baking technology. A complete set of bulletins is available on CD-ROM and on the Internet.

## Dairy Products

*Australian Journal of Dairy Technology*. 1946– . Melbourne, Australia: Dairy Industry Association of Australia. Twice a year. ISSN: 0004-9433. Official journal of the association.

*Bulletin of the IDF*. 1960– . International Dairy Federation Brussels, Belgium: International Dairy Federation. 10 issues/year. ISSN: 0250-5118. Includes articles, short monographs, proceedings, and IDF News. Mostly one subject per issue.

*International Dairy Journal*. 1991– . Amsterdam: Elsevier. Monthly. ISSN: 0958-6946. Available from the World Wide Web at www.elsevier.com/locate/idairyj. Focuses on applied research and the interface of the dairy and food industries. Incorporates *Netherland Milk* and *Dairy Journal*.

*Journal of Dairy Research*. 1929– . Cambridge: Cambridge University Press. Quarterly. ISSN: 0022-0299. Original research on all aspects of dairy science.

*Journal of Dairy Science*. 1917– . Champaign, IL: American Dairy Science Association. Monthly. ISSN: 0022-0302. Available from the World Wide Web at http://www.adsa.org/jds. Journal of the association.

*La documentation fromagère*. *Résumé*. 1988– . Rennes, France: Institut technique de Gruyère. Monthly. Publication that specializes in monitoring technical and scientific developments in the cheese industry. Fact sheets summarizing information on a given topic are classified under 20 headings such as milk bacteria, salting and ripening, lactic starters, and so forth. Keyword index. Articles can be ordered from the ITG documentation service in Rennes.

*Le Lait*. (International Journal of Dairy Science and Technology). 1921– . Lyon, France: Lons-Le-Saunier. Bimonthly. ISSN: 0023-7302. Available from the World Wide Web at http://www.edpsciences.org/docinfos/INRA-LAIT. Publishes English- and French-language original scientific articles and technical notices on the microbiology, biochemistry, and physico-chemistry of milk and its derivatives and on the transformation procedures.

*Milchwissenschaft (Milk Science International)*. 1945– . Munich, Germany: VV-GmbH. Monthly. ISSN: 0026-3788. Well-known German publication with articles in English and a bibliography in German.

## Food Engineering

*Food Engineering*. 1928– . Highlands Ranch, CO: Cahners Business Information. 11 issues/year. ISSN: 0193-323x. Available from the World Wide Web at http://www.foodengineeringmag.com. Trade journal with articles on processing developments, plant management, marketplace, events, and issues.

*Food Processing.* 1940– . Chicago, IL: Putman Publishing Company. Monthly. ISSN: 0015-6523. Available from the World Wide Web at www.foodprocessing. com. Trade journal on strategy, technology, and trends.

*International Journal of Food Science and Technology.* 1966– . Oxford, UK: Blackwell Science. Bimonthly. ISSN: 0950-5423. Journal of the Institute of Food Science and Technology Trust Fund, UK. Covers the basic food science and includes research papers, critical topical reviews, and short communications.

*Journal of Food Engineering.* 1982– . Amsterdam: Elsevier. Monthly. ISSN: 0260-8774. Available from the World Wide Web at www.elsevier.com/locate/ jfoodeng. Original research and review papers on any subject at the interface between food and engineering, particularly those with relevance to industry.

*Journal of Food Process Engineering.* 1977– . Westport, CT: Food and Nutrition Press. Bimonthly. ISSN: 0145-8876. Covers engineering aspects of post-production handling, storage, processing, packaging, and distribution of food.

*Journal of Food Processing and Preservation.* 1977– . Westport, CT: Food and Nutrition Press. Bimonthly. ISSN: 0145-8892. Discussion of the latest knowledge and advances in processing and preservation, with a balance between fundamental aspects and applied food-processing procedures.

*Process. Magazine des technologies alimentaires.* 1945– . Rennes, France: Process. 11 issues/year. ISSN: 0998-6650. Includes chronicles, training, new publications, and events to come along. Includes information on the ingredients, processes, and equipment.

## Food Microbiology

*Food Microbiology.* 1984– . New York: Academic Press. Bimonthly. ISSN: 0740-0020. Available from the World Wide Web at http://www.academicpress. com/foodmicro. Publishes primary research papers, short communications, reviews, reports of meetings, book reviews, and news items dealing with all aspects of the microbiology of foods.

*International Journal of Food Microbiology.* 1984– . Amsterdam: Elsevier. 24 issues/year. ISSN: 0168-1605. Available from the World Wide Web at www.elsevier.com/locate/ijfoodmicro. Official journal of the International Union of Microbiological Societies (IUMS) and the International Committee on Food Microbiology and Hygiene (ICFMH).

*Journal of Food Protection.* 1937– . Ames, IA: International Association for Food Protection. Monthly. ISSN: 0362-028x. Available from the World Wide Web at http://www.foodprotection.org. Journal of the association.

*Journal of Food Safety.* 1977– . Westport, CT: Food and Nutrition Press. 4 issues/year. ISSN: 0149-6085. Covers the microbiological areas of food safety with emphasis on mechanistic studies involving inhibition, injury, and metabolism of food poisoning microorganisms, and the regulation of growth and toxin production in both model systems and complex food substrates.

## Meat and Meat Products

*Bulletin de liaison du CTSCCV.* 1991– . Maisons-Alfort, France: Centre Technique de la Salaison, de la Charcuterie et des Conserves de Viandes (CTSCCV). Bimonthly. ISSN: 1162-0676. Includes review articles, a digest of recently published articles, and a column on recent legislation and regulations in France. Copies of articles can be obtained from the documentation service of the CTSCCV.

*Fleischwirtschaft International.* 1920– . Frankfurt, Germany: Deutscher Fachverlag Gmb H. Monthly. ISSN: 0015-363x. All aspects of meat processing are covered from the production and processing up to marketing and quality management. Regular topics are production, preservation, packaging, feed, and storage.

*Journal of Muscle Foods.* 1990– . Westport, CT: Food & Nutrition Press. Quarterly. ISSN: 1046-0756. Focused primarily in the areas of postmortem technologies involving fat reduction in processed meats, meat pigment chemistry, factors that affect meat tenderness, histology of fresh meat, food safety, and microbiology.

*Meat Science.* 1977– . Amsterdam: Elsevier. Monthly. ISSN: 0309-1740. Available from the World Wide Web at http://www.elsevier.com/locate/meatsci. Official journal of the American Meat Science Association.

## METHODS

Quality control of food relies on methods of analysis. These methods are reviewed and supported by a large number of qualified people so they represent a true consensus of expert opinion. They offer a valid and recognized basis for control, evaluation, and practices.

American Oil Chemist's Society. 1985. *Official Methods and Recommended Practices of the American Oil Chemists' Society.* 2 vols. Champaign, IL: The Society. Over 400 analytical methods, illustrated and published in a loose-leaf format for easy updating.

American Society of Brewing Chemists. 1992. *Methods of Analysis of the American Society of Brewing Chemists*. 8th ed. St. Paul, MN: The Society. 586 p. This has been a valuable resource for brewing, malting, hops, and other allied industries. Published since 1936, it is in a loose-leaf format for easy updating.

*Approved Methods of the American Association of Cereal Chemists*. 1983– . 2 vols. St. Paul, MN: The Society. Up-to-date techniques for scientists working with grain-based ingredients. Information on instruments, chemicals, and equipment needed for each method is included. In a loose-leaf format for easy updating, it is also kept current with annual supplements. The product is available on CD-ROM and on the Internet.

Association of Official Analytical Chemists. 1998. *FDA Bacteriological Analytical Manual (BAM)*. 8th ed. Arlington, VA: AOAC International. Methods currently in use in U.S. Food and Drug Administration Laboratories for the microbiological analysis of foods. Print is in loose-leaf format for updating and is also available on CD-ROM.

Association of Official Analytical Chemists. 1995. *Officials Methods of Analysis of AOAC International*. 2 vols. Arlington, VA: AOAC International. Authoritative source of analytical methods used worldwide. In loose-leaf format for easy updating with supplements and also available on CD-ROM.

Collins, C.H., P.M. Lyne, and J.M. Grange. 1995. *Collins and Lyne's Microbiological Methods*. 7th ed. Oxford: Butterworth-Heinemann. Bench reference on technical methods for the examination and identification of bacteria, molds, and yeasts, and the microbiological examination of pathological materials and foodstuffs.

Downes, F.P. and K. Ito, eds. 2001. *Compendium of Methods for the Microbiological Examination of Foods*. 4th ed. Washington, DC: American Public Health Association. 600 p. Sixty-four chapters including general laboratory procedures; laboratory quality assurance; environmental monitoring procedures; sampling plans; sample collection; shipment; preparation for analysis; microorganisms involved in processing and spoilage of foods; foods and the microorganisms involved in their safety and quality; indicator microorganisms and pathogens; microorganisms and food safety; foodborne illness; preparation of microbiological materials—media, reagents, and stains; and much more.

Institute of Medicine (U.S.) Committee on Food Chemicals Codex. 1996. *Food Chemicals Codex*. One volume plus supplement. 4th ed. Washington, DC: National Academy Press. Accepted standards for quality and purity in food chemicals, officially referenced by the U.S. Food and Drug Administration and many agencies in other countries.

Richardson, G.H., ed. 1996. *Standard Methods for the Examination of Dairy Products*. 16th ed. Washington, DC: American Public Health Association. 546 p. Illustrated compilation of microbiological, chemical, and physical methods for analyzing dairy and dairy food substitutes.

## PATENTS

Patents play an important role in food processing. Scientists who work for food companies may not publish in journals, but they often apply for patents to protect their invention. Besides protecting inventors' creativity, patents also serve as a means for disseminating technology. Each patent document describes a new technological innovation that is made public to promote the exchange of knowledge. Several journals such as *Food Technology* review recent patents, but coverage is not exhaustive. Databases provide even more current patent information. Many patent offices have Internet sites, which also facilitates access to these types of documents. In addition, industrialized countries have patent offices where copies of their patents can be obtained and some even include patents from other countries.

The British company Derwent reviews and processes more than 1 million patents a year from 40 international patent authorities. Derwent World Patents Index (DWPI), produced by Derwent Information, provides access to information from more than 18 million patent documents, giving details of over 10 million inventions. About 200,000 of these patents related to food science. A companion file, Derwent Patents Citation Index (DPCI) details examiner and author patent citations for patents from the 16 largest patent-issuing authorities. Each patent is processed into patent families with English-language abstracts. Their databases cover over 40 patent-issuing authorities, with coverage of technology going back to 1963, and extended coverage of those areas of industry associated with chemistry such as engineering and pharmaceuticals.

Contact information:
14 Great Queen Street
London WC2B 5DF United Kingdom
Telephone: +44 0 20 7344 2800
Fax: +44 0 20 7344 2900
URL: http://www.derwent.com

The European Patent Office gives access to over 30 million patents from 19 European countries. Patent information centers are located in every member country.

Contact information:
European Patent Office
Erhardtstrasse 27
D-80331 Munich, Germany
Telephone: +49 89 23 99-0

Fax: +49 89 23 99 -44 65
URL: http://www.european-patent-office.org/index_f.htm
Questel-Orbit offers a database called QPAT·WW, which has the full text of all
U.S. patents issued since January 1, 1974 and European patents issued from 1987
to the present. This database has powerful search features.
Contact information:
Questel-Orbit
8000 Westpark Dr.
McClean, VA 22102
Telephone: +1 703-442-0900
Fax: +1 703-893-4632
URL: http://www.questel.orbit.com
U.S. Patent and Trademark Office (PTO) with its large USPTO Web Patents
Database offers access to the full text of U.S. patents issued since 1790.
Contact information:
General Information Services Division
U.S. Patent and Trademark Office
Crystal Plaza 3, Room 2C02
Washington, DC 20231 USA
Telephone: +1 703-308-4357
Fax: +1 703-305-7786
URL: http://www.uspto.gov/patft/index.html.

## PROCEEDINGS

Important conferences or congresses such as IUFoSt World Congress of Food
Science and Technology (now held biannually) are regularly held in the food-
science sector. The International Association for Cereal Science and Technology
(ICC) also holds congresses and symposia at regular intervals. Proceedings result
from the International Dairy Federation World Dairy Congress, held every four
years, and from the International Congress of Meat Science and Technology,
held annually. The Institute of Food Technologists has the IFT Annual Meeting
and IFT Food Expo every year, with the results published in a book of abstracts.
Sponsored by Rhodia Company, the annual Marschall Cheese Seminar Proceedings are now available on CD-ROM and through the Web at http://
www.rhodiadairy.com/marschall/proceed/index.htm.

## REVIEWS

Reviews present a digest of material by acknowledged authorities in the many
specialized fields of food science. In bringing together new information, reviews
help practitioners keep abreast of developments in the field and identify an extensive list of references.

*Advances in Food and Nutrition Research.* 1970—. San Diego, CA: Academic Press. Each issue contains five to six articles on all aspects of food science from wines to frozen fruits. Formerly called *Advances in Food Research*.

American Association of Cereal Chemists. 1976–90. *Advances in Cereal Science and Technology.* 10 vols. St. Paul, MN: American Association of Cereal Chemists. Includes 68 chapters in 10 volumes written by 137 authors from 115 countries.

Davies, R. 1982–88. *Developments in Food Microbiology.* 4 vols. London: Applied Science Publishers. Topical developments in various aspects of food microbiology as well as an introduction of new concepts from current research.

Fox, P.F. 1992–7, *Advanced Dairy Chemistry.* 2nd ed. 3 vols. New York: Chapman & Hall. V. 1: Milk proteins; v. 2: Milk lipids; and v. 3: Lactose, water, salts, and vitamins.

Hudson, B.J.F. 1982. *Developments in Food Proteins.* 7 vols. London: Applied Science Publishers. Presentation of new topics concerned with food proteins.

Lawrie, R. 1980–91. *Developments in Meat Science.* 5 vols. London: Applied Science Publishers. Selection of topics covering scientific progress in meat science.

Thorne, S. 1981–9. *Developments in Food Preservation*, 5 vols. London: Applied Science Publishers. Review of the current state of food preservation.

## STANDARDS

An important aspect of food science and technology is quality control, which relies on proven testing methods. A number of international standards apply to the food sector. Most countries have established quality standards for their products and these standards can be ordered from their National Standards Office. Only the major internationally recognized standards are presented in this chapter.

### Association Française de Normalisation

The Association française de normalisation (AFNOR) has published more than 900 standards on all aspects of food technology. It is a state-approved organization supervised by the Ministry of Industry and works to prepare, approve, promote, and facilitate the use of standards. The Web site provides access to standards; a database of French, European, and ISO standards; reference texts; and an online search capability for the catalog, including ordering, payment, and delivery information.

Contact information:
AFNOR Association française de normalisation
Tour Europe
92049 Paris la Défense Cedex France
Telephone: +33 1 42 91 55 55
Fax: +33 1 42 91 56 56
URL: http://www.afnor.fr (French version); http://www.afnor.fr/index_gb.htm (English version)

## Codex Alimentarius

The *Codex Alimentarius* (Latin for food law or code) is a set of international food standards adopted by an international commission established in 1962 by the FAO and the WHO. These standards apply to most food products—processed, semiprocessed, and unprocessed. The *Codex* includes provisions concerning food hygiene and nutritional quality, especially microbiological specifications, provisions respecting food additives, pesticide residues, contaminants, labeling, presentation, and analysis and sampling methods. The *Codex* also contains reference documents on codes of practice, guidelines, and other recommended measures. The complete edition of the *Codex Alimentarius* consists of 13 volumes. An abridged version containing only the food standards also has been published. It is a quick reference work that complements the original edition. The CD-ROM version has retrieval software and text in English, French, and Spanish. Available from the World Wide Web at http://www.fao.org/waicent/faoinfo/economic/esn/codex/Default.htm.

## International Dairy Federation

The International Dairy Federation (IDF) is an organization made up of 36 national committees representing the various aspects of the dairy industry in their respective countries. It publishes the *Bulletin of IDF*, standards, books, and news items. The organization publishes over 300 bulletins and close to 200 standards, mostly in English and French, that are developed by teams of experts. An annual subscription to the *Bulletin of IDF* entitles the subscriber to the bulletins, standards, books, and news items published in the course of the year. The Federation holds the International Dairy Convention every four years.
Contact information:
International Dairy Federation
41 Square Vergote
1030 Brussels, Belgium
Telephone: +32 2 733 9888
Fax: +32 2 733 0413
URL: http://www.fil-idf.org

## International Organization for Standardization

The International Standards Organization, (ISO) is a worldwide federation of over 100 national standards bodies. ISO's work leads to international agreements that are published in the form of international standards. Export-oriented agri-food companies have long recognized that agreeing on international standards will facilitate world trade. ISO has published a series of food technology standards that can be obtained from the standardization agencies of ISO member countries. It is possible to search, order, and receive standards through the Web site at http://www.iso.ch.
Contact information:
ISO Central Secretariat
International Organization for Standardization
1, rue de Varembé, Case postale 56
CH-1211 Geneva 20, Switzerland
Telephone: +41 22 749 01 11
Fax: +41 22 733 34 30

## TRADE LITERATURE AND EXPOSITIONS

The content of trade literature is ephemeral, but it is an important source of current information on commercial aspects of food technology.

### Agra Europe

Agra Europe has been a leading publisher of newsletters, special reports, and market studies on the food industry since 1963. Their weekly publication *FOOD-NEWS* supplies market information and covers the world market for fruit juice concentrates; canned and frozen foods; tomato products; food ingredients; dried fruit and nuts; and dairy products. The publication also offers daily news services.
Contact information:
Agra Europe (London) Ltd.
80 Calverley Road, Tunbridge Wells
Kent, TN1 2UN United Kingdom
Telephone: +44 1892 533813
Fax: +44 1892 544895
URL: http://www.agra-europe.com

### Centre Français du Commerce Extérieur

The Centre français du commerce extérieur (CFCE) publishes several works on exporting that provide economic surveys on a number of countries. Titles such as *Le secteur agro-industriel au Maroc*, *La filière fruits et légumes transformés*

*en Roumanie*, or *Marché des produits laitiers de base dans 58 pays* provide information on the national markets. The bookshop "International Trade" distributes publications of such organizations as the FAO, the General Agreement on Tariffs and Trade (GATT), and the Organization for Economic Cooperation and Development (OECD).
Contact information:
La Librairie du Commerce International
BP 438, 75233 Paris Cedex 05 France
Fax: +33 1 4336 4798
URL: www.cfce.fr

## Euromonitor

Publishes over 200 publications on market analysis, production statistics, and industry studies. Several of the publications deal with food processing, such as *The World Market for Confectionery*; *The World Market for Dairy Products*; and *The European Food Databook*. These publications have a good reputation and while expensive, they should be considered.
Contact information:
Euromonitor
60-61 Britton St.
London, ECIM 5NA United Kingdom
Telephone: +44 0 171-251 1105
Fax: +44 0 17 1251 0985
URL: http://www.euromonitor.com

## Food and Agricultural Organization of the United Nations

The Division of the Statistics of FAO publishes several directories related to commodities marketing such as the *FAO Trade Yearbook* and the *FAO Production Yearbook*. These works comprise a collection of statistical data and other information on the international trade of basic agricultural produce for all countries and territories of the world. The two works are trilingual: English, Spanish, and French. In most countries there are one or more national distributors of FAO publications. FAO also offers the FAOSTAT database at http://apps.fao.org/lim500/agri_db.pl, which contains over 1 million worldwide time-series statistics dating from the most recent statistics back to 1961.
Contact information:
Sales and Marketing Group, Information Division
FAO, Viale delle Terme di Caracalla, 00100 Rome, Italy
Telephone: +39 06 5705 1
Fax: +39 06 5705 3360
URL: http://www.fao.org

## EXPOSITIONS

Fairs and expositions are significant sources of information. Such events are often where a company will choose to launch a new product. They also provide an opportunity to discuss food issues with providers or potential buyers in order to better know the competitors or the key stakeholders in a sector. A majority of trade journals announce upcoming events and some publish special issues for these occasions.

### ANUGA FoodTec—International Food Technology Fair

Held every three years in Cologne, Germany. Provides a venue for dialogue between the food technology sector and the food industry. Previous fairs have attracted 1,074 exhibitors and 40,000 visitors.
Contact information:
Messe und Ausstellungs Ges. mbH Koln
Messeplatz 1, 50679 Koln Germany
Telephone: +49 2 21 8210
Fax: +49 2 21 2574
URL: http://www.anugafoodtec.com

### FOODEX—International Food and Beverage Exhibition

Secretariat of FOODEX Japan

Held yearly by the Secretariat of FOODEX Japan, previous exhibitions have attracted 2,300 exhibitors with 90,000 visitors.
Contact information:
Japan Management Association
3-1-22, Shiba-Koen, Minato-ku
Tokyo 105-8522 Japan
Telephone: +81 3 3434 3453
Fax: +81 3 3434 8076
URL: http://www.jma.or.jp/foodex

### Institute of Food Technologists Annual Meeting and Food Expo

Held yearly. Previous expos have attracted 2,400 exhibitors with 24,000 visitors.
Contact information:
IFT—Meetings and Expositions Department
221 N. La Salle St.
Chicago, IL 60601 USA
Telephone: +1 312-782-8424

Fax: +1 312-782-8348
URL: http://www.ift.org

## SIAL (The Global Food Market Place)

Held every two years this is the largest international food exhibition with 5,000
exhibitors and approximately 135,000 visitors.
Contact information:
International Trade Exhibitions in France
39 Rue de la Bienfaisance
75008 Paris France
Telephone: +33-1-42-89-46-87
URL: http://www.sial.fr

## WORLD WIDE WEB SITES

This is a selected list of Internet sites. Many food science departments at academic
institutions and trade journals now have their own sites that link to many other
Web sites in the area of food science.

AgriWeb (Canada) (http://www.agr.gc.ca).   Directory of online resources in
Canadian Agriculture and Agri-Food. Click on ''Links.''

Food Navigator (http://www.foodnavigator.com/).   European-driven portal on
food ingredients, aims to provide a European view of the food technology industry. The site offers five divisions of expertise in the fields of food Internet resources, market trends, food technology papers, ingredients and industry databases, and daily food news. Sponsored by many leading ingredients companies,
it includes access to 12,000 Web sites.

Food and Nutrition Internet Index (http://www.fnii.ifis.org/).   Produced by the
International Food Information Service, the Food and Nutrition Internet Index
(FNII) is a searchable Web site describing and indexing food and nutrition resources available on the Internet. The main focus is on food science, food technology, and human nutrition, although there is also reference to food business and
company information. FNII has a worldwide perspective and provides informative summaries and comprehensive indexing. Access requires a fee and password.

INFOMINE      (http://infomine.ucr.edu/search/bioagsearch.phtml).   Contains
highly useful Internet/Web resources including databases, electronic journals,
electronic books, bulletin boards, listservs, online library card catalogs, articles,
directories of researchers, and many other types of information.

Institute of Food Science and Technology (http://www.easynet.co.uk/ifst/re-
source.htm).   Useful links to external food-related World Wide Web resources.

Institute of Food Technologists—Food Laws and Regulations Division (http://www.ift.org/divisions/food_law/jumpmain.htm). Contains links to a variety of resources on food law and regulations. International coverage.

Réseau Technologie et Partenariat en Agroalimentaire (France) (http://www.gret.org/tpa/liens/liens/kso.htm). The objectives are to promote exchanges and circulation of information by food production enterprises in developing countries, particularly Africa.

## BIBLIOGRAPHY

Bernard F. Importance stratégique de l'information. In: L'entreprise agro-alimentaire: assurer sa croissance. St-Jean sur Richelieu, Canada: Éditions du Monde Alimentaire. 1996, pp. 202–268.

Blanchfield J.R. Food Science and Technology on the Internet. Trends in Food Science & Technology 7: 1–8, 1996.

Davidson B. et al. Internet Access to Information on Food Trends and Technology. In: Proceedings of Food Ingredients Europe: Conference proceedings, London, November 4–6. 1997. Maarssen, The Netherlands: Miller Freeman, 1998, pp. 177–190.

Gazeay M. and C. Gougeon. L'Europe de l'Agro-Alimentaire. Paris: ABDS Editions, 1994.

Green S. Keyguide to Information Sources in Food Science and Technology. London: Mansell Publishing, 1985.

Green S. and G.P. Lilley, eds. Food Science. In: Information Sources in Agriculture and Food Science. London: Butterworths, 1981, pp. 477–499.

Hill S. The IFIS Food Science and Technology Bibliographic Databases. Trends in Food Science & Technology 11: 269–271, 1991.

Kernon JM. The Foodline Scientific and Technical, Marketing and Legislation Databases. Trends in Food Science & Technology 11: 276–278, 1991.

Metcalfe J. Food Information on the Internet—An Overview for FI Europe '97. In: Proceedings of Food Ingredients Europe: Conference proceedings, London, November 4–6, 1997. Maarssen, The Netherlands: Miller Freeman, 1998, pp. 182–184.

Rudge K. The Web That We Weave. International Food Ingredients 4: 20–23, 1998.

# 11

## Grey Literature and Extension Resources

**Sheila D. Merrigan**
University of Arizona, Tucson, Arizona, USA

**Tim McKimmie**
New Mexico State University, Las Cruces, New Mexico, USA

The inclusion of a separate grey literature chapter in this book attests to the value of this literature. This chapter is relevant to all the other chapters in this book because each agriculture field has an element of grey literature. In recent years, grey literature has attracted attention in the information professions. Although some may tend to think of grey literature as a minor component of information resources, in terms of quantity, grey literature may sometimes represent the majority of all information available on a given topic. In terms of quality, of course, there are no substitutes for scholarly articles and monographs. However, in many cases the sought-after information may not be found in scholarly sources. The information seeker must then decide whether to pursue the myriad of grey literature sources that may be the sole source of the needed information. This chapter suggests that information seekers who are willing to take the time to venture into grey literature may find highly relevant and useful information.

Grey literature is by nature problematic, having in fact, no generally agreed upon definition. Grey literature is usually difficult to identify and is often not indexed by traditional indexing services. Sometimes it may be identifiable but is not accessible or obtainable. Grey literature is often written for a specific audience or a specific purpose. For example, information on controlling a crop-damaging pest may be critical to farmers in a localized area, but of little interest to anyone else.

Grey literature has been described as fugitive, informal, nonconventional, and ephemeral, or in other words, literature that "falls through the cracks." For the purposes of this book, we define grey literature rather loosely, and include many traditional databases as sources of grey literature. Ideally, the producers of these traditional indexes would possess all of the indexed documents (whether grey or not) and would be willing to loan or copy such documents if access was not available elsewhere.

It is difficult to categorize different types of literature as being "grey" based on format, publisher, or subject matter. Grey literature may be in print or electronic format. Some examples of literature that commonly fall into the grey category include corporate documents, discussion papers, in-house journals and newsletters, surveys, working papers, technical reports, trade association publications, institutional or association reports, and bulletins. Some conference proceedings, academic reports, and government reports are also considered to be grey literature. However, many government reports and conference proceedings are published as books or journal articles, thereby eliminating those particular documents from grey literature.

Extension literature is included in this chapter because extension publications are often not accessible through traditional information tools such as library catalogs or bibliographic databases. However, extension publications may be accessed through specialized databases, extension personnel, or publication offices of land-grant universities. Many universities offer excellent points of access to extension literature. For others, access may be difficult. Therefore, extension literature from some states might be considered "grey," but could be considered "white" from others. This chapter will delineate the various ways to access extension literature.

Identification, accessibility, and utility are of primary importance to the information seeker. Our use of the term *accessibility* assumes that the information seeker has access to modern Internet technology and the services of a research library or information professional. Theoretically, the use of the World Wide Web, search engines, and other Internet technology increases our ability to identify and access grey literature. The Web provides access to a global sphere of information on any conceivable topic, often in full text and with links to related information. On the other hand, the Web also contains information of little or no value, reference to material that may not be available, and links to sites and information that may disappear. The promise of the Web as the great organizer and access tool for grey literature has not yet materialized. If agriculture information specialists, however, were to focus on making grey literature more available, the Web would be a natural forum.

There are literally thousands of points of access to grey and extension literature throughout the world and it would be impossible to produce a comprehensive list of sources. Corporate agricultural materials account for a large volume of literature. Because of the proprietary nature of this material, however, access may be difficult or impossible. Documents from nongovernmental and nonprofit organizations are also numerous although perhaps not as difficult to access. This chapter presents the major gateways and access tools to grey and extension literature. It also discusses strategies for identifying grey literature, some of which may not be readily apparent. Sources of information regarding access to grey literature are presented first, followed by sources for accessing extension information. Grey literature may be more difficult to locate than traditional literature, but for a given project, grey literature may often represent the most useful material available.

## STRATEGIES FOR FINDING GREY LITERATURE

Researchers, librarians, and other information seekers often do not search grey literature unless their need for information is not fulfilled by a search of traditional sources. When searching for agricultural information, for example, the information seeker will often first use traditional databases such as AGRICOLA, CAB Abstracts, Science Citation Index, and WorldCat. The search may end there, particularly if the researcher finds material that answers the information need(s). If no information is found, there may be various responses. The lack of information may be good news for it may signify that the researcher is indeed entering new, unexplored territory. By their nature, scientists seek to work on research topics on which little or no information has been published. Thus, it is possible that the literature review is not expected to retrieve results. Further, it may be the opinion of some researchers that refereed, indexed publications are the only valid source of published literature. Therefore, the researcher may be satisfied when no publications are located and may proceed with the investigation without taking the time to search the grey literature.

There are many reasons to take additional steps and search the grey literature, though. Many information seekers want to identify any available information, regardless of how obscure it is. One of the most common reasons for turning to grey literature is the seeker's conviction that there is information available on the topic. The seeker may have heard of a study or may feel that someone must have looked at the problem at hand and made the results available somewhere. Searching grey literature is likely to open the researcher to new horizons of information. Unindexed periodicals, agency reports, unpublished manuscripts, Web page reports, pamphlets, and limited-distribution conference proceedings are examples of grey literature that may have scholarly as well as practical value.

It is worthy of note that many seekers of grey literature may not be researchers at all. These users may be interested in applied information rather than research and may not be concerned with format or authority. Much information relevant to organic farming, for example, is buried in the grey literature of agriculture. Regardless of the needs of the clients, information providers such as reference librarians should be well versed in providing access to grey literature. Indeed, many librarians and other information professionals will have their own grey literature collections.

There are steps that one can take to facilitate the identification of grey literature sources. One of the first tools to use in identifying grey literature may be the telephone. If little or no information on a particular topic can be found using traditional sources, a telephone call to an organization or expert in that field may produce a plethora of information.

Personal contacts are an extremely useful source of knowledge on specific agricultural subjects. Extension agents are a good example of experts who have accumulated practical knowledge that may not exist on paper. Many experts keep extensive grey literature files. As with the telephone, e-mail to respected or knowledgeable colleagues and acquaintances might quickly provide details on sources of information.

Listservs on the Internet can be very useful in tracking grey literature. Listservs are, generally, discussion groups set up for a specific topic. Listservs are communication tools that may be used not only to identify literature and answer questions, but they may also be useful for brainstorming sessions that may result in solutions to problems. USAIN-L, IAALD-L, or STS-L are appropriate for agricultural information questions. USAIN-L is available from the United States Agricultural Information Network (USAIN), IAALD-L from the International Association of Agricultural Information Specialists (IAALD), and STS-L from the Science and Technology Section of the Association for College and Research Libraries (ACRL).

In the vast ocean of grey literature, the Web can be a useful resource for quickly identifying an expert or organization as well as providing contact information. Most of the major agricultural societies maintain Web pages. (Due to the large number of agricultural-related societies and the ease of locating these societies on the Web, individual societies are not listed in the following bibliographic section of this chapter.) Examples include the American Society of Animal Science, American Society for Horticultural Science, American Society of Agronomy, and the American Association of Family and Consumer Sciences. These sites often include sources of information on various topics including jobs, conferences, and publications. Web search engines are effective tools for locating information, sometimes in full text. Search strategies differ depending on the search engine, and irrelevant material may be a problem, but the patient searcher should be able to develop a degree of competence that will lead to successful searches.

Web sites such as AgNIC that compile links to Internet sites and documents, provide an enormous wealth of information, much of it grey literature. The goal of AgNIC is to eventually provide links to information for the entire spectrum of agriculture. AgNIC information includes not only factual and research-related information but also items such as directories of expertise, calendars of events, and conference information. Additionally, AgNIC participants agree to answer questions in their area of expertise using e-mail. One of the best examples of the potential of this approach is the Managing Rangelands site from the University of Arizona (http://ag.arizona.edu/agnic/). In addition to information links, the site includes a range-management textbook, maps and other research tools, a full-text journal, and practical range-management information.

Theses and dissertations should also be consulted as sources of grey literature. Graduate students often turn up obscure reports, personal communications, and other sources of information not indexed in traditional databases. Of course, some of this material may be very difficult to access and may require contact with the author or other persons or agencies.

*Directories in Print* and *The Encyclopedia of Associations* list associations by subject. These works provide contact information for persons with subject expertise and list association publications, which often fall into the category of grey literature. Many associations also sponsor conferences. They may also publish proceedings that are distributed to conference attendees, but not necessarily indexed. A directory of particular interest is the *Agriculture Information Resource Centers*. It describes nearly 4,000 sites worldwide that are potential sources of agricultural grey literature. The information professional can also take the opportunity to get on mailing lists of publishers in specialty areas.

Trade literature is another unique form of grey literature. It is generally the publications from commercial sources, small associations, societies, or organizations on a specific topic such as Arabian horses, pesticides, or roses. Trade literature often takes the form of magazines. The articles in these magazines may or may not be reviewed and generally are written by professionals or experts (but generally not academics) within the subject matter. Much of the trade literature is not indexed, but USAIN is making an attempt to bring some of these titles together through a list on their Web site (http://usain.org).

The bibliography section that follows, provides a starting point in searching the grey literature. There you will find references to indexes and databases, associations, directories, guides, journals, and Web sites. Just as it would be impossible to list all avenues of grey literature information it would also be presumptuous of us to suggest that there is any one formula for searching it. The wise seeker will use knowledge, intuition, technology, networking, and personal contacts to locate the best information available. Even then, the elusive nature of grey literature may leave us in doubt as to whether some stones were unturned. Fortunately we have a tool of enormous potential, the Internet, that promises to make our task much easier. Not only does it globalize information resources, regardless of their format, quality, quantity, or subject matter, but it also promises a semblance of equality of access regardless of economics.

## Abstracts, Indexes, and Databases

Many of the general abstracts, indexes, and databases of agriculture include access to some grey literature. AGRICOLA contains citations to grey materials such as technical reports, bulletins, conference proceedings, trade literature, and some extension publications that may not be easily obtained. AGRICOLA does not contain the materials, but it does identify and help locate them. AGRIS collects bibliographic references (to date, about 3 million) to either conventional (journal articles, books) or nonconventional materials (theses, reports) not available through normal commercial channels. AGRIS encourages the exchange of information among developing countries, whose literature is often not covered by other international systems. Dissertations often contain citations to grey literature so use of Dissertation Abstracts is also an important tool for identifying the grey literature. Biological Abstracts/RRM provides access to worldwide meeting literature as well as literature reviews, reports, patents, and CD-ROM and other software. CAB Abstracts includes learned, professional, and trade journals, reports, bulletins, monographs, conference proceedings, and theses. CARIS was created by the Food and Agriculture Organization of the United Nations (FAO) in 1975 to identify and to facilitate the exchange of information about current agricultural research projects being carried out in, or on behalf of, developing countries and CARIS is the USDA's documentation and reporting system for ongoing and recently completed research projects in agriculture, food and nutrition, and forestry. Finally, Science Citation Index/Web of Science indexes references to articles that include a great deal of grey literature.

The sources listed here provide indexed access to agricultural grey literature (and often to traditional agricultural literature as well). Each citation includes a description of the database content, publisher, and Web addresses, when available. When possible, the originating organization or agency information is given. Many of the databases are available free of charge and, if this is the case, it is so noted in the citation information. Those sources that are fee-based may be available from the database producer or from one or more database vendors. Prices may vary widely either for subscriptions or per use, even for the same product. Some sources are available in multiple formats including print, CD-ROM, tapeload, or Web access.

Agri2000 (http://orton.catie.ac.cr/agri2000en.htm). Agricultural Information and Documentation System for America (SIDALC). Free. Agri2000 contains Latin America and Caribbean agriculture literature from national and international research and educational institutions.

Agriculture and Environment for Developing Regions (TROPAG & RURAL) (http://www.kit.nl/information_services/html/agriculture_environment.asp). Koninklijk, The Netherlands: Instituut Voor de Tropen (Royal Tropical Institute). This database contains literature relating to the practical aspects of agriculture and rural development in tropical and subtropical regions. Regional coverage encompasses Africa, Asia and the Pacific, Latin America, and the Caribbean. Updated with 4,000 to 6,000 abstracts per year.

ANRO (http://www.infoscan.com.au/contents/index.html). Australian Agriculture and Natural Resources Online. Free. ANRO is a research information service with access to several different agency databases including Australian Rural Research in Progress (ARRIP), Australian Bibliography of Agriculture (ABOA), and Streamlines (Australia's Natural Resources). The databases contain citation information to reports, books, conferences, dictionaries, fact sheets, journal articles, legal information, maps, pamphlets, research projects, reviews, serials, and theses.

Current Web Contents (http://www.isinet.com/isi/products/cc/cwc/webselect.html). Institute for Scientific Information (ISI). Current Web Contents is an ever-growing collection of high-quality scholarly Web sites. ISI editorial experts review Web sites based on authority, accuracy, and currency. Agricultural Web sites such as AgNIC are included in this database.

ERIC (http://www.accesseric.org/searchdb/searchdb.html). National Library of Education. Washington, D.C. Free. ERIC is supported by the U.S. Department of Education's Office of Educational Research and Improvement and is administered by the National Library of Education (NLE). Included in the database are citations to agricultural and rural education literature. The ERIC index includes two types of publications: journals and ERIC documents on microfiche. The articles come from over 750 education-related journals. ERIC microfiche documents include materials that often are not published elsewhere such as curriculum guides, research reports, and conference papers. Abstracts are included and coverage is from 1966 to the present.

ICAR (Inventory of Canadian Agri-Food Research) (http://res2.agr.ca/icar/). Canadian Agriculture Research Council (CARC). Free. ICAR is a comprehensive database for agriculture and food research in Canada. It describes current research projects from industry, universities, and provincial and federal establishments.

Food Science and Technology Abstracts (FSTA) (http://www.ifis.org)/. International Food Information Service (IFIS). FSTA is an international database that covers food science, food technology, and food-related human nutrition for all commodities. FSTA is a bibliographic database with abstracts and contains information from scientific journals, patents, books, conference proceedings, reports, theses, standards, and legislation. Contains over 560,000 bibliographic records from 1969 to present.

GrayLIT Network: A science portal of technical reports. (http://www.osti.gov/graylit/). Developed by the U.S. Department of Energy's (DOE) Office of Scientific and Technical Information in collaboration with the U.S. Department of Defense (DOD), National Aeronautical and Space Administration (NASA), and Environmental Protection Agency (EPA). Free. The GrayLIT Network is a portal for technical report information generated through federally funded research and development projects. It provides access to grey literature (much of which is full text) of federal agencies such as DOD/DTIC, DOE, EPA, and NASA. Participation will be expanding as the site develops.

NTIS (http://www.ntis.gov/). U.S. National Technical Information Service. Free. NTIS offers bibliographic and full descriptive summaries of more than 2 million reports received from government agencies since 1964. It covers U.S. government and worldwide government-sponsored research. The NTIS database covers engineering; biotechnology; computers; the environment; health and safety; business; and the physical, biological, and social sciences. Types of materials include technical publications, electronic data files, audiovisuals, software, and standards. Many items found within the NTIS database are difficult to locate elsewhere. Most documents abstracted in the database can be ordered from NTIS.

Orton Online Library Database (http://www.catie.ac.cr/library/default.htm). Orton Commemorative Library. Free. The Orton Commemorative Library was founded in 1943 and is currently jointly administered by the Inter-American Institute of Cooperation for Agriculture (IICA) and the Tropical Agricultural Research and Higher Education Center. The Orton database contains information on general agriculture, forestry and agroforestry, animal husbandry, natural resources, biodiversity, sustainability, soil science, coffee, cacao, and related sciences, with an emphasis on nonconventional literature.

Rural Development Abstracts (http://www.cabi.org/publishing/products/journals/abstract/rda/index.asp). CAB International. First issued in 1978, Rural Development Abstracts covers all economic and social aspects of rural development in the developing world. Types of material include journal articles, reports, conference proceedings, and books. A subset of CAB Abstracts.

SIGLE (System for Information on Grey Literature in Europe) (http://www.kb.nl/infolev/eagle/frames.htm). European Association for Grey Literature in Europe (EAGLE). SIGLE is a multidisciplinary bibliographic database that covers technology, natural sciences, biology and medicine, economics, and social sciences and humanities. SIGLE is a unique source for grey literature and contains over 364,000 records. Several national information and documentation centers in European Community (EC) countries such as the United Kingdom, France, Germany, Belgium, Ireland, Italy, the Netherlands, Spain, and Sweden contribute to the database.

## Agencies, Associations, and Centers

There are many more agricultural agencies, associations, and centers throughout the world than are possible to list here. Most countries have some sort of department or ministry of agriculture and many have Web sites. It is the authors' desire that the resources listed here will provide a starting point for finding these types of agricultural entities.

Agricultural Communicators in Education (ACE) (http://www.aceweb.org). ACE is an international association of professionals who practice in all areas of agricultural communication.

Agricultural Information and Documentation System for America (SIDALC) (http://orton.catie.ac.cr/defaulten.htm). SIDALC was established in order to strengthen and support the different documentation units and centers in various countries and, in collaboration with them, create real networks at the national, regional, and hemispheric levels. SIDALC attempts to satisfy information needs of the scientific community, students, development institutions, and the business and rural communities. The following countries have been integrated in to the system: Argentina, Barbados, Bahamas, Belize, Bolivia, Brazil, Chile, Colombia, Costa Rica, Cuba, Dominica, Ecuador, El Salvador, Guatemala, Guyana, Haiti, Honduras, Jamaica, Mexico, Nicaragua, Panama, Peru, Dominican Republic, Santa Lucia, St. Kitts and Nevis, Surinam, S. Vincent and Grenadines, Trinidad and Tobago, Uruguay, and Venezuela.

Agriculture and Agri-Food Canada (http://www.agr.ca/index_e.phtml). Agriculture and Agri-Food Canada provides information, research and technology, and policies and programs to achieve security of the food system, health of the environment, and innovation for growth.

AGROPOLIS (http://www.agropolis.fr/). AGROPOLIS is an international center for research and higher education in agriculture. AGROPOLIS's principal objective is the economic and social development of Mediterranean and tropical regions. It is made up of more than 200 laboratories, 300 researchers in the Montpellier region, and 600 international scientific groups in 60 countries.

Alternative Farming Systems Information Center (AFSIC) (http://www.nalusda.gov/afsic/). National Agricultural Library (NAL) (U.S.). AFSIC is an information center at the NAL specializing in locating and accessing information related to alternative cropping systems including sustainable, organic, low-input, biodynamic, and community-supported agriculture.

Animal Welfare Information Center (AWIC) (http://www.nalusda.gov/awic/). NAL (U.S.). AWIC is an information center at the NAL specializing in information for improved animal care and use in research, teaching, and testing.

Appropriate Technology Transfer for Rural Areas (ATTRA) (http://www.attra.org/). National Center for Appropriate Technology. This center provides technical assistance to farmers, extension agents, market gardeners, agricultural researchers, and other agricultural professionals.

Asian Vegetable Research and Development Center (AVRDC) (http://www.avrdc.org.tw/). This center was established to improve the nutrition, health, and incomes of people in developing countries. Its goal is to develop environmentally safe and sustainable vegetable production technologies that can be adapted by national agricultural research systems.

Consultative Group on International Agricultural Research (CGIAR) (http://www.cgiar.org/). CGIAR's mission is to contribute to food security and poverty eradication in developing countries through research, partnership, capacity building, and policy support. CGIAR promotes sustainable agricultural development based on the environmentally sound management of natural resources. Sixteen international agricultural research centers make up the global network known as CGIAR. A map of the centers can be found at http://www.cgiar.org/centers.htm. Each center has publications or a database available.

CIAT: Centro Internacional de Agricultura Tropical (International Center for Tropical Agriculture) (http://www.ciat.cgiar.org/). Cali, Columbia.

CIFOR: Center for International Forestry Research (http://www.cifor.cgiar.org/). Bogar, Indonesia.

CIMMYT: Centro Internacional de Mejoramiento de Maiz y Trigo (International Maize and Wheat Improvement Center) (http://www.cimmyt.org/). Mexico City, Mexico.

CIP: Centro Internacional de la Papa (International Potato Center) (http://www.cipotato.org/). Lima, Peru.

ICARDA: International Center for Agricultural Research in the Dry Area (http://www.icarda.cgiar.org/) Aleppo, Syrian Arab Republic.

ICLARM: International Center for Living Aquatic Resources Management (http://www.cgiar.org/iclarm/). Penang, Malaysia.

ICRAF: International Centre for Research in Agroforestry (http://www. icraf.cgiar.org/). Nairobi, Kenya.

ICRISAT: International Crops Research Institute for the Semi-Arid Tropics (http://www.icrisat.org/). Andhra Pradesh, India.

IFPRI: International Food Policy Research Institute (http://www. ifpri.cgiar.org/). Washington, DC.

IITA: International Institute of Tropical Agriculture (http://www.cgiar.org/ iita/). Ibadan, Nigeria.

ILRI: International Livestock Research Institute (http://www.cgiar.org/ ilri/). Nairobi, Kenya.

IPGRI: International Plant Genetic Resources Institute (http://www. ipgri.cgiar.org/). Rome, Italy.

IRRI: International Rice Research Institute (http://www.cgiar.org/irri/). Makati City, Philippines.

ISNAR: International Service for National Agricultural Research (http:// www.cgiar.org/isnar/). The Hague, Netherlands.

IWMI: International Water Management Institute (http://www.cgiar.org/ iwmi/). Colombo, Sri Lanka.

WARDA: West Africa Rice Development Association (http://www. warda.cgiar.org/). Bouaké, Côte d'Ivoire.

CSREES State Partners (http://www.reeusda.gov/1700/statepartners/usa.htm). This is a directory of land-grant universities that are state partners of the Cooperative State Research, Education, and Extension Service (CSREES). Also included is the CSREES Online Directory of Professional Workers in Agriculture, the State Extension Service Directors and Administrators Directory, as well as links to the Web sites of the schools of forestry, higher education, human sciences, veterinary sciences, and state extension services and state experiment stations.

Deutsche Forschungsgemeinschaft (http://www.dainet.de/zbl/tausch.htm). National Agricultural Library of Germany. The National Agricultural Library of Germany has a 150-year tradition in collecting agricultural literature. Since WWII they have collected agricultural literature worldwide—especially grey literature.

European Association for Grey Literature in Europe (EAGLE) (http://www. kb.nl/infolev/eagle/frames.htm). EAGLE is a cooperative network for identification, location, and supply of grey literature. It is a nonprofit association formed by the national centers participating in SIGLE. There are national SIGLE centers in 15 countries. EAGLE produces and provides access to SIGLE, a multidisciplinary, bibliographic database. (See also SIGLE entry on p. 329.)

Food and Agriculture Organization of the United Nations (FAO) (http://www. fao.org/). The largest autonomous organization within the United Nations, the FAO has worked to alleviate poverty and hunger by promoting agricultural development, improved nutrition, and the pursuit of food security.

Food and Nutrition Information Center (FNIC) (http://www.nal.usda.gov/ fnic/). NAL (U.S.). One of several information centers at the U.S. NAL. FNIC provides resource lists, databases, and many other food and nutrition–related links.

Inter-American Institute for Cooperation on Agriculture (IICA) (http:// iicanet.org/). IICA is the specialized agency for agriculture of the inter-American system of the Organization of the American States. Its mission is to provide cooperative services for agriculture and to strengthen and facilitate inter-American dialogue.

Inter-American Reference Center for Information on Agriculture (CRIIA) (http:// iicanet.org/criia/). CRIIA facilitates the transmission and exchange of information for decision-making policies of the public and private organizations of the member states of the IICA. CRIIA's goal is to make information accessible and enable members to define their investment, production, and trade strategies. This will improve competition levels, increase incomes, and lessen poverty within the context of sustainable development of agriculture and the rural environment.

International Association of Agricultural Information Specialists (IAALD) (http://www.lib.montana.edu/~alijk/IAALD.html). IAALD facilitates professional development of, and communication among, members of the agricultural information community worldwide. IAALD's goal is to enhance access to and use of agriculture-related information resources.

Secretaría de Agricultura, Ganadería, Desarrollo Rural, Pesca y Alimentacion (Secretary of Agriculture, Animal Science, Rural Development, Fisheries, and Nutrition) (http://www.sagarpa.gob.mx/index3.htm). This site provides information on agricultural policies, research, and programs in Mexico.

Sustainable Agriculture Network (SAN) and Sustainable Agriculture Research and Education Program (SARE) (http://www.sare.org/). SARE is a USDA-funded initiative that sponsors competitive grants for sustainable agriculture research and education. SAN is the communications and outreach arm of the SARE program, and is dedicated to the exchange of scientific and practical information on sustainable agriculture systems.

Tropical Agricultural Research and Higher Education Center (CATIE) (http:// www.catie.ac.cr/catie/). CATIE is an international, nonprofit civil association. Its main purpose is research, higher education, and outreach in agricultural sciences, natural resources, and related subjects in the American tropics. CATIE provides access to two databases, the Orton Library's database and a tree seed database.

United States Agricultural Information Network (USAIN) (http://usain.org/ usain.html). USAIN is an organization for information professionals that provides a forum for discussion of agricultural issues, takes a leadership role in the formation of a national information policy as related to agriculture, makes recommendations to the NAL on agricultural information matters, and promotes cooperation and communication among its members. USAIN hosts biannual conferences focusing on agriculture information.

United States Department of Agriculture (USDA) (http://www.usda.gov/). USDA's mission is to enhance the quality of life for the American people by supporting production of agriculture. This is accompanied by ensuring a safe, affordable, nutritious, and accessible food supply; caring for agricultural, forest, and range lands; supporting sound development of rural communities; providing economic opportunities for farm and rural residents; expanding global markets for agricultural and forest products and services; and working to reduce hunger in America and throughout the world. Two vital sources of information within the USDA include the CSREES (http://www.reeusda.gov/) and the NAL (http:// www.nalusda.gov/).

Water Quality Information Center (WQIC) (http://www.nalusda.gov/wqic/). NAL (U.S.). WQIC is an information center at the NAL specializing in information about water and agriculture.

## Agriculture Data and Statistics

Finding data on items such as agricultural production, value, and exports has become much easier since much of it is government information and is available free of charge on the Web. The following section lists 6 sites of note for agriculture data and statistics.

Census of Agriculture (http://www.nass.usda.gov/census/). USDA. 1997. Free. The Census of Agriculture is a complete accounting of U.S. agricultural production. The census is the only source of uniform, comprehensive agricultural data for every county in the United States. The census is taken every five years.

Economic Research Service (ERS) (http://www.ers.usda.gov/). USDA. Free. The ERS is the main source of economic information and research from the USDA. The mission of ERS is to inform and enhance public and private decision-making on economic and policy issues related to agriculture, food, natural resources, and rural development.

FAOSTAT (http://apps.fao.org/). Food and Agriculture Organization of the United Nations. Free. FAOSTAT is multilingual and contains over one million time-series records covering international statistics on agriculture production, trade, food balance sheets, fertilizer and pesticides, land use and irrigation, forest products, fishery products, population, agricultural machinery, and food aid shipments.

Foreign Agricultural Service (FAS) (http://www.fas.usda.gov). USDA. Free. FAS works to improve foreign market access for U.S. products. FAS operates programs designed to build new markets and improve the competitive position of U.S. agriculture in the global marketplace.

National Agricultural Statistics Service (NASS) (http://www.usda.gov/nass). USDA. Free. NASS conducts hundreds of surveys and prepares reports covering virtually every facet of U.S. agriculture. NASS's 45 state statistical offices publish data about many of the same topics for local audiences. NASS publications cover a wide range of subjects, from traditional crops, such as corn and wheat, to specialties, such as mushrooms and flowers; from calves born to hogs slaughtered; from agricultural prices to land in farms.

USDA Economics and Statistical System (http://usda.mannlib.cornell.edu/). USDA and Cornell University. Free. The USDA Economics and Statistics System contains nearly 300 reports and datasets from the economic agencies of the USDA. These materials cover U.S. and international agriculture and related topics. Most reports are text files that contain time-sensitive information. Most data sets are in spreadsheet format and include time-series data that are updated yearly.

## Directories

Directories are valuable for locating persons and organizations. The latter will often produce publications of limited distribution but that target specific subject areas. The location of persons with expertise can save valuable research time. Either way, networking is an important part of uncovering valuable grey literature. (See also those listed in the introductory chapter on general resources.)

*Agricultural Research Centres: A World Directory of Organizations and Programmes.* 1995. 12th ed. London: Cartermill. 679 p. ISBN: 1860670067. More than 8,000 entries profile universities, research centers, corporations, trade associations, and consultants in all aspects of agricultural research.

*Directories in Print.* 2000. 19th ed. A.L. Darga, ed. 2 v. Detroit, MI: Gale Group. ISBN: 0787631175 (set). *Directories in Print* describes approximately 15,500 active rosters, guides, and other print and nonprint address lists published in the

United States and worldwide. Information about the directory is available from the World Wide Web at http://www.galegroup.com/.

*Encyclopedia of Associations.* 2001. 36th ed. T.E. Sheets, ed. 2 vols. Detroit, MI: Gale Group. ISBN: 0787635391 (set).    This reference covers approximately 20,800 multinational and national membership organizations from Afghanistan to Zimbabwe—including U.S.–based organizations with a binational or multi-national membership. Entries are arranged in 15 general subject chapters allowing users to browse in sections that interest them. There are three indexes—geographic, executive, and keyword. Information about the directory is available from the World Wide Web at http://www.galegroup.com.

*International Guide to Persons and Organisations in Grey Literature.* Farace, D.J. and J. Frantzen, eds. 2000. 5th edition. Amsterdam: GreyNet, Grey Literature Network Service. 71 p. ISBN: 9074854273.    The fifth edition of this guide contains roughly 300 records from 40 countries worldwide. The records include address information, as well as the names of persons and their respective organizations. Information on the products and services, as well as the (inter)national contacts that organizations maintain in the field of grey literature are entered in the records if available.

*Research Centers Directory.* 2001. 28th ed. Detroit, MI: Gale Group. ISBN: 0787642827.    Covers nearly 1,000 agriculture oriented institutions in the United States and Canada with in-depth coverage of university experiment stations and USDA research centers.

## Electronic Journals and Trade Literature

Many agricultural-related journals such as the trade literature, journals produced by small associations, and e-journals are not indexed. This information, however, is increasingly available through the World Wide Web. AgZines (formerly called Tomato Juice) is an index to grey agriculture periodicals available on the Web. The publication sections of Web guides (see Guides section) such as Agrisurfer and Not Just Cows also provide links to e-journals and trade literature.

Agrisurfer (Agrisurf—Publications) (http://www.agrisurf.com/).    Free. The publications section of Agrisurf provides links to agricultural newsletters, magazines, journals, and newsletters.

AgZines (http://usain.org/agzines.html). USAIN Communications Committee.    Free. AgZines is a full-text index of agriculture-related, Web-based serials. Items included in the index are related to the field of agriculture, accessible through a Web browser without charge, serial in nature, content-rich, with more than just access to table of contents or summary data. Currently Agzine indexes about 100 Web-based agricultural serials. Extension publications are not included in AgZines.

Not Just Cows (E-Journals) (http://www.morrisville.edu/~drewwe/njc/). Drew, W. Free. The e-journals section of this site provides links to agricultural-related electronic journals, magazines, reports, and newsletters.

## Guides

This section contains two types of guides: (1) those that refer to sources, primarily monographs, that describe the use of grey literature and (2) those, primarily Web sources, that refer to sources that provide access to agriculture information in any format. Many of these latter sources have been developed in the past five years and provide a critical service to agriculture information professionals, farmers, and others in need of agriculture information.

AgNIC (http://www.agnic.org/). NAL (U.S.). Free. AgNIC is a guide to quality agricultural information on the Internet as selected by agriculture subject specialists, the NAL (U.S.), land-grant university librarians, and other agricultural institutions, in cooperation with citizen groups and government agencies.

AgriGator (http://www.ifas.ufl.edu/AgriGator/ag.htm). Institute of Food and Agriculture, University of Florida. Free. AgriGator is a collection of Internet sites and resources that provide agricultural- and biological-related information.

Agrisurfer (http://www.agrisurfer.com/). Free. Agrisurfer is an online weekly guide to agricultural information. It provides a free news service sent through e-mail that describes Web sources of agriculture information. It also has a Web directory called Agrisurf that links to agricultural Web sites.

Armstrong, C.J., ed. 1996. *World Databases in Agriculture*. London; New Jersey: Bowker-Saur. 1130 p. ISBN: 1857390431. Lists more than 1,000 bibliographic, directorial, statistical, and textual databases in agriculture-related fields.

Auger, C.P. 1998. *Information Sources in Grey Literature*. London: Reed Business Information Ltd. 4th edition. Series: Guides to Information Sources. 177 p. ISBN: 1857391942. A guide on all aspects of grey literature, from what it is to where to find it. Provides specific information on aerospace, life sciences, business and economics, the European Community, education, energy, and science and technology.

Bush, E.A.R. 1974. 2 vols. *Agriculture: A Bibliographical Guide*. London: Macdonald and Jane's. ISBN: 0356045056. A bibliography of bibliographies covering the years 1958–71. Entries include books (which include reports, bulletins, guides, and lists) and periodical material. Contains an extensive subject index and an author index.

Drew, W. Not Just Cows (http://www.morrisville.edu/~drewwe/njc/). Free. The purpose of this guide is to direct users to the best resources in agriculture that are available on the Internet. This site is put together by a single individual but contains links to information in many areas of agriculture.

Drew, W., ed. 1995. *Key Guide to Electronic Resources*: *Agriculture*. Medford, NJ: Learned Information. 123 p. ISBN: 157387005. Includes descriptions of databases, bibliographies, directories, library catalogs, and other electronic services.

Farace, D.J. 1997. *Notebook on Grey Literature*: *A Systematic Approach to Grey Literature in Printed and Electronic Form*. Amsterdam: GreyNet, Grey Literature Network Service. 145 p. ISBN: 907485415. The first edition appeared in 1995 under the title: *State of the Art Seminar on Grey Literature*.

Farace, D.J. and J. Frantzen, eds. 2000. *Annotated Bibliography on the Topic of Grey Literature*. Amsterdam: GreyNet, Grey Literature Network Service. 4th ed. 164 p. ISBN: 9074854265. Many of the documents cited in this publication are relevant to agriculture and related fields. When available, abstracts and information on accessing the documents have been included. A complete index to all authors, as well as a keyword index is included.

The Farmer's Guide to the Internet (http://www.rural.org/favorites.html). TVA Rural Studies Program. Free. The TVA Rural Studies Program was created in November of 1994 by the Tennessee Valley Authority to research issues shaping the future of rural America. This site compiles nearly 2,000 different links to useful sites.

## Listservs

Listservs are electronic discussion groups. Members of the list post e-mail that may request assistance, advice, or information. They are also used for news items of interest to the group. Specific subject-oriented listservs can be an extremely efficient and helpful resource to members of the list. Some listservs of interest to agriculture information professionals follow.

IAALD-L (http://www.lib.montana.edu/~alijk/dislist.html). IAALD. This listserv gives information professionals around the world a forum to discuss agricultural issues and assist one another with questions.

STS-L (http://www.ala.org/acrl/sts/sts-l-gif.html). ACRL. Science and Technology Section. STS-L is the moderated discussion list of the Science and Technology Section (STS) of the ACRL. STS provides a forum through which librarians in scientific and technical subject fields can achieve and maintain awareness of the impact and range of information with which they work.

USAIN-L (http://www.usain.org/listserv.html). USAIN.    USAIN-L is the list-serv for the United States Agricultural Information Network (USAIN). This list-serv gives information professionals in the United States a forum to discuss ag-ricultural issues and is often used as a mechanism for requesting assistance or sharing materials.

## EXTENSION RESOURCES

Extension literature is produced primarily by people involved in Cooperative Extension. In the United States, Cooperative Extension is part of the USDA and is associated with land-grant universities in every state and territory in the United States. Therefore, in the United States extension literature is generally published at the state level. The audience varies, but includes farmers, agriculture marketers and businesses, homemakers, homeowners, gardeners, and agricultural organiza-tions. Much Cooperative Extension material falls into the grey literature category (poorly indexed and limited geographic distribution), although some falls into traditional publication forms such as journal articles and books. Citations to some extension literature can be found in AGRICOLA, but most extension publications are not indexed and the NAL no longer regularly indexes extension literature. WorldCat is also another source for extension literature but it is not cataloged in a comprehensive, systematic manner.

Extension literature often falls outside the subject boundaries of traditional agriculture. For example, research-based information is produced under the aus-pices of Cooperative Extension on topics as diverse as child development, com-munity development, consumer issues, ecology, economics, environment, 4-H, family issues, fishing, forestry, gardening, lawn and turf, public policy, range-lands, technology, water issues, and youth development.

Most state and territory Cooperative Extension offices have a list of publi-cations, and many publications are available online in full text. The CSREES Web site (http://www.reeusda.gov/1700/statepartners/usa.htm) provides a map and list with links to the Cooperative Extension Web sites in each state or terri-tory. These sites may provide links to online publications, often in full-text for-mat. E-Answers is an Internet source that provides access to extension publica-tions from around the United States. Additionally, some librarians have created retrospective indexes to extension titles. New Mexico State University, for exam-ple, provides an index to all of its extension documents dating to 1915 (see http://lib.nmsu.edu/subject/ag/aggietext.html).

### Abstracts, Indexes, and Databases

AGRICOLA.    Extension publications were routinely indexed in AGRICOLA by NAL staff until about 1997. Today, NAL staff are indexing very few current extension publications. The land-grant libraries in a few states have taken over these responsibilities. Currently 4.7% of the total number of records in the AGRI-COLA database are extension publications. NAL has just begun loading indexing records for the period 1970–1978 (Esman, personal communication, 2001).

CARIS (Current Agricultural Research Information System). CARIS was created by FAO in 1975 to identify and to facilitate the exchange of information about current agricultural research projects being carried out in, or on behalf of, developing countries. CARIS identifies projects dealing with all aspects of agriculture: plant and animal production and protection; post-harvest processing of primary agricultural products; forestry; fisheries; agricultural engineering; natural resources and the environment as related to agriculture; food and human nutrition; agricultural economics; rural development; and agricultural administration, legislation, information, education, and extension. Some 137 national and 19 international and intergovernmental centers participate in CARIS.

## Agencies, Associations, and Centers

Association for International Agricultural and Extension Education (http://ag.arizona.edu/aiaee/). The Association for International Agricultural and Extension Education (AIAEE) was established in 1984 to provide a professional association to network agricultural educators and extension personnel who share the common goal of strengthening agricultural and extension education programs and institutions worldwide.

Cooperative State Research, Education, and Extension Service (CSREES) (http://www.reeusda.gov). CSREES provides the focus to advance a global system of research, extension, and higher education in the food and agricultural sciences and related environmental and human sciences to benefit people and communities everywhere. CSREES programs increase and provide access to scientific knowledge; strengthen the capabilities of land-grant and other institutions in research, extension, and higher education; increase access to, and use of, improved communication and network systems; and promote informed decision making by producers, families, communities, and other customers.

Epsilon Sigma Phi (ESP) (http://espnational.org/). ESP is dedicated to fostering standards of excellence in the extension system and developing the extension profession and professional.

National Association of County Agricultural Agents (http://www.cas.psu.edu/docs/COEXT/regions/southeast/cumberland/nacaa/nacaa.html). The mission of this association is to provide communication and cooperation among extension educators and provide professional improvement opportunities for its members.

## Journals

The authors acknowledge that the following is not a comprehensive list of extension journals available throughout the world, but hope that this sampling of journals provides the reader with an appreciation for the diversity available.

*Bangladesh Journal of Extension Education.* 1973– . Mymensingh, Bangladesh:
Bangladesh Agricultural Extension Society. ISSN: 1011-3916. A semiannual
journal that began in 1986 and written in English. Covers agricultural extension
work in Bangladesh.

*Eastern Africa Journal of Rural Development.* 1973– . Kampala, Uganda: Eastern Africa Agricultural Economics Society. ISSN: 0377-7103. This journal
started in 1973 and is published in English. Since 1974 it has been published
jointly by the Eastern Africa Agricultural Economics Society and the Department
of Rural Economy and Extension, Makerere University. It focuses on agricultural
economic aspects of Eastern Africa.

*Indian Journal of Extension Education.* 1965– . New Delhi: Indian Society of
Extension Education. ISSN: 0537-1996. This quarterly journal started in 1965
and is written in English. It covers agricultural extension work in India.

*The Journal of Agricultural Education and Extension.* 1988– . Wageningen, The
Netherlands: Group Communication and Innovation Studies. LCCN: sn 98-
15322. Free. Available from the World Wide Web at http://www.agralin.nl/
ejae/. This journal was formerly titled *European Journal of Agricultural Education and Extension.* Started in 1998, the *Journal of Agricultural Education and
Extension* is a quarterly journal aiming to publish authoritative and well-refer-
enced articles on topical issues in agricultural higher and secondary education
and extension. The journal strives to stimulate the debate on the interrelationship
between education and extension services, and other social and economic institutions in the agricultural sector.

*Journal of Continuing Education and Extension.* 1991– . Morogoro, Tanzania:
Institute of Continuing Education, Sokoine University of Agriculture. ISSN:
0856-4094. This semiannual journal began in 1991 and is written in English.
It covers agricultural research in Tanzania.

Journal of Extension (JOE) (http://joe.org/index.html). 1994– . Eugene, OR: Extension Journal, Inc. ISSN: 1077-5315. Free. The Journal of Extension is an
electronic journal that is the peer-reviewed publication of the Cooperative Extension system. It seeks to expand and update the research and knowledge base for
extension professionals and other adult educators to improve their effectiveness.
JOE was first published electronically in 1994 after 30 years as a print journal.

*Journal of Extension Systems*. 1985– . Karnal, India: National Dairy Research Institute. LCCN: 87-909312. Available from the World Wide Web at http://www.jesonline.org/. A semiannual journal began in 1985 and written in English. Covers agricultural extension work in India.

*The Journal of International Agricultural Education and Extension (International Journal on Changes in Agricultural Knowledge and Action Systems)*. 1994– . Tucson, AZ: University of Arizona: Association for International Agricultural and Extension Education. ISSN: 1077-0755. Available from the World Wide Web at http://ag.arizona.edu/aiaee/Journal.htm. Started in 1994 and written in English. This semiannual journal is the official refereed publication of the AI-AEE. Its purpose is to develop a broad research and knowledge base on agricultural and extension education in developing countries.

*Khaosan Kasetsat*. 1956– . Krungthep, Thailand: Mahawitthavalai, Kasetsat. ISSN: 0125-104X. This bimonthly journal was first published in 1956 and is written in English and Thai. It covers agricultural information, including extension, in Thailand.

*The Legon Agricultural Research and Extension Journal*. 1991– . Legon, Ghana: Faculty of Agriculture, University of Ghana. This journal was first published in 1991 and is written in English. It covers agriculture in Ghana.

*The Nigerian Journal of Agricultural Extension*. 1981– . Zaria, Nigeria: Agricultural Extension and Research Liaison Service, Ahmadu Bello University. ISSN: 0331-7757. This semiannual journal began in 1981 and is written in English. It covers agriculture and agricultural extension work in Nigeria.

*The Nigerian Journal of Rural Extension and Development*. 1991– . Ibadan, Nigeria: Department of Agricultural Extension Services, University of Ibadan. ISSN: 0795-7432. This annual journal began in 1991 and is written in English. It covers agricultural extension work, rural extension, and rural development.

## World Wide Web Sites

AgNIC (http://www.agnic.org/). Free. AgNIC is a guide to quality agricultural information on the Internet as selected by the NAL (U.S.), land-grant universities, and other agricultural institutions, in cooperation with citizen groups and government agencies. AgNIC focuses on providing agricultural information in electronic format over the World Wide Web through the Internet. Many extension people are involved in the development of AgNIC.

Alternative Information in Agriculture (http://agebb.missouri.edu/mac/links/index.htm). Compiled by The Missouri Alternatives Center. Free. This site provides alternative forms of agricultural information. It includes a list of links by subject to Cooperative Extension publications from various university research centers around the world.

Cooperative Extension and Agricultural Experiment Station Publications and Video Tape Presentations List (http://www.oznet.ksu.edu/library/other_st/other_st.htm). Compiled by Kansas State University Agricultural Experiment Station and Cooperative Extension Service.    Free. This site contains a list of links by state to publications. Some states provide full-text publications.

E-answers (http://128.227.242.197/).    Free. E-answers is made possible by participating universities and ACE through a grant from the USDA, CSREES. E-answers is a searchable Web site that provides reliable, research-based information on a wide range of extension or outreach-oriented subjects. The practical, current, and unbiased information in this site represents the work of Cooperative Extension and Agricultural Experiment Station professionals at land-grant universities throughout the United States.

PEOPLE (Portable Extension Office for Program Literature Exchange) (http://www.uog.edu/cals/people/INDEX.HTM). University of Guam.    Free. This site provides full-text publications for sustainable extension efforts on tropical islands. This project is funded by USDA, CSREES.

Plant Facts (http://plantfacts.ohio-state.edu/). Compiled by Ohio State University.    Free. This site provides a fact-sheet database that includes guides for answering plant-related questions from 46 different universities and government institutions across the United States and Canada. Over 20,000 pages of extension fact sheets and bulletins provide a concentrated source of plant-related information.

## CONCLUSION

Using agricultural grey literature and extension resources may be a bit overwhelming at first, but the strategies and resources presented in this chapter are intended to make this task less daunting. Grey and extension literature represents a large portion of agriculture information and, while it has never been easy to identify or access, it is becoming more widely available. Admittedly, the Internet does not solve all of our access problems, but it does provide a very useful outlet for grey and extension literature. Should the information seeker take the time, the effort expended for identifying and accessing grey literature and extension resources will usually be worth the reward.

## ACKNOWLEDGMENTS

The authors would like to thank Doug Jones, University of Arizona Science and Engineering Library and Joanna Szurmak, New Mexico State University Library for reviewing this chapter. Their comments and insight are greatly appreciated.

# 12

## Horticulture

**Elaine A. Nowick**
University of Nebraska, Lincoln, Nebraska, USA

The reference works in this chapter deal with basic and applied research on ornamental flowers, fruits, vegetables, edible mushrooms, herbs and medicinal plants,
minor crops, turfgrass, ornamental shrubs, urban forestry, and amenity trees. The
field of horticulture is one of the broadest in several dimensions. While the body
of horticultural literature includes the traditional peer-reviewed journals and
books typical of other sciences, it also includes works directed toward home
gardeners, serious amateurs, and commercial growers, as well. Some of the more
popular literature may be quite scholarly and well-researched, and of value to
serious students and scientists. Some of the works that bridge the area between
serious amateur and professional have been included in this chapter, especially
if there is no equivalent research-level work available.

Horticultural crops span a wide range of uses. They may be grown for
human food, for medicinals, for aesthetics, or for other purposes. This chapter
has attempted to include the literature related to vegetables, fruits, medicinal
plants, and minor crops as well as ornamental flowers, tree, and shrubs, but has
not included the classic reference texts for garden design or landscape architec-
ture. Sources on information about medical applications of botanicals can be
found in guides to herbals and in the ethnobotanical literature and are not included
here.

The wide range of uses for horticultural crops is reflected in the wide taxonomic range of plant materials grown by horticulturists. These plant materials include fungi; mosses and other bryophytes; gymnosperms; and both monocotyledonous and dicotyledonous angiosperms. For this reason the taxonomic literature is of great importance to horticulturists. Readers are referred to Davis and Schmidt (1995) for coverage of the taxonomic literature as well as information sources about plant physiology, genetics, biochemistry, and more basic aspects of plant science. Guides to the botanical literature may also be useful (Davis and Schmidt, 1996).

Illustrations are important sources of information to the horticulturist of fruit and vegetable crops as well as ornamentals. There are a number of dictionaries, encyclopedias, and similar works with lists of the wide number of species that have been cultivated. When illustrations are included, they have been noted. Web sites and online resources have been incorporated into the appropriate sections of this chapter.

There are overlaps between crops that are considered to be horticultural and those that are grown as field or row crops. Additional relevant reference works can be found in other chapters in this book, especially those on field crops and soil science.

## ABSTRACTS AND INDEXES

The abstracting services that cover general agriculture such as AGRICOLA or CAB Abstracts include many of the publications dealing with horticulture and general plant sciences. The abstracting journals in the plant sciences published by CAB International (formerly the Commonwealth Agricultural Bureaux) such as *Plant Growth Regulator Abstracts*, *Plant Genetic Resources Abstracts*, *Plant Breeding Abstracts*, *Seed Abstracts*, or *Crop Physiology Abstracts* are good sources of information for horticulture. These abstracting journals are combined into the online or CD-ROM versions of CAB Abstracts. BIOSIS (*Biological Abstracts*) is an excellent index to the literature in basic research in the plant sciences.

*Applied Botany Abstracts*. v. 1– . 1981– . Lucknow, India: National Botanical Research Institute, Economic Botany Information Service. Quarterly. ISSN: 0970-2377. Focusing on phytochemical, biotechnological, ethnobotanical, and cultural research and related areas of botany and horticulture. Book reviews included. Emphasis is on Indian crops and plant uses. This abstracting journal is indexed by CAB International and included in their database.

*Flowering Plant Index: FPI*. v. 1– . 1991– . Chanhanssen, MN: Andersen Horticultural Library. Semiannually. ISSN:1061-9011. An index to illustrations of flowering plants. The print edition will be published periodically with the most up-to-date information available online for a fee at http://plantinfo.umn.edu/ default.asp.

*Garden Literature*: *An Index to Periodical Articles and Book Reviews*. v. 1– . 1992– . Boston, MA: Garden Literature Press. Annually. ISSN:1061-3722. Covers over 100 periodicals on gardening published in the United States and United Kingdom.

*Horticultural Abstracts*. v. 1– . 1931– . Wallingford, UK: CABI Publishing. Monthly/Annual Author and Subject Indexes. ISSN: 0018-5280. Also available on CD-ROM, through HORT CABWeb, and the online version of CAB Abstracts. Covers journal articles, reports, conferences, and books on fruits, vegetables, ornamental plants, nuts, plantation crops, and minor industrial crops from an international perspective. Aspects covered are cultivars and taxonomy; propagation and in vitro culture; agronomy and crop management; crop protection; crop botany; environmental aspects and stress; harvesting and postharvest technology; and product quality.

*Ornamental Horticulture*. v. 1. 1975– . Wallingford, UK: CABI Publishing. Bimonthly/Annual Author and Subject Indexes. ISSN: 0305-4934. This title is also available through Hort CABWeb, through the online version of CAB Abstracts, and as a CD-ROM. Covers journal articles, conferences, reports, and books on orchids; conifers; roses; trees and shrubs; bulbs and tubers; lawns and sports turf; aquatic plants; foliage and succulent plants; poinsettias; rhododendrons; and azaleas.

*Review of Aromatic and Medicinal Plants*. v. 1– . 1995– . Wallingford, UK: CABI Publishing. Bimonthly/Annual Author Subject Indexes. ISSN: 1356-1421. Also published as a part of HortWeb, through the Internet, or on CD-ROM. Covers the literature on cultivated and wild species of culinary herbs and spices and essential oil and medicinal plants. Aspects considered include botany; natural resources; crop production; chemistry; biotechnology; human and veterinary medicine; pharmacology; biological activity; herbal drugs; essential oils; culinary herbs and spices; and economics.

## ATLASES

Information on the distribution of plants at present or historically is important to horticulturists because the distribution gives an indication of plant adaptation in new and similar areas. Atlases also provide information on the distribution of wild species that are often the sources of genetic material for breeding of horticultural crops. Wild plants are still being domesticated as ornamentals, medicinals, and industrial crops. Regional floras and field guides have not been included here, but may also provide good information on the distribution of wild, weedy, and domesticated plants.

Davis, S.D., V.H. Heywood, and A.C. Hamilton. 1994. *Centres of Plant Diversity*. Cambridge, UK: World Wide Fund for Nature (WWF) and ICUN—World Conservation. 3 vols. ISBN: 283170197X (v. 1); 2831701988 (v. 2); 2831701996 (v. 3). Out of print. Datasheets for each region include geography, vegetation, flora, useful plants, social and environmental values, threats, conservation, and references. Regions are selected by the richness of endemic species.

Englander, C. and P. Hoehn. 1997. Checklist of Online Vegetation and Plant Distribution Maps (http://www.sul.stanford.edu/depts/branner/vegmaps4. html). (cited July 13, 2000) Stanford, CA: Stanford University Libraries. Links to an extensive list of plant distribution and ecoregion maps throughout the world.

Sauer, J.D. 1993. *Historical Geography of Crop Plants: A Selected Roster*. Boca Raton, FL: CRC Press. 320 p. ISBN: 0849389011. Origins, history, and geography of cultivars and their wild progenitors. Domestication, evolution, and spread are traced for fruits, vegetables, and ornamentals, as well as row crops.

USDA. World Agricultural Weather Facility. 1994. *Major World Crop Areas and Climatic Profiles*. Washington, DC: U.S. Government Printing Office. Agricultural Handbooks (USDA) no. 664. 279 p. Out of print. Crop zones, crops, and climate diagrams, charts, and maps.

## BIBLIOGRAPHIES

The following bibliographies provide guides to the literature in subject areas closely related to horticulture, such as taxonomy. Bibliographies can be especially useful in finding historical information from literature published before the time periods covered by electronic abstracts and indexes.

Davis, E.B. 1995. *Guide to Information Sources in the Botanical Sciences*. 2nd ed. Littleton, CO: Libraries Unlimited. 275 p. ISBN: 1563080753. Provides information on bibliographic sources; abstracts, indexes, and databases; dictionaries and encyclopedias; handbooks and methods; directories and groups; taxonomy; biographical and historical materials; textbooks; and important series on all aspects of botany.

Davis, E.B. and D. Schmidt. 1995. *Using the Biological Literature: A Practical Guide*. 2nd ed., rev. and expanded. New York: Marcel Dekker. 440 p. ISBN: 082479477X. Lists abstracts and indexes, dictionaries, encyclopedia, nomenclatural guides, Internet guides, handbooks, periodicals, and other reference works concentrating on the more theoretical aspects of biology.

*Quick Bibliography Series*. 1976– . Beltsville, MD: U.S. National Agricultural Library. Price varies. ISSN: 1052-5378. Individual volumes summarize citations from the AGRICOLA database on a wide range of topics related to agriculture and horticulture.

World Conservation Monitoring Center. 1990. *World Plant Conservation Bibliography*. Kew, UK: Royal Botanic Gardens. 645 p. ISBN: 0947643249. Over 10,000 citations on international, national, and local plant conservation. Listed by country.

Wyatt, H.V., ed. 1997. *Information Sources in the Life Sciences*. 4th ed. Providence, NJ: Bowker-Saur. 264 p. ISBN: 1857390709. Includes guides to statistics and software, major secondary sources including CD-ROMs, foreign-language literature, and translations. Library services in other countries, invisible colleges, and scientific fraud are also covered.

## CONFERENCES AND PROCEEDINGS

The list of conferences included in the section is not comprehensive, but includes those sponsored by major professional societies in horticulture and those whose proceedings are published on an ongoing basis. Many state horticultural societies publish annual proceedings, such as the Florida State Horticultural Society and the Oregon Horticultural Society. Proceedings from individual state societies have not been listed here. Information on conferences sponsored by these organizations on particular themes is available by directly contacting the organization. Web sites for a number of sponsoring organizations are listed in this chapter in the Societies section. Another good source of information on upcoming programs is the calendar section of AgNIC maintained by the U.S. National Agricultural Library (NAL).

American Society for Horticultural Science. *Program and Abstracts*. Alexandria VA: American Society for Horticultural Science. Abstracts from the annual meeting have been published yearly in the journal *HortScience* since 1989.

Canadian Horticultural Council. *Annual Meeting Reports*. 1922– . Ottawa, Canada: Canadian Horticultural Council. ISSN: 0068-8908. Proceedings of meetings on issues related to the horticultural industry in Canada including regulatory information and trade.

European Association for Research on Plant Breeding (EUCARPIA). 1956– . *Report of the Congress*. New York: Springer-Verlag. ISSN: 0071-2615. Proceedings of the general meeting held every three years.

International Plant Propagators' Society. 1950– . *Combined Proceedings of Annual Meetings*. Seattle, WA: International Plant Propagators' Society. ISSN: 0538-9143. Includes proceedings of nine regional meetings on plant propagation.

*Plants and Fern-Like Plants in Latin, Russian, English, and Chinese.* Koenigstein, Germany: Koeltz Scientific Books. 1,033 p. ISBN: 387429398X.   An extensive work listing common names of plants in several languages not well known in western countries. Names commonly used in pharmacological literature are included as well as an extensive bibliography.

Soule, J. 1985. *Glossary for Horticultural Crops.* New York: John Wiley. 898 p. ISBN: 0471884995. Out of print.   This reference work provides ready access to technical terms over a broad spectrum of the plant and plant-related sciences for students, research scientists, extension specialists, administrators, growers, and home gardeners.

Stearn, W.T. 1995. *Botanical Latin*: *History, Grammar, Syntax, Terminology and Vocabulary.* 4$^{th}$ ed. Portland, OR: Timber Press. 560 p. ISBN: 0881923214.   This extensive and useful reference includes geographical names, color terms, and a list of commonly used symbols and abbreviations.

Steinmetz, H. 1972. *Horticultural Techniques and Implements*: *Multilingual Illustrated Dictionary.* Woodstock, NY: American Agricultural News Service. 396 p. ISBN not listed. Out of print.   Names for tools and equipment commonly used for horticulture in German, English, French, Spanish, Italian, and Dutch with illustrations.

Trehane, P. 1996. *International Code of Nomenclature for Cultivated Plants.* Regnum Vegetable, no. 133. Champaign, IL: Balogh Scientific Books. ISBN: 094811701. Out of print.   Endorsed by the International Commission for the Nomenclature of Cultivated Plants (ICNCP). Revisions of the code are issued when necessary.

USDA, Natural Resources Conservation Service. 1999. Plants Database (http://plants.usda.gov/) (cited July 13, 2000). Baton Rouge, LA: National Plant Data Center.   An online source of information on phylogenetics, cultural significance, threatened status, economics, and geographic distribution of a wide range of vascular plants.

Wiersema, J.H. 2000. World Economic Plants in GRIN (http://www.ars-grin.gov/npgs/tax/taxecon.html) (cited July 17, 2000). Washington, DC: USDA-ARS.   This is the online version of *A Checklist of Names for 3,000 Vascular Plants of Economic Importance*, Agricultural Handbook (USDA), no. 505 by E.E. Terrell (1986).

Weiresema, J.H. and B. Leon. 1999. *World Economic Plants.* Boca Raton FL: CRC Press. 749 p. ISBN: 0849321190.   This book covers over 10,000 vascular plants of commercial importance. Scientific names, synonyms, economic uses, and geographic distribution are listed for each entry.

Wrobel, M. and G. Creber. 1999. *Elsevier's Dictionary of Fungi and Fungal Plant Diseases*. Elsevier Science. ISBN: 0444504508. Edible and poisonous fungi and plant pathogens listed by scientific name with common names in English, German, French, and Italian. Variants of English and French names in Australia, the United States, New Zealand, Canada, and other countries are included.

## DIRECTORIES

Works included in this section are guides to institutions and organizations, such as arboreta, botanic gardens, and research institutes. Also included are lists of nurseries, seed companies, and other sources for seeds and other propagules for horticultural plants.

Barrett, T.M., ed. 1998. *North American Horticulture: A Reference Guide*. 2nd ed. New York: MacMillan Library Reference. 427 p. ISBN: 0028970012. A directory listing thousands of horticultural organizations; plant societies; educational programs; public and historic gardens; scholarly organizations; botanical and horticultural libraries; conservation groups; garden club associations; and other information of interest to amateurs and professionals.

Barton, B. 1995. *Gardening by Mail: A Source Book*. 5th ed. New York: Houghton Mifflin. 404 p. ISBN: 0395877709. A directory of plant and seed companies, nurseries, garden accessories, services, libraries, societies, newsletters, books, and Internet resources cross-referenced by type of plant and state.

Brooks, R.M. and H.P. Olmo. 1997. *Brooks and Olmo Register of Fruit and Nut Varieties*. 3rd ed. Alexandria, VA: ASHS Press. 743 p. ISBN: 0961502746. This work covers cultivars in 80 commodities developed by controlled breeding since the 1920s. Each entry lists variety origin, discovery date, its inventor, the patent number, date, and assignee; the seedling type, length, description, and date of harvest; and any other relevant characteristics (such as flavor, disease resistance or sensitivity, and type of shell).

Facciola, S. 1998. *Cornucopia II: A Source Book of Edible Plants*. Vista, CA: Kampong Publications. 713 p. ISBN: 0962808725. Over 7,000 varieties in 3,000 species are listed with cultivation information, uses, and sources where the seed or plants can be obtained.

Heywood, V.H., C.A. Heywood, and P.W. Jackson. 1993. *International Directory of Botanical Gardens*. Koenigstein, Germany: Koeltz Scientific Books. 1,021 p. ISBN: 387429319X. Out of print. Sponsored by the WWF—World Wide Fund for Nature, Botanic Gardens Conservation and the Secretariat, International Association of Botanical Gardens.

International Society for Horticultural Science. 1993. *Horticultural Research International: Directory of Horticultural Research Institutes and Their Activities in 74 Countries.* 5th ed. Wageningen, The Netherlands: International Society for Horticultural Science. 1,000 p. ISBN: 9066051957. Out of print. Twenty-five thousand horticultural research institutes with addresses and names of 18,000 research workers, with their main fields of interest. The completely revised sixth edition will be published on the Internet.

Royal Horticultural Society. 2000. RHS Plantfinder (http://www.rhs.org.uk/rhsplantfinder/plantfinder.asp) (cited July 26, 2000). A searchable Web site listing 70,000 plants and 800 nurseries in the United Kingdom.

Whealey, K. 1999. *Garden Seed Inventory: An Inventory of Seed Catalogs Listing All Non-Hybrid Vegetables Seeds Still Available in the United States and Canada.* 5th ed. Decorah, IA: Seed Savers Publication. 806 p. ISBN: 1882424549. While providing a guide to seed sources of many rare or heirloom vegetables, the stated purpose of the work is to alert readers to cultivars no longer in commercial production and in need of preservation.

## ENCYCLOPEDIAS

### General Works

Included in this section are works that cover a wide range of plant species. The reader is also directed to the sections on dictionaries and handbooks. The annotations list the range of plants covered, any images included, and other information.

Armitage, A.M. 2000. *Armitage's Garden Perennials: A Color Encyclopedia.* Portland OR: Timber Press. 320 p. ISBN: 0881924350. Information on over 1,400 garden perennials with color photographs and suggested species for particular situations by one of the foremost authors in horticulture.

Armitage, A.M. 1997. *Herbaceous Perennial Plants: A Treatise on their Identification, Culture, and Garden Attributes.* 2nd ed. Champaign, IL: Stipes Publishers. 1,141 p. ISBN: 087563723X. Cultural specifics, propagation techniques, and suggested companions for plants hardy in USDA zones three to eight. Reviews of newly introduced cultivars and species are included. A CD-ROM version is also available.

Bailey, L.H., ed. 1963. *The Standard Cyclopedia of Horticulture: a Discussion for the Amateur, and the Professional and Commercial Grower, of the Kinds, Characteristics and Methods of Cultivation of the Species of Plants Grown in the Regions of the United States and Canada for Ornament, for Fancy, for Fruit and for Vegetables; with Keys to the Natural Families and Genera, Descriptions of the Horticultural Capabilities of the States and Provinces and Dependent Is-*

*lands, and Sketches of Eminent Horticulturists.* 2nd ed. New York: MacMillan. 3,639 p. ISBN not listed. Out of print. Chosen as one of the American Horticultural Societies 75 Great Garden Books, this classic reference includes over 4,000 illustrations, some in color.

Bailey, L.H. and E.Z. Bailey. 1976. *Hortus Third: A Concise Dictionary of Plants Cultivated in the United States and Canada.* New York: Macmillan Publishing Co. 1,312 p. ISBN: 0025054708. This classic and still outstanding plant encyclopedia lists scientific names alphabetically with botanical descriptions. An extensive common name index and a glossary are included.

Dirr, M. 1997. *Dirr's Hardy Trees and Shrubs: An Illustrated Encyclopedia.* Portland OR: Timber Press. 493 p. ISBN: 0881924040. Habitat, landscape use, availability from commercial sources, time of flowering, size, and flower color for trees and shrubs hardy in USDA zones 3 to 8 with over 1,600 color photographs. This work was chosen as a 1999 Outstanding Reference Source by the American Library Association.

Everett, T.H. 1980. *New York Botanical Garden Illustrated Encyclopedia of Horticulture.* New York: Garland Publishing. 3,064 p. ISBN: 0815302568. This is one of the most comprehensive works available for horticulture. Entries include taxonomic and nomenclatural information, as well as cultivation tips for each species with illustrations for many. Horticultural techniques and terms are included as well.

Graf, A.B. 1992. *Hortica: Color Cyclopedia of Garden Flora in All Climates and Exotic Plants Indoors.* Rutherford, NJ: Roehrs. 1,215 p. ISBN: 0911266259. This work includes over 8,100 high-quality color photographs with short plant descriptions.

Graf, A.B. 1992. *Tropica: Color Cyclopedia of Exotic Plants and Trees for Warm-Region Horticulture, in Cold Climate, the Summer Garden, or Sheltered Indoors.* 4th ed., rev. and enlarged. East Rutherford, NJ: Roehrs. 1,152 p. ISBN: 0911266240. This classic reference in horticulture includes over 7,000 photographs of ornamental, tropical, and house plants.

Graf, A.B. 1982. *Exotica Series IV International: Pictorial Cyclopedia of Exotic Plants from Tropical and Near Tropic Regions.* East Rutherford, NJ: Roehrs. 2 vols. 2,560 p. ISBN: 091126678. Out of print. Over 16,000 photographs of indoor ornamentals for home, office, greenhouse, or patio or for use as outdoor ornamentals in warm climates. Listed by plant family.

Jelitto, L. and W. Schacht. 1990. *Hardy Herbaceous Perennials.* Portland, OR: Timber Press. 2 vols. 721 p. ISBN: 0881921599. Describes 3,617 cultivars in 4,286 species, with 690 color photographs. Translated by Michael Epp.

Kohlein, F. and P. Menzel. 1994. *Color Encyclopedia of Garden Plants and Habitats*. Portland, OR: Timber Press. 320 p. ISBN: 0881922986. This unique work has information on habitats that are present or can be created in gardens and the plant species adapted to them along with descriptions and illustrations.

Krussman, G. 1986. *Manual of Cultivated Broad-leaved Trees and Shrubs*. Beaverton, OR: Timber Press. 3 vols. (v. 3 only). ISBN: 0917304780 (v. 1), 0881920053 (v. 2), 0881920061 (v. 3). This work was originally published in German. It gives cultivation and propagation information for over 6,000 cultivars in 5,000 species with taxonomic keys and illustrations. Volumes 1 and 2 are out of print, but Volume 3 is available through Timber Press.

Poor, J.M., ed. 1984. *Plants that Merit Attention: Trees*. Portland, OR: Timber Press. 349 p. ISBN: 0917304756. Poor, J.M., J. Meakes, and N. Brewster. 1996.

*Plants that Merit Attention: Shrubs*. Portland, OR: Timber Press. 364 p. ISBN: 0881923478. These two companion volumes focus on less familiar but noteworthy plants. Each entry has at least two color photos, a physical description, cultural information, landscape value, and a list of the public gardens where the plant can be observed. An appendix with source nurseries is included. Sponsored by the Garden Club of America.

Rocklein, J.C. and P. Leung. 1987. *Profile of Economic Plants*. New Brunswick and Oxford: Transaction Books. ISBN: 0887381677. An extensive listing of plant species grown worldwide by principal end use. Includes information on adaptation, common names, a summary of cultivation methods, and references.

## GUIDES TO SPECIFIC CROPS

Included here are works that are concerned with specific families, genera, or other more focused groups of plants. This list is not comprehensive, and there are other good guides to specific types of plants. This section tries to cover the most commonly grown plant groups. Other reference works dealing with the cultivation of specific genera or plant groups can be found in the handbooks section of this chapter.

Austin, D. 1998. *Botanica's Roses: The Encyclopedia of Roses*. New York: Mynah. 704 p., includes a CD-ROM. ISBN: 0091838037. Information on the rose's history and culture; 600 pages of photos and descriptions of varieties including wild roses. A reference table summarizes the main features of each rose.

Beckett, K.A. and C. Grey-Wilson, eds. 1993. *Encyclopaedia of Alpines*. Pershore, UK: AGS Publications. 2 vols. ISBN: 0900048638. Out of print. A comprehensive list of alpine and rock garden plants from the Alpine Garden Society (Great Britain).

Bryan, J.E. *Bulbs.* 1989. Portland, OR: Timber Press. 2 vols. 750 p. ISBN: 0881921017. Out of print. Propagation, history, botany, cultivation, forcing, landscape use, pests, and diseases for ''bulb'' species in a broad sense. Appendices list genera by height, color, flowering time, and other characteristics. A glossary and bibliography are included.

Burrell, C.C. 1997. *A Gardener's Encyclopedia of Wildflowers: An Organic Guide to Choosing & Growing over 150 Beautiful Wildflowers.* Emmaus, PA: Rodale Press. 216 p. ISBN: 087596723X. One of the 75 Great American Garden Books from the American Horticultural Society.

Darke, R. 1999. *Color Encyclopedia of Ornamental Grasses, Sedges, Restios, Cat-tails, and Selected Bamboos.* Portland, OR: Timber Press. 325 p. ISBN: 0881924644. The biology of grasses and grasslike plants, their use in garden design, native habitat, physical description, culture, cultivars, and color photographs.

Davidian, H.H. 1982. *The Rhododendron Species.* Portland, OR: Timber Press. 4 vols. ISBN: 0713416394 (v. 1), 0881921092 (v. 2), 0881921688 (v. 3), and 0881923117 (v. 4). A complete taxonomic and historical account of the cultivated species of Rhododendron, with keys for identification of the series and species.

Galle, F.C. 1997. *Hollies: The Genus Ilex.* Portland, OR: Timber Press. 619 p. ISBN: 088192380X. Descriptions of more than 800 species (30 deciduous and 780 evergreen) of hollies with information on landscape use; botany and morphological characteristics; cultivars; and techniques on hybridizing and propagation with color photographs and black-and-white illustrations.

Galle, F.C. 1988. *Azaleas.* Enlarged and rev. ed. Portland, OR: Timber Press. 627 p. ISBN: 0881920916. Species and cultivars, breeding, culture, pests, and diseases of azaleas.

Goulding, E. 1995. *Fuchsias: The Complete Guide.* Portland, OR: Timber Press. 208 p. ISBN: 0881923281. Includes 49 color photos of cultivars and species with growing and hybridizing information.

Greenlee, J. 1992. *Encyclopedia of Ornamental Grasses: How to Grow and Use over 250 Beautiful and Versatile Plants.* Emmaus, PA: Rodale Press. 182 p. ISBN: 0875961002. Color photographs, climatic adaptations, and cultivation tips for grasses and related plants from the most well-known spokesman for the ornamental grass movement. One of the 75 Great American Garden Books from the American Horticultural Society.

Grey-Wilson, C. 2000. *Clematis: The Genus: A Comprehensive Guide for Gardeners, Horticulturists and Botanists.* Portland, OR: Timber Press. 224 p. ISBN:

0881924288. Contains information on both wild and cultivated species, as well as cultivars. Color photographs and maps showing the distribution of species in the wild are also included.

Janick, J., J.E. Simon, and A. Whikey. 1998. *New Crop Compendium*: *Navigating New Crops*. West Lafayette, IN: Purdue University Center for New Crops & Plant Products; Rome, Italy: Food and Agriculture Organization of the United Nations. ISBN: 0931682703. Information on new, specialty, neglected, and underutilized crops for scientists, growers, marketers, processors, and extension personnel on CD-ROM. Companion volumes, *Perspectives on New Crops* and *Progress in New Crops* are also available.

Jaynes, R.A. 1998. *Kalmia*: *The Mountain Laurel and Related Species*. Portland, OR: Timber Press. 360 p. ISBN: 0881923672. Describes the 8 species and 80 cultivars of the Kalmia genus with color photographs.

Jones, D.L. 1987. *Encyclopedia of Ferns*. Portland, OR: Timber Press. 450 p. ISBN: 0881920541. A growing guide for ferns comprises the first half of the book. The second half includes descriptions for hundreds of species with color photographs and line drawings.

Kohlein, F. 1991. *Gentians*. Portland, OR: Timber Press. 189 p. ISBN: 0881921920. A guide to some of the 400 species on this genus adapted to all climates with color photographs. Translated from the German.

Krussman, G., G.S. Daniels, and H.D. Warder. 1985. *Manual of Cultivated Conifers*. 2^nd rev. ed. Portland, OR: Timber Press. 160 p. ISBN: 088192007X. Describes 607 species and 2,075 cultivars, extensively illustrated and includes references to illustrations in other works. A common name and invalid plant name index is included.

Macoboy, S. 1998. *Illustrated Encyclopedia of the Camellias*. Portland, OR: Timber Press. 304 p. ISBN: 0881924210. Separate sections focus on sasanquas, japonicas, higos, and reticulatas. Over 1,000 cultivars are described with color photographs.

Mickel, J. 1997. *Ferns for American Gardens*: *The Definitive Guide to Selecting & Growing More Than 500 Kinds of Hardy Ferns*. New York: MacMillan Publishing. 370 p. ISBN: 0028616189. More than 400 species of ferns are included. Each entry describes varieties and cultivars, habitat, landscape use, and ease of cultivation. Includes some 360 color photographs and 30 line drawings. The author is curator of ferns at the New York Botanic Garden.

Miller, D. 1996. *Pelargoniums*: *A Gardener's Guide to the Species and Their Hybrids and Cultivars*. Portland, OR: Timber Press. 208 p. ISBN:

088192363X. Includes information on more than 200 lesser known species of Pelargoniums in addition to the popular regal and zonal geraniums.

Mineo, B. 1999. *Rock Garden Plants: A Color Encyclopedia*. Portland, OR: Timber Press. 284 p. ISBN: 0881924326. Color photographs of rock garden plants from around the world including rare and unusual species as well as old favorites.

Pietropaolo, J. and P. Pietropaolo. 1996. *Carnivorous Plants of the World*. Portland, OR: Timber Press. 206 p. ISBN: 0881923567. Carnivorous plant types are combined into chapters on Venus's flytraps, pitcher plants, sundew types, butterworts, and bladderworts. Cultivated species are treated in depth with brief mentions of wild species. A growing guide is included.

Petit, T.L. and J.P. Peat. 2000. *Color Encyclopedia of Daylilies*. Portland, OR: Timber Press. 296 p. ISBN: 0881924881. Primarily focused on the cultivars in the genus *Hemerocallis* with over 1,200 color photographs. Also includes information on the characteristics breeders select in this genus.

Preston-Mafham, R. and K. Preston-Mafham. 1991. *Cacti: The Illustrated Dictionary*. Portland, OR: Timber Press. 224 p. ISBN: 0881924008. Concentrates on the genus *Mamillaria*, but includes some information on columnar forms of cacti. Over 1,000 color photographs.

Pridgeon, A. 1993. *Illustrated Encyclopedia of Orchids*. Portland, OR: Timber Press. 304 p. ISBN: 0881922676. Includes over 1,100 species and hybrids in a range of commonly cultivated genera. Each entry includes descriptions, taxonomy, names and synonyms, geographic distribution, notes on culture, many with photographs.

Rehder, A. 2001. *Manual of Cultivated Trees and Shrubs Hardy in North America, Exclusive of the Subtropical and Warmer Temperate Regions*. Caldwell, NJ: Blackburn Press. 996 p. ISBN: 1930665326. This is a recent reprint of the 1940 revised edition. It includes introduced species that have been successfully cultivated in western regions not treated by Gray or Britton. Taxonomic keys are included as well as indexes to common and Latin names and authors of original species descriptions.

Sajeva, M. and M. Costanzo. 2000. *Succulents II: The New Illustrated Dictionary*. Portland, OR: Timber Press. 234 p. ISBN: 0881924490. Intended to be a supplement to the authors earlier book listed in the following text, this volume includes an additional 900 species and photographs of 300 species included in the previous volume. Brief information on cultivation and Convention on International Trade in Endangered Species (CITES) status in the wild describe each entry.

Sajeva, M. and M. Costanzo. 1997. *Succulents: The Illustrated Dictionary*. Portland, OR: Timber Press. 240 p. ISBN: 0881923982. Includes information on

195 genera of succulents with over 1,200 photographs of species used for houseplants and ornamentals. Rare plants, some endangered in the wild, are included.

Schenk, G. 1991. *Complete Shade Gardener*. 2nd ed. Willmington, MA: Houghton Mifflin. 311 p. ISBN: 0395574269. Lists of plants adapted to low light, their descriptions, uses, and requirements, subdivided between trees, shrubs, perennials, and annuals. Landscape plans and information on the effects of tree species on undergrowth are included with color and black-and-white photographs and line drawings.

Schmidt, W.G. 1992. *The Genus Hosta*. Portland, OR: Timber Press. 468 p. ISBN: 0881922013. Descriptions of all species, varieties, forms, and registered cultivars, and a comprehensive list of the nonregistered classic hostas of historic and garden interest. Includes a taxonomic revision of the genus.

Slocum, P.D., P. Robinson, and F. Perry. 1996. *Water Gardening, Water Lilies and Lotuses*. Portland, OR: Timber Press. 434 p. ISBN: 0881923354. Part one is a guide to water gardening as well as an extensive illustrated encyclopedia of submerged, floating, marginal, and bog plants. Part two is an encyclopedia of the water lilies and lotuses.

van Gelderen, D.M. and J.R.P. van Hoey Smith. 1996. *The Illustrated Encyclopedia of Conifers*. Portland, OR: Timber Press. 2 vols. 706 p. ISBN: 0881923540. Each genus is surveyed overall, with brief notes on distribution, botanical classification and characteristics, and mention of species of particular interest. Photographs of individual species, hybrids, and cultivars follow, arranged by natural relationships and geographical ranges.

Vertrees, J.D. 1987. *Japanese Maples*. 2nd ed. Portland, OR: Timber Press. 180 p. ISBN: 0881920487. History, taxonomy, nomenclature, culture, and propagation of *Acer palmatum* and *Acer japonicum* with 250 color photographs.

## HANDBOOKS AND MANUALS

The emphasis in the following works is on the more practical aspects of horticultural crop and ornamental growing for both researchers and commercial growers. These works contain information on breeding; adaptation; seed production; propagation; nutrition; disease and insect control; induction of flowering; marketing; greenhouse construction and maintenance; and post-harvest handling of fruits, vegetables, cut flowers, bedding plants, and pot plants.

Alford, D.V. 1995. *Colour Atlas of Pests of Ornamental Trees, Shrubs, and Flowers*. London: Manson. 448 p. ISBN: 0470234946. Describes life history, damage caused, and control suggestions for insects, mites, and other pests with over 1,000 photographs. The emphasis is on pests of the British Isles and Europe.

Alford, D.V. 1984. *Colour Atlas of Fruit Pests*. Glasgow, Scotland: Wolfe Publishers 320 p. ISBN: 0723408165. Out of print. Insects are listed by family with descriptions, life histories, damage, and control measures with color photographs and line drawings. An index to scientific names of plants and insects and references are included.

Armitage, A.M. 1994. *Ornamental Bedding Plants*. Crop Production in Agriculture, v. 4. Wallingford, UK: CABI Publishing. 175 p. ISBN: 0851989012. Considers aspects of propagation and cultivation such as the role of nutrition and media; temperature; light; supplemental carbon dioxide and growth regulators; postproduction handling; diseases and pests; and mechanization for tender annuals and biennials.

Armitage, A.M. 1993. *Specialty Cut Flowers: The Production of Annuals, Perennials, Bulbs, and Woody Plants for Fresh and Dried Cut Flowers*. Portland, OR: Timber Press. 372 p. ISBN: 0881922250. A guide to growing, harvesting, and marketing both common and unusual species in the greenhouse and field for cut flowers, foliage, and fruit.

Atherton, J.G. and J. Rudich. 1986. *The Tomato Crop: A Scientific Basis for Improvement*. London: Chapman and Hall. 658 p. ISBN: 0412251205. Out of print. Genetics and breeding, physiology, crop production, diseases, pests, disorders, and greenhouse production are covered in this comprehensive work.

Ball, V. 1998. *Ball Redbook*. 16th ed. Batavia, IL: Ball Publishers 816 p. ISBN: 1883052157. This complete guide to floriculture crops includes detailed instructions for seed germination, nutrition, lighting, and marketing for 120 species.

Bates, D.M., R.W. Robinson, and C. Jeffrey. 1990. *Biology and Utilization of the Cucurbitaceae*. Ithaca, NY: Cornell University Press. 485 p. ISBN: 0801416701. Out of print. Focuses on systematics, breeding, sex expression, and utilization of some lesser known species.

Bechtel, H. and P. Cribb. 1992. *Manual of Cultivated Orchid Species*. 3rd ed. Cambridge, MA: MIT Press. 585 p. ISBN: 0262023393. A definitive reference originally published in German. Includes descriptions of 140 genera and over 800 color photographs.

Bosland, P.W. and E.J. Votava. 1999. *Peppers: Vegetable and Spice Capsicums*. Wallingford, UK: CABI Publishing. 216 p. ISBN: 0851993354. History, genetics, composition, production, harvesting, and diseases and pests of peppers used for food, flavorings, medicinals, and cosmetics.

Bowes, B.G. 1999. *A Colour Atlas of Plant Propagation and Conservation*. London: Manson. 224 p. ISBN: 1874545707. An illustrated work on the techniques of propagation.

Bunt, A.C. 1988. *Media and Mixes for Container-Grown Plants*. London: Unwin Hyman. 309 p. ISBN: 0046350160. Out of print. Physical and chemical characteristics of both soil containing and soilless mixtures, plant nutrition, irrigation, and sterilization.

Chase, A.R., M. Daughtrey, and G.W. Simone. 1995. *Diseases of Annuals and Perennials: A Ball Guide to Identification and Control*. Batavia, IL: Ball Publishers 202 p. ISBN: 1883052084. This work is a guide to physiological disorders and foliar, root, and stem diseases, including causal agents and susceptible species. Illustrations, some in color, are included.

Chemical and Pharmaceutical Press. 2000. *Turf and Ornamental Chemicals Reference*. 9th ed. New York: Chemical and Pharmaceutical Press. 3 vols. 1,100 p. ISBN: 1570090017. This reference lists product labels and Material Safety Data Sheets (MSDS) indexed by brand name, producer, plant, site, pest, product category, and common name. Adjuvants are included.

Davidson, H., R. Mecklenburg, and C.M. Peterson. 1999. *Nursery Management: Administration and Culture*. 4th ed. Upper Saddle River, NJ: Prentice Hall. 513 p. ISBN: 0138579962. Management issues covered include business planning, marketing, employee relations, legislation, shipping, and other topics. Cultural information on nutrition, growth regulation, irrigation, and soils.

Decoteau, D.R. 2000. *Vegetable Crops*. Upper Saddle River, NJ: Prentice Hall. 464 p. ISBN: 0139569960. History, taxonomy, environmental factors, cultivation, post-harvest care, and economics of vegetable growing for commercial growers.

Dirr, M. 1998. *Manual of Woody Landscape Plants: Their Ornamental Characteristics, Culture, Propagation, and Uses*. 5th ed. Champaign, IL: Stipes Publishers 250 p. ISBN: 0875637957. This is the essential guide to woody ornamentals and includes descriptions, hardiness, growth rate, culture, diseases, insects, landscape uses, cultivars, and propagation for each species described. A CD-ROM version is also available.

Dirr, M. and C. Heuser. 1987. *Reference Manual of Woody Plant Propagation: From Seed to Tissue Culture*. Athens, GA: Varsity Press, Inc. 240 p. ISBN: 0942375009. Chapters on propagating trees and shrubs by seed, cuttings, grafting, and tissue culture are followed by detailed instructions for each species. Includes numerous references to the literature.

Dole, H.M. and H.F. Wilkins. 1999. *Floriculture: Principles and Species*. Upper Saddle River, NJ: Prentice-Hall. 613 p. ISBN: 133747034. Covers a wide range of potted plant, cut flower, and bedding plant crops. Includes both production and postharvest aspects from both a U.S. and an international perspective.

Duke, J.A. 1988. *CRC Handbook of Nuts*. Boca Raton, FL: CRC Press. 368 p. ISBN: 0849336368. The text defines nuts and discusses their economic and nutritional value. Each nut is listed by species, arranged alphabetically by scientific name with illustrations, taxonomy, colloquial names, uses, folk medicine, chemistry, germplasm, distribution, ecology, and cultivation.

Foster, H.L. 1982. *Rock Gardening: A Guide to Growing Alpines and Other Wildflowers in the American Garden*. Portland, OR: Timber Press. 466 p. ISBN: 0917304292. An award-winning classic that includes rock-garden construction techniques as well as an illustrated catalog of 1,900 rock garden plants.

George, R.A.T. 2000. *Vegetable Seed Production*. 2$^{nd}$ ed. Wallingford, UK: CABI Publishing 328 p. ISBN: 0851993362. This work describes vegetable seed production for both developing and developed countries. Entries are listed by plant family.

Gill, S. and J. Sanderson. 1998. *Ball Identification Guide to Greenhouse Pests and Beneficials*. Batavia, IL: Ball Publishers. 143 p. ISBN: 1883052017. Contains color photographs of 450 insects and includes information on how to submit a sample for professional diagnosis.

Gist, J. and F.X. Jozwik. 2000. *Greenhouse and Nursery Handbook: A Complete Guide to Growing and Selling Ornamental Container Plants*. Mills, WY: Andmar Press. 806 p. ISBN: 0916781232. This work includes detailed information on growing and marketing for the commercial greenhouse operator.

Goldblatt, P. and D.E. Johnson, eds. 1999. *Index to Plant Chromosome Numbers 1994–1995*. St. Louis, MO: Missouri Botanic Gardens Press. 208 p. ISBN: 0915279592. This series indexes species by family with chromosome numbers in the gametophytic and sporophytic generations with references to the source literature.

Halevy, A.H. 1989. *CRC Handbook of Flowering*. Boca Raton, FL: CRC Press. 776 p. ISBN: 0849339162. Information is provided on all aspects of flower development; including sex expression, requirements for flowering initiation and development; photoperiod; light density; vernalization and other temperature effects; and interactions for species listed alphabetically.

Hall, R., ed. 1991. *Compendium of Bean Diseases*. St. Paul, MN: American Phytopathological Society. 114 p. ISBN: 0890541183. Jones, P., R.E. Stall, and T.A. Zitter, eds. 1991. *Compendium of Tomato Diseases*. St. Paul, MN: American

Phytopathological Society. 141 p. ISBN: 0890541205. Ritchie, D.F., J.M. Ogawa, E.I. Zehr, G.W. Bird, eds. 1995. *Compendium of Stone Fruit Diseases*. St Paul, MN: American Phytopathological Society. 128 p. ISBN: 0890541744. The Disease Compendium Series published by the American Phytopathological Society contains volumes describing diseases in a wide range of ornamentals, fruits, and vegetables. Each of the volumes has information on diagnosis and control using biological, chemical, and integrated pest-management techniques with color photographs.

Hanan, J.J. 1997. *Greenhouses: Advanced Technology for Protected Horticulture*. Boca Raton, FL: CRC Press. 720 p. ISBN: 0849316987. Addresses the major environmental factors of light, temperature, water, nutrition, and carbon dioxide, and features extensive discussions of greenhouse types, construction, and climate control. The book highlights technology such as hydroponics, computer control of environments, and advanced mathematical procedures for environmental optimization.

Handreck, K.A. and N.D. Black. 1994. *Growing Media for Ornamental Plants and Turf*. Randwick, Australia: University of New South Wales Press. 448 p. ISBN: 0868403334. Physical and chemical properties of growing media, effects on plant growth with an emphasis on fertilization, irrigation, and porosity.

Harris, R.W. and J.B. Ingram. 2000. *A Guide for Plant Appraisal*. 9th ed. Savoy, IL: International Society of Arboriculture. 144 p. ISBN: 1881956253. Information on how to evaluate landscape plants based on size, species, condition, and location.

Harris, R.W., J.R. Clark, and N.P. Matheny. 1999. *Arboriculture: Integrated Management of Landscape Trees, Shrubs, and Vines*. 3rd ed. Upper Saddle River, NJ: Prentice-Hall. 687 p. ISBN: 0133866653. Plant selection for specific regions and conditions; pruning; soil improvement; nutrition, irrigation; injury prevention and repair; pest and disease concerns for trees; and other woody species.

Jackson, R.S. 2000. *Wine Science: Principles and Applications*. 2nd ed. Orlando, FL: Morgan Kaufmann Publishers. 550 p. ISBN: 012379062X. A review of the world literature on viticulture and wine making.

Johnson, W.T. and H.H. Lyon. 1991. *Insects that Feed on Trees and Shrubs*. 2nd ed. Ithaca, NY: Cornell University Press. 560 p. ISBN: 0801426022. Insects are grouped by the types of trees or shrubs attacked and then by type of damage caused. Life-cycle information for each pest is included. Color photographs of plant damage and life stages of each insect are shown.

Jones, J.B., Jr. 1997. *Plant Nutrition Manual*. 2nd ed. Boca Raton, FL: CRC Press. 160 p. ISBN: 188401531X. Nutritional requirements are given for 143 plants grouped in 7 categories from food crop plants to ornamentals. Information on tissue sample analysis is also included.

Jones, J.B., Jr. and S.C. Andersen. 1997. *Hydroponics: A Practical Guide for the Soilless Grower*. Boca Raton, FL: CRC Press. 248 p. ISBN: 1884015328. Explains the basics of plant growth and development as related to hydroponics, the different methods of preparing and using hydroponic nutrient solutions, and hydroponic options for various environmental conditions. Charts, tables, and illustrations are included.

Kelley, A.F. 1998. *Encyclopaedia of Seed Production of World Crops*. New York: Wiley. 414 p. ISBN: 0471982024. A guide to the production, storage, and use of seeds of 300 food, industrial, and commercial crops.

Leslie, A.R. 1994. *Handbook of Integrated Pest Management for Turf and Ornamentals*. Boca Raton, FL: CRC Press. 672 p. ISBN: 0873713508. Describes benefits of using integrated pest management and techniques to minimize chemical use. Outlines situations when chemical use is needed. Includes bibliography and case studies.

MacDonald, B.A. 1986. *Practical Woody Plant Propagation for Nursery Growers*. Portland, OR: Timber Press. v. 1. ISBN: 0881920622. Propagation by seed or by vegetative techniques such as grafting, rooting of cuttings, and air layering are covered including facilities and tools needed.

Marschener, H. 1995. *Mineral Nutrition of Higher Plants*. 2$^{nd}$ ed. Academic Press. 889 p. ISBN: 0124735428. Covers the effects of mineral nutrition on all plant functions with an emphasis on horticultural crops. An extensive bibliography is included.

Maynard, D.N., G.J. Hochmuth, and J.E. Knott. 1997. *Knott's Handbook for Vegetable Growers*. 4$^{th}$ ed. New York: Wiley. 582 p. ISBN: 0471131512. This is the standard guide for commercial growers. Information on planting rates; schedules; spacing; soil; fertilization; insect and disease control; water management, germination; and weed control are included for each crop.

McKinlay, R.G. 1992. *Vegetable Crop Pests*. Boca Raton, FL: CRC Press. ISBN: 0849377293. A guide to the major insect pests of temperate vegetables. For each pest geographical distribution, description, life cycle, and plant damage are described with biological and chemical control measures.

McRae, E.A. 1998. *Lilies: A Guide for Growers and Collectors*. Portland, OR: Timber Press. 392 p. ISBN: 0881924105. Contains information on the biology, propagation, cultivation, and breeding of lilies. Includes a list of wild lily species and contemporary cultivars and hybrids.

Nau, J. 1999. *Ball Culture Guide: The Encyclopedia of Seed Germination*. 3$^{rd}$ ed. Batavia, IL: Ball Publishers. 243 p. ISBN: 188305219X. This volume included germination and scheduling information for 300 seed-grown crops including bedding plants, potted plants, foliage plants, herbs, cut flowers, perennials, vegetables, and ornamental grasses.

Nau, J. 1996. *Ball Perennial Manual*. Batavia, IL: Ball Publishers. 487 p. ISBN: 1883052106. This work covers 300 species in 149 genera. Information includes nomenclature, description, propagation, adaptation, cultivation, a source list, glossary, and color photographs.

Nelson, P.V. 1998. *Greenhouse Operation and Management*. 5$^{th}$ ed. Parasmus, NJ: Prentice-Hall. ISBN: 0133746879. This work is a guide to building design and operation for commercial greenhouses including information on environmental control systems, fertilization, water quality, material prices, cooling and heating, and post-production handling of crops. The sixth edition is in preparation.

Nieuwhof, M. 1969. *Cole Crops: Botany, Cultivation, and Utilization*. London: Leonard Hill Books. 353 p. No ISBN available. Out of print. Although dated, this book contain a great deal of information on sprouting, growing and breeding cabbage, broccoli, brussel sprouts, cauliflower, kale, and kohlrabi.

Ogawa, J.M. and H. English. 1991. *Diseases of Temperate Zone Tree Fruit and Nut Crops*. Oakland, CA: A N R Publications. 464 p. ISBN: 0931876974. Includes information on diseases of pome fruits, stone fruits, walnuts, pistachios, olives, figs, and some minor fruit and nut tree species. Color photographs are included.

Powell, C.C. and R.K. Lindquist. 1997. *Ball Pest and Disease Manual: Disease, Insect, and Mite Control on Flower and Foliage Crops*. 2$^{nd}$ ed. Batavia, IL: Ball Publishers 426 p. ISBN: 1883052130. This guide provides information on disease and insect control using Integrated Pest Management, chemicals, and biological controls. It also includes information on safety regulations, new products, and pest diagnosis and detection.

Rabinowitch, H.D. and J.L. Brewster. 1990. *Onions and Allied Crops*. Boca Raton, FL: CRC Press. 3 vols. ISBN: 0849363000 (v. 1), 0849363012 (v. 2) Volume 1 covers botany, physiology, and genetics. Volume 2 deals with agronomy, biotic interactions, pathology, and crop protection. The third volume of this title has not yet been published.

Robinson, R.W. 1997. *Cucurbits*. Wallingford, UK: CABI Publishing. 240 p. ISBN: 0851991335. Taxonomy, breeding, and crop production of cucumbers, gourds, muskmelons, pumpkins, squashes, and watermelons.

Rubatzky, V.E., C.F. Quiros, and P.W. Simon. 1999. *Carrots and Related Vegetable Umbelliferae*. Wallingford, UK: CABI Publishing. 294 p. ISBN: 0851991297. Botany, production, pest control, breeding, postharvest care, and utilization of root, foliage, seed, and herb crops in this plant family.

Ryder, E.J. 1999. *Lettuce, Endive, and Chicory*. Wallingford, UK: CABI Publishing 208 p. ISBN: 0851992854. Taxonomy, genetics and breeding, production, crop protection, marketing, and food safety concerns.

Salunkhe, D.K. and S.S. Kadam. 1998. *Handbook of Vegetable Science and Technology*. New York: Marcel Dekker. 721 p. ISBN: 0824701054. Includes production and post-harvest technology for 70 major and minor vegetable crops. Aspects considered are taxonomy, anatomy, adaptation, propagation, cultivation, disease and pest control, post-harvest handling, and processing.

Salunke, D.K. and S.S. Kadam. 1995. *Handbook of Fruit Science and Technology: Production, Composition, Storage, and Processing*. New York: Marcel Dekker. 611 p. ISBN: 0824796438. A guide to the botany, adaptation, propagation, culture, disease and pest control, harvesting, composition, storage, and processing of 60 fruits grown worldwide.

Schaffer, B. and P.C. Andersen. 1994. *Handbook of Environmental Physiology of Fruit Crops*. Boca Raton, FL: CRC Press. 2 vols. ISBN: 0849301793. Describes the effects of irradiance, temperature, water, and salinity on growth and reproduction of both tropical and temperate fruit.

Schenk, G.H. 1997. *Moss Gardening: Including Lichens, Liverworts, & Other Miniatures*. Portland, OR: Timber Press. 262 p. ISBN: 08819237. Techniques for transplanting, propagating, and growing mosses and other cryptogams as groundcovers, in containers, and in bonsai with color illustrations and a suggested reading list.

Sinclair, W.A., H.H. Lyon, and W.T. Johnson. 1987. *Diseases of Trees and Shrubs*. Ithaca, NY: Cornell University Press. 574 p. ISBN: 0801415179. This comprehensive work covers 350 disease agents and environmental factors affecting 250 plant species in the United States and Canada. It includes 247 pages of color photographs and 1,700 illustrations.

Small, E. 1997. *Culinary Herbs*. Ottawa, Canada: NRC Research Press. 710 p. ISBN: 0660166682. Discusses taxonomy, history, uses, cultivation, and phytochemistry of 125 culinary and medicinal herbs with numerous references to other works.

Snowdon, A.L. 1990. *A Colour Atlas of Post Harvest Diseases and Disorders of Fruits and Vegetables*. London: Wolfe Scientific. 2 vols. ISBN: 0723416362 (v. 1) out of print; ISBN: 0723416362 (v. 2). Volume 1 of this work covers a general introduction and is unfortunately out of print. Volume 2 is available and covers vegetable crops.

Young, J.A. and C.G. Young. 1986. *Collecting, Processing, and Germinating Seeds of Wildland Plants.* Portland, OR: Timber Press. 236 p. ISBN: 0881920576. A valuable resource for native plant horticulture.

## JOURNALS AND OTHER SERIAL PUBLICATIONS

This list of serials is not comprehensive, but aims to provide a guide to the most important serials in horticulture for colleges, universities, and research institutions. While popular garden magazines may provide a wealth of reliable and useful information, they are not included here. These journals cover topics related to fruits, nuts, vegetables, ornamentals, minor crops, turfgrass, and medicinals. Most of them are peer reviewed. There are a number of newsletters devoted to the genetics of specific crops or plant families that are not included here, but are valuable sources of information for breeders and geneticists.

*Advances in Economic Botany.* v. 1. 1983– . Bronx, NY: New York Botanic Gardens. Irregular. ISSN: 0741-8280. Contains original research and symposia on the use and management of plants.

*Advances in Horticulture & Forestry.* v. 1– . 1990– . Jodhpur, India: Scientific Publishers. Annually. ISSN: 0971-0507. Publishes applied and basic research in pomology, oleiculture, floriculture, fruit and vegetable preservation, and landscaping gardening.

*Advances in Horticultural Science.* 1987– . Firenze, Italy: Universita degli Studia di Firenze, Dipartimento di Ortoflorofrutticoltura. Quarterly. ISSN: 0394-6169. Refereed publication dealing with production, post-handling, and processing of horticultural products. Text and summaries in English.

*American Journal of Enology and Viticulture.* v. 10– . 1959– . Davis, CA: American Society for Enology and Viticulture. Quarterly. ISSN: 0002-9254. Focuses on research on the cultivation of wine grapes and wine making.

*Arboricultural Journal: The International Journal of Urban Forestry.* v. 2– . 1974– . London: Arboricultural Association. Quarterly. ISSN: 0307-1375. Research papers and review articles in urban and amenity forestry. Continues the *Arboriculture Association Journal.*

*Biological Agriculture & Horticulture: An International Journal for Sustainable Production Systems.* v. 1– . 1982– . Bicester, Oxfordshire, UK: AB Academic Publishers. Quarterly. ISSN: 0144-8765. Publishes research and reviews related to biological, integrated agriculture, and comparative studies.

*Economic Botany.* v. 1. 1947– . Bronx, NY: New York Botanic Gardens Press. Quarterly. ISSN: 0013-0001. Deals with the utilization of plants by people.

*Ethnobotany.* v. 1– . 1989. New Delhi, India: Deep Publications (Society of Ethnobotanists). Annual. ISSN: 0971-1252. A refereed publication focused on research in ethnobotany and related fields.

*Herbalgram.* v. 1. 1979– . Austin, TX: American Botanical Council. Quarterly. ISSN: 0899-5648. A peer-reviewed publication focusing on research and legislation related to medicinal plants.

*Horticultural Reviews.* v. 1. 1976– . Westport, CN: AVI Pub. Co. ISSN: 0163-7851. Presents reviews on topics in the horticultural sciences with an emphasis on applied research including the production of fruits, vegetables, nut crops, and ornamental plants of economic importance. Published in conjunction with the American Society for Horticultural Science and edited by Jules Janick.

*Horticulturist.* v. 1– . 1992– . London: Institute of Horticulture. Quarterly. ISSN 0964-8992. Publishes information of interest to the profession of horticulture.

*HortScience.* v. 1– . 1966– . Alexandria, VA: American Society for Horticultural Science. Eight times a year. ISSN: 0018-5345. An international publication reporting results of preliminary or ongoing research, progress reports, research notes, and brief announcements of new concepts, methods, applications and findings, and information related to horticultural education and extension. Publishes registrations of new horticultural varieties and germplam lines.

*HortTechnology.* v. 1. 1991– . Alexandria, VA: American Society for Horticultural Science. Quarterly. ISSN: 1063-0198. Provides peer-reviewed practical horticultural information with a scientific basis for professional horticulturists, practitioners, educators, and scientists.

*Journal of Arboriculture.* v. 1. 1975– . Champaign, IL: International Society of Arboriculture. Bimonthly. ISSN: 0278-5226. Publishes information on growth and maintenance of shade and ornamental trees.

*Journal of the American Pomological Society.* v. 54. 2000– . University Park, PA: American Pomological Society. Quarterly. ISSN: 1527-3741. Promotes fruit variety and rootstock improvement through breeding and testing. Publishes latest information on fruit variety introductions and performance of existing varieties. Continues *Fruit Varieties Journal.*

*Journal of the American Society for Horticultural Science.* v. 1– . 1903– . Alexandria, VA: American Society for Horticultural Science. Bimonthly. ISSN:0003–1062. Publishes original research in applied and basic horticulture.

*Journal of Environmental Horticulture*. v. 1– . 1983– . Washington, DC: Horticultural Research Institution. Quarterly. ISSN: 0738-2898. Publishes peer-reviewed reports with a nontechnical summary of each article.

*Journal of Herbs, Spices, and Medicinal Plants*. v. 1– . 1992– . Binghampton, NY: The Haworth Press. Four times a year (during the academic year). ISSN: 1049-6475. Features original articles and short reviews on the production and development of herbs, spices, and medicinal plants including physiology, breeding, productivity, commercial applications, and marketing.

*Journal of the Horticultural Association of Japan (Engei Gakkai Zasshi)*. v. 1– . 1925– . Japan: Engei Gakkai (Japanese Society for Horticultural Science). Quarterly. ISSN: 0013–7626. Articles published in both Japanese and English.

*Journal of Horticultural Science and Biotechnology*. v. 73. 1998– . Ashford, UK: The Invicta Press. Bimonthly. ISSN: 0022-1589. Results of original research on temperate and tropical fruit, perennial crops, vegetables, and flowers in Britain and overseas. Continues *Journal of Horticultural Research*.

*Journal of Tree Fruit Production*. v. 1– . 1996– . Binghampton, NY: Food Products Press. Semiannually. ISSN:1055–1387. Results of practical research on training, growth regulation, pest management, sustainable production, harvesting, handling, storage, and marketing of tree fruit.

*Journal of Turfgrass Management: Developments in Basic and Applied Turfgrass Research*. v. 1– . 1995– . Binghampton, NY: Food Products Press. Quarterly. ISSN: 1070–437X. Gathers and disseminates current advances in basic and applied turfgrass research for scientists and turfgrass managers.

*Journal of Vegetable Crop Production*. v. 1– . 1995– . Binghampton, NY: Food Products Press. Semiannually. ISSN: 1049–6467. Covers problems of vegetable crop management such as land preparation, seeding, or consumption with minimal dependence on chemical use.

*New Plantsman*. v. 1– . 1994– . London: Royal Horticultural Society. Quarterly. ISSN: 1352–4186. Taxonomy, physiology, conservation, and history of ornamental plants.

*New Zealand Journal of Crop and Horticultural Science*. v. 17– . 1989– . Wellington, New Zealand: SIR Publishing. Quarterly. ISSN: 0114–0671. Covers all aspects of horticulture and field crops.

*Scientia Horticulturae*. v. 1. 1973– . Amsterdam: Elsevier. Sixteen times a year (4 vols). ISSN:0304-4238. Deals with horticulture under moderate subtropical and tropical conditions, as well as with open and protected crop growing of vegetables, fruits, mushrooms, bulbs, and ornamentals.

*Small Fruits Review*. v. 1. 2000– . Binghampton, NY: Food Products Press. Quarterly. ISSN: 1522–8851. Focusing on new technologies and innovative techniques for the management and marketing of small fruits and berries.

## PATENTS

In the United States, patent protection for plants is provided by either the Plant Variety Protection office or the U.S. Patent Office, depending on the type of crop. Both of the offices listed in the following text provide links to similar offices in other countries.

Plant Variety Protection Office (http://www.ams.usda.gov/science/pvpo/pvp. htm) (cited July 12, 2000). Washington, DC: USDA, Agricultural Marketing Services. Intellectual property rights for plant breeders of seed- and tuber-propagated plants are protected by the Plant Variety Protection Act in the United States Cultivars and varieties of plants registered under the provisions of this act can be searched by crop, owner name, variety name, or number at the URL listed here. Links to relevant Web sites of other countries are provided through the Plant Variety Protection Office Web site.

United States Patent and Trademark Office: A Performance Based Organization. United States Patent Office (http://www.uspto.gov/) (July 12, 2000). Clonally propagated plants can be patented in the United States. At the U.S. Patent Office Web site, there is a database of U.S. patents that can be searched by keyword, assignee, or patent number. Patents may also apply to genetically engineered organisms or genes, and to some aspects of natural product utilization. The Web site includes links to intellectual property rights offices in many other countries.

## SOCIETIES

This list includes major professional societies concerned with research in horticulture. There are a number of associations devoted to a particular species or groups of species that were not included on this list. Trade associations also were not included, nor were associations with more general membership emphasizing home gardening.

American Association of Botanical Gardens and Arboreta (AABGA)
Contact information:
351 Longwood Rd.
Kennett Square, PA 19348-1807
Telephone: +1 610-925-2500
Fax: +1 610-925-2700
URL: http://www.aabga.org

Founded in 1940. 3,000 members. Professional society of directors and staff of botanical gardens, arboreta, institutions maintaining or conducting horticultural courses, and others. Promotes professional development and advocates for the interests of public gardens in political, corporate, foundation, and community arenas. Encourages gardens to adhere to professional standards. Publications: *American Association of Botanical Gardens and Arboreta—Newsletter*; *Internship Directory*. Annual conference.

American Horticultural Therapy Association (AHTA)
Contact information:
909 York St.
Denver, CO 80206
Telephone: +1 303-370-8087
Fax: +1 303-331-5776
URL: http://www.ahta.org/
Founded in 1973. 800 members. Promotes and encourages the development of horticulture and related activities such as a therapeutic and rehabilitative medium. Publications: *AHTA Annual Membership Directory*; *Journal of Therapeutic Horticulture*; *People Plant Connection Newsletter*. Annual conference.

American Pomological Society
Contact information:
102 Tyson Building
University Park, PA 16802
URL:http://garden.cas.psu.edu/aps/f-index.htm
Founded in 1848. 1,000 members. Promotes the study and culture of fruit and nuts, and disseminates information pertaining to fruit and nut cultivars. Publications: *Fruit Varieties Journal*. Annual conference.

American Society for Horticultural Science (ASHS)
Contact information:
113 S. West St., Ste. 200
Alexandria, VA 22314-2851
URL: http://www.ashs.org
Founded in 1903. 4,500 members. Promotes and encourages scientific research and education in horticulture within the United States and throughout the world. Publications: *ASHS Newsletter*; *HortScience*; *HortTecnology*; *Journal of the American Society for Horticultural Science*. Annual meeting, general assembly, and symposium. See Web site for schedule.

Canadian Society for Horticultural Science (CSHS)
Contact information:
141 Laura Ave. W, Set 1112
Ottawa ON, Canada K1P 5J3

URL: http://www.aic.ca/members/cshs.html
Founded in 1956. 100 members. Members are scientists, educators, students, extension agents, and industry personnel involved in research, teaching, information, and technology related to all horticultural crops such as fruits, vegetables, nuts, herbs, greenhouse-grown plants, flowers, nursery plants, and more. Publications: *Canadian Journal of Plant Science* (contributes). Annual meeting.

European Association of Plant Breeders (EUCARPIA)
Contact information:
P.O. Box 315
6700 AH
Wageningen, The Netherlands.
URL: http://www.eucarpia.org/
Founded in 1956. EUCARPIA has 12 sections focusing on specific crops including sections for vegetables, fruits, and ornamentals. The organization facilitates international exchange of information and materials. Triennial General Conference and section meetings. See Web site for schedule.

Horticultural Research Institute (HRI)
Contact information:
1250 Eye St., NW Ste. 500
Washington, DC 20005
Founded in 1962. 250 members. The research division of the American Nursery and Landscape Association. Focuses on research in nursery management, marketing, production, pest management, and water use. Publications: *New Horizons*, *Journal of Environmental Horticulture*. Annual meeting.

Institute of Horticulture
Contact information:
14-15 Belgrave Sq.
London SW1X 8PS, England
URL: http://www.horticulture.demon.co.uk/
Founded in 1984. 2,010 members. Promoting the profession of horticulture; acting as an authoritative body for the purpose of consultation with government and other policy-making bodies on matters of interest and concern to professional horticulturists; conferring status upon professionally qualified and experienced horticulturists; promoting educational and training opportunities; encouraging the development of all disciplines within horticulture; disseminating information on matters affecting the profession; providing opportunities for discussion among horticulturists through publications, conferences, seminars, lectures, and other appropriate media; and promoting the important part that horticulture plays in food production, improving the environment, and in supporting one of the main

leisure pursuits—gardening. Publications: *Horticulturist*. Annual residential conference. See Web site for seminar schedule.

International Society for Horticultural Science (ISHS) Societe
Internationale de la Science Horticole (SISH)
Contact information:
Kardinaal Mercierlaan 92
B-3001 Louvain, Belgium
URL: http://www.ishs.org/index 1.htm
Founded in 1959. 2,950 members. The leading worldwide organization of horticultural scientists. The aim of the ISHS is to promote and to encourage research in all branches of horticulture and to facilitate the cooperation of scientific activities and knowledge transfer on a global scale by means of its publications and events. Publications: *Acta Horticulturae*; *Chronica Horticulturae*; *Horticultural Research International*; *Scientia Horticulturae*. Quadrennial International Horticulture Congress; numerous symposia each year. Check Web site for schedule.

North American Plant Preservation Council (NAPPC)
Contact information:
HC 67 Box 539B
Renick, WV 24966
URL: http://www.gardenweb.com/orgs/nappc/nappc.html
Founded in 1990. 5,000 members. Promotes formation of local and regional collections of endangered indigenous plants; facilitates botanical and horticultural study; gathers and disseminates information on the care of endangered plants; conducts research, educational, and charitable programs. Publications: *Directory of Collections*. Annual board meeting.

Society for Economic Botany (SEB)
Contact information:
New York Botanical Garden
Bronx, NY 10458
URL: http://www.econbot.org
Founded in 1959. 1,200 members. Includes botanists, anthropologists, pharmacologists, and others interested in scientific studies of useful plants. Seeks to develop interdisciplinary channels of communication among groups concerned with past, present, and future uses of plants. Publications: *Economic Botany*; *SEB Newsletter*. Annual conference and symposium.

# 13

## Human Nutrition

**Amy L. Paster**
The Pennsylvania State University, University Park,
Pennsylvania, USA

**Heather K. Moberly**
Oklahoma State University, Stillwater, Oklahoma, USA

Few subjects elicit as much scientific research and emotional response as food. Human nutrition is a topic of study that spans many agricultural and medical fields and this diverse nature does not lend itself to an overall core literature. The semiannual "Brandon/Hill Selected Lists for Allied Health" [*Bulletin of the Medical Library Association* 2000 Jul; 88 (3): 218–33] and "Small Medical Libraries" [*Bulletin of the Medical Library Association* 1999 Apr; 87 (2): 145–69] provide insight into books and journals for some of the medical subdisciplines.

The published body of information about human nutrition is immense, varying from peer-reviewed primary literature to unsubstantiated supposition and professional to trade publications. Areas of research that may influence human nutritional literature in the near future include such topics as functional foods or nutraceuticals. Many dietary habits, such as Kosher or Halal, are based in tradition and faith. They do not necessarily have a *nutrition* component. Other patterns of consumption, such as vegetarianism are often included within the nutrition literature. This chapter focuses on current print reference works that emphasize general and clinical research in the areas of human nutrition, diet therapy, dietetics, nutrition and disease, food composition, nutrition through the life cycle, phytomedicines, and preventive nutrition.

For additional information covering Internet resources see:

Hedges, B., H.K. Moberly, and A.L. Paster. 1997. "Food, Agriculture, and Nutrition." In: *Internet Tools of the Profession: A Guide for Information Professionals*. 2nd ed. H. Tillman, ed. Washington, DC: Special Libraries Association, p. 109–25.

Moberly, H.K. and A.L. Paster. 1998. "Nutrition and Vegetarianism Sites to Really Sink Your Teeth Into." *College and Research Libraries News* 59(4): 265–268.

This chapter intentionally does not include information about food policy, law and legislation, cookbooks, or popular literature.

## ABSTRACTS AND INDEXES

Human nutrition research by its nature bridges many disciplines and their literature, such as biochemistry, biology, food science, microbiology, chemistry, food technology, plant science, and animal science. This makes it necessary to be both flexible and accurate while searching databases and online public access catalogs. There is no inclusive index dedicated to all of the aspects of human nutrition. The following sources cover the majority of the current literature.

Depending on the topic within human nutrition that is being searched, other abstracts and indexes to consider include those that emphasize a specific interdisciplinary component such as Chemical Abstracts, a specific format such as Dissertation Abstracts International, or a specific time frame such as Current Contents (primarily the Life Sciences section).

In general, the most effective search strategies will combine terms that describe a topic using two methods: free-text keywords and controlled vocabulary. Free-text keywords are the terms that are found in the title, abstract, and so forth of the resource that is being described. Controlled vocabulary terms, usually called *subject* or *descriptor* terms, are assigned by the creator of the database to assist you in locating similar items. For example, the author of one article about basil may use the term *basil* in the title and another author may use the term *Ocimum sanctum* in the title. Searching using the controlled vocabulary of a database should link all of the records with the term *basil* to the term *Ocimum sanctum* allowing you to locate both the articles. The help information about each database will list which controlled vocabulary is the authority for that database. Controlled vocabularies are most often arranged in a hierarchy of broader, narrower, and related terms. Combining controlled vocabulary with free-text key words is essential.

Scientific research is written and published all over the world. An awareness of variant spellings and their inclusion in your search strategy will also make your search more effective. Depending upon the database, terms that Americans may consider variant will be the standard term. For example, the *CAB Thesaurus* includes over 500 terms for which there are both British and American spellings. The British spelling is standard for entries in CAB Abstracts and the American spelling is standard for entries in AGRICOLA. Examples of common spelling variants are color versus colour, yogurt versus yoghurt, hemoglobin versus haemoglobin, soybean versus soyabean, and estrogen versus oestrogen. In addition to variant spellings, there are differences in terminology standards. Examples of common terminology variants are corn versus maize, peanut versus groundnut, and zucchini versus courgette.

## CABI Publishing Products

CAB Abstracts. [online] 1973– . Wallingford, UK: CABI Publishing. CAB Abstracts uses the *CAB Thesaurus* for its controlled vocabulary. CAB Abstracts (http://www.cabi.org) is the source file for the CABI Publishing (CAB) print indices, of which one specializes in human nutrition (please see next entry). The comprehensive coverage of the worldwide literature of agriculture and related sciences includes 9% human and animal nutrition and 7% horticulture including medicinal plants. The human nutrition component includes analytical methods, food composition, physiology, biochemistry, diet and effects on health, nutritional disorders, and therapeutic nutrition. The medicinal plant coverage includes pharmacology, animal studies, plant composition, allergens, and toxins. CAB Abstracts is available through a variety of vendors. Depending on the vendor, limiting by subset may be available.

Nutrition Abstracts and Reviews, Series A: Human and Experimental. (Print index) v. 47– . 1977– . Wallingford, UK: CABI Publishing. ISSN: 0309-1295. Monthly. Nutrition Abstracts and Reviews use the *CAB Thesaurus* for its controlled vocabulary. This focused index includes content derived from CAB Abstracts and CAB Health databases with the occasional added review article.

## International Food Information Service Products

*Food Science and Technology Abstracts.* (Print index and electronic database) v. 1– . 1969– . Shinfield, UK: The International Food Information Service (IFIS) under the direction of the Commonwealth Agricultural Bureaux (now CAB International), Institut für Dokumentationswesen, and the Institute of Food Technologists. ISSN: 0015-6574. Monthly.

*The Food Science and Technology Abstracts* (FSTA) uses the *Food Science and Technology Abstracts Thesaurus*. The FSTA includes coverage of the literature of food science, food products, and food processes. The print index is the equivalent of the electronic database. The electronic database is available through a variety of vendors.

## National Library of Medicine Products

The U.S. National Library of Medicine (NLM) in Bethesda, MD publishes a wide variety of citation databases and full-text sources covering the biomedical sciences. NLM databases, in general, use the Medical Subject Headings (MeSH) as their controlled vocabulary. Audience levels range from consumer to researcher. The development and evolution of these resources will outpace the descriptions in monographic literature such as in this chapter. Please visit the NLM Web sites for the latest information.

Index Medicus. (Print index) v. 1– . 1879– . Bethesda, MD: U.S. National Library of Medicine. ISSN: 0019-3879. Monthly.    Issued monthly with cumulative volumes. Coverage includes 3,220 journals encompassing the literature of biomedicine. Volumes include subject, author, and bibliography of biomedical literature reviews. Information is included in Medline.

Medline (MEDlars onLINE) (PubMed—http://www.pubmed.gov). 1966– . Bethesda, MD: U.S. National Library of Medicine.    Coverage includes over 4,500 biomedical journals and is comprised of article citations from *Index Medicus*, *Index to Dental Literature*, *International Nursing Index*, and other sources. Available through a variety of vendors or free of charge from NLM through PubMed.

Medlineplus (http://www.nlm.nih.gov/medlineplus/). Nutrition topics: (http://www.nlm.nih.gov/medlineplus/foodnutritionandmetabolism.html).    Bethesda, MD: U.S. National Library of Medicine.    Designed for a primarily consumer audience, Medlineplus provides information about and links to authoritative Web sources for medical information, preformulated Medline searches, and reference materials. Available free of charge from NLM.

NLM Gateway (http://gateway.nlm.nih.gov/gw/Cmd). Bethesda, MD: U.S. National Library of Medicine.    The Gateway targets searchers who are unfamiliar with NLM databases. It provides one-stop searching to NLM products. At the time of writing of this chapter, the Gateway includes Medline and PubMed (1966 to present), OLDMEDLINE (1958–1965), LOCATORplus (NLM online library catalog), AIDSLINE, HSRProj (Health Services Research Projects), and Medlineplus. Additional NLM products are scheduled to be migrated to the NLM Gateway during future phases of development. Available free of charge from NLM.

PubMed (http://www.pubmed.gov). Bethesda, MD: U.S. National Library of Medicine. Web-based search system superset including: Medline, Pre-Medline (bibliographic citations for articles too new to have full records), more complete coverage of many journals than their Medline coverage, and additional biomedical and life science journals. Available free of charge from NLM.

## National Agricultural Library Products

AGRICOLA (http://www.nal.usda.gov/Ag98/). (AGRICultural OnLine Access). Beltsville, MD: National Agricultural Library. AGRICOLA uses the *CAB Thesaurus* for its controlled vocabulary. Coverage includes agriculture and the related sciences. Database is comprised of two bibliographic data sets: library catalog of the National Agricultural Library (NAL) and the journal citation index. In addition to journal literature, AGRICOLA includes citations to conference proceedings, state experiment station publications, government reports, and other nontraditional sources. See also entry on USDA Food and Nutrition Information Center, part of NAL, in the General section of this chapter. AGRICOLA is available through a variety of vendors or free of charge from NAL's Web site. Depending on the vendor these two data sets may be searched together. The development and evolution of these resources will outpace the descriptions in monographic literature such as this chapter. Please visit the NAL Web site for the latest information.

USDA Food and Nutrition Information Center (http://www.nal.usda.gov/fnic). The Food and Nutrition Information Center (FNIC), a specialized information center at the NAL, provides a variety of information and services related to food and human nutrition including reference service, database access, full-text publications, and bibliographies. Services are available free of charge.

## BIBLIOGRAPHIES AND REVIEWS

Depending on the research topic there may not be an appropriate, thorough, or timely index. The best strategy may be to seek review articles. Reviews often have a temporal or topical focus while providing an overview and review of the published literature. This section highlights a variety of types of review publications and includes citations for several traditional monographic bibliographies.

*Advances in Food and Nutrition Research*. v. 33– . 1989– . San Diego, CA: Academic Press. ISSN: 1043-4526. Annual. Although listed as an annual, sometimes a second thematic volume is published. Each volume includes an index.

*Annual Review of Nutrition*. v. 1– . 1981– . Palo Alto, CA: Annual Reviews. ISSN: 0199-9885. Annual. Each annual volume includes a one-volume subject index, a five-year chapter title index, and a five-year contributing author index.

Introductory materials include a list of related articles in other Annual Reviews series titles (such as *Annual Review of Biochemistry*, or *Annual Review of Medicine*).

Brogdon, J. and W.C. Olsen, eds. 1995. *The Contemporary and Historical Literature of Food Science and Human Nutrition*. Ithaca, NY: Cornell University Press. 296 p. ISBN: 0801430968.   As part of the Literature of the Agricultural Sciences series, this volume includes extensive explanatory text in addition to annotated bibliographic entries. Topics include determination of core publications, characteristics of the literature, and primary historical literature. Includes references and index.

*Critical Reviews in Food Science and Nutrition*. v. 1– . 1970– . Boca Raton, FL: CRC Press. ISSN: 1040-8399. Bimonthly.   Each issue includes an average of three reviews. There is no inclusive indexing.

*Nestle Nutrition Workshop Series*. 1983– . New York: Raven Press. ISSN: 0742-2806. Irregular.   Based on Nestle Workshop series, these volumes are thematic and each has an individual title and ISBN.

Newman, J.M. 1993. *Melting Pot: An Annotated Bibliography and Guide to Food and Nutrition Information for Ethnic Groups in America*. 2nd ed. New York: Garland Publication 240 p. ISBN: 0824077563.   Annotated entries describe cultural and social aspects of food habits, general dietary issues, nutrition, and health concerns of the major ethnic groups of the United States.

*Nutrition Research Reviews*. v. 1– . 1988– . Wallingford, UK: CABI Publishing. ISSN: 0954-4224. Biannually.   Published on behalf of The Nutrition Society, this source averages eight reviews per volume. Each issue includes author index and the second issue includes a complete author and subject index for the volume.

*Nutrition Reviews*. v. 1– . 1942– . Washington, DC: International Life Sciences Institute. Monthly. ISSN: 0029-6643.   Although listed as a monthly publication, some issues include a separate *Part II* (conference proceedings from a wide variety of sponsoring organizations). The last issue in any volume includes subject index, author index, and tables of contents for the volume.

Szilard, P. 1987. *Food and Nutrition Information Guide*. Littleton, CO: Libraries Unlimited. 358 p. ISBN: 0872874575.   This traditional bibliography, with lengthy annotations, is devoted to reference materials about human nutrition, dietetics, food science and technology, and related subjects such as food service. Includes references and index.

Werbach, M.R. 1996. *Nutritional Influences on Illness*. 2nd ed. Tarzana, CA: Third Line Press. 698 p. ISBN: 0961855053.   This volume includes citations to nutritional literature related to 87 specific illnesses. Arranged by illness, each entry arranges citations with abstracts by specific nutrients and their proffered therapeutic affect. Includes appendices, references, and index.

*World Review of Nutrition and Dietetics.* v. 1– . 1959– . New York: Karger. ISSN: 0084-2230. Biannually. The number of reviews in each thematic volume varies. Each volume has an individual title, ISBN, and subject index. Some volumes are based on a conference and some include an author index.

## DICTIONARIES

Although some human nutrition terms will be defined in traditional or science dictionaries, the terminology often requires clarification that is easier to locate and comprehend in a subject specific dictionary. Intentionally, works that focus on food science and technology, policy and legislation, or culinary and hospitality topics have been excluded.

Anderson, Kenneth. 1993. *The International Dictionary of Food and Nutrition.* New York: John Wiley. 330 p. ISBN: 0471559571. This volume uses 7,500 definitions across 40 languages to present a comprehensive collection of terms about foods and cookery from around the world.

Bender, D.A. and A.E. Bender. 1999. *Benders' Dictionary of Nutrition and Food Technology.* 7th ed. Cambridge, UK: Woodhead Publishing. 463 p. ISBN: 1855734753. (Also available from CRC Press, ISBN: 0849300185) This source defines more than 5,000 terms with detail varying from a few sentences to half a page. Tables cover units of measure; reference values for food labeling; recommended nutrient intakes for the United States, United Kingdom, and European Union, food additives permitted in the European Union, and nomenclature of fatty acids. Includes cross-references and tables.

Berdanier, C.D. 1998. *CRC Desk Reference for Nutrition.* Boca Raton, FL: CRC Press. 358 p. ISBN: 0849396824. Descriptive tables are incorporated into the text of this reference. The subject coverage includes common medical terms, biochemical pathways, and physiological processes. Includes tables, diagrams, and figures.

Campbell-Platt, Geoffrey. 1987. *Fermented Foods of the World.* Boston, MA: Butterworths. 291 p. ISBN: 0407003134. This book includes both 3,500 individual entries and 250 groups of foods. These groups are ranked to indicate their importance (minor, medium, or major), regions of production, class of food, consumption, types, production, microbiology and biochemistry, composition, and nutritive value. Includes references. The second edition, published in 2000, was not available for review at the time of this writing.

Drummond, K.E. 1996. *The Dictionary of Nutrition and Dietetics*. Albany, NY: Van Nostrand Reinhold. 391 p. ISBN: 0442022255. Clearly written definitions are geared to undergraduate students. Appendices cover other sources for information, recommended dietary allowances, professional associations, and nutrition labels. Includes tables.

Lagua, R.T. and V.S. Claudio. 1996. *Nutrition and Diet Therapy Reference Dictionary*. 4th ed. New York: Chapman and Hall. 491 p. ISBN: 0412070510. More than 3,000 terms contain lengthy definitions in this source targeted at nutritionists, physicians, nurses, health care practitioners, and students. Includes tables.

Leitzmann, C. and U. Dauer. 1996. *Dictionary of Nutrition = Wörterbuch der Ernährung = Dictionnaire de la nutrition = Dizionario di nutrizione = Diccionario de nutrición*. Stuttgart: Ulmer. 516 p. ISBN: 3800121484. This reference contains approximately 3,000 nutritionally significant terms alphabetically listed and defined in English and translated into German, French, Italian, and Spanish. Additionally there are separate indices in German, French, Italian, and Spanish. Includes references.

## ENCYCLOPEDIAS

Ensminger, A.H., et al. 1995. *The Concise Encyclopedia of Foods and Nutrition*. Boca Raton, FL: CRC Press. 1,178 p. ISBN: 0849344557.

Ensminger, A.H., et al. 1994. *Foods and Nutrition Encyclopedia*. 2nd ed. 2 vols. Boca Raton, FL: CRC Press. 2,415 p. ISBN: 0849389801. These encyclopedias cover food, nutrition, and health. Writing style is direct and easy to comprehend. Food composition tables, including data from both Agricultural Handbook (USDA) no. 8–5 and international sources, are edged in black for quick access. Although the text of both encyclopedias is identical, the two-volume set includes additional detail for some entries. Also, the *Concise Encyclopedia of Foods and Nutrition* has smaller print and margins. Each includes references and index.

Macrae, R., R.K. Robinson, and M.J. Sadler, eds. 1993. *Encyclopaedia of Food Science, Food Technology, and Nutrition*. 8 vols. San Diego, CA: Academic Press. 5,365 p. ISBN: 0122268504. Designed to provide the most comprehensive coverage possible of food science, food technology, and nutrition. Contains approximately 1,000 articles; 3,000 words in length. Includes references and index.

Sadler, M.J., J.J. Strain, and B. Caballero, eds. 1999. 3 vols. *Encyclopedia of Human Nutrition*. San Diego, CA: Academic Press. 1,973 pages plus appendices and index. ISBN: 0122266943. This encyclopedia strives for comprehensive coverage within human nutrition. Major topics are nutrient physiology; nutritional assessment; food composition; associations between foods, lifestyle, and health; and medical diet therapy. Includes references and index.

## FOOD COMPOSITION TABLES

Food composition tables include the standard value of nutrients, vitamins, micronutrients, and macronutrients in food. Many tables include *generic* ingredient-level food values, such as fruits, meats, and vegetables. Others include brand name–prepared or ingredient-level foods either as examples or comparisons. Yet others include brand name–restaurant and fast-food values.

## United States and Canada

Although the United States and Canada have at times had different recommended daily intakes of nutrients for their populations, they are typically included in the same publications. With the advent of the *Dietary Reference Intakes*, (see the Nutrient Recommendations and Dietary Guidelines section of this chapter), the two countries now share a framework for recommendations. The United States Department of Agriculture (USDA) has long been the keeper of the raw data from which many of these works have drawn their values in the form of Agricultural Handbook (USDA) no. 8. See USDA Nutrient Data Laboratory entry for information about its evolution.

Hands, E.S. 2000. *Nutrients in Food*. Baltimore, MD: Lippincott, Williams and Wilkins. 315 p. With CD-ROM. ISBN: 0683307053.   In addition to food composition tables, this book includes detailed background information about nutrients and dietary guidelines for the United States and Canada. Each item in the tables includes basic components, additional fats, Vitamin A and components, vitamins, and minerals. Additionally, each item has a reference code, to be used with the CD-ROM, designed to interact with leading nutrition software packages. Supplemental tables include data for amino acids, alcohol, caffeine, phytosterol, pectin, and theobromine. Includes brand name, prepared, and restaurant foods. References are on the CD-ROM. Includes index.

Netzer, C.T. 1992. *Encyclopedia of Food Values*. New York: Dell Books. 903 p. ISBN: 0440503671.   This book presents an alphabetic listing of foods and basic components (calories, protein, carbohydrates, total fat, saturated fat, cholesterol, sodium, and fiber). Includes brand name, prepared and restaurant foods, references, and tables.

Pennington, J.A.T. 1998. *Bowes and Church's Food Values of Portions Commonly Used*. 17th ed. Philadelphia, PA: Lippincott, Williams and Wilkins. 481 p. ISBN: 0397554354.   This book is divided into 34 categories. Each item includes basic components, vitamins, and minerals. Supplemental tables include data for alcohol, amino acids, caffeine, calorie and carbohydrates in chewing gum, mints, candies, and medications; gluten; iodine; pectin; phytosterol; purines; salicylates; selenium; sugars; theobromine; Vitamin D; Vitamin E; Vitamin E as alpha-tocopherol; and Vitamin K. Covers brand name, prepared, and restaurant foods. Includes extensive references and index.

Souci, S.W., W. Fachmann, H. Kraut, H. Scherz, and F. Senser. 2000. *Food Composition and Nutrition Tables*. 6th ed. Stuttgart, Germany: Medpharm. 1,182 p. ISBN: 3887630769. (Also available from CRC Press, ISBN: 0849307570) This volume is written as a parallel text in German, English, and French. Each table has text in English and contains the nutritional (energy) values, information on waste, and concentration of food constituents. Includes list of sources for the data evaluated during the last decade, tables, appendices, and index.

USDA Nutrient Data Laboratory (http://www.nal.usda.gov/fnic/foodcomp). This Web site provides access to a large number of food composition databases. The USDA Nutrient Database for Standard Reference Release-13 (SR-13) contains the nutrient content of over 6,200 foods and up to 82 nutrients. Much of this information had been previously available in Agriculture Handbook (USDA) no. 8 (AH-8). AH-8 is no longer available in print, but may be downloaded from this site. SR-13 includes all information from and replaces AH-8. The electronic version supersedes the printed version. These databases and their documentation may be searched by keyword and downloaded in a variety of formats.

## International

Differences in both the types of foods and their preparation across the world dictate that there be composition tables for a number of regions, ethnic, and culinary differences. This section lists a sample of the types of tables that have been published. Additionally the *Directory of International Food Composition Tables* from the Food and Agriculture Organization of the United Nations (FAO) includes a variety of specific titles. Available from the World Wide Web at http://www.fao.org/infoods/fdtables/0fdtable.htm.

Arab-Kohlmeier, L., M. Wittler, and G. Schettler. 1987. *European Food Composition Tables in Translation*. New York: Springer-Verlag. 153 p. ISBN: 0387173935. This source compiles the translations of the introductory materials of food composition tables from European countries.

Burlingame, B.A., et al. 1997. *The Concise New Zealand Food Composition Tables*. 3rd ed. Palmerston North, New Zealand: New Zealand Institute for Crop and Food Research; Wellington, New Zealand: Public Health Commission. 167 p. ISBN: 0478108001.

Dignan, D.A. 1994. *The Pacific Islands Food Composition Tables*. Palmerston North, New Zealand: New Zealand Institute for Crop & Food Research. 147 p.

*Food Composition Tables for the Near East*. A research project sponsored jointly by the Food and Agriculture Organization of the United Nations, Food Policy and Nutrition Division, Rome, and U.S. Department of Agriculture, Human Nutrition Information Division, Consumer Nutrition Center, Hyattsville, MD. FAO Food and Nutrition paper no. 26, 1982. Rome, Italy: FAO. 265 p. ISBN: 9251012776.

*Food Composition Table for Use in East Asia*. A research project sponsored by the U.S. National Institute of Arthritis, Metabolism, and Digestive Diseases, and Nutrition Program, and the FAO, Food Policy and Nutrition Division. 1972. Bethesda, MD. National Institutes of Health. 334 p.

*Food Composition Tables for Use in the English-Speaking Caribbean*. 2nd ed. 1995. Kingston, Jamaica: The Caribbean Food and Nutrition Institute. ISBN: 9766260206.

*Food Composition Table Recommended for Use in the Philippines*. 3rd ed. 1964. Manila, Philippines: Food and Nutrition Research Center. 134 p.

Leung, W.-T. W. and M. Flores. 1961. *Food Composition Table for Use in Latin America*. Bethesda, MD: National Institutes of Health, Interdepartmental Committee on Nutrition for National Defense. 145 p.

McCance, R.A. and E.M. Widdowson, eds. 1991. *The Composition of Foods*. 5th rev. and extended ed. London: Royal Society of Chemistry; Ministry of Agriculture, Fisheries and Food. 462 p. ISBN: 0851863914. This volume includes information from the U.K. food composition tables. Includes references, appendices, and tables. Includes separately issued supplements.

Miller, J.B., K.W. James, and P.M.A. Maggiore. 1993. *Tables of Composition of Australian Aboriginal Foods*. Canberra, ACT: Aboriginal Studies Press. 256 p. ISBN: 0855752424.

Nobmann, E. 1993. Nutrient value of Alaska native foods. Anchorage, AL: Alaska Area Native Health Service. Available from the Health Sciences Information Service, University of Alaska, Anchorage. 31 p.

Visser F.R. and J.K. Burrows. 1983. *Composition of New Zealand Foods*. 5 vols. DSIR Bulletin no. 235. Wellington, New Zealand: New Zealand Department of Scientific and Industrial Research. ISSN: 0077-961X. This is an extremely comprehensive treatment.

## HANDBOOKS, MANUALS, AND TEXTBOOKS

This section is divided into ten categories representing the wide variety found within the broad subject of human nutrition. The subject areas include clinical practices and nutritional assessment, general, life cycle, medical/specific illness, military nutrition, nutrient recommendation and dietary guidelines, phytomedicinals, research methods, sport nutrition, and textbooks/references.

### Clinical Practices and Nutritional Assessment

These titles are representative of the types of nutritional information that clinicians are likely to need as at-their-fingertips tools. They are guides, literally to be used in a practical setting, that lead to other more in-depth resources.

Buchman, A. and W.J. Klish. 1997. *Handbook of Nutritional Support*. Baltimore, MD: Williams and Wilkins. 186 p. ISBN: 0683302388.   Intended as a pocket-sized guidebook rather than as an inclusive reference text, this handbook outlines nutritional assessment throughout the life cycle. Special emphasis is placed on disease-specific nutrition and pregnancy. Practical tabular data includes clinical signs of nutrient deficiencies, composition of milks and infant formulas, and measures of strength and growth. Includes tables, references, appendices, and index.

Heimburger, D.C. and R.L. Weinsier. 1997. *Handbook of Clinical Nutrition*. 3rd ed. St. Louis, MO: Mosby. 600 p. ISBN: 0815192746.   This pocket-sized, spiral-bound, ready-reference handbook is divided into three parts. Part one is Nutrition for Health Maintenance, which covers nutrition throughout life. Part two is Nutritional Support in Patient Management in the hospital setting. Topics covered include the nutritional support team, parenteral nutrition, and sample forms. Part three is Nutrition in Specific Clinical Situations. This is the largest section of the book and does a thorough job of covering nutritional concerns for a variety of diseases and illnesses. Does not include menus; more diagnostic in nature. Includes five basic appendices, references, and an extensive index.

Lee, R.D. and D.C. Nieman. 1996. *Nutritional Assessment*. 2nd ed. St. Louis, MO: Mosby. 689 p. ISBN: 0815153198.   This volume, designed to be a textbook for dietetic and public health curricula, covers nutritional assessment and standards, results from extant nutritional surveys, American dietary trends, reviews of nutritional assessment software, role of assessment in disease prevention, interpretation of laboratory test, and counseling theories. Includes appendices (nearly 100 pages of assessment tools and reference tables), glossary, references, and index.

*Manual of Clinical Dietetics*. 2000. 6th ed. Chicago, IL: American Dietetic Association. 950 p. ISBN: 0880911875.   This edition, developed through a cooperative effort of American and Canadian dietitians, is substantially expanded and reorganized. It opens with a discussion of adult and pediatric nutrition assessment. Guidelines are broadly divided into life cycle (curiously this includes the vegetarian nutrition section), nutrition management (adverse reactions to food, burns, cancer, cardiovascular disease, diabetes, eating and weight disorders, gastrointestinal disease, HIV/AIDS, renal disease, transplant, and other), nutrition support [variety of enteral (tube feeding) and parenteral situations], and diets (modified consistency, therapeutic, modified mineral, and others including test diets and Kosher diet). Includes extensive and varied appendices, references, and index. The American Dietetic Association also publishes a separate *Pediatric Manual of Clinical Dietetics*. 1998. 678 pages. ISBN: 0880911603.

Matarese, L.E. and M.M. Gottschlich. 1998. *Contemporary Nutrition Support Practice: A Clinical Guide*. Philadelphia, PA: W.B. Saunders Company. 694 p. ISBN: 0721659993. Designed as both reference and text, this book covers nutrition support and the nutritional support team through the life cycle (pregnancy through geriatrics), disorders of specific systems (such as renal or hepatic failure), disorders of general systems (such as cancer or diabetes), physiologic stress (such as organ transplantation or burns), and management and professional issues (such as home care, ethics, research, and economics). Includes appendices, references, and index.

Sauberlich, H.E. 1999. *Laboratory Tests for the Assessment of Nutritional Status*. 2nd ed. Boca Raton, FL: CRC Press. 486 p. ISBN: 0849385067. This volume, a complete update of the 1974 edition, outlines the biochemical laboratory techniques available for gathering data for nutritional assessment. Chapters include information specific to water-soluble vitamins, fat-soluble vitamins, semi- or quasi-vitamins, body electrolyte minerals, macrominerals, trace elements, and ultratrace elements. Additionally there are extensive sections covering protein-energy malnutrition and essential fatty acid deficiencies. Part of the CRC Series in Modern Nutrition. Includes tables, references, and index.

## General

Many nutrition resources attempt to present a true overview of the subject. General topic resources often gather summaries of information whose detailed sources will be strewn throughout other literature. Although they include more substantial information than traditional directories, these works quickly become dated because in addition to providing topical summaries they include directions for seeking additional information. A good example is the following: Frank, R.C. *Directory of Food and Nutrition Information for Professionals and Consumers*. 1992. Phoenix, AZ: Oryx Press. 332 p. ISBN: 0897746899.

Bellenir, K. ed. 1999. *Diet and Nutrition Sourcebook*. 2nd ed. Detroit, MI: Omni-graphics. 650 p. ISBN: 0780802284. Divided into 7 parts (60 chapters), this general sourcebook covers all aspects of nutrition. Divisions include nutrition fundamentals, nutrition needs throughout life, nutrition for people with specific medical concerns, weight control, dietary supplements and food additives, nutrition research, and additional help and references. Part of the Health Reference Series. Includes references and index.

Bender, D.A. and A.E. Bender. 1997. *Nutrition: A Reference Handbook*. New York: Oxford University Press. 573 p. ISBN: 0192623680.    This is a handbook for individuals working in the broad fields of diet and health, food, and nutrition. Brings together in one volume information that is usually scattered among various reference sources. Coverage includes food science and technology, labeling legislation, biochemistry, physiology, clinical medicine, and pediatrics. Each chapter includes a list of further reading books and review articles. There is an extensive index and a full bibliography that refers the reader back to the original research literature. Includes references and index.

*Chapman & Hall Nutrition Handbooks* (Chapman & Hall—numbered monographic series)

> *Geriatric Nutrition Handbook*. 1998. v. 5 ISBN: 0412136414.
> *Obstetrics/Gynecology Nutrition Handbook*. 1996. v. 1. ISBN: 0412075016.
> *Pediatric Nutrition Handbook*. 1996. v. 3. ISBN: 0412075113.
> *Preventive and Therapeutic Nutrition Handbook*. 1996. v. 2. ISBN: 0412074915.
> *Surgery Nutrition Handbook*. 1996. v. 4. ISBN: 0412075210.

*Clinical Nutrition in Health and Disease* (Marcel Dekker—unnumbered monographic series)

> *Folate in Health and Disease*. 1995. ISBN: 0824792807.
> *Handbook of Nutritionally Essential Minerals and Elements*. 1997. ISBN: 0824793129.

*CRC Series in Modern Nutrition* or *Modern Nutrition* (CRC Press—unnumbered monographic series)

> *Advanced Human Nutrition*. 2000. ISBN: 0849385660.
> *Advanced Nutrition: Macronutrients*. 2000. 2nd ed. ISBN: 0849387353.
> *Advanced Nutrition: Micronutrients*. 1998. ISBN: 0849326648.
> *Antioxidants and Disease Prevention*. 1997. ISBN: 0849385091.
> *Calcium and Phosphorus in Health and Disease*. 1996. ISBN: 0849378451.
> *Gender Differences in Metabolism: Practical and Nutritional Implications*. 1999. ISBN: 0849381940.

*Handbook of Dairy Foods and Nutrition.* 2000. 2ⁿᵈ ed. ISBN: 0849387310.
*Laboratory Tests for the Assessment of Nutritional Status.* 1999. ISBN: 0849385067.
*Manganese in Health and Disease.* 1994. ISBN: 0849378419.
*Nutrients and Foods in AIDS.* 1998. ISBN: 084938561X.
*Nutrients and Gene Expression: Clinical Aspects.* 1996. ISBN: 0849394856.
*Nutrition and AIDS.* 1994. ISBN: 0849378427.
*Nutritional Care for HIV Positive Persons: A Manual for Individuals and Their Caregivers.* 1995. ISBN: 0849378435.
*Nutritional Concerns of Women.* 1996. ISBN: 0849385024.
*Nutritional and Environmental Influences on the Eye.* 1999. ISBN: 0849385652.
*Nutrition: Chemistry and Biology.* 1999. ISBN: 0849385040.
*Nutrition and Health: Topics and Controversies.* 1995. ISBN: 0849378494.
*Nutrition in Spaceflight and Weightlessness Models.* 2000. ISBN: 0849385679.
*Practical Handbook of Nutrition in Clinical Practice.* 1994. ISBN: 0849378478.

Lysen, L.K. 1997. *Quick Reference to Clinical Dietetics.* Gaithersburg, MD: Aspen Publishers. 292 p. ISBN: 0834206293. Designed for use as a text or reference book, this work emphasizes the biological bases of human nutrition at molecular, cellular, tissue, and whole-body levels. Information is included about clinical dietetics in a hospital or home setting following a progressive order from the management of patients with various diseases/conditions to nutrition support, through discharge planning, to management of the patient outside the hospital setting. Thorough section about meal planning (objective, nutritional adequacy, summary of guidelines). Includes appendices, references, and index.

## Life Cycle

Among research specialities, differentiation in nutrition throughout the human life cycle is well represented in the literature. Many professional medical associations, for example the American Academy of Pediatrics, have nutrition divisions or participate in cooperative research and information dissemination in human nutrition. Among the most heavily represented life-cycle stages in the literature are pregnancy, lactation, pediatric including adolescent, adult, menopause, and geriatric.

Bendich, A. and R.J. Deckelbaum. 1997. *Preventive Nutrition: The Comprehensive Guide for Health Professionals.* Totowa, NJ: Humana Press. 579 p. ISBN: 0896033511. The second edition, published in 2001, was not available for review at the time of this writing. Objectives of the book are to provide physicians and other health care professionals with the new and up-to-date research indicating that the risk of many of the major diseases affecting middle-aged adults can be prevented, or at least delayed, with simple nutritional approaches. The second objective is to examine key research linking nutritional status with the prevention of birth defects and optimization of birth outcomes. Part one is "Public Health Implications of Preventive Nutrition." Part two is "Prevention of Major Disabilities." Part three is "Optimal Birth Outcomes." Part four is "Benefits of Preventative Nutrition in the United States and Europe." Part five is "Implications of Prevention Nutrition for the Far East, South America, and Developing Areas." Part six is "Nutrition-Related Resources." Includes references and index.

Chernoff, R. 1999. *Geriatric Nutrition: The Health Professional's Handbook*. 2nd ed. Gaithersburg, MD: Aspen Publishers. 518 p. ISBN: 0834210827. The opening chapters address issues facing our society that are related to shifting demographics, present knowledge in nutrient requirements for macronutrients, vitamins, minerals, and trace metals. There is a large section that examines the impact of aging on organ systems. The closing chapter is about health promotion for the elderly. An entire chapter is devoted to polypharmacy, with related drug-drug and drug-nutrient interactions. There is also a chapter dealing with exercise and aging. Includes references and index.

Kleinman, R.E. ed. 1998. *Pediatric Nutrition Handbook 4*[th] ed. Elk Grove Village, IL: American Academy of Pediatrics. 833 p. ISBN: 1581100051. Designed as a ready reference tool for physicians and other health professionals working with the health of children from infants through adolescents. Major subject areas include feeding normal infants and older children; micronutrients and macronutrients; nutrition delivery systems; disorders of metabolism and nutrition; and diet and prevention of disease. Includes tables, appendices, references, and index.

Worthington-Roberts, B.S. and S.R. Williams, eds. 2000. *Nutrition Throughout the Life Cycle*. 4th ed. Boston, MA: McGraw-Hill. 469 p. ISBN: 0072927321. This text provides an overview of the role of nutrition in each stage of life. Each chapter includes basic concepts, nutrient requirements, special considerations, a summary, and review questions. Includes glossary, tables, appendices, references, and index.

## Medical/Specific Illness

The interaction between nutrition and wellness is another area of well-defined cooperation among specialists. A tremendous body of literature is devoted to the nutritional influences in the maintenance of wellness and the prevention of, recovery from, or curing of disease. Many resources will focus on a particular disease and its interaction with nutrition in general or with a particular nutrient. These sources illustrate the wide variety of information that is available: general, diagnostic, and disease-specific information.

Escott-Stump, S. 1998. *Nutrition and Diagnosis-Related Care*. 4ᵗʰ ed. Philadelphia, PA: Lippincott, Williams and Wilkins. 785 p. ISBN: 0683301209. Designed to supplement other references, this source lists nutritional acuity ranking; definitions and notation; objectives; dietary and nutritional recommendations; profile; common drugs used; and potential side effects and patient education for each disease, disorder, or situation discussed. Includes tables, references, and index.

Gershwin, M.E., J.B. German, and C.L. Keen, eds. 2000. *Nutrition and Immunology: Principles and Practice*. Totowa, NJ: Humana Press. 505 p. ISBN: 0896037193. This highly authoritative volume provides an extensive review of the role of nutrition in health. The bulk of this work focuses on clinical issues. Additionally, nutrition assessment, requirements for specific populations, and nutrient immune interactions are covered. Heavily referenced, this book borders on being a review of the literature. Includes tables, references, and index.

Kris-Etheron, P. and J.H. Burns. 1998. *Cardiovascular Nutrition: Strategies and Tools for Disease Management and Prevention*. Chicago, IL: American Dietetic Association. 307 p. ISBN: 088091159X. Intended for physicians, dietitians, nurses, and exercise physiologists, this volume is divided into four sections: Introduction, Background and Assessment, Management, and Implementation Strategies. Among the useful tools for practitioners are gender-specific risk tables and detailed treatment algorithms. Includes tables, references, and index.

Powers, M.A. 1996. *Handbook of Diabetes Medical Nutrition Therapy*. 2ⁿᵈ ed. Gaithersburg, MD: Aspen Publishers. 711 p. ISBN: 0834206315. In its second edition, this book is a comprehensive reference guide for dietitians and health care professionals. It is divided into seven parts: understanding diabetes; setting and achieving management goals; selecting a nutrition education approach; macronutrient influence on blood glucose and health; making food choices; life stages; nutrition and specific clinical conditions; and making it all work. Appendices are at the end of chapters to which they refer. Includes tables, references, and index.

Snetselaar, L.G. 1997. *Nutrition Counseling Skills for Medical Nutrition Therapy*. Gaithersburg, MD: Aspen Publishers, Inc. 409 p. ISBN: 0834207559. This text guides clinical dietetic students and dietitians through nutritional counseling of specific health conditions including: coronary heart disease, obesity, diabetes, hypertension, renal disease, and cancer risks. Includes tables, references, and index.

Zappia, V., et al. eds. 1999. *Advances in Nutrition and Cancer 2.* Advances in Experimental Medicine and Biology, no. 471. New York: Kluwer. 313 p. ISBN: 0306463067. Contributions representing diverse scientific disciplines illustrate recent research about the interaction between nutrition and cancer. Broad subjects include the relationship between risk factors and diet, molecular epidemiology, and nutrition intervention as prevention. Includes tables, references, and index.

## Military Nutrition

There are a number of special areas in nutrition that are often overlooked. Military nutrition is one of these areas. The Committee on Military Nutrition Research, a part of the U.S. Institute of Medicine, is the major producer of research in this field. The following is a list of some of the titles currently available.

> *Body Composition and Physical Performance*: *Applications for the Military Services.* 1992. ISBN: 030904586X.
>
> *Committee on Military Nutrition Research*: *Activity Report, December 1, 1994, through May 31, 1999.* 1999. ISBN: 0309065852.
>
> *Emerging Technologies for Nutrition Research*: *Potential for Assessing Military Performance Capability.* 1997. ISBN: 0309057973.
>
> *Food Components to Enhance Performance*: *An Evaluation of Potential Performance-Enhancing Food Components for Operational Rations.* 1994. ISBN: 030905088X.
>
> *Military Strategies for Sustainment of Nutrition and Immune Function in the Field.* 1999. ISBN: 0309063450.
>
> *Not Eating Enough*: *Overcoming Under-Consumption of Military Operational Rations.* 1995. ISBN: 0309053412.
>
> *Nutritional Needs in Cold and in High-Altitude Environments*: *Applications for Military Personnel in Field Operations.* 1996. ISBN: 0309054842.
>
> *Nutrition Needs in Hot Environments*: *Applications for Military Personnel in Field Operations.* 1993. ISBN: 0309048400.
>
> *The Role of Protein and Amino Acids in Sustaining and Enhancing Performance.* 1999. ISBN: 0309063469.

## Nutrient Recommendations and Dietary Guidelines

From 1941 to 1989 the U.S. Institute of Medicine, Food and Nutrition Board published the *Recommended Dietary Allowances* providing quantitative data about nutrient intakes for healthy individuals. Since 1989 the Dietary Reference Intakes (DRI) have expanded the structure of government recommendations for nutrient intake. The DRI will be published as a multivolume series. Some volumes will provide DRI for nutrients and other volumes will describe methodologies. To date three volumes of DRI and one methodology volume have been published. *Dietary Reference Intakes for Vitamin A, Vitamin K, Arsenic, Boron, Chromium, Copper, Iodine, Iron, Manganese, Molybdenum, Nickel, Silicon, Vanadium, and Zinc* was published in 2002. These National Academy Press titles are freely available in full text at http://www.nap.edu/.

*Dietary Reference Intakes for Calcium, Phosphorus, Magnesium, Vitamin D, and Fluoride.* Standing Committee on the Scientific Evaluation of Dietary Reference Intakes, Food and Nutrition Board, Institute of Medicine. Washington, DC: National Academy Press. 1997. 432 p. ISBN: 0309063507.

*Dietary Reference Intakes: A Risk Assessment Model for Establishing Upper Intake Levels of Nutrients.* Food and Nutrition Board, Institute of Medicine. Washington, DC: National Academy Press. 1998. 71 p. ISBN: 0309063485. The DRI values replace the previously published United States RDA and Canadian Recommended Nutrient Intakes (RNI). The DRI are comprised of several reference values: RDA, Adequate Intake (AI), Tolerable Upper Intake Level (UL), and Estimated Average Requirement (EAR). This volume describes the risk assessment model that is used to develop ULs. Includes references and appendices.

*Dietary Reference Intakes for Thiamin, Riboflavin, Niacin, Vitamin B6, Folate, Vitamin B12, Pantothenic Acid, Biotin, and Choline.* A Report of the Standing Committee on the Scientific Evaluation of Dietary Reference Intakes and its Panel on Folate, Other B Vitamins, and Choline and Subcommittee on Upper Reference Levels of Nutrients, Food and Nutrition Board, Institute of Medicine. Washington DC: National Academy Press. 1998. 564 p. ISBN: 0309064112.

*Dietary Reference Intakes for Vitamin C, Vitamin E, Selenium, and Carotenoids.* A Report of the Panel on Dietary Antioxidants and Related Compounds, Subcommittees on Upper Reference Levels of Nutrients and Interpretation and Uses of Dietary Reference Intakes, and the Standing Committee on the Scientific Evaluation of Dietary Intakes, Food and Nutrition Board, Institute of Medicine. Washington, DC: National Academy Press. 2000. 506 p. ISBN: 0309069491. These volumes include information about the function of the nutrient, methodologies for determining the requirements, external factors that may influence function, and the relationship between the nutrient and chronic disease or developmental abnormality. The DRI values replace the previously published U.S. RDA and Canadian RNI. The DRI are comprised of several reference values: RDA, AI, UL, and EAR. Each includes references, appendixes, and index.

National Research Council, Food and Nutrition Board. 1989. *Recommended Dietary Allowances.* 10th ed. Washington DC: National Academy Press. 284 p.

ISBN: 0309040418. This is the last publication of the RDA before their replacement by the DRI. Includes references.

Nutrition and Your Health: Dietary Guidelines for Americans, 5th ed. (http://www.usda.gov/cnpp/Pubs/DG2000/Index.htm) 2000. This site provides links to comprehensive explanatory information about the USDA Dietary Guidelines. The most recent guidelines, revised in September 2000, are available full-text as HTML (including graphics and tables) and PDF files.

## Phytomedicinals

Phytopharmaceuticals or phytomedicinals are plants or plant constituents that are consumed to produce a therapeutic effect in the human body. Although they do not necessarily have a nutritive value, they are currently regulated in the United States under the Dietary Supplement Health and Education Act (DSHEA, 1994) as dietary supplements rather than as medicines. See the Office of Dietary Supplements at the National Institutes of Health at http://odp.od.nih.gov/ods/ for more information. Locating consistent and reliable information, written in or translated into English, about these substances is often difficult. Listed here is a brief sample of references whose information is widely regarded as reliable.

Bisset, N.G. and M. Wichtl. 2001. *Herbal Drugs and Phytopharmaceuticals*: A *Handbook for Practice on a Scientific Basis*. Boca Raton, FL: CRC Press. 566 p. ISBN: 0849310113. This work was originally published in two German editions, edited by Wichtl, (MedPharm GmbH Scientific Publishers) in 1984 and 1989. The second German edition has been translated to English, edited by Bisset and Wichtl, and published in two editions (CRC Press) in 1994 and 2001. The 2001 edition, although listed in many sources as second English-language edition, is a reprint.

Blumenthal, M. 2000. *Herbal Medicine*: *Expanded Commission E Monographs*. Newton, MA: Integrative Medicine Communications. 519 p. ISBN: 0967077214. This updated reference, based on the *German Commission E Monographs*, contains 107 expanded monographs and color photographs. Also contains expanded information about chemistry, pharmacology, dosage, and administration. Extensive references.

Blumenthal, M. and W.R. Busse. 1998. *The Complete German Commission E Monographs*, *Therapeutic Guide to Herbal Medicines*. Austin, TX: American Botanical Council; Boston, MA: Integrative Medicine Communications. 685 p. ISBN: 096555550X. Botanical medicines are strictly regulated in Germany. Translated by the American Botanical Council, this volume contains translations of 380 monographs (pharmaceutical inserts). It is the source from which many other works derive their information.

Chevallier, A. 1996. *The Encyclopedia of Medicinal Plants*. New York: DK Pub. 336 p. ISBN: 0789410672.

Schulz, V., R. Hansel, and V.E. Tyler. 2001. *Rational Phytotherapy: A Physician's Guide to Herbal Medicine*. 4th ed. New York: Springer-Verlag. 383 p. ISBN: 3540670963.

## Research Methods

Although these titles are about designing research, they are specifically about research in human nutrition and its challenges. Most other research books cover generic study design or the statistical manipulations rather than the actual design of a human nutrition study. An excellent general research design text is the following: Portney, L.G. and M.P. Watkins. 2000. *Foundations of Clinical Research: Applications to Practice*. 2nd ed. Upper Saddle River, NJ: Prentice Hall Health. 752 p. ISBN: 0838526950.

Dennis, B.H. 1999. *Well-Controlled Diet Studies in Humans: A Practical Guide to Design and Management*. Chicago, IL: American Dietetic Association. 418 p. ISBN: 0880911581. This book addresses the difficulties in standardization during human dietary studies. It is divided into five main sections: study design, human factors, dietary intervention, research kitchen, and enhancing the outcome of dietary studies. The focus of this volume is feeding studies rather than other types of nutrition research. The audience is researchers in the field. Includes references and index.

Greenfield, H. and D.A.T. Southgate. 1992. *Food Composition Data, Production, Management and Use*. London; New York: Elsevier Applied Science. 243 p. ISBN: 1851668810. This volume, a product of the International Network of Food Data Systems (INFOODS) initiative, leads the researcher through the creation of a comprehensive food composition database. Includes references, appendices, and index. Reprinted with corrections in 1994 as Greenfield, H. and D.A.T. Southgate. 1992. *Food Composition Data, Production, Management and Use*. London: Chapman & Hall. 243 p. ISBN: 0412537508.

Ireton-Jones, C.S., M.M. Gottschlich, and S.J. Bell. 1998. *Practice-Oriented Nutrition Research: An Outcome Measurement Approach*. Gaithersburg, MD: Aspen Publishers, Inc. 260 p. ISBN: 0834208857. Comprised of two sections: how to conduct nutrition research in clinical practice, and case studies. The nine chapters of section one provide an overview of general research methodology including outcomes research in nutrition support. The case studies section describes four types of research: nurse-led, pharmacist-led, dietitian-led, and physician-led. Includes references and index.

Monsen, E.R. 1992. *Research: Successful Approaches*. Chicago, IL: American Dietetic Association. 449 p. ISBN: 0880910925. This is a publication from the Nutrition Research Dietetic Practice Group of the American Dietetic Association and is more advanced than *Practice-Oriented Nutrition Research*. The audience is practicing dietitians. It is divided into eight parts, the first four parts have to do with research methodology. The next two parts present the unique content of this volume—key research techniques and statistical analyses focusing on nutrition and dietetics (e.g., meta-analysis, sensory evaluation methods, and dietary intake methodology). The final parts describe presentation and application of the research in practice. Includes reference and index.

## Sport Nutrition

Another very specialized area of nutrition is its relationship with athletic performance. There is a great volume of trade and popular literature about sport nutrition, however, there is a growing body of research-based materials as well.

Dorfman, L. 2000. *Vegetarian Sports Nutrition Guide: Peak Performance for Beginner to Gold Medalist*. New York: John Wiley. 270 p. ISBN: 0471348082. Includes references and index.

Maughan, R., ed. 2000. *Nutrition in Sport*. Malden, MA: Blackwell Science. 680 p. ISBN: 0632050942. Volume 7 of *Encyclopaedia of Sports Medicine*. Includes references and index.

McArdle, W.D., F.I. Katch, and V.L. Katch. 1999. *Sports and Exercise Nutrition*. Philadelphia, PA: Lippincott, Williams & Wilkins. 750 p. ISBN: 0683304496.

*Nutrition in Exercise and Sport Series* (CRC Press—unnumbered monographic series)

> *Amino Acids and Proteins for the Athlete: The Anabolic Edge*. 1997. ISBN: 0849381932.
> *Body Fluid Balance: Exercise and Sport*. 1996. ISBN: 0849379180.
> *Energy-Yielding Macronutrients and Energy Metabolism in Sports Nutrition*. 2000. ISBN: 0849307554.
> *Exercise and Disease*. 1992. ISBN: 0849379121.
> *Exercise and Immune Function*. 1996. ISBN: 0849389108.
> *Nutrients as Ergogenic Aids for Sports and Exercise*. 1993. ISBN: 0849342236.
> *Nutrition Applied to Injury Rehabilitation and Sports Medicine*. 1995. ISBN: 084937913X.
> *Nutrition and Exercise Immunology*. 2000. ISBN: 0849307414.
> *Nutrition in Exercise and Sport*. 3rd ed. 1998. ISBN: 0849385601.
> *Nutrition and the Female Athlete*. 1996. ISBN: 0849379172.

*Nutrition, Physical Activity, and Health in Early Life.* 1996. ISBN: 0849379199.
*Nutrition for the Recreational Athlete.* 1995. ISBN: 0849379148.
*Sports Drinks: Basic Science and Practical Aspects.* 2001. ISBN: 0849370086.
*Sports Nutrition.* 2000. ISBN: 0849381975.
*Sports Nutrition: Minerals and Electrolytes.* 1995. ISBN: 0849379164.
*Sports Nutrition: Vitamins and Trace Elements.* 1997. ISBN: 0849381924.

## Textbooks/References

Many of the overviews of human nutrition are published as university-level textbooks or as both a text and a reference book. They are frequently revised, contain overviews of timely/hot topics, and include bibliographies, charts, and reference values. Listed here is a brief sampling of some of the most respected titles in this group.

Desal, B.D. 2000. *Handbook of Nutrition and Diet.* New York: Marcel Dekker. 797 p. ISBN: 0824703758. This volume, part of the *Food Science and Technology Series*, is intended to be a broad reference for students, teachers, and professionals in the study of both food science/technology and nutrition/health. This volume emphasizes nutrition and health covering both traditional topics (such as diabetes and cardiovascular disease) and nontraditional topics (such as mental health, metabolic, bone, skin, teeth, and hair). Includes references and index.

Guthrie, H.A. 1995. *Human Nutrition.* St. Louis, MO: Mosby. 806 p. ISBN: 0815140436. This basic undergraduate text is designed to provide students with an overview of nutrition principles, macronutrients, life cycle, physical fitness, and food safety. Includes references and index.

Mahan, K. and S. Escott-Stump, eds. 2000. *Krause's Food, Nutrition, and Diet Therapy.* 10th ed. Philadelphia, PA: W.B. Saunders Company. 1,194 p. ISBN: 0721679048. This book is designed to serve as a both a text and reference. This edition has an increased emphasis on pathophysiology, incorporated the new DRI, and the ICD-9 codes. Includes tables, appendices, references, and index.

Owen, A.L., P.L. Splett, and G.M. Owen. 1999. *Nutrition in the Community: The Art And Science of Delivering Services.* 4th ed. Boston, MA: McGraw-Hill. 654 p. ISBN: 0815133111. Intended for a broad audience this text emphasizes the development of community nutrition programs and services. Includes references and index.

Shils, M.E., et al., eds. 1999. *Modern Nutrition in Health and Disease.* 9th ed. Philadelphia, PA: Lippincott, Williams and Wilkins. 2,161 p. ISBN: 068330769X. This book is designed to serve as both a text and a reference covering the history, scientific base, and practice of nutrition. New to this edition is a discussion of the role of nutrition in integrated biological systems. Only drawback is the weight of the volume. Includes appendices, tables, references, and index.

Stipanuk, M.H. 2000. *Biochemical and Physiological Aspects of Human Nutrition*. Philadelphia, PA: W.B. Saunders Company. 1,007 p. ISBN: 072164452X. This book is designed to serve as both a text and a reference emphasizing the biological bases of human nutrition at the molecular, cellular, tissue, and whole-body levels. It is intended for readers already competent in organic chemistry, biochemistry, molecular biology, and physiology. Includes tables, references, and index.

Williams, S.R. 1999. *Essentials of Nutrition and Diet Therapy*. 7th ed. St. Louis, MO: Mosby. 729 p. ISBN: 0323003982. This book is an abridged version of *Nutrition and Diet Therapy* designed for entry-level undergraduate nutrition classes. See next entry.

Williams, S.R. 1997. *Nutrition and Diet Therapy*. 8th ed. St. Louis, MO: Mosby. 850 p. ISBN: 0815192738. Designed as an undergraduate text, this book provides the student with an overview of nutrition, nutrition in the community, nutrition throughout the life cycle, and nutrition in clinical care. Includes tables, appendices, references, and index.

## JOURNALS

*American Journal of Clinical Nutrition*. v. 1– . 1952– . Bethesda, MD: American Society for Clinical Nutrition. ISSN: 0002-9165. Monthly. Indexed in *Food Science and Technology Abstracts* and *Index Medicus/Medline*. Official journal of the American Society for Clinical Nutrition. Publishes basic and clinical studies that relate to human nutrition.

*Annals of Nutrition and Metabolism*. v. 25– . 1981– . Basel: Karger. ISSN: 0250-6807. Bimonthly. Indexed in *Current Contents*, *Food Science and Technology Abstracts*, and *Index Medicus/Medline*. Official journal of the Federation of European Nutrition Societies (FENS). Provides information relating to human nutrition and metabolic diseases. Continues *Nutrition and Metabolism*.

*British Journal of Nutrition: An International Journal of Nutritional Science*. v. 1– . 1947. Wallingford, UK: CABI Publishing. ISSN: 0007-1145. Monthly. Indexed in CAB Abstracts, *Current Contents*, and *Index Medicus/Medline*. Published on behalf of The Nutrition Society. Subject coverage includes research relating to human nutrition, clinical nutrition, general nutrition, and animal nutrition.

*EJCN: European Journal of Clinical Nutrition.* v. 55– . 2001– . Basingstoke, UK: Nature Publishing Group, a division of Macmillan Publisher. ISSN: 0954-3007. Monthly. Indexed in CAB Abstracts, *Current Contents*, and *Index Medicus/Medline.* Official Journal of the European Academy of Nutritional Sciences. Emphasizes the theoretical aspects of the science. Does not include reports of animal research unless a parallel study on human subjects is included. Continues *European Journal of Clinical Nutrition.*

*International Journal of Food Sciences and Nutrition.* v. 43– . 1992– . Basingstoke, UK: Carfax Publishing, Taylor and Francis. ISSN: 0963-7486. Bimonthly. Indexed in *Food Science and Technology Abstracts, Index Medicus/Medline,* and *Nutrition Abstracts and Reviews.* Focuses on the integration of food science with nutrition. Continues *Food Sciences and Nutrition.*

*International Journal of Sport Nutrition and Exercise Metabolism.* v. 10– . 2000– . Champaign, IL: Human Kinetics. ISSN: 1050-1506. Quarterly. Indexed in CAB Abstracts, *Index Medicus/Medline,* and *Nutrition Abstracts and Reviews.* Publishes information directly relating to the nutritional aspects of sport and exercise.

*Journal of the American College of Nutrition.* v. 1– . 1982– . New York: American College of Nutrition ISSN: 0731-5724. Bimonthly. Indexed in *Chemical Abstracts, Index Medicus/Medline,* and *Nutrition Abstracts.* Official journal of the American College of Nutrition. Publishes primarily reports of research with application to researchers and critical reviews that highlight key teaching points. October issue includes abstracts from the annual meeting.

*Journal of the American Dietetic Association.* v. 1– . 1925– . Chicago, IL: American Dietetic Association. Monthly. ISSN: 0002-8223. Indexed in CAB Abstracts and *Food Science and Technology Abstracts.* Official research publication of the American Dietetic Association. Publishes research relating to the broad aspects of dietetics, nutrition, diet therapy, community nutrition, education and training, and administration.

*Journal of Applied Nutrition.* v. 1– . 1947– . Addison, TX: International and American Associations of Clinical Nutritionists. ISSN: 0021-8960. Quarterly. Indexed in *Biological and Agricultural Index* and *Nutrition Abstracts and Reviews.* An official publication of the International and American Associations of Clinical Nutritionists. Focuses on micronutrients and macronutrients in relation to health and disease. Continues *International Journal of Sport Nutrition.*

*Journal of Clinical Biochemistry and Nutrition.* v. 1– . 1986– . Mitake, Japan: Institute of Applied Biochemistry. ISSN: 0912-0009. Bimonthly. Indexed in *Biological Abstracts* and *Chemical Abstracts.* Publishes research and review articles about clinical biochemistry, and clinical nutrition based on biochemistry.

*Journal of Human Nutrition and Dietetics*. v. 1– . 1988– . Oxford: Blackwell Science. ISSN: 0952-3871. Bimonthly. Indexed in *Current Contents, Food Science and Technology Abstracts*, and *Nutrition Abstracts*. Official journal of the British Dietetic Association. Focuses on the relationship between human nutrition and health and disease emphasizing therapeutic and preventive applied nutrition and dietetics.

*Journal of Medicinal Food*. v. 1– . 1998– . Larchmont, NY: Mary Ann Liebert. ISSN: 1096-620X. Quarterly. Indexed in *Chemical Abstracts* and *Excerpta Medica*. Focuses on the uses of foods and their components for the treatment and prevention of disease in humans.

*Journal of Nutrition*. v. 1– . Sept. 1928– . Bethesda, MD: American Society for the Nutritional Sciences. ISSN: 0022-3166. Monthly. Indexed in *Food Science and Technology Abstracts, Index Medicus/Medline*, and *Nutrition Abstracts*. Official publication of the American Society for the Nutritional Sciences. Emphasizes the biochemical and metabolic aspects of nutrition; community and international nutrition; nutrient requirements; interactions; and toxicity.

*Journal of Nutritional Biochemistry*. v. 1– . 1990– . New York: Elsevier. ISSN: 0955-2863. Monthly. Indexed in *Chemical Abstracts, Food Science and Technology Abstracts*, and *Nutrition Abstracts*. Emphasizes research in nutrition where it overlaps biochemistry, molecular biology, neurochemistry, toxicology, and pharmacology.

*Journal of Nutrition Education*. v. 1– . 1969– . Hamilton, Canada: BC Decker ISSN: 0022-3182. Bimonthly. Indexed in *Current Contents* and *Food Science and Technology Abstracts*. Official publication of the Society for Nutrition Education. Emphasis is on applied nutritional sciences and the dissemination of this information to educators.

*Journal of Nutrition for the Elderly*. v. 1– . 1980– . Binghamton, NY: Haworth Press. ISSN: 0163-9366. Quarterly. Indexed in CAB Abstracts, *Food Science and Technology Abstracts*, and *Index Medicus/Medline*. Includes research articles, reports of literature, and book reviews for a primarily clinician audience rather than researchers.

*Journal of Nutrition, Health and Aging*. v. 1– . 1997– . New York: Springer Publishing. ISSN: 1279-7707. Three times a year. Indexed in *Index Medicus/Medline*. Emphasizes nutrition research as it relates to the life cycle.

*NMCD: Nutrition, Metabolism and Cardiovascular Disease*. v. 1– . 1991– . Milan: Medikal Press. ISSN: 0939-4753. Quarterly. Indexed in *Current Contents, Index Medicus/Medline*, and *Nutrition Abstracts*. Associated with the Italian Society for Atherosclerosis (SISA) and the Canadian Society of Atherosclerosis, Thrombosis and Vascular Biology (CSATVB). Focuses on the relationship between nutrition and cardiovascular diseases.

*Nutrition and Cancer: An International Journal.* v. 1– . 1978– . Mahwah, NJ: Lawrence Erlbaum Associates. ISSN: 0163-5581. Bimonthly. Indexed in *Cambridge Scientific Abstracts, Food Science and Technology Abstracts,* and *Index Medicus/Medline.* Publication emphasizes effects of nutrition on the etiology, therapy, and prevention of cancer.

*Nutrition and Health.* v. 1– . 1982– . Berkhamsted, UK: A. B. Academic Publishers. ISSN: 0260-1060. Quarterly. Indexed in *Food Science and Technology Abstracts, Index Medicus/Medline,* and *Nutrition Abstracts.* Associated with the McCarrison Society. Covers the effect of food on the human body, health maintenance, and prevention of and recovery from illness.

*Nutrition: An International Journal of Applied and Basic Nutritional Sciences.* v. 3. 1987– . New York: Elsevier. ISSN: 0899-9007. Monthly. Indexed in CAB Abstracts, *Food Science and Technology Abstracts,* and *Index Medicus/Medline.* Includes reports of nutrition research with special emphasis on new technologies, outcomes research and meta-analyses, policy, and practice. Continues *Nutrition International.*

*Nutrition Research.* v. 1– . 1981– . New York: Elsevier. ISSN: 0271-5317. Monthly. Indexed in *Current Contents, Food Science and Technology Abstracts,* and *Science Citation Index.* Emphasizes rapid publication of research in applied and clinical nutrition.

*Plant Foods for Human Nutrition.* v. 37– . 1987– . Dordrecht, The Netherlands: Kluwer. ISSN: 0921-9668. Quarterly. Indexed in *Food Science and Technology Abstracts, Index Medicus/Medline,* and *Nutrition Abstracts.* Focuses on research and critical reviews about the quality of plant foods for humans. Continues *Qualitas Plantareum, Plant Foods for Human Nutrition.*

*Proceedings of the Nutrition Society.* v. 1– . 1994– . Wallingford, UK: CABI Publishing. ISSN: 0029-6651. Bimonthly. Indexed in AGRICOLA, CAB Abstracts, and *Food Science and Technology Abstracts.* Published on behalf of The Nutrition Society. Publishes full papers and short communications from the scientific meetings of The Nutrition Society.

*Public Health Nutrition.* v. 1– . 1998– . Wallingford, UK: CABI Publishing. ISSN: 1368-9800. Bimonthly. Indexed in CAB Abstracts and *Index Medicus/Medline.* Published on behalf of The Nutrition Society. Focuses on the promotion of good health through nutrition.

*Topics in Clinical Nutrition.* v. 1– . 1986– . Frederick, MD: Aspen Publishers, Inc. ISSN: 0883-5691. Quarterly. Indexed in AGRICOLA, *Food Science and*

*Technology Abstracts*, and *Nutrition Abstracts*. Emphasizes issues and research relevant to the continuing education and clinical practice of dietitians and nutritionists. Continues *Nutrition Clinics*.

## ASSOCIATIONS AND ORGANIZATIONS

American College of Nutrition (ACN) (http://www.am-coll-nutr.org). Founded in 1959. Publishes *The Journal of the American College of Nutrition*. Provides educational opportunities (including continuing education), facilitates information dissemination, advocates nutrition education in medical schools, and advises physicians.

American Dietetic Association (ADA) (http://www.eatright.org). Founded in 1917. Publishes *The Journal of the American Dietetic Association*. Offers educational accreditation through the Commission on Accreditation for Dietetics Education, individual credentialing through the Commission on Dietetic Registration, and national dietitian referral service. Provides electronic access to many full-text resources, such as position papers, food pyramids, bibliographies, and nutrition fact sheets.

American Society for Nutritional Sciences (ASNS) (http://www.faseb.org/ain/). Founded in 1928 as the American Institute of Nutrition (AIN). Publishes the *Journal of Nutrition*. Supports research in human and animal nutrition, disseminates and archives research results, and nurtures education and training. Clinical division: American Society for Clinical Nutrition (ASCN) (http://www. faseb.org/ascn/). Founded in 1959. Publishes the *American Journal of Clinical Nutrition*.

British Dietetic Association (BDA) (http://www.bda.uk.com/). Founded in 1936. Publishes the *Journal of Human Nutrition and Dietetics*. Comprised of professional registered dietitians. Acts as a liaison between dietitians in the United Kingdom and the rest of the world. Sponsors educational opportunities, arranges meetings, and promotes the science and practice of dietetics.

Federation of American Societies for Experimental Biology (FASEB) (http:// faseb.org). Founded in 1912. Publishes the FASEB journal. Comprised of scientific societies including the American Society for Nutritional Sciences and their clinical division American Society for Clinical Nutrition. Researches and disseminates information in the fields of biomedical sciences and life sciences.

Food and Agriculture Organization of the United Nations (FAO) (http:// www.fao.org). Founded in 1945. Largest autonomous agency within the United Nations. Collects, analyses, and distributes nutritional and agricultural statistical data including FAOSTAT, *Codex Alimentarius* (international food standards), food composition tables, and other FAO publications.

International and American Associations of Clinical Nutritionists (IAACN) (http://www.iaacn.org). Founded in 1971. Publishes the *Journal of Applied Nutrition*. Sponsors the Certified Clinical Nutritionist (CCN) credential under the responsibility of the Clinical Nutrition Certification Board (CNCB). Encourages research (particularly in the relationship between nutrition and disease), promotes study of nutrition and complementary therapies, provides referral services, and sponsors symposia.

International Food Information Council (IFIC) (http://ificinfo.health.org). Founded in 1985, incorporated as a nonprofit organization in 1990. Facilitates communication about food safety, nutrition, and health and translates research into a form that is useful to opinion leaders and consumers. Their Web site includes full-text information, a directory of organizations, and a glossary.

International Union of Nutritional Sciences (IUNS) (http://www.monash.edu. au/IUNS/). Founded in 1946. Affiliated with the International Council of Scientific Unions—France. Promotes cooperation among nutrition societies, encourages research and information exchange.

International Vegetarian Union (IVU) (http://www.ivu.org). Founded in 1893, successor to the Vegetarian Federal Union. Comprised of vegetarian societies. Web site includes articles produced by members, links to members, recipes, news, events, and glossary.

The Nutrition Society (NS) (http://www.nutsoc.org.uk/). Founded in 1941. Publishes the *British Journal of Nutrition*, Nutrition Research Reviews, *Proceedings of the Nutrition Society*, and *Public Health Nutrition*. Promotes the scientific study of nutrition in humans and animals and disseminates information. Maintains two registers: one of accredited nutritionists and the other public health nutritionists. Sponsors a main annual meeting and smaller meetings.

Society for Nutrition Education (SNE) (http://www.sne.org). Founded in 1967. Publishes the *Journal of Nutrition Education*. Dedicated to promoting healthy, sustainable food choices and represents the unique professional interests of nutrition educators in the United States and worldwide.

Vegetarian Resource Group (VRG) (http://www.vrg.org). Founded in 1982. Publishes the *Vegetarian Journal*. Supports education of the public about vegetarianism and interrelated issues. Assists in the development of publications and answers media queries. Publishes a variety of documents, many available in full text at the Web site.

# 14

## Rural Development and Sociology in the United States

**M. Louise Reynnells**

National Agricultural Library, ARS-USDA, Beltsville, Maryland, USA

Finding rural, small town, or nonmetropolitan information is not as easy as one would think. Many resources do not distinguish between *rural* and *urban* and that can be frustrating when looking for specific data, research, books, or articles relating to rural issues. When dealing with *rural*, possible subject areas are numerous and the lines of what is and what is not rural can be ambiguous, at best. *Rural* issues are everyday life in small towns. Thus, this chapter is limited to more general rural resources with a focus on the United States and is intended as a starting point in locating and using the literature for rural areas.

Besides the more traditional resources of bibliographies, books, and databases, this chapter also includes nontraditional items from a variety of media. Planning or visioning tools; case studies or best practices; ''how-to'' handbooks; manuals and guides; organizations and associations; and periodicals, including newsletters and World Wide Web sites, are also valuable resources to communities, even if all of these resources are not specifically rural focused. Some sources may be broader in scope to provide a framework for ideas to assist rural communities.

To better understand rural information, it is best to start with a definition. The first discussion in this chapter will discuss definitions of *rural*. The second section is a more general look at rural resources. Next is the largest section that focuses on economic/community development resources, including recreation and tourism as they are used as rural development tools. Finally, the last category takes a look at rural change and social issues in nonmetropolitan areas.

## DEFINING RURAL

*Rural* is defined in numerous ways. This section provides resources for the main federal definitions: U.S. Census Bureau, U.S. Department of Agriculture (USDA), U.S. Department of Health and Human Services, and the U.S. Office of Management and Budget. It also includes Web sites with resource links that contain *rural* as a way of life versus a statistical definition.

### Government Documents

Government documents are important to the researcher identifying rural issues and definitions. Statistical information, research, and policy all play a role in how rural communities operate and are sustained. Government documents provide many of these resources, which are easily accessible, and in many cases considered the ''standard'' in the topic and provide a basis for defining policy at the local level based on national and state legislation, regulations, and research.

Goldsmith, H.F., D.S. Puskin, and D.J. Stiles. 1993 with minor updates made in 1997. *Improving the Operational Definition of "Rural Areas" for Federal Programs*. Rockville, MD: Federal Office of Rural Health Policy, U.S. Department of Health and Human Services, Health Resources and Services Administration. 10 p. Available from the World Wide Web at http://www.nal.usda.gov/orhp/Goldsmith.htm (cited June 25, 2001). This publication provides a methodology for defining *rural* called the Goldsmith Rural Modification for Metropolitan Counties designated by the U.S. Office of Management and Budget. This is a definition used to assist with rural health concerns.

Ricketts, T.C., K.D. Johnson-Webb, and P. Taylor. 1998. *Definitions of Rural: A Handbook for Health Policy Makers and Researchers*. Rockville, MD: U.S. Department of Health and Human Services, Federal Office of Rural Health Policy, Health Resources and Services Administration. 20 p. Provides a comprehensive look at federal definitions of *rural* and looks at the concept of frontier areas. Also includes maps, a bibliography, and listing of U.S. counties and their populations.

U.S. General Accounting Office (GAO). 1993. *Rural Development: Profile of Rural Areas, GAO/RECD-93-40FS*. Washington, DC: GAO. 32 p. This report is based on a congressional request to develop a demographic and economic profile of rural areas. It provides information on trends in population, age, and per capita income in nonmetropolitan areas and the geographic distribution of farm program payments. Provides Office of Management and Budget definitions of metropolitan and nonmetropolitan areas.

## World Wide Web Sites

When looking for any information starting with the World Wide Web is now becoming the norm instead of the exception. Everything on the Web is not, however, considered good or reliable information. In order to obtain better information, sticking to government (.gov or state.xx.us—where xx is the state abbreviation), university or educational (.edu), and nonprofit (.org) information is usually a safer bet then commercial (.com or .net) or personal (Wayne's Web site, etc.) Web sites that can be biased in content and slanted to focus on their product or services and their likes or dislikes. Using search engines is a must and using search features on individual Web sites is also useful to identify information. Using the ''links'' section of a particular Web site is helpful in identifying additional resources similar to the topic at hand.

Rural Areas (http://www.rhc.universalservice.org/eligibility/ruralareas.asp) (cited June 25, 2001). Washington DC: Rural Health Care Division, Rural Health Care Program, Universal Service Administration Company. This resource includes an automated listing of rural areas by county for each state. It also lists urban counties that have exceptions within the county considered rural by census tract. A quick resource for finding rural areas.

Selected Historical Census Data. Urban and Rural Definitions and Data (http://www.census.gov/population/www/censusdata/ur-def.html) (cited June 25, 2001). Washington, DC: U.S. Census Bureau, Department of Commerce. Formal definition of urban and rural as defined by the U.S. Census Bureau is provided here along with tables of urban and rural population: 1900–90. Source of items on this Web site are from the 1990 Census of Population and Housing, ''Population and Housing Unit Counts,'' CPH-2-1.

What is Rural? (http://www.ers.usda.gov/briefing/rurality/WhatisRural/) (cited June 25, 2001). Washington, DC: Economic Research Service, USDA. Looks at nonmetropolitan definitions in terms of research. Provides a table with totals of rural and urban residents by county and it links to a data set called the rural-urban continuum code and the urban influence code. This provides a listing that is ranked by rurality according to its proximity to a large urban area.

What is Rural? Rural Information Center, Frequently Asked Questions (FAQ) (http://www.nal.usda.gov/ric/faqs/ruralfaq.htm) (cited June 25, 2001). Baltimore, MD: Rural Information Center, USDA National Agricultural Library. Provides information on the major definitions of *rural* and includes bibliographic citations and links to electronic publications, lists of rural counties, and other rural definition Web sites.

What is Rural? The Rural Womyn Zone (http://www.ruralwomyn.net/rural.html) (cited June 25, 2001). This Web site has links to formal definition sites and also provides links to nontraditional information including statistics, maps, and personal perspectives that define rural areas and their social issues.

## RURAL INFORMATION RESOURCES

This section is a general look at rural areas. From historical and reference text resources to online abstracting databases and statistical information this section reviews sources that give the end user an idea of where to look for rural information.

### Abstracts, Indexes, Databases, and Other Tools

When starting a "search" for rural literature it is best to organize your thoughts for the subject area, identify specific jargon used in that area, and finally focus your search on *rural* and geographic location. For instance, when dealing with local planning issues the term *visioning* is used in setting planning goals in community development. However, this can be used in larger urban settings when searching by such terms as *rural*, *small town*, and/or *nonmetropolitan*, and then using specific database codes for geographic focus to the United States produces better results. A broader search without limiting to *rural* can still identify appropriate literature, but may require the review of more citations that are not applicable and should only be used when the search results using rural terms provide no hits or a small number of hits.

Databases such as AGRICOLA and CAB Abstracts have subject headings and thesaurus terms for rural issues. AGRICOLA and CAB Abstracts are described in earlier chapters and are not repeated here. When a search for rural information is focused on the United States, both of these databases will produce international citations and should be reviewed carefully for applicable use. Limiting for geographic focus to the United States and English for language can help. CAB results tend to produce more hits on foreign and developing countries even with limitations. AGRICOLA will also include international citations but produces more pertinent hits with a U.S. focus. The Educational Resources Information Center (ERIC) has a separate database dedicated to rural education literature and in searching that database using the term *rural* to narrow the search is not necessary because all items are already in that category. This broader search technique should also be considered when using the Local Government Information Network (LogIn) database. Searching with rural terms provides zero or limited hits and is better if not identified using other more specific jargon instead.

LogIn has several search modes such as beginning, advanced, and expert. Librarians will find the beginning mode much too cumbersome. The advanced mode can be used easily by novice searchers because of its form or fill-in-the-parameters ability. However, expert mode requires in-depth knowledge of the database to be successful. Using WilsonSelect database provides current and archival periodical resources and should be limited to rural. In all databases, keyword searching using Boolean or indexed phrases and limiting to the title and/or identifier fields usually produces the hits needed for pertinent documents.

The Educational Resources Information Center (ERIC) (http://www.ael.org/eric/about.htm) (cited June 25, 2001). Washington, DC: U.S. Department of Education, National Library of Education. Sixteen clearinghouses, one of which focuses specifically on rural education called the ERIC clearinghouse on Rural Education and Small Schools. This database is located on the Web at http://www.ael.org/eric/rural.htm (cited June 25, 2001). Subjects include all aspects of education and related issues. ERIC also produces the *Rural Education Directory* and other publications specific to rural education.

H.W. Wilson's Select Full Text (WilsonSelect). Bronx, NY: H.W. Wilson. With over 320,000 records and updated weekly, WilsonSelect provides full-text articles in science, humanities, and business from over 800 sources.

Local Government Information Network (LogIn) (http://services.login-inc.com/login/) (cited June 25, 2001). St. Paul, MN: LogIn Contains over 17,000 documents in its main database on concerns, issues, policy, and programs for local governments. This online Web-based (for fee) service provides access to end-user provided case studies and full-text items from *Governing and American City and Counties* magazine, all emphasizing government management and practices. Also has U.S. government legislative and funding tracking systems, information from the Conference of U.S. Mayors, and the ability for end users to ask LogIn users for information, programs, and regulatory examples.

W.K. Kellogg Collection of Rural Community Development Resources (http://www.unl.edu/kellogg/main.html) (cited June 25, 2001). Lincoln, NE: Heartland Center for Leadership Development. A collection of rural development resources compiled by the Heartland Center for Leadership Development with a grant from the W.K. Kellogg Foundation. Divided into seven categories with annotations and availability listed for each citation. Categories include community development, strategic planning, telecommunication/education, leadership development, economic development, land use/natural resources, and health care.

## Bibliographies

This section provides bibliographies used to help locate topical literature references. Bibliographies are a great place to look for supporting and background materials. They also provide boundaries to topics and current jargon used in specific fields.

McKearney, S.J. 1994. *Rural Studies Bibliography*. Beltsville, MD: USDA, NAL Rural Information Center. 60 p. This publication is a collaborative effort between the U.S. Rural Information Center of the NAL and the U.S. National Rural Studies Committee. An extensive list of monographs and articles that is intended to represent the "best" of the rural studies literature is included. The committee selected citations based on suggestions and recommendations from its committee members and from rural studies professionals and scholars throughout the United States.

## Books

These resources focus on rural topics and can be historical, in-depth, and statistical in nature on broad rural-America information or very specific rural topics.

Danbom, D.B. 1995. *Born in the Country: A History of Rural America*. Baltimore, MD: The John Hopkins University Press. 306 p. This is a historical review of rural areas from rural Europe and Pre-Columbian America to rural life at the end of the twentieth century. Bibliography and index included.

Luloff, A.E. and L.E. Swanson, eds. 1990. *American Rural Communities*. Boulder, CO: Westview Press. 276 p. Provides a look into nonmetropolitan America and the struggles it has overcome and has yet to address. There is no attempt to perpetuate the "myth" of rural life in America. A close examination of small towns by the collection of papers included in this publication indicates that rural areas have many hurdles to jump in order to survive the present and the future. Bibliography and index provided.

USDA, Economic Research Service. 1995. *Understanding Rural America*. Washington, DC: The Service. 25 p. This publication is a small but informative, statistical glance at rural America by researchers at the USDA. Includes tables and maps.

## Encyclopedias

These publications are normally more substantial in content and provide definitions and examples of rural cultures and populations through small summaries.

Davis, M.B., ed. 1996. *Native America in the Twentieth Century: An Encyclopedia*. New York: Garland Publishing. 787 p. This reference work is a useful tool that looks at the past and present issues, contributions, policies, and resources of Native Americans. Signed chapters by historians, anthropologists, and other specialists, arranged alphabetically and covering subject areas such as art, economic conditions, education, languages, health, religion, and life on reservations. Much of this reference is devoted to articles on twentieth-century Native nations. Includes an index.

Gorham, G.A., ed. 1997. *Encyclopedia of Rural America: The Land and People.* 2 vols. Santa Barbara, CA: ABC-CLIO. 861 p. Formatted in traditional encyclopedia fashion, this resource takes a look at rural areas and people. Historical information, policy, an extensive look at rural communities, networks, the rural town experience, social change, and definitions of rural are all included in this two-volume resource. A selected bibliography and index are also provided.

## Periodicals

These publications reflect research, calendars, policy papers, and concerns that are timely in nature. Many periodicals are refereed journals and are used by professionals in this field to develop research on particular aspects of rural areas. Some periodicals are produced by organizations and professional associations as a means of keeping current in the literature of their fields.

*Rural America.* v. 16, no. 1– , 2000– . Washington, DC: USDA Economic Research Service. Available from the World Wide Web at http://www.ers. usda.gov/publications/ruralamerica/about.htm (cited June 25, 2001). Continues as *Rural Development Perspectives* and is published three times a year. Provides individually authored articles on current rural issues. Includes book reviews.

*Rural Voices: The Magazine of the Housing Assistance Council.* v. 2– , 1996/ 1997– . Washington, DC: Housing Assistance Council. Available from the World Wide Web at http://www.ruralhome.org/pubs/ruralvoc.htm (cited June 25, 2001). Produced quarterly with each edition focused on a special rural topic through individually authored articles.

*Small Community Quarterly: Newsletter of the National Center for Small Communities.* Washington, DC: National Association of Towns and Townships. Available from the World Wide Web at http://www.natat.org/ncsc/pubs/ newsletter/scqintro.html (cited June 25, 2001). This newsletter on small community concerns provides in-depth interviews, information on upcoming meetings; training and workshops; resources through its Tool Box section; and subject-oriented articles.

## Rural Research Groups

This section provides major research groups that contribute to the literature on many rural topics. These groups stay abreast of ''hot topic'' areas and provide insight based on the research by staff.

Aspen Institute. One Dupont Circle, NW, Suite 700, Washington, DC 20036-1133, telephone: +1 202-736-5800. Available from the World Wide Web at http://www.aspeninstitute.org/ (cited June 25, 2001). A nonprofit organization that includes a Community Strategies Group that focuses on rural and small town community economic development issues. It publishes the *Rural Update* newsletter and has several program initiatives in rural development.

The Assest-Based Community Development Institute. Institute for Policy Research, North Western University, 2040 Sheridan Road, Evanston, IL 60208-4100, telephone: +1 847-491-8712. Available from the World Wide Web at http://www.nwu.edu/IPR/abcd.html (cited June 25, 2001). Community Development research organization that includes capacity-building community development, training opportunities, and workbooks and guides on neighborhood capacity building.

Center for Rural Affairs. 101 S. Tallman Street, P.O. Box 406, Walthill, NE 68067, telephone: +1 402-846-5428. Available from the World Wide Web at http://www.cfra.org/ (cited June 25, 2001). An unaffiliated nonprofit corporation that serves to build responsible communities with ''social justice, economic opportunity, and environmental stewardship.'' Conducts research and provides opportunities through various outreach programs from beginning farmers to its Rural Enterprise Assistance Project (REAP). Publishes topical publications, annual reports, and a monthly newsletter.

Center for Rural Studies. 207 Morrill Hall, University of Vermont, Burlington, VT 05405. Available from the World Wide Web at http://crs.uvm.edu/ (cited June 25, 2001). A nonprofit, fee-based, research organization. Subject areas covered include economic, resource-based, and social concerns of rural people and communities. Research areas are divided into four main topics: agriculture, human services and education, rural community and economic development, and Vermont community data.

Community and Rural Development Institute. Cornell University, 43 Warren Hall, Ithaca, NY 14853, telephone: +1 607-255-9510. Available from the World Wide Web at http://www.cardi.cornell.edu/ (cited June 25, 2001). Provides research, education, and policy analysis on rural issues. Brings together extension and university professionals with local government officials to link these areas with information to assist those in rural communities. Publications, conferences, and workshops are among the information resources available.

National Center for Small Communities. National Association of Towns and Townships (NATaT), 444 N. Capital Street, NW, Suite 208, Washington, DC 20001-1202, telephone: +1 202-624-3550. Available from the World Wide Web at http://www.natat.org/ncsc/Default.htm (cited June 25, 2001). A national nonprofit organization that serves small town leaders to maintain community character, protect environmental resources, and expand economic opportunities. Publishes a newsletter and small town guides on various topics, and provides technical assistance and resource information on economic development for small communities.

Rural Policy Research Institute (RUPRI). 135 Mumford Hall, University of Missouri, Columbia, MO 65211-6200, telephone: +1 573-882-0316. Available from the World Wide Web at http://ww.rupri.org/ (cited June 25, 2001). Focuses on public policy research and assistance to decision makers in rural areas to understand the impacts of rural policy. Publishes research reports and working papers.

## Regional Rural Development Centers

There are four regional centers in the United States that coordinate rural research and extension education programs regionally and nationally. Each center publishes its own newsletter, topical publications, and fact sheets; and sponsors workshops or conferences on rural topics. The centers are listed:

North Central Regional Center for Rural Development. Iowa State University, 108 Curtiss Hall, Ames, IA 50011-1050, telephone: +1 515-294-1329. Available from the World Wide Web at http://www.ncrcrd.iastate.edu/ (cited June 25, 2001).

Northeast Regional Center for Rural Development. The Pennsylvania State University, 7 Armsby Building, University Park, PA 16802-5602, telephone: +1 814-863-4656. Available from the World Wide Web at http://www.cas.nercrd. psu.edu (cited June 25, 2001).

Southern Rural Development Center. Box 9656, 410 Bost Extension Building, Mississippi State, MS 39762, telephone: +1 662-325-3207. Available from the World Wide Web at http://ext.msstate.edu/srdc/ (cited June 25, 2001).

Western Rural Development Center. Utah State University, 8335 Old Main Hill, Logan, UT 84322-8335, telephone: +1 435-797-9732. Available from the World Wide Web at http://extension.usu.edu/WRDC/ (cited June 25, 2001).

## Rural Statistics

Locating statistics for rural areas can be a challenge. Many times review of urban statistics is needed to find small portions dedicated to rural areas. Sometimes it is necessary to develop your own statistical data sets based on which definition of *rural* you choose. County-level data is provided in most cases as a general statistical source. By defining which counties are rural, data sets can be determined and general statistical sources can be used. The USDA's Economic Research Service (ERS) does much of this, including providing a list of rural-urban continuum codes for defining rural by county in every state.

*American Fact Finder*. 1999. Washington, DC: U.S. Department of Commerce, Census Bureau. Available from the World Wide Web at http://factfinder. census.gov/ (cited June 25, 2001). This is a comprehensive data dissemination system for demographic and economic information. This sophisticated data system allows the user to customize information on communities that includes tables, maps, and socioeconomic data. Community profiling using U.S. Census Bureau data sources is a resource tool now available to anyone doing research in rural or urban areas.

*American Statistical Index (ASI): A Comprehensive Guide and Index to the Statistical Publications of the U.S. Government*. 1974– . Bethesda, MD: Congressional Information Service. Abstracts and index of U.S. federally produced statistical resources that are divided by subject, type of data (such as metropolitan or nonmetropolitan), publication title, and report number. Available electronically through the Statistical Universe at http://www.lexisnexis.com/academic/1univ/ stat/default.htm (cited March 6, 2002).

Buse, R.C. and J.L. Driscoll, eds. 1992. *Rural Information Systems: New Directions in Data Collection and Retrieval*. Proceedings of a two-day symposium held in July 1989. Ames, IA: Iowa State University Press. 458 p. These proceedings provide information from the two-day symposium held to discuss future data needs for rural America. This publication was developed to help rural professionals better understand and serve rural areas. Subjects include institutions of public data, advances in survey data collection methods, uses of public data, USDA dissemination of data, information systems, and future data availability.

Christenson, J.A., R.C. Maurer, and N.L. Strang, eds. 1994. *Rural Data, People, and Policy: Information Systems for the 21st Century*. Rural Studies Series. Boulder, CO: Westview Press. 232 p. Individually authored chapters that cover three main areas: The Politics of Data; Data for Rural Information Systems; and Technologies for Rural Information Systems. Developed to help sort out problems existing in rural data collection, it also serves as an instrument for creating unique data systems that can help to clarify the needs of people and places in rural America.

Courtright, J. and A. Reamer. 1998. *Socioeconomic Data for Understanding Your Regional Economy: A User's Guide*. Washington, DC: U.S. Department of Commerce, Economic Development Administration. 98 p. Available from the World Wide Web at http://www.doc.gov/eda/pdf/socio.pdf (cited June 29, 2001). Provides information on regional sources of data used for socioeconomic needs.

Identifies Web sources and print resources as well as pitfalls and the "Ten Habits of Highly Effective Data Analysts." State data centers, national and regional associations, and state labor market information are listed within the appendices. Includes index.

EconData.Net (http://www.econdata.net/) (cited June 29, 2001). Cambridge, MA: Andrew Rearner. An online guide to regional economics with more than 400 links to socioeconomic data sources, arranged by subject and providers.

Rural Briefing Room (http://www.ers.usda.gov/emphases/rural/) (cited June 30, 2001). Washington, DC: USDA Economic Research Service. Provides access to rural-specific data resources such as the rural-urban continuum codes, state data fact sheets, and publications of the ERS that deal with rural areas. Many items are full text in .pdf format. This site is keyword searchable.

Rural by the Numbers (http://www.rupri.org/policyres/rnumbers/index.html) (cited June 30, 2001). Columbia, MO: Rural Policy Research Institute. Subject-oriented statistical information specific to rural America. It indicates only current and well-documented resources are used at this site.

Salant, P. and A.J. Waller. 1995. *Guide to Rural Data*. Washington, DC: Island Press, 140 p. This is a revised edition of *A Community Researcher's Guide to Rural Data*, 1990. The manual was developed to provide researchers and community planners with data resources to better understand rural areas through planning and policy development. Includes an overview of the resources; the types of data available and how to use them; and the basic understanding of the characteristics and economics of rural communities. Bibliographic references, glossary, and subject index provided.

Statistical Reference Index (SRI). 1980– . Bethesda, MD: Congressional Information Service. Provides information on statistical resources for state, university, and the private sectors. Available online in the Lexis/Nexis database Statistical Universe at http://www.lexisnexis.com/academic/1univ/stat/default.htm (cited March 6, 2002).

U.S. Department of Commerce, Bureau of Economic Analysis (BEA) (http://www.bea.doc.gov/) (cited June 30, 2001). Washington, DC: Bureau of Economic Analysis. "The mission of BEA is to produce and disseminate accurate, timely, relevant, and cost-effective economic accounts statistics that provide government, businesses, households, and individuals with a comprehensive, up-to-date picture of economic activity. BEA's national, regional, and international economic accounts present basic information on such key issues as U.S. economic growth, regional economic development, and the Nation's position in the world economy (Web site)."

U.S. Department of Commerce, Census Bureau. *Statistical Abstracts of the United States 1878–*. Washington, DC: Government Printing Office. Annual. Available from the World Wide Web at (http://www.census.gov/prod/www/statistical-abstract-us.html) (cited June 25, 2001). This standard reference library resource is now available over the Internet in .pdf format. It provides statistical information from social, political, and economic aspects of American life for regions, states, and Metropolitan Statistical Areas with limited information for nonmetropolitan areas. This is also in print through the U.S. Government Printing Office.

## RURAL COMMUNITY AND ECONOMIC DEVELOPMENT

The references under this section provide informational resources for general economic/community development efforts in rural areas.

### Bibliographies

Brown, D.M. and O.L. Flake. 1999. *Rural Transportation: An Annotated Bibliography*. Washington, DC: USDA and the U.S. Department of Transportation. 65 p. Organized into five categories to provide an overview of different aspects in rural transportation and other important issues in community development efforts. These areas consist of availability, demand, and condition of rural transportation and infrastructure; transportation and rural development, which includes the impact of transportation infrastructure and employment in rural areas; federal and local government issues; other issues not categorized in the other sections including social services, safety, and environmental issues and transportation; and additional resources such as maps, atlases, bibliographies, status reports, index, and nonannotated references are also included.

Estes, R.J. 1998. *Resources for Social and Economic Development: A Guide to the Scholarly Literature*. Fort Washington, PA: Communication Graphics. 175 p. Developed as a guide to the vast amount of literature that exists on social and economic development. The author suggests this publication could be used as a supplementary reference tool in macro-level courses in social policy, planning, administration, and research.

Stierman, J.K. 1999. *Finding Rural Development Resources on the World Wide Web: Tips and Techniques for Efficient Searches*. Macomb, IL: Illinois Institute for Rural Affairs. 36 p. Available from the World Wide Web at http://www.iira.org/pubs/PDF/ruraldev.pdf (cited June 30, 2001). Provides an insight into searching the World Wide Web for rural development issues. The introduction provides a detailed account of Web search engines, how to use them, and the differences between them. The publication also furnishes relevant Web sites for rural development. References included. Online version has Web sites enabled.

Wright, E., ed. 1999. *An Annotated Bibliography For Faith-Based Community Economic Development.* Washington, DC: National Congress for Community Economic Development, Faith-Based Community Development Program. 41 p. Describes references to faith-based community development through efforts or the work of churches and church charities in community programs.

## Books

Clark, C. 1997. *101 More Ideas on Economic Development.* Omaha, NE: Utili-Corp United. 128 p. Derived from a weekly series in a newspaper column by Cal Clark, this publication is an update of the original, *101 Ideas on Economic Development,* which came from earlier versions of the same weekly column. Subject areas include characteristics of successful communities; why have economic development? Or what is its importance?; the role of community leadership; community resources and foundations, visioning, and strategic planning efforts; trends; and how to organize for economic development, industry, entrepreneurs and target marketing, health care, and housing issues.

Cornell, S. and J.P. Kalt., eds. 1995. *What Can Tribes Do?: Strategies and Institutions In American Indian Economic Development.* Los Angeles, CA: University of California, American Indian Studies Center. 336 p. A collection of papers based on research through the "Harvard Project on American Indian Economic Development." This publication is directed toward the efforts of Indian tribes to develop their own solutions to economic development and tribal well being. Based on what works, where, and why, this resource provides information and insights to tribal leaders working to develop their own economic opportunities.

Corporation for Enterprise Development. 1999. *Ideas in Development: 20 Years; Growing Assets, Expanding Opportunities, 1979–1999.* Washington, DC: Corporation for Enterprise Development 100 p. Produced in response to the Corporation for Enterprise Development's (CFED's) twentieth anniversary, this collection of ideas range from entrepreneurship to trade agreements, and from capacity building to taxation. Some are formal ideas and solutions, while others portray a more spiritual concept. This collection is put forth to inspire a positive step toward assisting everyone in a community to be self-sufficient.

Dalton, L.C., C.J. Hoch, and F.S. So, eds. 2000. *The Practice of Local Government Planning.* 3rd ed. Washington, DC: International City/County Management Association (ICMA), ICMA Training Institute. 496 p. This document is a comprehensive resource for local and regional planning efforts. Includes information on the impact of technology, diversity, citizen participation, and planning issues in housing, transportation, community development, and urban/rural design. Includes bibliographic references and an index.

Galston, W.A. and K.J. Baehler. 1995. *Rural Development in the United States: Connecting Theory, Practice, and Possibilities.* Washington, DC: Island Press. 353 p. This book focuses on providing a framework for bringing together policies, research, and practical models for rural development. It is meant to act as a catalyst in promoting rural development, based on new policies that are effective in revitalizing and sustaining rural areas. Bibliography and index included.

Gringeri, C.E. 1994. *Getting By: Women Homeworkers and Rural Economic Development.* Lawrence, KS: University Press of Kansas. 200 p. Describes a research project, which was designed to explore how industrial homework has become an integral facet of economic development in two Midwestern towns. Data was collected by personal, in-depth interviews. Includes bibliography and index.

Phillips, P.D. 1990. *Economic Development for Small Communities and Rural Areas.* Urbana, IL: University of Illinois at Urbana-Champaign, Office of Continuing Education and Public Service, Community Information and Education Service Programs. 180 p. Intended for economic developers in rural areas, this document provides definitions, factors, and step-by-step procedures in creating successful economic development programs in rural areas. Includes an annotated bibliography.

Ratner, S. 2000. *The Informal Economy in Rural Community Economic Development.* Lexington, KY: TVA Rural Studies Program 24 p. Available from the World Wide Web at http://www.rural.org/publications/Ratner00-03.pdf (cited June 30, 2001). This report defines an informal economy as one that includes unpaid labor and labor exchanges, unreported business transactions, subsistence production, volunteer work, household production and consumption, interhousehold bartering, sharing, and care giving to young and old. This everyday life informal economy is the one that usually is not counted when looking at economic/community development efforts. The author believes that this informal economy should be considered in the planning process of economic development efforts. Bibliographic references included.

Reese, L.A. 1997. *Local Economic Development Policy: The United States and Canada.* New York: Garland Publishing. 161 p. Presents policy in a cross-national context. Compares Canadian and U.S. policies using examples from cities in the state of Michigan in the United States and the Ontario Province in Canada. Chapters include comparisons of cross-border policies, economic development models, and decision-making processes. References and index included.

Rowley, T.D., D.W. Sears, G.L. Nelson, J.N. Reid, and M.J. Yetley, eds. 1996. *Rural Development Research: A Foundation for Policy.* Westport, CT: Greenwood Press. 248 p. This book is a summary of presentations of rural development experts from academia and government who participated in a series of six conferences over a three-year period. There are two parts to this book: Part one deals with four components of rural development: education, entrepreneurship, physical infrastructure, and social infrastructure. Part two examines analytic methods of measuring rural development efforts, models, and case studies.

Sears, D.W. and J.N. Reid. 1995. *Rural Development Strategies*. Chicago, IL: Nelson-Hall Publishers. 304 p. This is a collection of essays on development planning. Coverage includes bottom-up models, worker cooperatives, community investment in education, business recruitment and expansion, transportation importance to local businesses, prisons and tribal communities, natural resources, and successful ventures. Index included.

## Directories

These resources include specialized organization member listings, more general rural economic development contacts, Native American resource lists, and funding program information directories.

American Economic Development Council. 1986– . *Who's Who in Economic Development*. Schiller Park, IL: American Economic Development Council. This annual publication is based on the American Economic Development Council (AEDC) membership directory.

*Directory of Community Development Investments*: Bank Holding Companies, State Member Banks. 1998. Washington, DC: Board of Governors of the Federal Reserve System. 192 p. A directory of federal reserve institutions for state member banks and bank holding companies. Index provided.

*Economic Development Directory*. 1999. Washington, DC: Public Works & Economic Development Association. 153 p. Prepared by the Public Works & Economic Development Association for the Economic Development Administration (EDA), U.S. Department of Commerce. Lists EDA national and regional offices and contacts by programs.

*National Guide to Funding for Community Development*. 1996– . New York: Foundation Center. Biennial. Publication with information on charities, trusts, and foundations along with Community Development Corporations and other economic/community development assistance information for the United States.

*Native American Directory*: Alaska, Canada, United States. 1996. San Carlos, AZ: National Native American Cooperative. 600 p. Includes tribal information from over 2,000 tribes, organizations, associations, media, events, a buyers guide to Native American art, and information on researching Indian ancestry.

Reynnells, M.L. 2000. *Federal Funding Sources for Rural Areas For FY2001.* *Rural Information Center Publication Series.* Beltsville, MD: USDA, NAL, Rural Information Center. 139 p. Annual. Available from the World Wide Web at http://www.nal.usda.gov/ric/ricpubs/funding/federalfund/ff.html (cited June 25, 2001). This annual publication is a directory of federal funding programs for which rural researchers, small town officials, or individuals living in rural areas are eligible to apply. Based on information from the *Catalog of Federal Domestic Assistance,* this listing of over 200 programs provides insight on available federal funding programs for rural areas. Includes subject index. Also available as a keyword searchable online database through the Rural Information Center Web site.

Rochin, R.I. and E. Marroquin. 1997. *Rural Latino Resources: A National Guide.* East Lansing, MI: Julian Samora Research Institute. 152 p. Available from the World Wide Web at http://www.jsri.msu.edu/RandS/books/rlr/ (cited June 25, 2001). Developed as a resource guide in the field of Latino Studies. Includes a listing of specialists, and details of resources available on rural and Latino issues.

Rural Local Initiatives Support Corporation. 1998. *Rural America: Communities Creating Opportunity. The First Directory of Rural Community Developers.* Washington, DC: Rural Local Initiatives Support Corporation (LISC), Stand Up for Rural America Campaign. 313 p. loose-leaf. Available from the World Wide Web at http://www.ruralamerica.org/directory.htm (cited June 25, 2001). This publication is based on a U.S. Census Bureau survey done in 1998 on community-based development organizations (CBDOs). All the CBDOs listed in this directory indicated that they served nonmetropolitan or mixed areas and gave permission by a written response or telephone interview to be listed. An estimated 1,000 CBDOs were surveyed and over 800 replied.

State Specific Economic Development Links (http://www.hhh.umn.edu/centers/slp/edweb/state.htm) (cited June 30, 2001). Minneapolis, MN: Hubert H. Humphrey Institute of Public Affairs, State and Local Policy Program. University of Minnesota. A state-by-state hyperlinked table that provides direct access to state economic development agencies nationwide.

*Working Together: Directory of State Coordination Programs, Policies and Contacts 1999–2000.* Washington, DC: American Public Works Association, with assistance from the Coordinating Council on Mobility and Access, Ecosometrics, Inc., and the National Transportation Consortium of States under a cooperative agreement with the U.S. Department of Transportation, Federal Transit Administration. 50 p. Listing is by state with the State Coordination Contact listed first.

## Government Documents

Aldrich, L. and L. Kusmin. 1997. *Rural Economic Development: What Makes Rural Communities Grow?* Agriculture Information Bulletin (USDA) no. 737.

Washington, DC: USDA, Economic Research Service, Food and Rural Economic Division. 7 p. Available from the World Wide Web at http://www.ers.usda.gov/ epubs/pdf/aib737/ (cited June 30, 2001). Presents an analysis of factors that encourage local and regional economic growth. Attractiveness to retirees, right-to-work laws, excellent high-school completion rates, good public-education expenditures, and access to transportation networks, were among those factors that strengthen rural growth rates. Reviews of the literature in this area and other data are provided with maps and tables.

Reamer, A. and J. Cortright. 1999. *Socioeconomic Data for Economic Development. An Assessment.* Washington, DC: U.S. Department of Commerce, Economic Development Administration. 184 p. An in-depth study of the nation's system for developing regional socioeconomic data finds that there is a need for better coordination among federal agencies to produce and disseminate regional data for use by practitioners.

U.S. General Accounting Office. *Rural Development: Rural America Faces Many Challenges.* 1992. Washington, DC: U.S. General Accounting Office. 126 p. To assist Congress in strengthening rural development policy, the U.S. GAO was asked to identify the challenges rural America faces in dealing with current economic realities. Viewpoints from a symposium on rural America held in June of 1992 are synthesized in this report.

## Guides

Guides are an important tool used by communities, organizations, and individuals in rural areas to develop programs, projects, and decision-making documents to assist in the economic development process.

Ames, S.C., P. Coppel, and C. Rains. 1998. *Guide to Community Visioning: Hands-On Information for Local Communities.* Rev. ed. Salem, OR: Oregon Chapter, American Planning Association, Oregon Visions Project. 41 p. Describes a step-by-step process of what visioning is and how to do it. Also provides resources, helpful hints, and case studies of Oregon communities that have undergone this process.

Guyette, S. 1996. *Planning for Balanced Development: A Guide for Native American and Rural Communities.* Sante Fe, NM: Clear Light Publishers. 312 p. A unique guide developed to provide planning methods that could apply to any community, but which places a special emphasis on the cultural resources that a community has to offer. The guide explains the planning process, community needs assessment, cultural revitalization, cultural tourism, and funding issues for projects developed. Contact resources, business plan guidelines, and articles of incorporation for an enterprise corporation are among the appendices. Bibliographic references and an index included.

International City/County Management. 1998. *Catalog of Data Sources for Community Planning*. IQ Service Report. Washington, DC: ICMA. 19 p. Reference guide to federal, state, and local administrative sources of data needed in community development and used in planning efforts. Bibliographic references included.

Mantell, M.A., S.F. Harper, and L. Propst. 1990. *Resource Guide For Creating Successful Communities*. Washington, DC: Island Press. 209 p. Provides programs from large and small communities concerning growth management, conservation of farmland, wetlands, rivers, historic and cultural areas, and open spaces. Includes sample articles of incorporation, bylaws for local organizations, and growth management tools and techniques.

Walter, T. 1996. *On-Line Resources For Rural Community Economic Development*. (Draft Version IV) Washington, DC: The Aspen Institute. 24 p. Available from the World Wide Web at http://aspeninstitute.org/csg/csg_online.html (cited June 30, 2001). This online guide is in constant draft stage and revisions are made through the Web. A definition of rural community economic development leads the resources that also include Internet provider information, e-mail group lists, bulletin board and newsgroups listings, rural Web sites, groupware resources, and real-time meetings.

### Handbooks and Manuals

These resources prove to be helpful in guiding rural communities through the economic development process. Many include example surveys, model programs, and sample forms for handling development projects. Local ordinances, regulations, economic indicators, and case studies provided in these handbooks are useful tools in planning and implementing community projects.

Burchell, R.W. 1994. *Development Impact Assessment Handbook*. Washington, DC: Urban Land Institute. 326 p. Includes computer disk (3 1/2″). Presented are methods for analysis, models, and impact assessments for community economics, physical infrastructure, and social concerns. Data sources are reviewed in each area and models provided. Also included is a computer disk (IBM compatible) containing a model of hypothetical proposals.

Daniels, T.L. 1995. *The Small Town Planning Handbook*. Chicago, IL: American Planning Association, Planners Press. 305 p. This information resource gives examples of how to create a small town plan, community profiling, data needed, citizen participation, and zoning laws. Bibliographic references and an index included.

Davies, A. 1997. *Managing for Change*: *How to Run Community Development Projects*. London: Intermediate Technology Publications. 164 p. A project-management resource guide that provides the management process, chapter-by-chapter, starting with problem identification and ending with post-project activities. This handbook includes questions and exercises after each chapter to reinforce the lessons learned.

North Central Regional Center for Rural Development (NCRCRD). 1999. *Measuring Community Success and Sustainability: An Interactive Workbook*. Ames, IA: NCRCRD. 77 p. Available from the World Wide Web at http://www.ag.iastate.edu/centers/rdev/RuralDev.html (cited June 25, 2001). Developed as a guide for communities, nonprofit organizations and agency personnel for learning how to measure local and/or regional impacts of economic and community development. This workbook provides information on how to measure community successes for long-term sustainability. Community assessments and case studies included.

Rypkema, D.D. and E. Jackson. 1996. *Community Initiated Development: Coming to the Table With Credibility*. Washington, DC: National Main Street Center, National Trust for Historic Preservation. 579 p. Includes a 63-page booklet. A step-by-step manual for redevelopment efforts of existing structures. A workbook with case studies and "fill-in-the-blank planning."

## Organizations and Associations

Organizations and associations provide numerous resources to rural communities. These organizations are many times the catalyst behind development programs and projects. These groups often produce publications including referred journals, research reports, specialized fact sheets, handbooks, and more. Some develop their own databases on best practices and development tools and should not be overlooked by those searching for rural information. This list is a selection of U.S. groups that represents a portion of those that assist rural communities. There are many groups at the state and local level that should be reviewed when working on or with specific area projects.

American Economic Development Council. One of the oldest economic development organizations, since 1926 it has served the profession with assistance in creating sustainable local economic development capabilities that are globally competitive. Provides support in building knowledge, forming alliances, and managing key issues. Publishes subject-oriented publications and reports, a monthly digest, a quarterly journal, and a who's who in economics.
Contact information:
1030 Higgins Road, Suite 301
Park Ridge, IL 60068
Telephone: +1 847-692-9944
URL: http://www.aedc.org/

American Planning Association. Provides publications, research, technical assistance, and educates policy makers on land use issues. The American Planning Association's (APA's) professional and education component is the American Institute of Certified Planners. The Small Town and Rural Planning Division (STaR) is the rural component of the APA.
Contact information:
1776 Massachusetts Ave., NW
Washington, DC 20036-1904
Telephone: +1 202-872-0611
URL: http://www.planning.org/

Community Development Society. An international organization established in 1969 as a professional association for community development practitioners and citizen leaders. Publishes a newsletter and the *Journal of the Community Development Society*. Has member chapters and provides certification for community developers.
Contact information:
1123 N. Water Street
Milwaukee, WI 53202
Telephone: +1 414-276-7106
URL: http://comm-dev.org/

International City/County Management Association. A professional and educational organization representing appointed administrators and managers in local government. Provides informational resources, publications, specific programs, and networks based on enhancing the quality of local government. Some of those programs are the Smart Growth Network and the Base Reuse Consortium.
Contact information:
777 North Capital Street, NE, Suite 500
Washington, DC 20002
Telephone: +1 202-289-4262
URL: http://www.icma.org/

National Association of Counties (NACO). Created in 1935 to represent county governments in the United States. Provides legislative, research, and technical services to member counties and acts as a liaison with other levels of government. Has a Rural Information Clearinghouse through their Web site. Publishes newsletters and subject-oriented publications.
Contact information:
440 First Street, NW, Suite 800
Washington, DC 20001
Telephone: +1 202-393-6226
URL: http://www.naco.org/

National Association of Development Organizations (NADO). Public interest group founded in 1967 to provide training, information, and representation for regional development organizations in small metropolitan and rural America. Produces publications including the series, Economic Development Digest, special reports, and *NADO News* newsletter. The NADO Web site provides rural resources.
Contact information:
400 N. Capital Street, NW, Suite 390
Washington, DC 20001
Telephone: +1 202-624-8813
URL: http://www.nado.org

National Association of Regional Councils (NARC). This national nonprofit organization provides information, publications, and reports on comprehensive regional planning and coordination and serves regional councils nationwide.
Contact information:
1700 K Street, Suite 1300
Washington, DC 20006
Telephone: +1 202-457-0710
URL: http://www.narc.org/

National Center for Small Communities. National nonprofit organization directing information to small community leaders. Publishes a newsletter and other subject-oriented publications. A part of the National Association for Towns and Townships.
Contact information:
444 N. Capital Street, NW, Suite 208
Washington, DC 20001-1202
Telephone: +1 202-624-3550
URL: http://www.natat.org/ncsc/Default.htm

National Rural Development Partnership. This national organization works to strengthen rural America through collaborative efforts from federal, state, local, tribal governments, and for-profit and nonprofit private sector. There currently are 37 State Rural Development Councils. The National Council consists of over 40 federal agencies, as well as national representatives from public interest, community-based, and private-sector organizations.
Contact information:
1400 Independence Ave., SW, Room 4225-S
Washington, DC 20250-3205
Telephone: +1 202-690-2394
URL: http://rurdev.usda.gov/nrdp/

Rural Local Initiatives Support Corporation. Established in 1995 to support community development corporations (CDCs) with training, technical assistance, and financially to increase their production and impact in rural areas. Has status in 37 states and encourages investments in rural CDCs. Also developed the Stand Up for Rural America Campaign.
Contact information:
1825 K Street, NW, Suite 1100
Washington, DC 20006
Telephone: +1 202-739-0882
URL: http://www.ruralisc.org/

Rural Sociological Society. Established in 1937, this society was created to promote the development of rural sociology through teaching, research, and extension. Publishes a journal, *Rural Sociologist*.
Contact information:
Department of Sociology
510 Arntzen Hall
Western Washington University
Bellingham, WA 98225-9081
Telephone: +1 360-650-7521
URL: http://ruralsociology.org/

Smart Growth Network International City/County Management Association. Developed to help create national, regional, and local coalitions to encourage metropolitan development that is environmentally, fiscally, and economically smart. This network is a partnership organization that supports collaborative efforts in smart-growth practices in neighborhoods, communities, and regions across the United States. The online Web site provides annotated bibliographic resources and links to full-text information on all smart-growth related topics.
Contact information:
777 North Capital Street, NE, Suite 500
URL: Washington, DC 20002-4201
Telephone: +1 202-962-3591
URL: http://www.smartgrowth.org/

## Periodicals

*American City and County*. v. 1– . 1975– . Pittsfield, MA: Morgan-Grampian Pub. Co. A monthly publication that concentrates on cities and counties. Municipal governing and services are the focus.

*Community Development Digest*. v. 1– . 1961– . Silver Spring, MD: CD Publications. A semimonthly digest that provides up-to-date information on community development programs, planning issues, legislation, regulations, and financial topics.

*Economic Development Digest.* v. 1– . 1992– . Washington, DC: National Association of Development Organizations Research Foundation. Available from the World Wide Web at http://www.nado.org/pubs/digest.html. Monthly report to help promote economic development in small towns and rural areas.

*Economic Development Review.* v. 1– . 1983– . Schiller Park, IL: American Economic Development Council. Quarterly. Publication of the American Economic Development Council that includes practitioner-oriented information and research articles.

*Governing.* v. 1– . 1987– . Washington, DC: Congressional Quarterly. Monthly. Magazine for state and local governments that covers current issues concerning governing practices, management, and trends. Also available on the LOGIN database.

*Housing and Development Reporter*: *Current Developments in Housing, Community Development, Finance and Taxation.* v. 1– . 1973– . Eagan, MN: West Group. Weekly. Publication that provides current information on housing and community development issues. This is not rural specific but does have rural housing and community development sections relevant to nonmetropolitan areas. Available in print or CD.

*Journal of the Community Development Society.* v. 1– . 1970– . Columbia, MO: Community Development Society (CDS). Published twice a year, the journal focuses on the dissemination of knowledge and practice in the field of community development and change.

*Journal of Economic Literature.* v. 1– . 1969– . Nashville, TN: American Economic Association. Quarterly. Publication of the association that contains survey and review articles, book reviews, and bibliographic indexes to current literature.

*Journal of Extension.* 1993– . Madison, WI: Extension Journal. Available from the World Wide Web at http://www.joe.org/ (cited June 25, 2001). Also known as the *Electronic Journal of Extension*, this electronic journal provides Extension Service programs and projects information in rural development and other issues involving family and farm.

*The Main Street Economist.* v. 1– . 1999– . Kansas City, MO: Center for the Study of Rural America, Federal Reserve Bank of Kansas City. Monthly. Publication that provides insight to rural economies and conditions with a focus on the Midwestern states. Available from the World Wide Web at http://www.kc.frb.org/RuralCenter/mainstreet/MainStMain.htm.

*Rural Conditions and Trends.* v. 1– . 1990– . Washington, DC: USDA, Economic Research Service. Available from the World Wide Web at http://www.ers.usda.gov/publications/rcat/rcat112/contents.htm. Provides statistical analysis of economic trends, in rural America. Lists in-depth data resources to back up predictions, trends and conditions in rural employment, industries, poverty, population, and addresses topics that change per issue.

## Proceedings

These documents contain the workings of conferences, meetings, and workshops. Includes information on the presenter, their research or findings on the topic at hand and sometimes specific policy recommendations are presented at the conclusion of these documents. This resource is valuable snapshot of current research and programs in progress that can be followed up on through the presenter listed.

*Building Economic Self-Determination in Indian Country. An Interagency Conference Sponsored by The White House (Office of Intergovernmental Affairs and Domestic Policy Council) and the Departments of Agriculture, Commerce, Defense, HHS, HUD, Interior, Justice, Labor, Transportation, Treasury, and the Indian Health Service, the Comptroller of the Currency and the Small Business Administration.* 1998. Washington, DC: The Sponsors. 1 vol. unpaged. This interagency-sponsored conference proceedings provides resources, both for funding and technical assistance, that are available for and specifically focused on Native American populations.

Schmidt, J. ed. 1997. *Rural Infrastructure as a Cause and Consequence of Rural Economic Development and Quality of Life. Proceedings of a Regional Workshop, Birmingham, Alabama, February 1997.* Mississippi State, MS: Southern Rural Development Center. 149 p. These proceedings address rural needs in the southern region of the United States. The underlying concern of this workshop was the ability of rural communities to be sustainable. Discusses modernization of infrastructure that assists future development. References are included with chapters.

## World Wide Web Sites

Rural Community Assistance Corporation (http://www.rcac.org/) (cited June 28, 2001). Dedicated to improving the quality of life for rural communities and disadvantaged people through partnerships, technical assistance, and access to resources. The Rural Community Assistance Corporation (RCAC) strives to help community-based organizations and rural governments increase their own capacity to implement solutions to their problems.

Rural Development at a Glance. USDA Economic Research Service (http://www.ers.usda.gov/briefing/rural/gallery/) (cited June 28, 2001). Provides research reports, statistical data, and maps on a variety of rural socioeconomic issues.

Rural Information Center. USDA, NAL (http://www.nal.usda.gov/ric) (cited June 28, 2001). This Web site provides a vast resource of information on rural topics with a comprehensive listing of rural and rural-related subject area resources. Includes online publications, Frequently Asked Questions addressing rural issues of concern, and the ability to "Ask A Question" online to one of the rural information specialists at the Rural Information Center (RIC).

Rural Development Web Site. USDA (http://www.rurdev.usda.gov) (cited June 28, 2001). Provides access to all three rural development agencies: Rural Business and Cooperative Service, the Rural Housing Service, and the Rural Utilities Service. National programs with information on special departmental initiatives.

## Yearbooks

These resources provide a year-at-a-glance information giving both statistics and trends in rural development issues and concerns.

*Development Report Cards for the States 2000: Economic Benchmarks for State and Corporate Decisionmakers.* 2000. 14th ed. Washington, DC: Corporation for Enterprise Development. Available from the World Wide Web at http://www.drc.cfed.org/ (cited June 25, 2001). This is an annual that assesses each state's economy and then analyzes its potential for growth based on over 70 data measurements. States are compared and rated by performance in regards to difference categories relevant to the data collected and the well-being of all citizens within the state. This online Web edition allows for complete download of data or customized versions.

Municipal Yearbook. v. 1– . 1934– . Washington, DC: International City/County Management. Annual. An annual resource that provides information on municipal management issues such as government employee salaries, technology trends, and management practices. Includes a variety of directories that include state municipal leagues, state agencies for community affairs, state associations of counties, and ICMA-recognized directors of councils of government. Also has a reference section with basic and specialized information resources and a cumulative index.

## SELECTED TOURISM AND RECREATION RESOURCES

This section is provided as an enhancement to the previous rural development section. Tourism efforts in rural areas have proven to be a good way to increase the economies in small towns. Many rural areas have natural resource areas that make excellent recreational attractions. Historical and cultural heritage regions also play an important role in many rural towns and can be used to expand economic opportunities for development efforts.

Baud-Bovy, M. and F. Lawson. 1998. *Tourism and Recreation Handbook of Planning and Design*. Oxford; Boston: Architectural Press. 289 p. A tourism planning handbook with a practical approach. Includes data resources, impact assessments, investment methods, bibliographic references, and an index.

Bruce, D. and M. Whitla, eds. 1993. *Tourism Strategies for Rural Development*. Sackville, Canada: Rural and Small Town Research & Studies Programme, Mount Allison University. 52 p. Authored chapters that include Canadian and U.S. regional tourism strategies, tourism and economic development case studies, tourism employment examples, and direct-marketing techniques for attracting tourists to rural areas. Bibliographic references and an index included.

## Books

Butler, R., C.M. Hall, and J. Jenkins, eds. 1998. *Tourism and Recreation in Rural Areas*. New York: John Wiley. 261 p. A collection of essays divided into four major groups: Continuity and Change in Rural Tourism; Tourism and Recreation Policy Dimensions; Image and Reimaging of Rural Areas; and Social and Economic Dynamics. Bibliographic references and an index are included.

Capalbo, S. 1996. *Cultural Heritage, and Environmental Tourism*. Washington, DC: Management Information Service Report, International City/County Management Association. 14 p. A how-to report to assist communities with cultural, heritage, and ecotourism efforts. Includes cases studies.

Kotler, P., J. Bowen, and J. Makens. 1998. *Marketing for Hospitality and Tourism*. 2nd ed. Upper Saddle River, NJ: Prentice Hall. 800 p. This colorful and comprehensive reference provides marketing, planning and promotion information, cases study examples, Internet sites, a glossary, and an index.

McGranahan, D.A. 1999. *Natural Amenities Drive Rural Population Change*. Agriculture Economic Report (USDA) no. 781. Washington, DC: USDA Economic Research Service, Food and Rural Economics Division. 24 p. Available from the World Wide Web at http://www.ers.usda.gov/publications/aer781/ (cited June 30, 2001). A natural amenities index is described using data from the past 25 years that indicates rural county population change over that time period has been based on climate, topography, and water areas. Information on recreation and retirement communities indicates that those rural counties specializing in these amenities have higher rates of population growth than other rural counties. References, figures, and data files are included.

U.S. Department of Commerce, United States Travel and Tourism Administration (USTTA). 1994. *Rural Tourism Handbook: Selected Case Studies and Development Guide.* Washington, DC: U.S. Department of Commerce, USTTA. 188 p. Provides how-to information and resources for tourism development with rural case-study examples. Reference bibliography included.

## Periodicals

*Journal of Hospitality & Tourism Research: the Professional Journal of the Council on Hotel Restaurant and Institutional Education.* v. 21, no. 1– . 1997– . Thousand Oaks, CA: Sage Publications. Continues *Hospitality Research Journal.* Three issues yearly featuring conceptual, empirical, and applied research articles in the tourism and hospitality fields.

*The Journal of Travel & Tourism Marketing.* v. 1, 1992– . Binghamton, NY: Haworth Press. A quarterly journal serving researchers and managers in the field of travel and tourism.

## Proceedings

*Tourism Innovations: Developments, Policy & Markets.* Proceedings of the National Extension Tourism Conference, May 17–19, 1998, Hershey, PA. 1998. University Park, PA: Northeast Regional Center for Rural Development. 134 p. Sponsored by all four Regional Rural Development Centers, the USDA's Cooperative State Research Education Extension Service, the USDA's Natural Resource Conservation Service, The National Association of Resource Conservation Districts (RC&Ds), and the National Tourism Foundation. Provides proceedings on developing ecotourism; case studies in agritourism; cultural and heritage tourism opportunities; collaborative efforts for rural areas; and policy, planning, and promoting recreation and tourism for rural areas.

## World Wide Web Sites

National Tourism Database (http://www.msue.msu.edu/msue/imp/modtd/mastertd.html) (cited June 28, 2001). Lansing, MI: Michigan State University Extension. This is a database of resources that includes full-text articles and experts in tourism. Agritourism and rural tourism areas are covered.

The National Tourism Education Clearinghouse (http://www.cas.nercrd.psu.edu/Tourism/main.html) (cited June 28, 2001). University Park, PA: Northeast Regional Center for Rural Development. This Web site is a compilation of resources for tourism educators.

Ten Factors for a Successful Tourism Program (http://www.ag.uiuc.edu/~lced/resources/factsheets/10factors.html) (cited June 28, 2001). Champaign, IL: University of Illinois at Urbana-Champaign, Community & Economic Development, College of Agricultural, Consumer and Environmental Sciences, Laboratory for Community and Economic Development. A listing of ten ''to-dos'' for successful rural tourism ventures.

Tourism Center (http://www.tourism.umn.edu/zRTD.html) (cited June 28, 2001). St. Paul: MN: The University of Minnesota's Extension Service. This Web site includes informational resources for tourism professionals and educators. It also provides agritourism resources.

Tourism, Hospitality, & Leisure Journals (http://omni.cc.purdue.edu/~alltson/journals.htm) (cited June 28, 2001). West Lafayette, IN: Purdue University, Purdue Tourism & Hospitality Research Center. Prepared by A.M. Morrison for Tourism Research. Comprehensive list of refereed academic journals in tourism, hospitality, leisure, and recreation. Hyperlinked to those with Web sites.

Tourism Resources (http://www.nal.usda.gov/ric/ruralres/tourism.htm) (cited June 12, 2001). Baltimore, MD: NAL, Rural Information Center. This is a Web site that includes hyperlinked resources for tourism with a focus on rural areas. Divided into sections: general resources; funding and program assistance; statistics; data resources; and publications.

## RURAL SOCIOLOGY AND CHANGE

This section is focused on the changes in rural America, both social and economic. How have rural areas adapted to technology, urbanization, out-migration, and what effect does this have on the quality of life of rural people? These resources provide viewpoints of ''rural change.''

### Abstracts, Indexes, Databases, and Other Tools

Sociological Abstracts. 1992– . San Diego, CA: Cambridge Scientific Abstracts. This is a database with a sociological focus. Subjects cover community development, demography, economics, political science, and social psychology. Items are pulled from over 2,500 journals in sociology and related disciplines.

### Bibliographies

Social Indicators: An Annotated Bibliography on Trends, Sources and Developments, 1960–1998. 1998. (http://www.ag.iastate.edu/centers/rdev/indicators/contents.html) (cited June 30, 2001). Ames, IA: North Central Regional Center for Rural Development. This annotated bibliography covers a 40-year span. It provides an overview of social indicators, how they are used and how they can be used to facilitate ''wise community management and development.'' Nontraditional social indicators are also presented to include sustainable communities, healthy communities, and healthy ecosystems. Developed with assistance from the United States Environmental Protection Agency, it focuses on social indicators that promote environmentally safe community and watershed management practices.

## Books

Brown, D.L., D.R. Field, and J.J. Zuiches, eds. 1993. *The Demography of Rural Life*: *Current Knowledge and Future Directions for Research*. NRCRD Publication no. 64. University Park, PA: Northeast Regional Center for Rural Development. 211 p.   A collection of papers presented at the October 18–19, 1991 conference in Madison, Wisconsin that honored rural demographer, Glenn V. Fuguitt. Identifies migration as the major demographic component for population redistribution at the time of this conference.

Castle, E.N., ed. 1995. *The Changing American Countryside*: *Rural People and Places*. Lawerence, KS: University Press of Kansas. 563 p.   This reference tool attempts to alleviate the normal stereotypes of *rural* and to identify true problematic concerns to rural areas and address them with realistic solutions. Bibliographic references and index included.

Conger, R.D. et al., eds. 1994. *Families in Troubled Times*: *Adapting to Change in Rural America*. New York: Walter de Gruyter. 303 p.   This publication is based on a study called, "The Iowa Youth and Families Project." This series of articles on change in rural areas and the impact to families was developed to encourage more effective social policies designed to reduce the adverse impacts of economic decline on individual lives, as experienced in the 1980s farm crisis. Author and subject index, as well as bibliographic references included.

Elder, G.H. Jr. and R.D. Conger. 2000. *Children of the Land*: *Adversity and Success in Rural America*. Chicago, IL: University of Chicago Press. 373 p.   Farm and nonfarm families were compared to assess factors that aid children to develop into successful, productive individuals. The major conclusions drawn from the research were that family matters in the lives of children. Other relevant factors are extended family, such as grandparents; fathers and good fathering; and social capital or family ties to local communities, churches, and schools.

Furuseth, O.J. and M.B. Lapping, eds. 1999. *Contested Countryside*: *The Rural Urban Fringe in North America*. Brookfield, VT: Ashgate Publishing. 291 p.   This book is a collection of essays from the United States and Canada. It lists perspectives in four themes: the examination of spatial structure and organization of the North American rural-urban fringe areas; planning issues and concern for the loss of agricultural land; rural-urban relationships; and the expansion of nongovernmental organizations and their role in environmental issues.

Henderson, D.A. 1997. *Urbanization of Rural America*. Commack, NY: Nova Science Publishers. 216 p.    Divided into three parts, this publication investigates the evolution of cities, urbanization in the twentieth century, and what urban centers may be like in the twenty-first century where technology plays an ever-increasing role in both urban and rural America.

Howell, F.M., Y. Tung, and C. Wade-Harper. 1996. *The Social Cost of Growing-Up in Rural America*: *Rural Development and Social Change during the Twentieth Century*. Social Research Report Series 96-5. Mississippi State, MS: Mississippi Agricultural and Forestry Experiment Station. 89 p.    A research report that identifies the most "prominent transformation in U.S. social history has been the farm-to-city migration pattern." This research review estimates trends in the effects of rural origins on socioeconomic attainments of adults in the United States during the twentieth century.

Ilbery, B., ed. 1998. *The Geography of Rural Change*. London: Addison Wesley Longman. 277 p.    Individually authored chapters that focus on positive and negative economic, physical and social changes of rural communities. Subject areas covered include agricultural productivism, agri-environmental policy, rural migration, industrialization, rural restructuring, and land-use issues. Chapter references and an index are provided.

Rogers, E.M., et al. 1988. *Social Change in Rural Societies*: *An Introduction to Rural Sociology*. 3rd ed. Englewood Cliffs, NJ: Prentice-Hall. 395 p.    Organized into three parts, this textbook examines the many changes that rural areas have undertaken, problems associated with those changes, and social consequences to incorporating those changes (and what happens when they are not addressed.) A look at rural life throughout the world gives a broader base of understanding of rural sociological issues. Glossary of concepts and an index are provided.

Salant, P. and J. Marx. 1995. *Small Towns, Big Picture*: *Rural Development in a Changing Economy*. Washington, DC: The Aspen Institute. 113 p.    A report on rural change, this publication describes the survival requirements for rural areas in a time of change. Areas addressed include the global economy, natural amenities, information technology, and rural poverty trends. Notes and references included.

## Periodicals

Many of those periodicals under the other sections in this chapter also address sociological issues in rural areas and should be reviewed as well.

*Rural Sociology*. v. 1– . 1936– . University Park, PA: Rural Sociology Society. Quarterly. The journal of the Rural Sociological Society provides informational research articles on current issues in rural sociology.

Small Town. v. 1– . 1969– . Eleensberg, WA: Small Towns Institute. Bimonthly. Provides, articles, research reports, and case studies of ''small town'' life.

## World Wide Web Sites

Social Resources (http://www.nal.usda.gov/ric/ruralres/social.htm) (cited June 28, 2001). Baltimore, MD: USDA National Agricultural Library, Rural Information Center. This is a Web site that includes linked resources for rural social issues. Includes family, poverty, homelessness, and minority resources.

The Social Statistics Briefing Room (http://www.whitehouse.gov/fsbr/ssbr.html) (cited June 28, 2001). Washington, DC: The White House. The purpose of this service is to provide easy access to current U.S. social statistics. It provides links to information produced by a number of U.S. agencies.

Texas Rural Church Network (http://rsse.tamu.edu/) (cited June 28, 2001). College Station, TX: Texas A&M University, College of Rural Sociology. Nonprofit organization developed for rural community leaders. Provides education, information, and support for rural community and churches. Focuses on three major programs, rural communities, rural churches, and rural families.

# 15

# International Rural Development and Rural Sociology

**Anita L. Hayden**
Arid Lands Information Center, Office of Arid Lands Studies, University of Arizona, Tucson, Arizona, USA

This chapter is offered as an addendum to the previous Rural Development and Sociology chapter, but from an international angle. A truly global approach to this topic would be overwhelming, so to keep the content manageable, the author chose to focus on resources available through the Internet, especially those resources that provide a good selection of materials available for download at no cost. These materials may include papers and other publications, databases, statistics, images, or audio files.

The topic of international rural development and sociology is inherently interdisciplinary. Just like all interdisciplinary fields, it is difficult, and sometimes inappropriate, to draw lines dividing what is or is not relevant to the topic. This chapter may reflect that dilemma by neglecting some aspects of rural life that have profound impacts on development. This is not meant to be an intentional statement about the topic's worth, but rather a misfortunate causality of abbreviation. As atonement for such oversights, a list of keywords is provided that may help in articulating the reader's research focus.

A parameter to keep in mind when designing research is that ''development'' may happen on several levels: international, national, local, as well as household and individual levels. The data must be extracted and separated into *urban* and *rural*, even though the two geographic areas may be completely dependent on each other. The availability, quality, and reliability of data may not be adequate in all countries. Also, the data that are available will not likely be comparable from one country to another. Rural development should not be tied exclusively to agriculture either, because nonagricultural enterprises, such as tourism, mining, or energy production may have a larger contribution to the rural economy than agriculture. Finally, development should not be measured exclusively by the monetary wealth of the community, but should include measures of other variables, such as equity and access to basic services such as health care and education. With these caveats in mind, it is hoped that the following information will be of use to the reader, and that, someday, the phrase *rural development* may be an old-fashioned and out-dated concept.

## SEARCHING THE LITERATURE

Bibliographic databases compiling abstracts of publications can be very good sources of information. Generally, journal articles focus on narrow topics, although occasionally comprehensive review articles and broad overviews of a field may be published. Searches in these databases may lead to improved search strategies. For example, a literature search performed in August 2001 using the Social Sciences Citation Index (one of the Institute for Scientific Information's Web of Science bibliographic databases) yielded references to many pertinent documents in this topic area. Specifically, the search term *rural development* yielded over 2,000 documents; *rural sociology* yielded 170, and *international development* yielded 425 documents. These documents appeared in more than 15 major journals, indicating that the topics were being discussed across multiple disciplines. There are many other appropriate keywords for targeting the best information. CAB International publishes the *CAB Thesaurus*, a useful tool when conducting literature searches or indexing information. The following is a list of commonly used keywords when searching topics pertaining to international rural development and rural sociology:

> ageing
> agriculture
> aid
> assets
> capacity building
> community health
> conservation

co-operatives
debt
demographics
development
ecology
education
environment
farming
food
gender
geography
health
housing
international development
land use
microcredit
minorities
natural resources
nutrition
planning
population
poverty
recreation
resource management
rural
sanitation
social justice
social migration
sociology
statistics
subsistence
sustainability
tourism
water
women

Most of the abstract databases mentioned here provide electronic access to the abstracts or even the full-text articles on a subscription basis. The subscription fees can be very expensive, however, many research university libraries have included these resources in their collections and offer the public free access to the databases and electronic journals through their library computer terminals. The following is a listing of the abstract and index databases that specifically emphasize international rural development and international rural sociology materials. Other general agricultural databases such as AGRICOLA, AGRIS, and CAB Abstracts will also contain relevant resources.

## ABSTRACTS, INDEXES, DATABASES, AND SEARCH ENGINES

AgriFor (http://agrifor.ac.uk). Nottingham, UK: BIOME. AgriFor is an Internet search engine providing links to selected Internet resources in agriculture, food, and forestry. AgriFor is created by a team of information and subject specialists based at the University of Nottingham, in partnership with international organizations. This database of Internet resources can lead students and researchers to many sites offering information on rural development and rural sociology.

Contact information:
BIOME
Greenfield Medical Library
Queens Medical Centre
Nottingham, NG7 2UH, United Kingdom
Telephone: 440-115-849-3251
Fax: 440-115-849-3265

Electronic Development and Environment Information System—ELDIS (http://www.eldis.org). Brighton, UK: British Library for Development Studies. An electronic directory and gateway to information resources on development and the environment through the Internet. Includes a variety of information sources, including online documents, links to Web sites, databases, library catalogs, bibliographies, and e-mail discussion lists. Hosted by the Institutes of Development Studies in Sussex, United Kingdom. This free service provides easy, well-organized access to a comprehensive collection of information at no charge to the user. Includes links to nearly 100 development centers worldwide, as well as a directory of libraries specializing in development. The database is free.

Contact information:
Eldis Project
British Library for Development Studies
Institute of Development Studies
University of Sussex
Falmer, Brighton, BN1 9RE, United Kingdom
Telephone: 440-1273-606261
Fax: 440-1273-621202

GEOBASE. 1980. Norwich, UK: Elsevier Science. A unique multidisciplinary database supplying bibliographic information and abstracts for development studies, the earth sciences, ecology, geomechanics, human geography, and oceanogra-

phy. The database provides current coverage of over 1,800 journals and archives coverage of several thousand additional titles. The material referenced includes refereed scientific papers; trade journal and magazine articles; product reviews, directories and any other relevant material. GEOBASE contains over one million records from 1980, with 74,000 records added annually. The database is available electronically either online by subscription or on CD-ROM.
Contact information:
Elsevier Geo Abstracts
The Old Bakery 111
Queens Road
Norwich NR1 3PL, United Kingdom
Telephone: 44 0 1603-626327
Fax: 44 0 1603-667934

*International Development Abstracts.* v. 1– . 1982– . New York: Elsevier Science. ISSN: 0262-0855. A bibliographical reference journal published six times per year covering over 500 core journals as well as fringe and popular literature in numerous languages. Entries are grouped into subject areas, including agriculture and rural development, environment and development, social policies, aid, international relations, and politics. Available both in print and electronically.
Contact information:
Elsevier Science
655 Avenue of the Americas
New York, NY 10010
Telephone: 1-212-633-3730
Fax: 1-212-633-3680

*Rural Development Abstracts.* v. 1– . 1978– . Wallingford, UK: CAB International. ISSN: 0140-4768. Rural Development Abstracts is issued quarterly in print, with weekly updates on the subscription Internet version. Covers journal articles, reports, conferences, and books concerning all economic and social aspects of Third World rural development.
Contact information:
CAB International
Nosworthy Way
Wallingford, Oxon OX10 8DE, United Kingdom
Telephone: 44 0 1491-832111
Fax: 44 0 1491-833508

*World Agricultural Economics and Rural Sociology Abstracts.* v. 1– . 1959– . Wallingford, UK: CAB International. ISSN: 0043-8219. This combined set of abstracts is available both in print (monthly updates) and through the Internet by

subscription (with weekly updates). Features abstracts from journal articles, reports, conferences, and books covering agricultural economics and rural sociology.
Contact information:
CAB International
Nosworthy Way
Wallingford, Oxon OX10 8DE, United Kingdom
Telephone: 44 0 1491-832111
Fax: 44 0 1491-833508

## ORGANIZATIONS AND ASSOCIATIONS

This listing is not intended to be a comprehensive list of organizations and associations involved in international rural development and sociology, but rather a list of governmental and nongovernmental organizations that provide a significant and varied collection of materials on their Web sites that may be of interest or assistance to a researcher.

Center for Indigenous Knowledge for Agriculture and Rural Development (http://www.iastate.edu/~anthr_info/cikard/). An academic and research center focusing on preserving and using the knowledge of farmers and rural people around the globe to facilitate participatory and sustainable approaches to development. Its goal is to record indigenous knowledge and make it available to local communities, development professionals, scientists, and scholars. The Center for Indigenous Knowledge for Agriculture and Rural Development (CIKARD) concentrates on four areas: indigenous innovations, knowledge systems (such as taxonomies), decision-making systems (such as what crops to grow on certain soils), and organizations (such as farmers' groups).
Contact information:
CIKARD
324 Curtiss Hall
Iowa State University
Ames, IA 50011
Telephone: 1-515-294-9503

Centre for Rural Social Research (http://www.csu.edu.au/research/crsr/). The Centre for Rural Social Research (CRSR) conducts applied multidisciplinary social research relevant to policy development and rural services, with a focus on rural Australia. Publishes papers, proceedings, and technical reports, as well as the *Rural Society* journal three times per year.
Contact information:
Charles Sturt University

Locked Bag 678
Wagga Wagga, NSW 2678, Australia
Telephone: 61-2-6933-2778
Fax: 61-2-6933-2293

Consultative Group on International Agricultural Research (http://www.cgiar.org/). The Consultive Group on International Agricultural Research (CGIAR) is a network of agricultural scientists who work toward achieving food security and poverty eradication in developing countries through research, partnerships, capacity building, and policy support, thus promoting sustainable agricultural development. Publications include the *CGIAR Annual Reports*, the *CGIAR News* (a quarterly newsletter), meeting documents, *Crawford Lectures*, *Issues in Agriculture*, *Gender Program* papers, *Study Papers*, and others.
Contact information:
CGIAR Secretariat, The World Bank
MSN G6-601, 1818 H Street NW
Washington, DC 20433
Telephone: 1-202-473-8951
Fax: 1-202-473-8110

Department for International Development (http://www.dfid.gov.uk). The Department for International Development (DFID) is the British agency responsible for promoting international development and reducing poverty (formerly called the Overseas Development Administration).
Contact information:
Department for International Development
1 Palace Street
London, SW1E 5HE, United Kingdom
Telephone: 44-0-20-7023-0000
Fax: 44-0-20-7023-0016

Institute of Development Studies (http://www.ids.ac.uk/ids/). The Institute of Development Studies (IDS) is an international center for research and teaching on development, and is closely linked with the University of Sussex and the United Kingdom's DFID. The IDS Internet bookshop gives access to all IDS publications, including many bibliographies and abstract compilations. IDS hosts ELDIS, an information gateway for international development described earlier in this chapter, as well as other development-related projects and Web sites. IDS also provides easy access to the British Library for Development Studies (BLDS), housing a large collection of current materials in international development.
Contact information:
Institute of Development Studies
University of Sussex

Brighton BN1 9RE, United Kingdom
Telephone: 44-0-1273-606261
Fax: 44-0-1273-621202/691647

International Development Research Centre (http://www.idrc.ca). The International Development Research Centre (IDRC) is a Canadian public corporation whose mandate involves helping developing countries find long-term solutions to the social, economic, and environmental problems they face. Many of the books published by the IDRC are available online in full text at no cost.
Contact information:
International Development Research Centre
P.O. Box 8500
Ottawa, ON K1G 3H9, Canada
Telephone: 1-613-236-6163, ext. 2075
Fax: 1-613-563-2476

OneWorld International (http://www.oneworld.net). An international nonprofit umbrella organization working as a network of centers and a community of over 950 organizations using the Internet to work toward sustainable development and social justice. The Web site offers a host of up-to-date information on development issues, including One World publications, special reports, and links to current news topics, as well as discussion guides, and audio files of interest. A very good portal for monitoring global activities in social justice.
Contact information:
OneWorld International
Floor 17
89 Albert Embankment
London SE1 7TP, United Kingdom
Telephone: 44-020-7735-2100
Fax: 44-020-7840-0798

Oxfam International (http://www.oxfam.org/). A confederation of autonomous nongovernmental organizations, Oxfam International (OI) is committed to fighting poverty and injustice around the world. Each organization maintains its own structure and identity, however the Web site contains links to all the independent Oxfam organizations, as well as a valuable collection of outside links.
Contact information:
Oxfam America
26 West Street
Boston, MA 02111
Telephone: 1-617-482-1211
Fax: 1-617-728-2594

Society for International Development (http://www.sidint.org). The Society for International Development (SID) is a global network of individuals and institutions concerned with issues surrounding development and social change. The SID's journal, *Development*, is published quarterly by Sage Publications. The Web site has many in-house publications available for download, as well as discussion forums.

Contact information:
Society for International Development
Via Panisperna, 207
00184 Rome, Italy
Telephone: 39-064872172
Fax: 39-064872170

United Nations (http://www.un.org/english). The United Nations (UN) is a global organization whose members are sovereign nations. Several agencies within the UN are dedicated to international development and rural issues. One of the misconceptions about the UN is that its primary activity is peacekeeping, however, over 70% of UN activities are devoted to development and humanitarian assistance. The Economic and Social Council (ECOSOC) is the principal organ for addressing international economic, social, health, and related problems. All UN agencies publish materials ranging from books and monographs to discussion papers, briefing papers, occasional papers, conference reports, and newsletters.

Contact information:
United Nations Publications
Room DC2-0853
2 UN Plaza
New York, NY 10017
Telephone: 1-800-253-9646; 1-212-963-8302
Fax: 1-212-963-3489

Food and Agriculture Organization of the United Nations (http://www.fao. org). The Food and Agriculture Organization of the United Nations (FAO) is one of the autonomous agencies within the ECOSOC, with a specific mandate that includes bettering the conditions of rural populations. The Sustainable Development (SD) department oversees many projects directly related to rural development, and hosts electronic discussion forums on current topics. The FAO publishes a wide range of materials, including over 3,000 photos with detailed captions in the digital photo archive known as MediaBase.

Contact information:
Food and Agriculture Organization of the United Nations
Via delle Terme di Caracalla

00100 Rome, Italy
Fax: 39-06-5705-3360

United Nations Research Institute for Social Development (http:// www.unrisd.org/). The United Nations Research Institute for Social Development (UNRISD) is another of the autonomous, multidisciplinary organizations that conducts research on the social dimensions of contemporary problems affecting development. It is not associated with any single specialized agency. The UNRISD has many publications available pertaining to rural development and sociology.
Contact information:
UNRISD
Palais des Nations
1211 Geneva 10, Switzerland
Telephone: 41-22-917-3020
Fax: 41-22-917-0650

United States Agency for International Development (http:// www.usaid.gov/). The United States Agency for International Development (USAID) is an agency of the federal government that extends foreign assistance and humanitarian aid in order to advance the political and economic interests of the United States. USAID has working relationships with more than 3,500 American companies and over 300 United States–based private voluntary organizations. The Web site has some pertinent development publications, including a database of statistics relevant to international development.
Contact information:
U.S. Agency for International Development Information Center
Ronald Reagan Building
Washington, DC 20523
Telephone: 1-202-712-4810
Fax: 1-202-216-3524

Wageningen University (http://www.sls.wau.nl/crds/Index.html). One of the oldest research groups to focus on rural development sociology, Wageningen University in The Netherlands continues to produce topical research in the area. The Web site contains a very interesting history of rural sociology. The group manages the editorship of *Sociologia Ruralis* journal, as well as a book series titled European Perspectives on Rural Development.
Contact information:
Rural Sociology Group
Hollandseweg 1
6707 KN Wageningen, The Netherlands

World Bank Group (http://www.worldbank.org). The World Bank is the

world's largest source of development assistance. The institution provided more than U.S.$15 billion in loans in the year 2000 to developing nations. The Web site contains useful links, statistics, maps, and discussion forums, including a Web-based database called World Bank Sources, containing over 6,000 reports.
Contact information:
The World Bank
1818 H Street, NW
Washington, DC 20433
Telephone: 1-202-477-1234
Fax: 1-202-477-6391

World Health Organization (http://www.who.int/home-page/). The World Health Organization (WHO) is involved in projects related to many of the specific issues affecting international rural development and rural sociology. A network of scientists called the Operation and Maintenance Working Group (OMWG) maintains an office called Water, Sanitation, and Health (WSH), which focuses on promoting cooperation between external support agencies and developing countries for establishment of sustainable water supplies and sanitation. The WHO maintains a large library of information available for free through the Internet.
Contact information:
WHO Headquarters Office in Geneva
Avenue Appia 20
1211 Geneva 27, Switzerland
Telephone: 00-41-22-791-21-11
Fax: 00-41-22-791-3111

## PUBLISHERS

It is not practical to attempt a compilation of books published in this ever-growing field. Nor is it practical to list all the publishers. However, there are some note-worthy publishers whose primary focus include social issues, international issues, or development. A few of those publishers, as well as some of the larger journal publishers who offer several journals related to the field, are listed.

### Blackwell Publishers

Focusing on publishing for the social sciences and humanities, Blackwell Publishers produces books and journals in areas of agriculture, development, and sociology, including the journal *Sociologia Ruralis*.
Contact information:
Blackwell Publishers Ltd.
108 Cowley Rd.

Oxford, OX4 1JF, United Kingdom
Telephone: 44-1865-791100
Fax: 44-1865-791347
URL: http://www.blackwellpublishers.co.uk

## Elsevier Science

Publishes numerous journals relevant to international rural development and rural sociology, including *International Development Abstracts*, *Food Policy*, *Journal of Rural Studies*, and *Rural Development*.
Contact information:
Elsevier Science
P.O. Box 211
1000 AE Amsterdam, The Netherlands
Telephone: 31-20-485-3757
Fax: 31-20-485-3432
URL: http://www.elsevier.nl/

## Island Press

Island Press is a nonprofit publisher with a focus on environmental issues and sustainability.
Contact information:
Island Press
1718 Connecticut Avenue, NW
Suite 300
Washington, D.C. 20009
Telephone: 1-202-232-7933
Fax: 1-202-234-1328
URL: www.islandpress.org

## Kumarian Press

Kumarian Press is an independent publisher specializes in books and other media that focus on international development and related topics.
Contact information:
Kumarian Press
1294 Blue Hills Avenue
Bloomfield, CT 06002
Telephone: 1-860-243-2098
Fax: 1-860-243-2867
URL: www.kpbooks.com

## Sage Publications

Sage is an international publisher, originally focusing on the social sciences but now publishing works in a wide range of fields. One noteworthy book published by Sage is titled *Collaboration in International Rural Development: A Practitioner's Handbook* by G. H. Axinn and N. W. Axinn (1997).
Contact information:
Sage Publications, Inc.
2455 Teller Road
Thousand Oaks, CA 91320
Telephone: 1-805-499-0721
Fax: 1-805-499-0871
URL: http://www.sagepub.com/

## Significant Journals

The following is a listing of the most popular journals addressing issues in international rural development. This list is not intended to be a complete listing, but rather those that appeared most frequently in sample literature searches.

*Community Development Journal.* v. 1– . 1966. Oxford, U.K.: Oxford University Press. ISSN: 0010-3802. Published four times per year by the Oxford University Press. Articles cover a wide range of topics including community action, regional planning, community studies, and rural development.
Contact information:
Oxford University Press
Great Clarendon Street
Oxford, OX2 6DP, United Kingdom
Telephone: 440-1865-556767
Fax: 440-1865-556646
URL: http://www3.oup.co.uk/cdj/

*Development.* v. 1– . 1957– . Thousand Oaks, CA: Sage Publications. ISSN: 1011-6370. Published four times per year for the Society for International Development by Sage Publications. Available in several languages.
Contact information:
Sage Publications, Inc.
2455 Teller Road
Thousand Oaks, CA 91320
Telephone: 1-805-499-0721
Fax: 1-805-499-0871
URL: http://www.sagepub.com/

*Development and Change.* v. 1– . 1969– . Oxford, U.K.: Blackwell Publishers. ISSN: 0012-155X. Published for the Institute of Social Studies at The Hague by Blackwell Publishers. Currently offering five issues per year, this journal focuses on development studies and social change.
Contact information:
Blackwell Publishers, Ltd.
108 Cowley Rd.
Oxford, OX4 1JF, United Kingdom
Telephone: 44-1865-791100
Fax: 44-1865-791347
URL: http://www.blackwellpublishers.co.uk

*Food Policy.* v. 1– . 1975– . New York: Elsevier Science. ISSN: 0306-9192. A bimonthly journal publishing original research and critical reviews on issues in the formulation, implementation, and analysis of policies for the food sector in developing, transitional, and advanced economies.
Contact information:
Elsevier Science
655 Avenue of the Americas
New York, NY 10010
Telephone: 1-212-633-3730
Fax: 1-212-633-3680
URL: www.elsevier.nl

*Journal of Development Studies.* v. 1– . 1964– . London: Frank Cass. ISSN: 0022-0388. An interdisciplinary approach to international development published six times per year. A focus on issues relevant to development economics, politics, and policy.
Contact information:
Frank Cass & Company Ltd.
Crown House
47 Chase Side
Southgate, London, N14 5BP United Kingdom
Telephone: 44-020-8920-2100
Fax: 44-020-8447-8548
URL: http://www.frankcass.com/jnls/jds.htm

*Journal of Rural Studies.* v. 1– . 1985– . New York: Elsevier Science. ISSN: 0743-0167. Published quarterly, the *Journal of Rural Studies* is both international and interdisciplinary in scope, with articles relating to various rural issues such as society, employment, land use, recreation, agriculture, and conservation.
Contact information:
Elsevier Science

655 Avenue of the Americas
New York, NY 10010
Telephone: 1-212-633-3730
Fax: 1-212-633-3680
URL: www.elsevier.nl

*Sociologia Ruralis.* v. 1– . 1960– . Oxford: Blackwell Publishers. ISSN: 0038-0199. Published quarterly, this journal covers a wide range of subjects, ranging from farming, natural resources, and food systems to rural communities, rural identities, and the restructuring of rurality. Published on behalf of the European Society for Rural Sociology by Blackwell Publishers.
Contact information:
Blackwell Publishers, Ltd.
108 Cowley Rd.
Oxford, OX4 1JF, United Kingdom
Telephone: 44-1865-791100
Fax: 44-1865-791347
URL: http://www.blackwellpublishers.co.uk

*World Development.* v. 1– . 1973– . New York: Elsevier Science. ISSN: 0305-750X. A monthly journal addressing issues such as poverty, unemployment, malnutrition, environmental degradation, access to information, trade, debt, discrimination, militarism, and lack of popular participation in economic and political life. Published by Elsevier Science.
Contact information:
Elsevier Science
655 Avenue of the Americas
New York, NY 10010
Telephone: 1-212-633-3730
Fax: 1-212-633-3680
URL: www.elsevier.nl

## CONCLUSION

Searching the literature for interdisciplinary topics relating to global issues is a never-ending endeavor. This chapter was compiled in an effort to provide some starting points for the researcher, as well as to emphasize the role of the Internet in disseminating information that was, historically, largely inaccessible to most people.

# 16

## Soil Science

**Carla Long Casler**
The University of Arizona, Tucson, Arizona, USA

**Karl R. Schneider**
National Agriculture Library, ARS-USDA, Beltsville, Maryland, USA

The study of soil science covers complex processes and morphology. The Soil Science Society of America has 11 "divisions of interest" that illustrate the scope of soil science: soil physics, soil chemistry, soil biology and biochemistry, soil fertility and plant nutrition, pedology, soil and water management and conservation, forest and range soils, nutrient management and soil and plant analysis, soil mineralogy, wetland soils, and soils and environmental quality. Additional concepts that appear frequently in recent literature are soil quality and soil health. Soil science is dynamic, expanding, and changing with each new technology applied to it.

The earliest record of the study of soils dates from 4,000 years ago in China, where soils were used as a basis for taxation and categorized in 9 broad classes and given descriptive names. Greeks such as Aristotle (384–22 B.C.) and his student Theophrastes differentiated soils by qualities affecting plant growth and recognized layers in the soil, now called horizons. Roman philosophers, Cato the Elder (234–149 B.C.), Varro (116–27 B.C.), and Columella (about 45 A.D.), discussed soils differences that caused variations in plant growth and ranked soils according to their suitability for plant growth (McCracken and Helms, 1994).

From 1700 through 1850, the field was mainly comprised of "gentleman practitioners" sharing ideas in meetings of agricultural societies, although some chemists and geographers also contributed to the study of soils. From 1850 through 1910 there was more intensive experimentation with organic chemistry applied to agriculture, including soil processes. Concern about exhausted, eroded lands resulted in discussions over whether loss of productivity was chemical or physical. Studies in the subdisciplines of soil science, except soil mineralogy, were conducted in this period. *La Pédologie* began publication in Russia in 1899, changing the title to *Pochvovedenie* in 1928. Many other national and international journals began in the early 1900s. From 1910 through 1945 the emphasis changed from geology and crop relations to soil genesis and classification based more on landscape characteristics influenced by studies in Russia in the late 1800s. In 1938, the Commonwealth Agricultural Bureaux (now CAB International) began publishing *Soils and Fertilizers*, a journal abstracting publications on soils. From 1940 through 1980 there was a greater emphasis on experimental studies in soils science, changing in the late 1970s to the application of general soils knowledge, and focusing on the landscape context. It was discovered that many experimental results were not applicable because processes occurring in the field had not been duplicated in the experiments. National societies for soil science were developed in many countries after the 1950s. The number of publications grew dramatically, but the increasing costs of journals combined with the growth of new specialized soils journals prevented most libraries from collecting comprehensively in the area (Warkentin, 1994).

The following quotations illustrate the complexity and vital importance of soils and soils science:

"Soils are geological bodies that take thousands to millions of years to develop. And, unlike living species, they do not reproduce nor can they be recreated" (Amundson, 1998).

"If all the elephants in Africa were shot, we would barely notice it, but if the nitrogen-fixing bacteria in the soil, or the nitrifiers, were eliminated, most of us would not survive for long because the soil could no longer support us. . . . Observing soils, studying them, and reflecting on them induces respect if not wonder. All of us relate to soil unconsciously in our daily nourishments that make us participants in the continuous flow of nutrient atoms that originate in the soil" (Stuart and Jenny, 1984).

"Soil is an amalgamation of sand, silt, and clay particles, combined with water, air, and many different microorganisms. The formation of different kinds of soils is influenced by temperature, climate, vegetation and other factors. Soil serves as a medium for plant growth.

It also helps to clean water, regulate climate, and purify wastes. Soil is not dirt, but rather a body of plant, animal, mineral and other matter that, in combination, becomes the 'ecstatic skin of the earth.''' [Hans Jenny] (Soil and Water Conservation Society, 2000).

"Natural processes can take more than 500 years to form one inch of topsoil. Soil scientists have identified over 70,000 kinds of soil in the United States" (USDA, Natural Resources Conservation Service, 2000).

Given that factors of climate, topography, organisms, parent material, and time influence the development of soils and that researching soils includes use of Geographic Information Systems (GIS) and remote sensing technologies, the study of soil science incorporates aspects of meteorology, computer science, geology, microbiology, chemistry, and physics in defining, describing, and mapping soils. The literature of soil science is equally diverse. For a comprehensive study of soil science publishing, see P. McDonald (1994). *The Literature of Soil Science*. Ithaca, NY: Cornell University Press. It covers the international history of soil science and illustrates the continuing evolution of soil science and its literature.

## SEARCHING FOR SOILS INFORMATION

Just as soils are complex, the variety of databases available for soil topics presents a broad range of data types, content, and retrieval options. The first choice in accessing soils information through databases is deciding between bibliographic databases to find publications, or nonbibliographic databases for data, maps, graphs, and nontext information.

For any specific search, database selection depends upon background knowledge available, needs and goals of the inquirer, and whether the information may be among known, standard properties of specific soils or soil systems. In the latter case, where particular soils properties or characteristics are sought, they may be found among the range of nonbibliographic databases now available. Many of these may be accessed using the Internet. Examples range from graphic maps and datasets in recent soil surveys, to GIS data for precision farming, hydrologic, or other applications, to soil-pesticide interaction data, and many others.

## ABSTRACTS, INDEXES, AND DATABASES

### Nonbibliographic Databases and Internet Accessible Resources

AgNIC (Agriculture Network Information Center), available from the World Wide Web at http://www.agnic.org, is one valuable resource for reviewing and selecting nonbibliographic databases for soils topics. A broad soil search (soil or soils) retrieved almost 100 items. Items retrieved included (as of June 18, 2001): "Aridic Soils of the United States and Israel," "USDA-NRCS Official Soil Series Descriptions," and "Diagnosis and Improvement of Saline and Alkali Soils." Narrowing the search to "soils and databases" retrieved 11 items, including the Soil Series Descriptions mentioned previously, "Worldwide Organic Soil Carbon and Nitrogen Data," and "X-ray Diffraction Analysis Database of Selected Soils and Sediments of the Southern Region." It is important to note that AgNIC is a dynamic resource, changing frequently as additional resources become available.

Because Internet soils resources listed on AgNIC are reviewed and selected for inclusion, items found there are likely to be the best of such resources. However, these are by no means the total of nonbibliographic soils databases. Internet search engines such as Google and AltaVista (among numerous others) will retrieve great numbers of soils items. Indexes are created, structured, and used in different ways by different search engines. Valuable features that help identify additional soils database resources include field limits for terms, truncation, parenthetical nesting, and Boolean term combination. AltaVista (http://www.altavista.com) allows these controls and was used to illustrate access to soil database resources through the Internet. In July 2001, using the AltaVista search system, over 54,000 items were found through a search using "soil and database." More specific focus by adding another concept set, (with terms such as *test\**, *fertil\**, *engineer\**, or *microb\**) still produced very large results sets. Retrievals with these terms ranged from about 8,000 pages to more than 35,000.

Because it is impractical to review such large numbers of perhaps unedited and unreviewed resources, initial use of AgNIC is suggested for nonbibliographic database selection. If exhaustive review of Internet resources is needed, it may be practical to add other concepts, with terms to further limit results. Concepts may be added with geographic or other specific terms that identify parameters of interest. Using AltaVista, the results noted previously were reduced considerably when other terms were added. Searching for *+soil +database +microb\* +rhizob\** showed 418 items. Adding *+irrigation* with the *+engineer* search gave just over 5,000 items. Adding *+micronutrients* to the search for *+soil +database +test\** search brought the results down to 541 items. Similarly, *+soil +database +fertil\* +ohio* showed over 2,100 items, and changing the added term from *+ohio* to *+micronutrient\** produced a results set of 425 items.

These refined Internet searches, although resulting in more manageable numbers, do not always indicate many useful items. Internet searches retrieve items based on unregulated term inclusion from a variety of fields. Greater precision may be obtained through specific field searching that is available from some search engines. Also, more efficient searches for soil database resources may be completed by using traditional bibliographic databases. These indexes include both bibliographic records for publications that describe specific database products or projects, and in some cases, records for the database products themselves.

To further explore the scope of nonbibliographic soils database resources, a search was conducted using prominent soils database files available through Dialog, a gateway system providing access to hundreds of databases. Over 1,000 items were identified in more than 40 databases. Many cited items were in CD, magnetic tape, and other nonprint formats. Because the bibliographic indexes often include records for published bibliographies, a strategy was used to eliminate the majority of that type of record. Note that such publications may at times add value. Examples of nonbibliographic soils database items from bibliographic database sources are shown in the following text. The list comprises selected titles of items found in the AGRICOLA database with a search using [(database or databases) in title combined with (soil or soils or the category code j)]:

> SALTDATA: a Database of Plant Yield Response to Salinity
> From Soil Survey to a Soil Database for Precision Agriculture
> Canada's Soil Organic Carbon Database
> Thermal Gravrimetric Analysis Database of Selected Soils and Sediments of the Southern Region
> Ecological Database Development and Analyses of Soil Variability in Northern New England
> The UNSODA Unsaturated Soil Hydraulic Database: User's Manual

Many of these record titles refer to publications about databases, but some are actual computerized database systems.

## Bibliographic Databases

While a large amount of material on soils from database sources has been described, the bulk of available information comes from published literature (articles, books, symposium proceedings, and so forth). Effective access to this information involves careful searching among the bibliographic databases. For complex searching and use of specialized database features, consult a local librarian.

Review of selected files available from the Dialog database vendor reveals extensive coverage of soils. Dialog's ''DialIndex'' system (file 411) is a useful tool for conducting a preliminary search to determine appropriate databases because it allows simultaneous review of multiple databases' contents and reports record counts for each database. From a search on *allfiles* (over 600 databases) for *soil* or *soils* as a title or descriptor term, the report reveals that almost 500 of these files have soil(s) terms posted with more than 4 million records identified. Many of these databases are not directly focused on agricultural science studies, including news, financial, business and industry data, patents, medical and other nonagriculture topics. Selected databases include almost 2 million soils records among more than 60 files particularly suited to agricultural research topics. Databases covering soils include the major bibliographic agricultural databases. About 25 specific bibliographic databases are listed here. Others are not described because of space limits. An excellent overview of soils databases is found in the extensive treatment of soils literature by McDonald (1994), and in some other monographs for soils, listed in that section of this chapter.

Soils bibliographic databases are listed in the following text in decreasing order by number of soils related subject records. Notes are given about specific codes and fields for soil subjects. Prefixes for Dialog use are noted. Numbers of soil records for each database follow each entry, reflecting counts (rounded to the nearest 1,000 records) for updates on Dialog in the last week of June 2000. The search strategy used for these counts was: select soil or soils in title and descriptor.

Access notes are also provided. Online sources (direct or through the Internet) may include some or all of these vendors: COS, DataStar, Dialog, DIMDI, NERAC, OCLC, Ovid, SilverPlatter, or STN. CD versions may be offered by database producers, or from vendors such as OVID or SilverPlatter. Printed bibliographic lists are published by many database producers, such as CAB International, and may include a single title, or comprise several specific indexing and abstracting publications. Internet links to free sites, and some commercial sites are shown.

AGRICOLA. 1970– . Beltsville, MD: USDA National Agricultural Library.   A-GRICOLA covers all aspects of agricultural soil science. Books, journal articles, proceedings and seminar papers, reports, and other types of print publications are included. AGRICOLA also indexes audio-visual and microform media, and some computer software and files. HTML links to online sources are sometimes shown. The U.S. National Agricultural Library (NAL), part of the U.S. Department of Agriculture (USDA) Agricultural Research Service produces the database. AGRICOLA uses one general and eight specific subject codes to differentiate records for soils topics, searchable in the "Section Heading" field (using: "SH=J---"). Access: online vendors, CD, print, local sites with the database locally loaded. Available from the World Wide Web at http://www.nal.usda.gov/ag98, or http://agricola.cos.com./. 282,000 soils records.

CAB Abstracts. 1972– . Wallingford, UK: CAB International.   CAB Abstracts also covers all aspects of soil science for agriculture. It has a generally broad global scope, and may be noted as the most comprehensive database for publications about soils in arid regions. CAB uses one general soil subject code and nine specific codes for particular soil science topics, searchable using "CABI-Codes" in the format "CC=JJ---". Access: online vendors, CD. Available from the World Wide Web at http://www.cabdirect.org/. 260,000 soils records.

TOXLINE. 1965– . Bethesda, MD: U.S. National Library of Medicine.   TOX-LINE covers the toxicological, pharmacological, biochemical, and physiological effects of drugs and other chemicals. It covers pesticides, waste disposal, and chemical impacts and interactions with the environment. Many TOXLINE records are also in the MEDLINE database. TOXLINE uses no specific subject codes for soils. Focused soil topic records may be selected by use of the /MAJ limit with basic index searches for soil(s) terms, to limit retrieval to records with *soil* as a major descriptor. Access: online vendors, CD. Available from the World Wide Web at http://igm.nlm.nih.gov/. 230,000 soils records.

BIOSIS Previews. 1969– . Philadelphia, PA: BIOSIS.   BIOSIS Previews provides comprehensive worldwide coverage of research in the biological and biomedical sciences. Because this database covers all aspects of all life sciences, it has significant coverage of soils subjects through the interdependence of terrestrial life with soils. BIOSIS uses four specific codes for access to particular types of soils records, and several codes covering specific organism types.

PASCAL. 1977– . Vandoeuvre-les-Nancy, France: Institute de l'Information Scientifique et Technique.   PASCAL is produced by the Institute de l'Information Scientifique et Technique (INIST) of the French National Research Council (CNRS). This database is multidisciplinary. The principal subject areas are physics and chemistry, life sciences, applied sciences and technology, earth sciences, and information sciences. A complex set of cascaded codes is used. Notes on these codes are available from INIST, but these are not found online. Access: online vendors, CD. Available from the World Wide Web at http://services.inist.fr/public/eng/conslt.htm. 174,000 soils records.

AGRIS. 1975– . Rome: Food and Agriculture Organization of the United Nations.   AGRIS database covers both economic and technical or scientific aspects of agriculture. Soils-related topics include general agriculture and forestry, plant and animal production, natural resources, and pollution. AGRIS employs one general and six specific codes for soils, in the format ''SC=P---''. These are all within a broad group searchable as ''SC=P'' (including all natural resources: water, land, soil, and climate). Access: online vendors, CD. Available from the World Wide Web at http://www.fao.org/agris/default32.htm (for 1996– ). 160,000 soils records.

GeoRef. 1785– . Alexandria, VA: American Geological Institute.   GeoRef, the database of the American Geological Institute (AGI), covers worldwide technical literature on geology and geophysics. About 60% of the records are from non-U.S. sources. GeoRef includes citations for items from thousands of professional society serials and other publications from member organizations of AGI. Maps, theses, and dissertations are also included. Detailed codes within soil topics are not used, but one general code for soil is available by a search with:

"SH=SOILS". Access: online vendors, print, CD. Available from the World Wide Web at http://www.georef.org/. 128,000 soils records.

Current Contents. 1990– . Philadelphia, PA: Institute for Scientific Information. Current Contents database provides access to the tables of contents from current issues of leading journals in seven broad groups. Soils topics are found mainly in four subfile groups, including: life sciences; engineering, technology, and applied sciences; agriculture, biology, and environmental sciences; and physical, chemical, and earth sciences. This database is produced by the Institute for Scientific Information (ISI). Current Contents uses "Journal Subject Code" to indicate general soil topic coverage, searchable using: "SC=SOIL". This only retrieves one fourth of the records with *soil(s)* as title or descriptor term. Access: online vendors, CD, print. Available from the World Wide Web at http://www.isinet.com/isi/products/cc/ccconnect/cccind.html. 83,000 soils records.

JICST-EPlus. 1985– . Tokyo, Japan: Japan Information Center for Science and Technology. JICST-EPlus (Japanese Science & Technology) includes citations to literature published in Japan from all fields of science, technology, and medicine. Agriculture, forestry, environment, geology, construction, and engineering are all included subjects. The database is produced by Japan Science and Technology Corporation. A complex set of alphanumeric codes is used for subject classification, available from the database producer. An online thesaurus may be used to identify needed terms. Access: online vendors. Available from the World Wide Web at http://jois.jst.go.jp./enjoy-jois/jois/nl0s5010.cgi (only in Japanese language). 80,000 soils records.

Ei Compendex. 1970– . Hoboken, NJ: Engineering Information, Inc. Ei Compendex database provides abstracted information from significant engineering and technological literature worldwide. COMPENDEX PLUS is produced by Engineering Information, Inc. This database includes journal article citations as well as records from published meeting reports and proceedings, and includes some soil-related topics such as pollution, geology, waste management, and others. Classification codes (CC) may be used to specify general category groups. These may be searched using: "CC=term". Access: online vendors, CD, print. Available from the World Wide Web at http://www.ei.org/. 62,000 soils records.

Energy Science and Technology. 1974– . Oak Ridge, TN: U.S. Department of Energy. Energy SciTech covers a broad range of topics for both basic and applied research in relation to energy sources, production, and use. Because energy activity often indirectly involves soils, especially with regard to generation, storage, and wastes, soils literature is abundant in this database. The U.S. Department of Energy produces Energy SciTech, in cooperation with other federal and international agencies and groups. Both numeric Subject Codes (SC) and corresponding text Subject Headings (SH) are used to identify selected topics, but neither include soil-specific terms or concepts. *Earth science* and *Geosciences* are both SH terms. The online thesaurus does not show any related terms for either of these. Access: online vendors, print. 58,000 soils records.

GEOBASE. 1980– . Norwich, UK: Elsevier Science. GEOBASE covers world-wide literature on geography, geology, ecology, and their related disciplines. GEOBASE international coverage includes citations to journals in English and foreign languages. Books, conference papers, and reports also included. No subject codes are used in this database. Access: online vendors, CD, print. Available from the World Wide Web at http://www.elsevier.com/inca/publications/store/ 4/2/2/5/9/7/index.htt. 56,000 soils records.

Ingenta. 1998– . Bath, UK: Ingenta PLC. Ingenta is a freely searchable database covering more than 20,000 journals, with coverage back to 1988. Articles are available in electronic form payable by deposit account or credit card. Libraries may register existing subscriptions for free full-text access through Ingenta. Ingenta acquired UnCover in 2000 and incorporated that database into the Ingenta database. Access: Available from the World Wide Web at http://www.ingenta.-com. 46,000 soils records.

NTIS: National Technical Information Service. 1964– . Springfield, VA: U.S. National Technical Information Service. The National Technical Information Service (NTIS) database consists of summaries of U.S. government–sponsored research, development, and engineering, plus analyses prepared by federal agencies, their contractors, or grantees. Some state and local government agencies also contribute summaries of their reports to the database. NTIS also provides access to the results of government-sponsored research and development from countries outside the United States, including the United Kingdom, Germany, and France, and many more. NTIS is produced by the National Technical Information Service, U.S. Department of Commerce. A few broad subject codes that cover soils-related topics are found in the filed called ''Section Heading'' (SH). Access: online vendors, CD, print. Available from the World Wide Web at http:// www.ntis.gov/. 42,000 records.

SciSearch. 1974– . Philadelphia, PA: Institute for Scientific Information. Sci-Search is a cited reference in science that is an international, multidisciplinary index to the literature of science, technology, biomedicine, and related disciplines, including agricultural sciences topics. SciSearch offers citation indexing, which permits searching by cited references. Only one general code covers soils in the SciSearch ''Journal Subject Category'' field entries, (''SC=AGRICULTURE, SOIL SCIENCE''). This code retrieves only a fraction of the total soils records in the database (compared to the *soil(s)* keyword). Access: online vendors, print, CD. Available from the World Wide Web at http://www.isinet.com/ isi/products/citation/wos. 41,000 soils records.

Environmental Bibliography. 1973– . Santa Barbara, CA: Environmental Studies Institute.   Environmental Bibliography provides access to the contents of periodicals covering all aspects of the environment. Coverage includes periodicals from major publishers on water, air, soil, and noise pollution; solid waste management; health hazards; urban planning; global warming; and other specialized subjects of environmental consequence. No subject codes are used. Access: online vendor, print, CD. Available from the World Wide Web at http://www.iasb.org/epb/. 40,000 soils records.

Water Resources Abstracts. 1968– . Bethesda, MD: Cambridge Scientific Abstracts.   Water Resources Abstracts offers a comprehensive range of water-related topics in the life and physical sciences, as well as the engineering and legal aspects of the conservation, control, use, and management of water. Though the focus of the database is water, there is considerable information about soils in arid regions related to irrigation, erosion, pollution, soil moisture, and evapotranspiration. The database was initiated in 1968 and produced by the U.S. Geological Survey until 1994. It is now produced by Cambridge Scientific Abstracts. Although subject codes (''SH='') are used, only a few are specific to soil topics so that basic index searches for soils terms is more productive. Access: online vendors, CD, print. Available from the World Wide Web at http://www.csa. com/. 37,000 soils records.

Inside Conferences. 1993– . Wetherby, West Yorkshire, UK: The British Library Document Supply Centre.   Inside Conferences is produced by The British Library. The database includes bibliographic details for papers given at congresses, symposia, conferences, expositions, workshops, and meetings received by the British Library Document Supply Centre since 1993. No subject codes are found in this database. Access: online vendors, print. Available from the World Wide Web at http://www.bl.uk/online/inside/intrials.html. 35,000 soils records.

GeoArchive. 1974– . Didcot, UK: Geosystems.   GeoArchive is a comprehensive database covering all types of information sources in geoscience, hydroscience, and environmental science. Serials, books, news items, maps, and dissertations are included. Numeric subject codes are used, in a field called Descriptor Code (DC), but these codes seem to be numeric versions of the subject terms from the Descriptor field (DE). Access: online vendors, CD (as GeoSEARCH, from NISC/COS), print. 35,000 soils records.

Biological and Agricultural Index. 1983– . Bronx, NY: The H.W. Wilson Company.   Biological and Agricultural Index includes citations from both popular and scientific journals. Selected sources include materials that are feature articles, biographical sketches, reports of symposia and conferences, review articles, abstracts and summaries of papers, and selected letters to the editor. No subject codes are used for this database. Access: online vendors, CD, print. Available from the World Wide Web at http://www.hwwilson.com/databases/bioag.cfm. 29,000 soils records.

Life Sciences Collection. 1982– . Bethesda, MD: Cambridge Scientific Abstracts. Life Sciences Collection database includes worldwide research literature in major areas of biology, medicine, biochemistry, biotechnology, ecology, and microbiology, and some aspects of agriculture and veterinary science. This database uses numeric subject classification codes (Section Headings) for both general (''SH=04600'') and numerous specific soil concepts. Codes are found for many aspects of soils and soil properties, soil organisms and particular types of organism involvement with soils or soil conditions. Text descriptor terms seem diverse, specific, and easier to use. Access: online vendors, CD, print, Internet. 27,000 soils records.

Enviroline. 1975– . Bethesda, MD: Congressional Information Service, Inc. Enviroline is the online version of the print *Environmental Abstracts*, produced by Congressional Information Service, Inc., with international scope, including journal articles, conference papers and proceedings, and other key sources of information on all aspects of the environment. Management, technology, planning, law, political science, economics, geology, biology, and chemistry as they relate to environmental issues are all covered. No specific codes are used for subject differentiation. Access: online vendors, CD, print. 20,000 soils records.

Pollution Abstracts. 1970– . Bethesda, MD: Cambridge Scientific Abstracts. Pollution Abstracts includes citations to references for environmentally related literature on pollution, pollution sources, and control. Air pollution, environmental quality, noise pollution, pesticides, radiation, solid wastes, and water pollution are all included in this database. No subject-specific code fields are found in this database. Access: online vendor, CD, print. 15,000 soils records.

INSPEC. 1969– . Stevenage, UK: Institution of Electrical Engineers. INSPEC database covers physics, electronics, and computing, all of which may be directly or indirectly a part of current soils science research efforts. Sources for records include journal articles, books, reports, and dissertations. INSPEC offers an online thesaurus with soil(s) as general terms and about 40 specific soil physics (and related) phrases. Several alphanumeric ''Class Codes'' are also used to identify records for general and specific major soil concepts. Access: online vendor, print, CD. Available from the World Wide Web at http://www.umi.com/hp/Features/Inspec/. 11,000 soils records.

CRIS. 1969– . Washington, DC: USDA. Current Research Information System (CRIS) is the USDA documentation and reporting system for ongoing agricultural, food and nutrition, and forestry research. CRIS is not a typical "biblio-graphic" database. It contains over 30,000 descriptions of current (most recent 4 years), publicly supported research projects of the USDA agencies, the State Agricultural Experiment Stations, the State land-grant colleges and universities, State schools of forestry, cooperating schools of veterinary medicine, and USDA grant recipients. Records for CRIS projects appear in the DIALOG database file called FEDRIP (Federal Research in Progress). CRIS reports on federally funded research, and has a pragmatic and results-focused tone. Publications of project results may be included as part of CRIS records. A number of codes are used in the database and these may enable more precise access to needed information. Codes and fields of subject significance include "Research Problem Area," "Activity," "Commodity," and "Field of Science." Soil topics are included among all of these code and topic groups. Two versions of the classification system exist. Older ongoing projects may use version V, newer projects use version VI. Documentation and search assistance is available on the CRIS Web site. Access: online vendors. Available from the World Wide Web at http://cris.csrees. usda.gov/. 9,000 soils records.

## Specific Soils Topics and Duplication in Bibliographic Databases

The following section provides an overview of topics within the soils realm. Ranks of databases for coverage within each topic area will be provided, with notes about the duplication among databases for each specific soil subject.

Although soil is taken for granted by most people—we walk on it, drive on it, pave over it, dig in it, plant in it, dump trash on it, and occasionally regret, when we think of it, that tons of fertile soil are lost to erosion every year—soil is a complex and dynamic material, serving as an integral part of living processes. The Soil Science Society of America (SSSA) has 11 divisions of specialties within soil science. In order to more effectively assist readers in determining databases suitable for soils use, a similar set of categories of soils information was constructed.

Nine topic areas are defined here. These include: (1) soil physics, (2) soil chemistry (with analysis and testing), (3) soil fertility (nutrients and yields), (4) soil genesis and pedology (with mapping, surveys, and classification), (5) soil and environment (including reclamation or remediation), (6) soil water (hydrology), (7) soil erosion, conservation, and management, (8) soil microbiology and soil-dwelling organisms, and (9) soil—plant relations.

Each topic is described, followed by a listing of the 10 databases with the greatest coverage for that topic and the number of citations in each database (as of June 2000) in parentheses. Databases are listed in decreasing coverage order. For the unrestricted search for any soils subjects, the number of records in each database was shown with the descriptive text preceding it. The topics were searched in databases previously described through the Dialog system. Dialog is a database vendor that provides access to over 600 databases from around the world. Duplication among the top five databases was also determined. This included recent updates only, because Dialog's Remove Duplicates (RD) feature is limited to comparing 500 records per search.

For the general search, items retrieved from the databases in the June 2000 update: CAB Abstracts (1,079 records), AGRICOLA (1,055), AESIS (Australia's Geoscience, Minerals, and Petroleum Database [not listed with the twenty-five previously mentioned databases because of lower overall coverage (25,000 soils records) and its Australia-only focus]) (855 records), GEOBASE (681), AGRIS (468), Biological and Agricultural Index (452), and GeoRef (361 records). It is critical to note that for these 5 databases, 4,951 records were added in the most recent updates at the end of June 2000. Among these items, only 51 records were found as duplicates within these 5 databases! This amount of overlap seems extremely small and may reflect different production schedules by the producers. It is important to note that for comprehensive searches of the most recent literature available, there is a need to search a number of databases.

Very specific searches can be conducted using Dialog's features for truncation and its special operators. Truncation is used to allow variations in word length or spelling. Unrestricted truncation allows any number of characters to follow the root (PHYSIC?). Restricted truncation limits to one additional character or a specified number of characters. (SOIL? ? limits to one additional character to allow an abbreviated way to include *soil* or *soils*, MECHANIC??? limits up to three additional characters, thus allowing for *mechanics* and *mechanical*). The command *S* means *select* or *search*. The special operator "(F)" means that the search terms must be found in the same field. This would minimize the number of false hits from terms pulled from different fields.

Soil physics was the first specific concept searched. The search strategy was: S SOIL? ? (F) (PHYSIC? OR MECHANIC??? OR ENGINEER?). Files with the greatest retrieval were: AGRICOLA (48,002), TOXLINE (44,710), GeoRef (44,205), PASCAL (29,636), CAB Abstracts (29,087), Ei Compendex (25,460), JICST-EPlus (22,463), NTIS (16,387), GeoArchive (15,630), and AGRIS (13,944). Record duplication was examined for this topic, among records added to the database on this topic within the year 2000. From the total of 1,536 records added to these 10 databases within this period, 1,370 were unique. This 10% overlap is rather small, and highlights the need for careful search and selection procedures.

Soil chemistry was searched using the strategy: S SOIL? ? (F)(CHEMI? OR TEST??? OR ANAL?). The top ten retrieval databases for this subject were CAB Abstracts (85,250), TOXLINE (76,721), AGRICOLA (57,107), AGRIS (51,910), BIOSIS (44,451), PASCAL (40,433), Current Contents (38,496),

GeoRef (34,291), SciSearch (34,210), and Energy SciTech (30,930). The last database listed was not described previously. A total of 1,842 records were found in these databases for the most recent update. Of these 1,711 were found to be unique items, giving a duplication rate of less than 10%.

Soil fertility was searched by S SOIL? ?(F)(FERTIL? OR NUTRI? OR YIELD??? OR PRODUCTIV?) as a strategy. This resulted in the following retrievals: CAB Abstracts (146,741), AGRICOLA (92,925), BIOSIS (72,704), AGRIS (57,188), Current Contents (33,016), SciSearch (32,637), TOXLINE (31,015), PASCAL (29,458), GEOBASE (16,963), and JICST-EPlus (12,398). Among this group, the search for duplicate coverage included items added during the year 2000 (as of June 2000). To enable the examination of a larger group of records, the "PRODUCTIV?" term shown was omitted. From this, a final total of 4,889 records were found. Out of this group, only 1,832 were shown as unique. This great reduction in unique records is partially explained by the fact that Dialog's RD command does not process all Current Contents records well. If a record is found to have "incomplete bibliographic data," it is not retained in the results set. For this search, almost 1,400 Current Contents records were reported this way.

Soil genesis, classification, and mapping were searched with the following results. The strategy: S SOIL? ?(F)(GENESIS OR PEDOLOG? OR CLASSIF? OR MAP???? OR SURVEY???) produced the following database results: GEOREF (34,749), CAB Abstracts (26,857), AGRICOLA (17,851), BIOSIS (13,741), PASCAL (13,230), TOXLINE (11,118), Current Contents (9,467), AGRIS (9,292), AESIS (8,857), and GEOBASE (8,682). Within this group, 1,463 records were added after 1999, and 797 of these were unique. This 50% duplicate ratio reflects both the loss of many Current Contents records, as well as significant overlap in the coverage within this soils topic area.

Soil and the environment was searched by the use of the strategy: S SOIL? ? (F)(ENVIRONMENT? OR RECLA? OR REMEDIAT? OR POLLUT?). The databases with the greatest coverage were: TOXLINE (202,940), BIOSIS (65,825), CAB Abstracts (45,914), PASCAL (37,610), AGRICOLA (35,710), AGRIS (35,144), GEOREF (32,429), Energy SciTech (32,207), Current Contents (22,889), and NTIS (20,737). For the duplication check, records added for this concept after 1999 were counted. The total number of records was 4,972 and 3,190 of these were unique. More than 800 of the discounted records were found in Current Contents, so the duplication factor among other databases was about 20%.

Hydrology was searched by the strategy: S SOIL? ?(F)(WATER? OR HYDROLOG? OR MOISTUR?). This showed the following coverages for the highest posted databases: TOXLINE (191,512), CAB Abstracts (114,852), BIOSIS (61,658), PASCAL (43,052), Current Contents (42,427), Water Resources Abstracts (41,965), AGRICOLA (40,110), SciSearch (39,968), GeoRef (36,323), and AGRIS (33,495). For duplicate checking, items added in the last update (as of late June 2000) were included. For this group, a total of 1,445 records were found. Of this group, 1,329 appeared to be unique, and only 17 of the original 71 records from Current Contents were excluded.

Soil conservation was searched and included soil erosion and management, by use of this strategy: S SOIL? ?(F)(CONSERV? OR BMP OR MANAG? OR EROD??? OR EROSION). From this search the most records were found in these databases: CAB Abstracts (48,770), AGRICOLA (47,727), BIOSIS (40,882), TOXLINE (24,777), AGRIS (23,543), JIST-EPlus (17,918), PASCAL (17,614), Water Resources Abstracts (17,422), Current Contents (16,169), and GeoRef (15,243). Duplication checking showed a total of 2,577 records on this topic added after 1999. Records shown as unique totaled 1,739. Over 250 of the excluded items were found in Current Contents.

Soil microbiology, including soil fauna, was searched using the strategy: S SOIL? ?(F)(MICROB? OR ANIMAL? ? OR FAUNA OR ORGANISM? ? OR MICROORGANISM? ?). Record counts in the top ten databases (June 2000) were: TOXLINE (82,134), CAB Abstracts (32,946), BIOSIS (29,259), JICST-EPlus (19,309), SciSearch (15,228), PASCAL (15,224), Current Contents (14,774), Life Sciences Collection (14,077), AGRIS (13,716), and AGRICOLA (13,276). For duplicate checking, records added in the year 2000 were used. A total of 2,629 records were found for this subject area in that period. There were only 1,079 unique records. Current Contents lost over 800 records during Dialog's duplicate checking. "Incomplete bibliographic data" was noted for most of these exclusions.

Soil-plant relations was the final concept searched. This employed the strategy: S SOIL? ? (F)(PLANT? ? OR CROP??? OR HARVEST??? OR VEGETAT? OR FLORA). Databases with the greatest number of postings for this topic were: CAB Abstracts (178,119), BIOSIS (83,780), AGRICOLA (83,342), TOXLINE (80,145), AGRIS (69,976), Current Contents (45,245), PASCAL (43,378), SciSearch (43,027), GEOBASE (25,040), and Energy SciTech (23,254). Removal of duplicates showed a total of 2,349 records added during the most recent update, and 2,107 were shown as unique records.

While these results are approximate, they demonstrate that particular topics are more successfully searched in certain databases. It is useful to note that while AGRICOLA, CAB Abstracts, TOXLINE, BIOSIS, PASCAL, AGRIS, Georef, Current Contents, JICST-Eplus, and SciSearch were shown to have the greatest number of soils records overall, not all of these databases are included in the "top ten" for each concept search. CAB Abstracts appears as the most-posted database for five of the nine concept searches, and is either second or third in three others.

This strongly supports Peter Mcdonald's report of CAB Abstracts being the best soil science database. These figures not withstanding, it should be recognized that AGRICOLA, TOXLINE, and PASCAL all appear in the "top ten" for each soil concept search. BIOSIS, Current Contents, GeoRef, and JISCT-EPlus appear in more than half of these concept ranking results. While databases such as GEOBASE, Water Resources Abstracts, Life Sciences Collection, Ei Compendex, and GeoArchive are found in only few of the subject-specific tests for soils data, they do have large numbers of overall soils postings, and should be included whenever possible, certainly when exhaustive results are required.

## JOURNALS

A variety of journals include articles about soils; such as *Agronomy Journal*, *Agricultural and Forest Meteorology*, *Journal of Agricultural and Food Chemistry*, *Nature*, and *Science*. Rather than repeat journal titles that are more general in scope and could appear in other chapters, we list those that are largely about soils or contain significant coverage of soil topics.

A number of publishers provide journal table of contents and/or abstracts on their Web sites; some offer full text of selected articles. Publishers' policies on providing access change as frequently as marketing strategies. In this chapter Web sites for journal titles are given for those journals, which provide at least tables of contents for issues.

*Acta Agriculturae Scandinavica Section B-Soil and Plant Science*. v. 1– . [1950?– ] London: Taylor & Francis Group. ISSN: 0906-4710. Quarterly. Available from the World Wide Web at http://www.tandf.co.uk/journals/tfs/09064710.html.

*Agrochimica*. v. 1– . 1956– . Pisa, Italy: Industrie Grafiche V. Lischi & Figli. Bimonthly. ISSN: 0002-1857. Available from the World Wide Web at http://www.infolab-it.com/agrisite/infautl.htm. In English, French, German, Italian, or Spanish with summaries in each of five languages.

*Applied and Environmental Microbiology*. v. 31– . 1976– . Washington, D.C.: American Society for Microbiology. ISSN: 0099-2240. Monthly. Available from the World Wide Web at http://aem.asm.org/. Continues *Applied Microbiology*.

*Applied Soil Ecology*. v. 1– . 1994– . Amsterdam: Elsevier Science. ISSN: 0929-1393. Quarterly. Available from the World Wide Web at http://www.elsevier.com/inca/publications/store/5/2/4/5/1/8/.

*Arid Soil Research and Rehabilitation*. v. 1– . 1987– . New York: Taylor and Francis. ISSN: 0890-3069. Quarterly. Available from the World Wide Web at http://www.journals.tandf.co.uk/tf/08903069.html.

*Australian Journal of Soil Research*. v. 1– . 1963– . Melbourne, Australia: Commonwealth Scientific and Industrial Research Organization (CSIRO). ISSN: 0004-9573. Six issues per year. Available from the World Wide Web at http://www.publish.csiro.au/journals/ajsr/index.html.

*BIOCYCLE*: *Journal of Composting and Recycling*. v. 22– . 1981– . Emmaus, PA: JG Press. ISSN: 0276-5055. Bimonthly. Available from the World Wide Web at http://www.environmental-expert.com/magazine/biocycle/index.htm, http://www.jgpress.com/. Continues *Compost Science/Land Utilization*

*Biology and Fertility of Soils*. v. 1– . 1985– . Berlin: Springer International. ISSN: 1432-0798. Quarterly. Available from the World Wide Web at http://link.springer.de/link/service/journals/00374/tocs.htm.

*Canadian Journal of Microbiology*. v. 1– . 1954– . Ottawa, Canada: National Research Council. ISSN: 0008-4166. Monthly. Available from the World Wide Web at http://www.cisti.nrc.ca/cgi-bin/cisti/journals/rp/rp_desy_e?cjm. English, includes some text in French.

*Canadian Journal of Soil Science*. v. 37– . 1957– . Ottawa, Canada: Agricultural Institute of Canada. ISSN: 0008-4271. Quarterly. Available from the World Wide Web at http://www.nrc.ca/aic-journals/cjss.html. Published in cooperation with the Canadian Society of Soil Science. Continues *Canadian Journal of Agricultural Science*.

*CATENA*. v. 1– . 1973– . Amsterdam: Elsevier. ISSN: 0341-8162. Bimonthly. Available from the World Wide Web at http://www.elsevier.nl/inca/publications/store/5/2/4/6/0/9/index.htt. A cooperative journal of the International Society of Soil Science, in English, French, or German with summaries in English, French, or German.

*Ciencia del Suelo*: *Revista de la Asociación Argentina de la Ciencia del Suelo*. v. 1– , 1983– . Buenos Aires, Argentina: La Asociación. ISSN: 0326-3169. Two issues per year. In Spanish.

*Clays and Clay Minerals*. v. 16– . 1968– . Lawrence, KS: Allen Press. The Clay Minerals Society. ISSN: 0009-8604. Bimonthly. Available from the World Wide Web at http://cms.lanl.gov/journal.html. Continues *Clay and Clay Minerals Proceedings of the Conference*.

*Communications in Soil Science and Plant Analysis*. v. 1– . 1970– . New York: Dekker. ISSN: 0010-3624. Semimonthly. Available from the World Wide Web at http://www.dekker.com/e/p.pl/0010-3624.

*Eurasian Soil Science*. v. 24– . 1992– . Silver Spring, MD: Scripta Technica, Inc. Moscow, Russia: Pochvovedenie. ISSN: 1064-2293. Ten issues per year.

Available from the World Wide Web at http://maik.rssi.ru/journals/soilsci.htm. Supersedes *Soviet Soil Science*; English translation of two Russian periodicals, *Agrokhimiia* and *Pochvovedenie*.

*European Journal of Soil Science*. v. 45. 1994– . Oxford: Blackwell Scientific and the British Society of Soil Science. ISSN: 1351-0754. Quarterly. Available from the World Wide Web at http://www.blacksci.co.uk/~cgilib/jnlpage.bin?-Journal=ejss&File=ejss&Page=aims. Formed by the union of *Journal of Soil Science*, *Pédologie*, and *Science du sol*.

*European Journal of Soil Biology*. v. 29– . 1993– . Amsterdam: Elsevier. ISSN: 1164-5563. Monthly. Available from the World Wide Web at http://www. elsevier.com/locate/ejsobi. Continues *Revue d'ecologie et de biologie du sol*.

*Geoderma*. v. 1– . 1967– . Amsterdam: Elsevier Science. ISSN: 0016-7061. Monthly. Available from the World Wide Web at http://www.elsevier.com/inca/ publications/store/5/0/3/3/3/2/.

*Irrigation and Drainage Abstracts*. v. 1– . 1975– . Wallingford, UK: CAB International. ISSN: 0306-7327. Quarterly. Available from the World Wide Web at http://www.cabi.org/catalog/journals/absjour/7s.htm. Abstracting journal. The records from this publication form a subfile of CAB Abstracts, and SOILCD.

*Irrigation Science*. v. 1– . 1978– . Berlin: Springer International. ISSN: 0342-7188. Bimonthly. Available from the World Wide Web at http://link. springer.de/link/service/journals/00271/index.htm.

*Journal of Environmental Quality*. v. 1– . 1972– . Madison, WI: American Society of Agronomy, Crop Science Society of America, and Soil Science Society of America. ISSN: 0047-2425. Quarterly. Available from the World Wide Web at http://www.agronomy.org/journals/jeq/.

*Journal of Soil and Water Conservation*. v. 1– . 1946– . Ankeny, IA: Soil and Water Conservation Society. ISSN: 0022-4561. Bimonthly.

*Journal of Soil Contamination*. v. 1– . 1992– . Boca Raton, FL: CRC. ISSN: 1058-8337. Six issues per year. Available from the World Wide Web at http:// www.aehs.com/jsc/jschome.htm. Sponsored by the Association for the Environmental Health of Soils (AEHS).

*Journal of the Indian Soil Science Society*. v. 1– . 1953– . New Delhi, India: Indian Society of Soil Science. ISSN: 0019-638X. Quarterly.

*Land Degradation and Rehabilitation*. v. 1– . 1989. Chichester, England: John Wiley & Sons. Online ISSN: 1099-145X; Print ISSN: 1085-3278. Quarterly. Available from the World Wide Web at http://www3.interscience.wiley.com/ cgi-bin/jtoc?ID=6175.

*Nutrient Cycling In Agroecosystems.* v. 46. 1996– . Dordrecht, Boston: Kluwer. ISSN: 1385-1314. Six issues (2 vols.) per year. Available from the World Wide Web at http://www.wkap.nl/jrnltoc.htm/1385-1314. Supersedes *Fertilizer Research.*

*Oikos.* v. 1– . 1949– . Copenhagen: Munksgaard. ISSN: 0030-1299. Monthly. Available from the World Wide Web at http://www.oikos.ekol.lu.se/Oikosjrnl.html. Chiefly in English; some French and German; summaries in English and Russian. Issued by the Nordic Society Oikos.

*Pedobiologia.* Bd. 1– . 1961– . Jena, Germany: Urban & Fischer. ISSN: 0031-4056. Six issues per year. Available from the World Wide Web at http://www.urbanfischer.de/journals/pedo/frame_template.htm?/journals/pedo/pedobiol.htm.

*Physiologia Plantarum.* v. 1– . 1948– . Copenhagen: Munksgaard. ISSN: 0031-9317. Monthly. Available from the World Wide Web at http://www.blackwell-synergy.com/journals/issuelist.asp?journal=ppl. Official publication of Societas Physiologiae Plantarum Scandinavica (Scandinavian Society for Plant Physiology).

*Phytopathology.* v. 1– . 1911– . St. Paul, MN: American Phytopathological Society. ISSN: 0031-949X. Monthly. Available from the World Wide Web at http://www.scisoc.org/journals/phyto/.

*Plant and Soil.* v. 1– . 1948– . Dordrecht: Kluwer Academic Publishers. ISSN: 0032-079X. Twenty issues per year. Available from the World Wide Web at http://www.wkap.nl/journalhome.htm/0032-079X.

*Plant Physiology.* v. 1– . 1926– . Bethesda, MD: American Society of Plant Physiologists. ISSN: 0032-0889. Monthly. Available from the World Wide Web at http://www.plantphysiol.org/.

*Soil & Tillage Research.* v. 1– . 1980– . Amsterdam: Elsevier. ISSN: 0167-1987. Quarterly. Available from the World Wide Web at http://www.elsevier.com/inca/publications/store/5/0/3/3/1/8/. In collaboration with the International Soil Tillage Research Organization. Absorbed *Soil Technology.* ISSN: 0933-3630.

*Proceedings/Soil and Crop Science Society of Florida.* v. 16– . 1956– . Hollywood, FL: Soil and Crop Science Society of Florida. ISSN: 0096-4522. Annual. Available from the World Wide Web at http://i.am/scssf. Continues *Proceedings/Soil Science Society of Florida.*

*Soil Biology and Biochemistry.* v. 1– . 1969– . Oxford: Pergamon Press. ISSN: 0038-0717. Monthly. Available from the World Wide Web at http://www.elsevier.com/inca/publications/store/3/3/2/.

*Soil Science*. v. 1– . 1916– . Baltimore, MD: Lippincott Williams and Wilkins. ISSN: 0038-075X. Monthly. Available from the World Wide Web at http:// www.soilsci.com/.

*Soil Science and Plant Nutrition*. v. 7– . 1961– . Tokyo, Japan: Society of Soil Science and Plant Nutrition, Japan. ISSN: 0038-0768. Quarterly. Available from the World Wide Web http://wwwsoc.nacsis.ac.jp/jssspn/index.html. In English. Continues *Soil and Plant Food*.

*Soil Science Society of America Journal*. v. 40– . 1976– . Madison, WI: Soil Science Society of America. ISSN: 0361-5995 (printed version); ISSN: 1435-0661 (electronic version). Monthly. Available from the World Wide Web at http://link.springer.de/link/service/journals/10089/index.htm. Continues *Proceedings of the Soil Science Society of America*.

*Soil Use and Management*. v. 1– . 1985– . Wallingford, UK: CAB International for British Society of Soil Science. ISSN: 0266-0032. Quarterly.

*Soils and Fertilizers*. v. 1– . 1938– . Wallingford, UK: CAB International. ISSN: 0038-0792. Monthly. Available from the World Wide Web at http://www. cabi.publishing.org/JOURNALS/Abstract/SF/Index.asp. Abstracting journal. Records from this publication form a subfile of CAB Abstracts and SOILCD.

*South African Journal of Plant and Soil*. v. 1– . 1984– . Pretoria, South Africa: Bureau for Scientific Publications. ISSN: 0257-1862. Quarterly. Text in English and Afrikaans. Issued in collaboration with South African Society of Crop Production; the Soil Science Society of South Africa; and the Southern African Weed Science Society.

*Transactions of the ASAE*. v. 1– . 1958– . St. Joseph, MI: American Society of Agricultural Engineers. ISSN: 0001-2351.

*Zeitschrift für Pflanzenernahrung und Bodenkunde*. Bd-123. 1969– . Weinheim: Wiley-VCH. Bimonthly. Online ISSN: 1522-2624; Print ISSN: 1436-8730. *Journal of Plant Nutrition and Soil Science* is available from the World Wide Web at http://www.wiley-vch.de/vch/journals/2045.html. Continues *Zeitschrift fur Pflanzenernahrung und Bodenkunde*.


## MONOGRAPHS

### General Soils Works

Brady, N.C. and R.R. Weil. 1999. *The Nature and Properties of Soils*. 12th ed. Upper Saddle River, NJ: Prentice Hall. 881 p. This classic soils text has been the backbone of soil science learning since its first 1922 edition. Coverage is comprehensive in 20 chapters that describe all aspects of soils and human interaction with them. A complete glossary, multitone illustrations and figures, and photographs in black and white (some in color) make it a pleasure to use.

Collins, M.E. 1995. The SOILdisc [interactive multimedia]. A multimedia approach to teach pedology. Edition: V1.0. Gainesville, FL: University of Florida, Soil and Water Science Department. This computer laser optical disc (CD-ROM) offers a unique way to explore soils learning through a pleasant and practical introduction to the science of soils.

Mcdonald, P., ed. 1994. *The Literature of Soil Science*. Ithaca, NY: Cornell University Press. 448 p. This fine work gives historical background for soil science and modern perspective to the information resources available in soils subjects. Detailed discussion in specific topics is a highlight for those interested in these areas, including such things as tropical soils literature, soil information systems, and primary journals for soils.

Miller, R.W., D.T. Gardiner, and J.U. Miller. 1998. *Soils in Our Environment*. 8th ed. Upper Saddle River, NJ: Prentice Hall. 649 p. This latest textbook edition (previously called *Soils: An Introduction to Soils and Plant Growth*) offers 19 chapters covering agricultural soils topics. A global perspective on soils and their use, conditions, and problems is provided. The book is well indexed and easy to read, with a glossary and many drawings and photographs to illustrate concepts.

Plaster, E.J. 1997. *Soil Science and Management*. 3rd ed. Albany, NY: Delmar Publishers. 402 p. This easy-reading book provides a clear illustrated introduction to soil science basics. Practical aspects of soil management are covered in the 20 chapters, with appendices including basic chemistry, sedimentation texture testing, soil classification in the United States, land evaluation, and soil characteristics for selected trees.

Sumner, M.E., ed. 2000. *Handbook of Soil Science*. Boca Raton, FL: CRC Press. This complete technical work describes and discusses current soils knowledge and explains and illustrates the application of science and mathematics to soils. Nine major sections include 57 chapters covering physics, chemistry, biology and biochemistry, fertility, pedology, mineralogy, and interdisciplinary studies. In a final section soils databases are discussed, with data applications as a key final chapter. This key book is illustrated with figures, tables, and formulae explaining modern soil-science techniques.

USDA. 1938. *Soils and Men. Yearbook of Agriculture*. Washington DC: U.S. Government Printing Office. 1,232 p. and *Soil*. 1957. Washington, DC: U.S. Government Printing Office. 784 p. These two USDA classics are fine reading. Filled with photographs and drawings, they show both the evolution of our understanding of soils, and the complexity of the issues we confront and attempt to grasp and understand. Problems from our unwise uses of soils and efforts toward their solution are highlighted. Regional soils issues are noted. Glossaries, indexes, and extensive bibliographic lists are included in both yearbooks.

## Soil Physics Monographs

Hillel, D. 1998. *Environmental Soil Physics*. San Diego: Academic Press. 771 p. The unique approach taken in this work does not limit its value as a text or reference work for all aspects of applied soil physics. Seven sections of 21 chapters cover soil basics, solid, liquid and gas phase factors, composite phenomena, the field water cycle, and soil-plant-water relations. Appendices cover spacial variation, soil physics in remediation, and the land in climate modeling. Detailed but clear text is aided by numerous illustrations, and mathematics is included when essential for complete subject treatment.

Roberts, J. 1996. *Understanding Soil Mechanics*. Albany, NY: Delmar Publishers. 255 p. This book introduces the reader to principles and practices in the uses and interactions of soils in construction. Clear text without mathematics includes numerous photographs and figures providing knowledge of the use and interpretation of tests, reports, and specific engineering practices. A glossary and index enable easy reading.

Scott, H.D. 2000. *Soil Physics: Agricultural and Environmental Applications*. Ames, IA: Iowa State University Press. This well-illustrated text introduces this advanced topic clearly in 12 chapters, from an introduction with historical overview through coverage of soil texture and structure, mass and energy, temperature and aeration, chapters on water and solute flow and transport, and soil-plant-water relations. Appendices offer geostatistics, mathematics, and physics reviews. Extensive reference lists follow each chapter.

Whitlow, R. 1995. *Basic Soil Mechanics*. 3rd ed. Harlow, UK: Longman Scientific & Technical; New York: Wiley. 559 p. Building and civil engineering applications, impacts, and considerations are this book's focus. Success in three editions is due to the mathematically developed and well-illustrated examples. A site investigations and in situ testing chapter ends the book, followed by a bibliography, Eurocode foundations specifications and geotechnical design safety factors, answers to exercises, and an index.

USDA, Soil Conservation Service. 1971. *SCS National Engineering Handbook*. Washington, DC: USDA. 1 v. (looseleaf). An overview of specific engineering operations involving soils. Sections now available online include: Part 630— Hydrology, available from the World Wide Web at http://www.wcc.nrcs. usda.gov/water/quality/common/neh630/4content.html; Part 642—Specifications for Construction Contracting, available from the World Wide Web at http://

www.ftw.nrcs.usda.gov/neh642.html; and Part 651—Agricultural Waste Management Field Handbook, available from the World Wide Web at http:// www.ftw.nrcs.usda.gov/awmfh.html.

*Earth Manual: Part 1.* 1998. Earth Sciences and Research Laboratory, Geotechnical Research, Technical Service Center. 3rd ed. Denver: U.S. Dept. of the Interior, Bureau of Reclamation (Washington, DC: Superintendent of Documents. U.S. G.P.O., distributor). 329 p. After extensive introductory discussions of soil physical properties, this book covers standards for investigations, analysis, and methods in earth systems construction for waterways, dams, roadways, and the like. Filled with photographs and figures, it provides an essential resource for understanding earth engineering requirements.

## Soil Chemistry Monographs

Hood, T.M., and J.B. Jones, Jr., eds. 1997. *Soil and Plant Analysis in Sustainable Agriculture and Environment.* New York, NY: Marcel Dekker. 864 p. This work comprises text, figures, and tables from 56 presentations at the 1995 International Symposium on Soil Testing and Plant Analysis, originally published in the journal *Communications in Soil Science and Plant Analysis* 27 (3–8), 1996. Included are key papers in four subject areas: quality assurance and reliability, soil testing methods, methods of soil and plant analysis, and recommendations from tests. Advancement of test reliability and communication consistency are among the goals of this work, with emphasis on practical application of results to cropping systems and environmental management.

Klute, A. et al. 1994, 1996. *Methods of Soil Analysis.* 3rd ed. 3 vols. Madison, WI: Soil Science Society of America. Issued as Soil Science Society of America Book Series no. 5; Variation: Agronomy no. 9. Table of contents include physical and mineralogical methods, microbiological and biochemical properties, and chemical methods. This American Society of Agronomy (ASA) and Soil Science Society of America (SSSA) series has been published since 1965. This latest edition has expanded the work to three volumes from the previous two-volume format. Discussion, references, and directions are given for each analysis. Various methods are shown for each data need. Critical research knowledge for those working in laboratories or devising and conducting their own studies is readily available here, and made usable by the excellent text and added value within these works. Each type of test is introduced by a section with an overview of that aspect of soil science.

Tan, K.H. 1998. *Principles of Soil Chemistry.* 3rd ed., rev. and expanded. New York: Marcel Dekker. 521 p. A good basic overview of soil chemical principles is presented in this book. It will serve well for a grower or producer with a problem soil (or a soil problem), or a beginning researcher who needs a solid foundation in this topic. While not extremely advanced, this text covers the multi-faceted nature of soils and soil components' interactions fully, from clay matrix ion exchange to organic chelates and metal complexes. It includes an extensive list of unnumbered references in a single section at the end.

Westerman, R.L., ed. 1990. *Soil Testing and Plant Analysis*. 3rd ed. Madison, WI: Soil Science Society of America. 784 p.   This work presents 27 individually authored chapters reviewing tests of soils and plant tissues for the management of soil chemistry in crop production. Basic principles are reviewed for various soil test types. Both general and crop-specific details are reviewed for plant analysis. Test methods are not presented, but sampling and test-results interpretation and use are covered. Extensive bibliographies follow each chapter.

## Soil Fertility Monographs

Black, C.A. 1993. *Soil Fertility Evaluation and Control*. Boca Raton, FL: CRC Press. 741 p.   In eight extensive chapters Dr. Black reviews soil fertility and its management, including principles of nutrient supply and yield response, economics, fertilizer evaluation, residual effects, placement factors, soil tests, and lime requirements. The book is indexed and includes an extensive list of cited works.

Cramer, C., ed. 1986. *The Farmer's Fertilizer Handbook: How to Make Your Own NPK Recommendation—and Make Them Pay*. New, expanded national edition. Emmaus, PA: Regenerative Agriculture Association. 208 p.   Practical self-help guide to understanding and managing soil fertility, with an emphasis on sustainability and do-it-yourself solutions. The book contains techniques and methods for producers to evaluate and manage the fertility status of their lands.

Foth, H.D. and B.G. Ellis. 1997. *Soil Fertility*. 2nd ed. Boca Raton, FL: CRC Lewis. 290 p.   An illustrated text in soil nutrient management for crop production, this book covers the basics of the topic in clear discussion of key points in 14 chapters. After an overview section, it covers soil ionic charges and their interactions with nutrients. It continues in coverage of individual nutrient elements and groups. The interplay of total versus available nutrient levels are stressed in relation to specific soil types. Fertilizers themselves are also discussed, including forms, application, and mixtures.

Tisdale, S.L., et al. 1993. *Soil Fertility and Fertilizers*. 5th ed. New York: Macmillan. 634 p.   This classic textbook presents in 15 chapters an understanding of soil fertility in terms of plant nutrition and nutrient management. Plant growth nutrient needs and essential and beneficial nutrient element requirements are covered, and the interactions of these elements in soils systems. A practical emphasis is maintained throughout, from notes on deficiency symptoms to cultural practice impacts and economic aspects of fertility management.

## Soil Genesis and Classification Monographs

Buol, S.W., et al. 1997. *Soil Genesis and Classification*. 4th ed. Ames, IA: Iowa State University Press. 527 p. As a classic title in this field, this book continues a tradition of excellent instruction in the science of soil formation and description. Soil genesis principles and processes are described in detail. The development and value of soil classification systems is described. Each soil order has a chapter with complete discussion and illustration of the suborders and great groups within each order. An index and extensive bibliography follow the last two chapters, describing soil cover, soil mapping units, and their understanding and use.

Fanning, D.S. and M.C.B. Fanning. 1989. *Soil: Morphology, Genesis and Classification*. New York: John Wiley. 395 p. This excellent textbook on soil formation and classification covers the topic in 44 short chapters with 4 appendices. The text is well written and easy to read, considering the complexity of the topics and interactive processes discussed. Major soil-forming processes are covered with each in separate chapters. Illustrations, tables, and photographs are included, and a list of references appears at the end of each chapter.

## Soils and Environment Monographs

Adriano, D.C., et al., eds. 1999. *Bioremediation of Contaminated Soils*. Madison, WI: American Society of Agronomy; Crop Science Society of America; Soil Science Society of America. 820 p. Extensive overview of aspects of soil bioremediation is found in 28 individually authored chapters. Explosives, radionuclides, petroleum products, aromatics, halogenated compounds, and preservatives are classes of contaminants covered. Principles of soil-microbe-contaminant interaction introduce the work. Specific problems and methodologies include enzymatic systems, anaerobic systems, prepared beds, screening soil organisms, mycorrhizae, and microbial enhancement. An integrated case study from Savannah River is presented. Reference lists end each chapter.

Brown, D., et al. 1986. *Reclamation and Vegetative Restoration of Problem Soils and Disturbed Lands*. Series. Park Ridge, NJ: Noyes Data Corporation. 560 p. *Pollution Technology Review*, no. 137. This book groups reclamation needs and options in two categories: mine land and riparian watershed areas. Mining methods impacts and specific soil problems and types are discussed, with equipment needs and uses covered extensively. Numerous photographs and illustrations aid understanding. Eight appendices provide valuable information in a glossary, site survey methods, details of specific soil stabilization measures, plant materials tables, a mulch guide, information resource lists, and lime needs estimation for acid sulfate soils.

Committee on Innovative Remediation Technologies, Water Science and Technology Board, Board on Radioactive Waste Management, Commission of Geosciences, Environment and Resources, National Research Council. 1997. *Innovations in Ground Water and Soil Cleanup*: *from Concept to Commercialization*. Washington, DC: National Academy Press. 292 p. P.S.C. Rao's (authoring committee chair) poem about the bridging needed between ideal actions and outcomes and real situations provides a positive keynote for this work. Covering challenges, market characteristics, state of practice, success measures, technology testing, and cost comparisons, this book provides detailed knowledge of the current status of remediation needs and options, and needed actions to enable market-based success. An executive summary includes recommendations in each area. It includes photographs, illustrations and tables, and lists of references. A list of databases including remediation technology information is found in an appendix.

Committee on Technologies for Cleanup of Subsurface Contaminants in the DOE Weapons Complex, Board on Radioactive Waste Management, Commission on Geosciences, Environment, and Resources, National Research Council (NRC). 1999. *Groundwater & Soil Cleanup*: *Improving Management of Persistent Contaminants*. Washington, DC: National Academy Press. 285 p. This book focuses on the cleanup (contaminant mass removal or immobilization) of soils contaminated with metals, radionuclides, and dense nonaqueous phase liquids (DNAPLs) contaminants from nuclear energy and weapons production sites and commercial sites related to these activities. These Department of Energy (DOE) operations comprise the most expensive environmental remediation in U.S. history. An update on the regulatory framework is presented, followed by sections reviewing the science and cleanup methods for each contaminant class, followed by discussion of the history and future directions in DOE remediation technology development and recommendations of the NRC. Tables and illustrations expand and elaborate points in the text. Appendices offer a list of DOE-responsible sites by state, (Subsurface Contaminants Focus Area SFCA) technology deployment report, biographies of the authors, and a list of acronyms. Bibliographies follow each chapter and an index is included.

Dickey, E.C., D.P. Shelton, and P.J. Jasa. 1997. Residue Management for Soil Erosion Control (computer file) NebGuide; no. G 81-544-A. Lincoln, NE: Cooperative Extension, Institute of Agriculture and Natural Resources, University of Nebraska—Lincoln. Available from the World Wide Web at http://www.ianr.unl.edu/pubs/fieldcrops/g544.htm http. This useful short document has photographs, tables, and text discussions to illustrate the value of cover and residue management for soil erosion control. Residue management, estimation of cover, and some potential problems are discussed to complete this very valuable, quickly read item.

Gregorich, E.G. and M.R. Carter, eds. 1997. *Soil Quality for Crop Production and Ecosystem Health. Developments in Soil Science*, no. 25. Amsterdam; New York: Elsevier. 448 p. Eighteen separately authored chapters treat soil quality in three major sections, each illustrating the complex interactivity of the minerals, water, air, plants, and other organisms comprising soil systems. A broad review of soil quality concepts is presented, followed by specific discussions of key factors and discussions of concepts and methods for soil quality evaluation. Case studies illustrate practical applications in soil quality assessment. Tables and graphs display empirical soil quality factors, processes, and data.

Purves, D. 1985. *Trace-Element Contamination of the Environment*. Rev. ed. Amsterdam; New York: Elsevier. 243 p. This work gives a comprehensive overview of the status of heavy metals and other trace elements on the land and in the water. The book gives extensive discussion of specific elements and evaluates options for the remediation of polluted sites and the management of wastes and waste-production processes.

## Soil Water Monographs

Boyer, J.S. 1995. *Measuring the Water Status of Plants and Soils*. San Diego: Academic Press. 178 p. This book gives a detailed overview of the methods, theories, techniques, and technologies for various methods of soil and plant water status measurements. Photographs and figures illustrate theories and specific equipment use.

Kramer, P.J. and J.S. Boyer. 1995. *Water Relations of Plants and Soils*. San Diego: Academic Press. 495 p. This overview of soil and plant relations with water includes detailed treatment of the physics of soils and water, and plant growth and metabolism with respect to water. Basic understanding of physical chemistry and plant physiology are key to full use of this advanced text. Chapters are illustrated and each concludes with a supplementary reading list. Almost 80 pages of references and an index conclude the book.

Maidment, D.R., ed. 1993. *Handbook of Hydrology*. New York: McGraw-Hill. (various paging). This work provides a comprehensive update and overview of the subject in 29 chapters prepared by over 50 subject experts. Four major subject areas include the hydrological cycle, water transport, hydrological data and statistics, and technology for hydrology. An understanding of mathematics, physics, and physical chemistry enable the most complete use of this valuable text. It includes an extensive index and numerous illustrations, photographs, tables, figures, and formulae.

Soils and their Role in Protecting Water Quality. 1996. Videorecording produced by Media Services, Cornell University; producer/director, Gary Ingraham. Ithaca,

NY: Dept. of Soil, Crop and Atmospheric Sciences, Cornell University, c1996. 1 videocassette (20 min., 55 sec.). A brief video providing an overview of major impacts of soils on water quality with a major focus of modern hydrological study.

Ward, R.C. and M. Robinson. 2000. *Principles of Hydrology.* 4th ed. London; Burr Ridge, IL: McGraw-Hill. 450 p. This revised text offers an in-depth introduction for the study of Earth's water. After preliminary facts and figures, precipitation, interception, evaporation, groundwater, soil water, runoff, and water quality are chapter themes, developed in text with limited mathematics. Figures expand concepts, review problems, and exercises and chapters. Varying theories of hydrology are compared. A new final chapter, "The Drainage Basin and Beyond," integrates knowledge and concepts presented in the book. An extensive current bibliography and author and subject indexes end the book.

## Soil Conservation and Management Monographs

Agassi, M., ed. 1996. *Soil Erosion, Conservation, and Rehabilitation.* New York: Marcel Dekker. 402 p. This book offers in-depth analysis of soil erosion factors and processes and describes common conservation practices. The last 4 of the 16 chapters (each by separate authors) describe modern rehabilitation practices, and include salt-affected soils reclamation. Reference lists conclude each chapter.

Fisher, R.F. and D. Binkley. 2000. *Ecology and Management of Forest Soils.* 3rd ed. New York: John Wiley. 489 p. This book covers general soils concepts with particular reference to forest soils. It describes major world forests and soil associations, highlighting the limited value of much crop agriculture soil management practice to these soils and situations. In addition to basic soils principles, hydrological factors, general forest soil biology, mycorrhizae, and tree physiology are specifically addressed. Site productivity, problem sites, anthropogenic factors, fire, nutrient needs and cycles, and intensive management versus long-term productivity are also covered in separate chapters. An extensive bibliography and index complete this book.

Pierce, F.J. and W.W. Frye. 1998. *Advances in Soil and Water Conservation.* Chelsea, MI: Ann Arbor Press. 239 p. This work, developed from seminars celebrating the 50th anniversary of the Soil and Water Conservation Society, reviews and updates major research areas critical to scientific understanding and effective soil resource use and management. From historical perspectives to current practice in technology use to social and governmental program impacts, key topic areas are fully described and discussed. A final chapter highlights future needs and developments.

Renard, K.G., et al. (1997). *Predicting Soil Erosion by Water: a Guide to Conservation Planning with the Revised Universal Soil Loss Equation (RUSLE).* Washington, DC: USDA, Agricultural Research Service. 384 p. Note: Supersedes Agricultural Handbook (USDA) no. 537, titled *Predicting Rainfall Erosion Losses: A Guide to Conservation*. This work gives details of tested and established procedures for use of the universal soil loss equation. Erosion factors are discussed at length, with specific data solution examples. The RUSLE computer program is described and a user's guide is presented in a final chapter. This software is available for use on personal computers. Available from the World Wide Web at http://www.sedlab.olemiss.edu/Rusle/index.html.

## Soil Microbiology Monographs

Paul, E.A. and F.E. Clark, eds. 1996. *Soil Microbiology and Biochemistry*. 2nd ed. San Diego: Academic Press. 340 p.   Biochemistry and microbiology in soils are treated in terms of both organisms and their interactions, and their impacts on and transformations of soil carbon, nitrogen, other macronutrient elements, and metals. Chemical changes and cycles and their kinetics are presented graphically and symbolically. References and selected additional readings end each chapter.

Subba Rao, N.S. 1999. *Soil Microbiology*. 4th ed. Enfield, NH: Science Publishers. 407 p.   The fourth edition of this work has been renamed (from *Soil Microorganisms and Plant Growth*). This reflects more general and comprehensive treatment of soil microbiology topics. Plant growth impacts of soil microbes are thoroughly covered, but current material on biodegradation of pesticides and pollutants and on biotechnology is added. Photographs, figures, tables, and charts illustrate important points, and references end chapters. An appendix on growth media preparation is also included.

Sylvia, D.M., et al. eds. 1998. *Principles and Applications of Soil Microbiology*. Upper Saddle River, NJ: Prentice Hall. 550 p.   This book comprises 23 chapters authored by subject experts, all current researchers. Three sections cover organisms and their habitat, transformations in soil, and applied and environmental topics. Modern techniques and current problems are both specifically addressed. The textbook format includes illustrations and boxes highlighting major principles and concepts. Further exploration of specific details is assisted by lists of cited and general references ending each chapter.

## Soil-Plant Relations Monographs

*Advances in Agronomy*. New York: Academic Press. v. 1– . 1949– .   This important monographic series covers key issues in agronomy with chapters in each volume providing scientific treatment and review. Examples of recent chapters with particular focus on soils include: Crop Residues and Management Practices:

Effects on Soil Quality: Soil Nitrogen Dynamics: Crop Yield, and Nitrogen Recovery (from no. 68, 2000); Surface Charge and Solute Interactions in Soils (from no. 67, 1999); Transfer of Phosphorus from Agricultural Soils (from no. 66, 1999); The Effects of Cultivation on Soil Nitrogen Mineralization (from no. 65, 1999).

*Agronomy*. Madison, WI: American Society of Agronomy. 1949– . This series of monographs includes extensive treatment of particular aspects of the management of crops and soils. Some specific soil-focused and recent crop-focused titles include: no. 4, Soil and Fertilizer Phosphorus in Crop Nutrition, 1953; no. 9, Methods of Soil Analysis. 1982-1; no. 10, Soil Nitrogen, 1965; no. 11, Irrigation of Agricultural Lands. 1967; no. 31, Modeling Plant and Soil Systems, 1991; no. 32, Turfgrass, 1992; no. 33, Oat Science and Technology, 1992; no. 34, Cool-Season Forage Grasses, 1996; no. 35, Sunflower Technology and Production, 1997; no. 36, Plant and Nematode Interactions, 1998; no. 37, Bioremediation of Contaminated Soils, 1999; no. 38, Agricultural Drainage, 1999.

Kinsey, N. and C. Walters. 1995. *Neal Kinsey's Hands-On Agronomy*. Metairie, LA: Acres U.S.A. 352 p.   Concern for soil-system health and efficient, sustainable production is the emphasis for this book, with a background perspective focused on avoiding unneeded or unhealthy applications of agricultural chemicals. Self-help practical advice and procedures are highlighted in discussions of basic soil chemistry and nutrient management options.

Magdoff, F. 1992. *Building Soils for Better Crops: Organic Matter Management*. Lincoln, NE: University of Nebraska Press. 176 p.   In a book for farmers, soil science basics, soil ecology and crop management, and a section on evaluating soil status and integrating production practices are provided in 21 chapters of well-written, plain text. Photographs and figures provide a clear explanation of topics to enable full understanding of critical soil-crop interactions and options for best land management.

*Managing Cover Crops Profitably*. 1998. 2nd ed. Beltsville, MD: Sustainable Agriculture Network. 212 p.   This book provides valuable information for farmers and educators to understand and utilize the benefits of cover crops for soil and production practice improvement. Discussion is augmented by illustrations, tables, and charts. Advantages and disadvantages of species are given, along with regional effectiveness rankings.

Prihar, S.S., et al. 2000. *Intensive Cropping: Efficient Use of Water, Nutrients, and Tillage*. New York: Food Products Press. 264 p.   This book is focused on developing world needs and options with intensive crop production required by population pressures and resource limits. Discussions of water, fertilizer, and energy use include coverage of key scientific principles for each. Final chapters describe and discuss sustainability needs and argue for a systems approach to modeling and holistic system development for crop-specific application at different levels of spatial aggregation. An index and more than 50 pages of references are provided.

## SOIL MAPPING AND SURVEYING

Mapping is an important part of soil science, particularly in compiling soil surveys. The definition of soil survey is "the systematic examination, description, classification, and mapping of soils in an area. Soil surveys are classified according to the kind and intensity of field examination." Soil Science Society of America, Glossary of Soil Science Terms, (http://www.soils.org/sssagloss/index.html) (cited September 19, 2000). The glossary site can be searched or browsed or a Microsoft Word version can be downloaded.

Soil surveys are conducted by the USDA's Natural Resources Conservation Service (NRCS, formerly Soil Conservation Service). A soil survey area may consist of a county, multiple counties, or parts of multiple counties. The print versions may be requested free from the state offices of NRCS. The considerable amount of detail in the text in addition to the maps makes these useful tools for land managers and scientists.

USDA, NRCS, Soil Survey Division, State Soil Geographic Database (STATSGO) (http://www.ftw.nrcs.usda.gov/stat_data.html) (cited September 19, 2000). State soil maps made by generalizing the detailed soil survey data. The mapping scale for STATSGO map is 1:250,000 (with the exception of Alaska, which is 1:1,000,000). The scale of mapping is designed for broad planning and management uses covering state, regional, and multistate areas. STATSGO data are designed for use in a Geographic Information System (GIS). Data can be downloaded from the Web site or ordered on CD-ROM.

USDA, NRCS, Soil Survey Division, National SSURGO Database (http://www.ftw.nrcs.usda.gov/ssur_data.html) (cited September 19, 2000). The soil maps in the Soil Survey Geographic (SSURGO) Database uses mapping scales ranging from 1:12,000 to 1:63,360; SSURGO is the most detailed level of soil mapping done by the NRCS. This level of mapping is designed for use by landowners, townships, and county natural resource planning and management. The complexity of the data requires users to be knowledgeable of soils data and their characteristics. Data can be downloaded from the Web site or ordered on CD-ROM.

Bullock, P., R.J.A. Jones, and L. Montanarella. *Soil Resources of Europe*. Luxembourg: Office for Official Publications of the European Communities. 1999. 202 p. Available from the World Wide Web at http://esb.aris.sai.jrc.it/documents/

1999-esb-research-report-6/ (cited September 19, 2000). The European Soil Bureau has published an overview of soil resources of Europe and country reports on soil survey and soil data by country. This book can be downloaded from the Web site.

FAO Digital Soil Map of the World and Derived Soil Properties. 1998 Rome: FAO. CD-ROM (version 3.5, November 1995) is based on the FAO/UNESCO Soil Map of the World, original scale 1:5,000,000 published from 1974 to 1978.

## Tools for Soil Classification and Surveying

Rossiter, D.G. A Compendium of On-Line Soil Survey Information. (http:// www.itc.nl/~rossiter/research/rsrch_ss.html) (cited September 19, 2000). Compiled by D.G. Rossiter of the Soil Science Division, International Institute for Aerospace Survey & Earth Sciences, in Enschede, The Netherlands. This is an excellent site providing annotated links to international and national organizations conducting soil surveys and to information on different soils classification systems, methods of surveying, and digital soil databases.

Soil Survey Division Staff of USDA, NRCS. 1998. *Keys to Soil Taxonomy*. 8th ed. Washington, DC: U.S. Government Printing Office, 326 p. Available from the World Wide Web at http://www.statlab.iastate.edu:80/soils/keytax/ (cited September 19, 2000). The Keys serves as an abridgement to *Soil Taxonomy*, which is easily used in the field and contains all the approved revisions and amendments to *Soil Taxonomy*.

Soil Survey Division Staff of USDA, NRCS. 1999. *National Soil Survey Handbook*. rev. ed. Washington, DC: U.S. Government Printing Office. Available from the World Wide Web at http://www.statlab.iastate.edu/soils/nssh/ (cited September 20, 2000). The *National Soil Survey Handbook* provides the standards, guidelines, definitions, policy, responsibilities, and procedures for conducting soil surveys in the U.S. It covers more of the operational procedures than the *Soil Survey Manual*.

Soil Survey Division Staff of USDA, NRCS. 1993. *Soil Survey Manual*. 3rd ed. Washington, DC: U.S. Government Printing Office. Available from the World Wide Web at http://www.statlab.iastate.edu/soils/ssm/gen_cont.html (cited September 20, 2000). The *Soil Survey Manual* explains the major principles and concepts for making and using soil surveys and the standards and conventions for describing soils, includes soil series and pedons.

Soil Survey Division Staff of USDA, NRCS. 1999. *Soil Taxonomy: a Basic System of Soil Classification for Making and Interpreting Soil Surveys*. 2nd ed. Washington, DC: U.S. Government Printing Office. 870 p. Available from the World Wide Web at http://www.statlab.iastate.edu/soils/soiltax/ (cited September 20,

2000). This describes the common base for the organization of knowledge about soils and the standards for their classification.

Soil Survey Division Staff of USDA, NRCS. Soil Series Classification Database. (http://www.statlab.iastate.edu/soils/sc/) (cited September 20, 2000). Ames, IA: Iowa State University. This database provides classification category details (order, suborder, great group, subgroup, and family) for each of the nearly 20,000 soil series named in U.S. soil survey reports. The USDA-NRCS Soil Series Classification Database contains the taxonomic classification of each soil series identified in the United States. Along with the taxonomic classification, the database contains other information about the soil series, such as office of responsibility, series status, dates of origin and establishment, and geographic areas of usage. Additions and changes are continually being made, resulting from ongoing soil survey work and refinement of the soil classification system. As the database is updated, the changes are immediately available to the user, so the data retrieved is always the most current. Series names for soil pedons may be searched to identify taxonomic characteristics. Searches of specific taxonomic characteristics identify matches among named soil series, with their locations.

Soil Survey Division Staff of USDA, NRCS. Official Soil Series Descriptions Database. (http://www.statlab.iastate.edu/soils/osd/) (cited September 20, 2000). Ames, IA: Iowa State University. The Official Soil Series Descriptions (OSD) is a national collection of more than 18,000 detailed soil series descriptions, covering the United States. The descriptions, in a text format, serve as a national standard. The soil series is the lowest category of the national soil classification system. The name of a soil series is the common reference term, used to name soil map units and serve as a major vehicle to transfer soil information and research knowledge from one soil area to another. OSD defines specific soil series in the United States, giving full taxonomic class names for each series, profile details, notes on variations in characteristics, and typical uses and limitations.

Schoeneberger, P.J. 1998. *Field Book for Describing and Sampling Soils* (http://www.statlab.iastate.edu/soils/nssc/field_gd/field_gd.pdf) (cited September 20, 2000). Lincoln, NE: USDA National Soil Survey Center, Natural Resources Conservation Service. This covers site description, soil profile description, geomorphology, geology, soil taxonomy, and others.

Welcome to the World Soil Resources Web Site by the World Soil Resources Office of the USDA, NRCS, Soil Survey Division. (http://www.nhq.nrcs.usda.gov/WSR/) (cited September 26, 2000). This includes an Online Soil Education Series with slide collections about soil classification and a full-text document *Soil Taxonomy Keys for Finland*, as well as a great deal of information about soils in Africa and Albania.

Outside of the United States, other systems of soil classification may be used. Some examples are as follows:

Australian Collaborative Land Evaluation Program, CSIRO, The Australian Soil Classification (http://www.cbr.clw.csiro.au/aclep/asc/soilhome.htm) (cited September 26, 2000). Includes a Key to Soil Orders, a glossary, references, and a table entitled "Approximate correlations between the Australian and other soil classifications."

Agriculture and Agri-Food Canada, Canadian Soil Information System (CanSIS) (http://res.agr.ca/cansis/) (cited September 22, 2000). Includes online maps, data, glossary, and full-text documents including the *Manual for Describing Soils in the Field*.

Arocena, J.M. and M. Abley. Soils of Canada (http://quarles.unbc.edu/nres/soc/soc.htm) (cited September 22, 2000). A collection of selected soil profiles representing the Great Groups within the Canadian System of Soil Classification. The properties for each Great Group are briefly described including soil profile description.

## WORLD WIDE WEB SITES

Throughout this chapter, Web sites have been provided for databases, journals, and government publications. The Web is widely recognized as a cost effective way to distribute information without geographic or time constraints. The main problem with the Web is the transitory nature of the electronic medium. These are not physical items to own and loan. A strategy for maintaining a list of reliable Web resources for the next five years would be to develop a "core" group of sites that provide information and links to other sites. In general, these would be sites of educational institutions, governmental agencies, and professional organizations, because these would have a commitment to maintaining a site beyond the interest span or tenure of an individual.

Some sites that are helpful starting points are:

Department of Soil Science, Soil and Agriculture Resources, Universiti Putra Malaysia, Dr. Soil Surfs and Soil Information/Tutorials (http://www.agri.upm. edu.my/jst/resources.html) (cited September 20, 2000). Originally compiled by Christopher Teh Boon Sung with over 1,000 links, now maintained by M. Hanif and K.B. Siva. This long list is organized alphabetically by Web site name.

Casler, C.L., E. Ben-Dor, and M. Haseltine. Soils of Arid Regions of the U.S. and Israel (http://ialcworld.org/soils/home.html) (cited September 20, 2000). The focus is about soils of arid regions, but there are many resources provided that cover all soils, including a page of links for "Educational Sites for Kids and Teachers."

Koch, J. Steenbock Memorial Library, University of Wisconsin-Madison. Soil Science Resources on the Internet (http://www.library.wisc.edu/libraries/Steenbock/electron/soilsci.htm) (cited October 3, 2000). This resource organizes Web sites into categories such as databases, directories, organizations, journals, jobs, and pesticide information.

Mcdaniel, P. Soil Taxonomy—The Twelve Soil Orders (http://soils.ag. uidaho.edu/soilorders/Index.htm) (cited September 20, 2000). This Web site provides photos of soil profiles and landscapes with several examples of each soil order.

Sathrum, R. Humboldt State University Library. General Soil Science (http://library.humboldt.edu/~rls/gensoil.htm) (cited July 24, 2001). This site serves as a guide to print and electronic soil science sources.

Many professional organizations maintain Web sites that offer substantive information, such as publications or educational information, besides information about the organization:

British Society of Soil Science (http://www.bsss.bangor.ac.uk/) (cited July 23, 2001). This site provides pages with extensive links on teaching, societies, and universities with soil science programs.

International Soil Reference and Information Centre (ISRIC) (http://www. isric.nl/) (cited September 21, 2000). Based in Wageningen, The Netherlands, this organization collects information on soils worldwide. This statement appears on the Web site: "In the near future ISRIC intends to use the Internet as the main outlet for its products and services (datasets, documents, maps, programs, etc.)." Currently, it provides full text of a few documents and downloadable archives of Soil and Terrain (SOTER) databases, one covering Kenya and one covering Latin and Central America and the Caribbean.

Soil Science Society of America. (http://www.soils.org/) (cited September 21, 2000). This site holds a wealth of information, including a searchable glossary of Soil Science terms, full-text publications of the Council of Soil Science Examiners, full-text access to *Soil Science Society of America Journal*, and well-organized links to other sites. Divisions within SSSA also have useful Web sites including one providing interactive tools for teaching within the Soil Physics Division, called Soil Physics Teaching Tools. Available from the World Wide Web at http://soilphysics.okstate.edu/toolkit/ (cited October 3, 2000). This has a variety of computational modules, some that require downloading a Java plug-in.

IFA: International Fertilizer Industry Association (http://www.fertilizer.org/index.htm) (cited September 29, 2000). Based in France, this organization provides full-text documents, conference papers, technical reports, and manuals on this Web site.

Several Web sites by governmental organizations supply considerable resources about soils, such as the following:

USDA, Natural Resources Conservation Service (NRCS) (http://www.nrcs.usda.gov/) (cited September 28, 2000). Besides information on plants, conservation, soil surveys, irrigation, soil salinity, this site includes full-text publications and downloadable databases.

European Soil Bureau (http://esb.aris.sai.jrc.it/) (cited September 19, 2000). Created in 1996, the European Soil Bureau (ESB) is a network of national soil science institutions, managed through a Secretariat located at the Joint Research Centre (JRC), Ispra, Italy. This site includes information about the organization as well as full-text documents and information about CD-ROMs and books about soils in Europe.

Technical reports are available on a variety of Web sites, including:

US Army Corps of Engineers, Cold Regions Research & Engineering Laboratory (http://www.crrel.usace.army.mil/) (cited September 28, 2000). Many recent (1995+) reports covering northern soils and environment topics, particularly decontamination of soils, are available in pdf format through the Reports and Products section.

NASA Visible Earth (http://visibleearth.nasa.gov/) (cited September 28, 2000). This site provides a searchable visual catalog of earth science-related visualizations and animations, including satellite images. Topics covered include soils, erosion, and sedimentation.

U.S. Department of Defense, The Defense Technical Information Center (DTIC). Scientific and Technical Information Network, STINET Technical Reports Collection (http://stinet.dtic.mil/str/tr4_fields.html) (cited October 3, 2000). This site provides access to all unclassified, unlimited citations to technical reports added into DTIC from late December 1974 to present; full-text pdf files are available for many documents.

U.S. Department of Energy. Digital Library (http://nattie.eh.doe.gov/library/library.html) (cited September 28, 2000). This site provides access to full-text reports largely involving contamination and remediation of soils and groundwater.

U.S. Department of Energy, Office of Scientific and Technical Information. DOE Information Bridg (http://www.osti.gov/bridge/) (cited October 3, 2000). This site is broader in scope than the previous site and also supplies full-text reports.

## REFERENCES

Amundson R. Do Soils Need Our Protection? Geotimes 43(3):16–20. 1998. Reprinted with permission on Vernal Pools and Endangered Species versus the Proposed U.C. Merced Campus, 5/3/2000 (http://www.vernalpools.org/UCMerced/geotimes.htm) (cited September 18, 2000).

McCracken RJ and Helms D. Soil Survey and Maps. In: McDonald P, ed. The Literature of Soil Science, Ithaca, NY: Cornell University Press, 1994.

Soil and Water Conservation Society. Soil: A Critical Environmental Resource (http://www.swcs.org/t_resources_critical_fact.htm) (cited September 18, 2000).

Soil Science Society of America. Glossary of Soil Science Terms (http://www.soils.org/sssagloss/index.html) (cited September 19, 2000).

Stuart K and Jenny H. My Friend, the Soil: A Conversation with Hans Jenny. Journal of Soil and Water Conservation 39(3):158–161, 1984.

USDA, Natural Resources Conservation Service. "What on Earth is Soil?" U.S. Environmental Protection Agency, Gulf of Mexico Program, August 28, 2000 (http://pelican.gmpo.gov/edresources/soil.html) (cited September 18, 2000).

Warkentin BP. Trends and Developments in Soil Science. In: McDonald P, ed. The Literature of Soil Science. Ithaca, NY: Cornell University Press, 1994.

# Index